Methods in Enzymology

Volume 378

QUINONES AND QUINONE ENZYMES

Part A

METHODS IN ENZYMOLOGY

Methods in Enzymology

Volume 378

Quinones and Quinone Enzymes

Part A

EDITED BY

Helmut Sies

INSTITUT FÜR BIOCHEMIE UND MOLEKULARBIOLOGIE I
HEINRICH-HEINE-UNIVERSTÄT DÜSSELDORF
DÜSSELDORF, GERMANY

Lester Packer

SCHOOL OF PHARMACY
UNIVERSITY OF SOUTHERN CALIFORNIA
LOS ANGELES, CALIFORNIA

ELSEVIER
ACADEMIC
PRESS

AMSTERDAM • BOSTON • HEIDELBERG • LONDON
NEW YORK • OXFORD • PARIS • SAN DIEGO
SAN FRANCISCO • SINGAPORE • SYDNEY • TOKYO

Academic Press is an imprint of Elsevier

Elsevier Academic Press
525 B Street, Suite 1900, San Diego, California 92101-4495, USA
84 Theobald's Road, London WC1X 8RR, UK

This book is printed on acid-free paper. 

For all information on all Academic Press publications visit our Web site at www.academicpress.com

ISBN: 0-12-182782-8

PRINTED IN THE UNITED STATES OF AMERICA
04 05 06 07 08 9 8 7 6 5 4 3 2 1

Dedicated to the Pioneers of this field: Frederick L. Crane, Lars Ernster, Karl Folkers, and Andrés O. M. Stoppani

Table of Contents

Contributors to Volume 378 xi

Preface xv

Volumes in Series xvii

Section I. Coenzyme Q: Detection and Quinone Reductases

1. Investigation of Regulatory Mechanisms in Coenzyme Q Metabolism — Jacob Grünler and Gustav Dallner — 3

2. Methods for Characterizing TPQ-Containing Proteins — Jennifer L. DuBois and Judith P. Klinman — 17

3. Aldo-Keto Reductases and Formation of Polycyclic Aromatic Hydrocarbon *o*-Quinones — Trevor M. Penning — 31

4. Redox Cycling of β-Lapachone and Structural Analogues in Microsomal and Cytosol Liver Preparations — Silvia Fernández Villamil, Andrés O. M. Stoppani, and Marta Dubin — 67

5. Quinone Chemistry and Melanogenesis — Edward J. Land, Christopher A. Ramsden, and Patrick A. Riley — 88

6. Quinoids Formed from Estrogens and Antiestrogens — Judy L. Bolton, Linning Yu, and Gregory R. J. Thatcher — 110

7. Ubiquinone and Plastoquinone Metabolism in Plants — Ewa Swiezewska — 124

8. Extramitochondrial Reduction of Ubiquinone by Flavoenzymes — Mikael Björnstedt, Tomas Nordman, and Jerker M. Olsson — 131

9. Tissue Bioavailability and Detection of Coenzyme Q — Igor Rebrin, Sergey Kamzalov, and Rajindar S. Sohal — 138

10. Coemzyme Q and Vitamin E Interactions — Rajindar S. Sohal — 146

11. Preparation of Tritium-Labeled 3-Methyl-3-buten-1[^{3}H]-yl Diphosphate (^{3}H-Isopentenyl Diphosphate) — TADEUSZ CHOJNACKI — 152

12. High-Performance Liquid Chromatography–EC Assay of Mitochondrial Coenzyme Q_9, Coenzyme Q_9H_2, Coenzyme Q_{10}, Coenzyme $Q_{10}H_2$, and Vitamin E with a Simplified On-Line Solid-Phase Extraction — MAURIZIO BATTINO, LUCIANA LEONE, AND STEFANO BOMPADRE — 156

13. Simultaneous Determination of Coenzyme Q_{10}, Cholesterol, and Major Cholesterylesters in Human Blood Plasma — CRAIG A. GAY AND ROLAND STOCKER — 162

14. Assay of Coenzyme Q_{10} in Plasma by a Single Dilution Step — GIAN PAOLO LITTARRU, FABRIZIO MOSCA, DANIELE FATTORINI, STEFANO BOMPADRE, AND MAURIZIO BATTINO — 170

Section II: Plasma Membrane Quinone Reductases

15. Quinone Oxidoreductases of the Plasma Membrane — D. JAMES MORRÉ — 179

16. Regulation of Ceramide Signaling by Plasma Membrane Coenzyme Q Reductases — PLÁCIDO NAVAS AND JOSÉ MANUEL VILLALBA — 200

17. Stabilization of Extracellular Ascorbate Mediated by Coenzyme Q Transmembrane Electron Transport — ANTONIO ARROYO, JUAN C. RODRÍGUEZ-AGUILERA, CARLOS SANTOS-OCAÑA, JOSÉ MANUEL VILLALBA, AND PLÁCIDO NAVAS — 207

Section III. Quinones, Cellular Signaling, and Modulation of Gene Expression

18. Regulation of Antioxidant Response Element–Dependent Induction of Detoxifying Enzyme Synthesis — ANIL K. JAISWAL — 221

19. Antioxidant Responsive Element Activation by Quinones: Antioxidant Responsive Element Target Genes, Role of PI3 Kinase in Activation — JIANG LI, JONG-MIN LEE, DELINDA A. JOHNSON, AND JEFFERY A. JOHNSON — 238

20. Signaling Effects of Menadione: From Tyrosine Phosphatase Inactivation to Connexin Phosphorylation — KOTB ABDELMOHSEN, PAULINE PATAK, CLAUDIA VON MONTFORT, IRA MELCHHEIER, HELMUT SIES, AND LARS-OLIVER KLOTZ — 258

21. Unique Function of the Nrf2–Keap1 Pathway in the Inducible Expression of Antioxidant and Detoxifying Enzymes — AKIRA KOBAYASHI, TSUTOMU OHTA, AND MASAYUKI YAMAMOTO 273

22. Role of Protein Phosphorylation in the Regulation of NF-E2–Related Factor 2 Activity — PHILIP J. SHERRATT, H.-C. HUANG, TRUYEN NGUYEN, AND CECIL B. PICKETT 286

23. Analysis of Transcription Factor Remodeling in Phase II Gene Expression with Curcumin — DALE A. DICKINSON, KAREN E. ILES, AMANDA F. WIGLEY, AND HENRY JAY FORMAN 302

24. Quinones and Glutathione Metabolism — NOBUO WATANABE, DALE A. DICKINSON, RUI-MING LIU, AND HENRY JAY FORMAN 319

25. Doxorubicin Cardiotoxicity and the Control of Iron Metabolism: Quinone-Dependent and Independent Mechanisms — GIORGIO MINOTTI, STEFANIA RECALCATI, PIERANTONIO MENNA, EMANUELA SALVATORELLI, GIANFRANCA CORNA, GAETANO CAIRO 340

26. Oxidant-Induced Iron Signaling in Doxorubicin-Mediated Apoptosis — SRIGIRIDHAR KOTAMRAJU, SHASI V. KALIVENDI, EUGENE KONOREV, CHRISTOPHER R. CHITAMBAR, JOY JOSEPH, AND B. KALYANARAMAN 362

AUTHOR INDEX 383

SUBJECT INDEX 411

Contributors to Volume 378

Article numbers are in parentheses and following the names of contributors.
Affiliations listed are current.

KOTB ABDELMOHSEN (20), Institut für Biochemie und Molekularbiologie I, Heinrich-Heine-Universität Düsseldorf, D-40225 Düsseldorf, Germany

ANTONIO ARROYO (17), Centro Andaluz de Biología del Desarrollo, Universidad Pablo de Olavide, E-41013 Sevilla, Spain

MAURIZIO BATTINO (12, 14), Institute of Biochemistry, Università Politecnica delle Marche, 60131 Ancona, Italy

MIKAEL BJÖRNSTEDT (8), Division of Pathology, Department of Laboratory Medicine, Karolinska Institutet, Huddinge University Hospital, SE-141 86 Stockholm, Sweden

JUDY L. BOLTON (6), Department of Medicinal Chemistry and Pharmacognosy, College of Pharmacy, University of Illinois at Chicago, Chicago, Illinois 60612-7231

STEFANO BOMPADRE (12, 14), Institute of Microbiology and Biomedical Sciences, Università Politecnica delle Marche, 60131 Ancona, Italy

GAETANO CAIRO (25), Institute of General Pathology, University of Milan School of Medicine, 20133 Milan, Italy

CHRISTOPHER R. CHITAMBAR (26), Division of Neoplastic Diseases, Medical College of Wisconsin, Milwaukee, Wisconsin 53226

TADEUSZ CHOJNACKI (11), Institute of Biochemistry and Biophysics, Polish Academy of Sciences, 02-106 Warszawa, Poland

GIANFRANCA CORNA (25), Institute of General Pathology, University of Milan School of Medicine, 20133 Milan, Italy

GUSTAV DALLNER (1), Department of Biochemistry and Biophysics, Stockholm University, S-106 91, Stockholm, Sweden

DALE A. DICKINSON (23, 24), Department of Environmental Health Sciences, School of Public Health, and Center for Free Radical Biology, University of Alabama at Birmingham, Birmingham, Alabama 35294

MARTA DUBIN (4), Bioenergetics Research Centre, National Research Council, School of Medicine, University of Buenos Aires, 1121-Buenos Aires, Argentine

JENNIFER L. DUBOIS (2), Department of Chemistry, University of California, Berkeley, Berkeley, California 94720

DANIELE FATTORINI (14), Institute of Biology and Genetics, Università Politecnica delle Marche, 60131 Ancona, Italy

SILVIA FERNÁNDEZ VILLAMIL (4), Bioenergetics Research Centre, National Research Council, School of Medicine, University of Buenos Aires, 1121-Buenos Aires, Argentine

HENRY JAY FORMAN (23, 24), Department of Environmental Health Sciences, School of Public Health, and Center for Free Radical Biology, University of Alabama at Birmingham, Birmingham, Alabama 35294*

Current affiliation: Division of Natural Sciences, University of California, Merced, Merced, California 95344

CRAIG A. GAY (13), Centre for Vascular Research, School of Medical Sciences, University of New South Wales, Sydney, New South Wales 2052, Australia

JACOB GRÜNLER (1), Department of Molecular Medicine, Karolinska Hospital, Karolinska Institutet Medical School, S-106 91, Stockholm, Sweden

H.-C. HUANG (22), Schering-Plough Research Institute, Kenilworth, New Jersey 07033-1300

KAREN E. ILES (23), Department of Environmental Health Sciences, School of Public Health, and Center for Free Radical Biology, University of Alabama at Birmingham, Birmingham, Alabama 35294

ANIL K. JAISWAL (18), Department of Pharmacology, Baylor College of Medicine, Houston, Texas 77030

JEFFREY A. JOHNSON (19), University of Wisconsin-Madison, School of Pharmacy, Madison, Wisconsin 53705-2222

DELINDA A. JOHNSON (19), University of Wisconsin-Madison, School of Pharmacy, Madison, Wisconsin 53705-2222

JOY JOSEPH (26), Biophysics Department, Medical College of Wisconsin, Milwaukee, Wisconsin 53226

SHASI V. KALIVENDI (26), Biophysics Department, Medical College of Wisconsin, Milwaukee, Wisconsin 53226

B. KALYANARAMAN (26), Biophysics Department, Medical College of Wisconsin, Milwaukee, Wisconsin 53226

SERGEY KAMZALOV (9), Department of Molecular Pharmacology and Toxicolgy, University of Southern California, Los Angeles, California 90089-9121

JUDITH P. KLINMAN (2), Department of Chemistry, University of California, Berkeley, Berkeley, California 94720

LARS-OLIVER KLOTZ (20), Institut für Biochemie und Molekularbiologie I, Heinrich-Heine-Universität Düsseldorf, D-40225 Düsseldorf, Germany

AKIRA KOBAYASHI (21), Center for Tsukuba Advanced Research Alliance, Institute of Basic Medical Sciences and JST-ERATO Environmental Response Project, University of Tsukuba, Tsukuba 305-8577, Japan

EUGENE KONOREV (26), Biophysics Department, Medical College of Wisconsin, Milwaukee, Wisconsin 53226

SRIGIRIDHAR KOTAMRAJU (26), Biophysics Department, Medical College of Wisconsin, Milwaukee, Wisconsin 53226

EDWARD J. LAND (5), School of Chemistry and Physics, Keele University, Keele, Staffordshire ST5 5BG, United Kingdom

JONG-MIN LEE (19), University of Wisconsin-Madison, School of Pharmacy, Madison, Wisconsin 53705-2222

LUCIANA LEONE (12), Institute of Microbiology and Biomedical Sciences, Università Politecnica delle Marche, 60100 Ancona, Italy

JIANG LI (19), University of Wisconsin-Madison, School of Pharmacy, Madison, Wisconsin 53705-2222

GIAN PAOLO LITTARRU (14), Institute of Biochemistry, Università Politecnica delle Marche, 60131 Ancona, Italy

RUI-MING LIU (24), Department of Environmental Health Sciences, School of Public Health, and Center for Free Radical Biology, University of Alabama, Birmingham, Alabama 35294

IRA MELCHHEIER (20), Institut für Biochemie und Molekularbiologie I, Heinrich-Heine-Universität Düsseldorf, D-40225 Düsseldorf, Germany

PIERANTONIO MENNA (25), Department of Drug Sciences and Centro Studi Invecchiamento, G. d'Annunzio University School of Medicine, 66013 Chieti, Italy

GIORGIO MINOTTI (25), Department of Drug Sciences and Centro Studi Invecchiamento, G. d'Annunzio University School of Medicine, 66013 Chieti, Italy

D. JAMES MORRÉ (15), Department of Medicinal Chemistry and Molecular Pharmacology, Purdue University, Lafayette, Indiana 47907

FABRIZIO MOSCA (14), Institute of Biochemistry, Università Politecnica delle Marche, 60131 Ancona, Italy

PLÁCIDO NAVAS (16, 17), Centro Andaluz de Biología del Desarrollo, Universidad Pablo de Olavide, E-41013 Sevilla, Spain

TRUYEN NGUYEN (22), Schering-Plough Research Institute, Kenilworth, New Jersey 07033-1300

TOMAS NORDMAN (8), Division of Pathology, Department of Laboratory Medicine, Karolinska Institutet, Huddinge University Hospital, SE-141 86 Stockholm, Sweden

TSUTOMU OHTA (21), Genomics Division, National Cancer Center Resarch Institute, Tokyo 104-0045, Japan

JERKER M. OLSSON (8), Division of Pathology, Department of Laboratory Medicine, Karolinska Institutet, Huddinge University Hospital, SE-141 86 Stockholm, Sweden

PAULINE PATAK (20), Institut für Biochemie und Molekularbiologie I, Heinrich-Heine-Universität Düsseldorf, D-40225 Düsseldorf, Germany

TREVOR M. PENNING (3), Department of Pharmacology, University of Pennsylvania School of Medicine, Philadelphia, Pennsylvania 19104-6084

CECIL B. PICKETT (22), Schering-Plough Research Institute, Kenilworth, New Jersey 07033-1300

CHRISTOPHER A. RAMSDEN (5), School of Chemistry and Physics, Keele University, Keele, Staffordshire ST5 5BG, United Kingdom

IGOR REBRIN (9), Department of Molecular Pharmacology and Toxicology, University of Southern California, Los Angeles, California 90089-9121

STEFANIA RECALCATI (25), Institute of General Pathology, University of Milan School of Medicine, 20133 Milan, Italy

PATRICK A. RILEY (5), Gray Cancer Institute, Mount Vernon Hospital, Northwood HA6 2JR, United Kingdom

JUAN C. RODRÍGUEZ-AGUILERA (17), Centro Andaluz de Biología del Desarrollo, Universidad Pablo de Olavide, E-41013 Sevilla, Spain

EMANUELA SALVATORELLI (25), Department of Drug Sciences and Centro Studi Invecchiamento, G. d'Annunzio University School of Medicine, 66013 Chieti, Italy

CARLOS SANTOS-OCAÑA (17), Centro Andaluz de Biología del Desarrollo, Universidad Pablo de Olavide, E-41013 Sevilla, Spain

PHILIP J. SHERRATT (22), Schering-Plough Research Institute, Kenilworth, New Jersey 07033-1300

HELMUT SIES (20), Institut für Biochemie und Molekularbiologie I, Heinrich-Heine-Universität Düsseldorf, D-40225 Düsseldorf, Germany

RAJINDAR S. SOHAL (9, 10), Department of Pharmacology and Toxicology, University of Southern California, Los Angeles, California 90089-9121

ROLAND STOCKER (13), Centre for Vascular Research, School of Medical Sciences, University of New South Wales, Sydney, New South Wales 2052, Australia

ANDRÉS O. M. STOPPANI (4), Bioenergetics Research Centre, National Research Council, School of Medicine, University of Buenos Aires, 1121-Buenos Aires, Argentine

EWA SWIEZEWSKA (7), Institute of Biochemistry and Biophysics, Polish Academy of Sciences, 02-106 Warszawa, Poland

GREGORY R. J. THATCHER (6), Department of Medicinal Chemistry and Pharmacognosy, College of Pharmacy, University of Illinois at Chicago, Chicago, Illinois 60612-7231

JOSÉ MANUEL VILLALBA (16, 17), Departamento de Biología Celular, Fisiología e Inmunología, Universidad de Córdoba, E-14071 Córdoba, Spain

CLAUDIA VON MONTFORT (20), Institut für Biochemie und Molekularbiologie I, Heinrich-Heine-Universität Düsseldorf, D-40225 Düsseldorf, Germany

NOBUO WATANABE (24), Department of Environmental Health Sciences, School of Public Health, and Center for Free Radical Biology, University of Alabama, Birmingham, Alabama 35294

AMANDA F. WIGLEY (23), Department of Environmental Health Sciences, School of Public Health, University of Alabama at Birmingham, Birmingham, Alabama 35294*

MASAYUKI YAMAMOTO (21), Center for Tsukuba Advanced Research Alliance, Institute of Basic Medical Sciences and JST-ERATO Environmental Response Project, University of Tsukuba, Tsukuba 305-8577, Japan

LINNING YU (6), Department of Medicinal Chemistry and Pharmacognosy, College of Pharmacy, University of Illinois at Chicago, Chicago, Illinois 60612-7231

Current affiliation: Department of Pathology, University of Alabama at Birmingham, Birmingham, Alabama 35294

Preface

Developments in genomics and proteomics rapidly generated focus on new *-omics,* particularly metabolomics and phenomics. Quinones, hydroquinones, semiquinones, and their metabolites are naturally-occurring compounds that serve as wonderful examples for this new paradigm of interdigitating *-omics;* in addition to a role as substrates and products in metabolism, quinone compounds are intermediates in many pathways of gene regulation, enzyme protein induction, feedback control, and waste product elimination. Quinones play a pivotal role in energy metabolism (Peter Mitchell's protonmotive *"Q cycle"*), many other key processes, and even in chemotherapy, where *redox cycling* drugs are utilized.

The present volume of *Methods in Enzymology* on quinones and quinone enzymes serves to bring together current methods and concepts on this topic. It focuses on the role in the so-called Phase II of drug metabolism (xenobiotics), but includes aspects on Phase I (CYP, cytochromes P-450) and Phase III (transport systems) as well. This volume of *Methods in Enzymology* (Part A) focuses on quinones and quinone enzymes in terms of coenzyme Q (detection and quinone reductases), plasma membrane quinone reductases, and the role of quinones in cellular signaling and modulation of gene expression, whereas Part B addresses mitochondrial ubiquinone and reductases, anticancer quinones, and the role of quinone reductases in chemoprevention and nutrition, as well as the role of quinones in age-related diseases. Phase II Enzymes, Part C, will focus on glutathione, glutathione S-transferases, and other conjugation enzymes.

The enzyme, NAD(P)H: quinone oxidoreductase, is the subject of a major section in this volume. This enzyme, discovered in 1958 in Stockholm by Lars Ernster and named DT-Diaphorase by him, has multiple roles, some of which were only recently discovered.

There are human polymorphisms in these enzymes relating to variations in cancer risk, and enzymes targeted by quinones are being investigated. Modern methods in assaying quinone reactions and, indeed, various quinones themselves, are also included in this volume.

Ubiquinone (Coenzyme Q_{10}) as a major naturally-occurring quinone became a highlight of scientific interest following its discovery in 1957, and was discovered to have an established role in mitochondrial electron transport by Frederick Crane. Fundamental contributions were made by Karl Folkers on its supplemental use for health benefits in disease prevention and by Andrés O. M. Stoppani, a pioneer of Argentinian biochemistry, in utilizing quinones for the treatment of Chagas disease.

We thank the Advisory Committee (Enrique Cadenas, Los Angeles; Gustav Dallner, Stockholm; Tom Kensler, Baltimore; Lars-Oliver Klotz, Düsseldorf; David Ross, Denver) for their valuable suggestions and wisdom in selecting the contributions for this volume.

Helmut Sies
Lester Packer

METHODS IN ENZYMOLOGY

Volume I. Preparation and Assay of Enzymes
Edited by Sidney P. Colowick and Nathan O. Kaplan

Volume II. Preparation and Assay of Enzymes
Edited by Sidney P. Colowick and Nathan O. Kaplan

Volume III. Preparation and Assay of Substrates
Edited by Sidney P. Colowick and Nathan O. Kaplan

Volume IV. Special Techniques for the Enzymologist
Edited by Sidney P. Colowick and Nathan O. Kaplan

Volume V. Preparation and Assay of Enzymes
Edited by Sidney P. Colowick and Nathan O. Kaplan

Volume VI. Preparation and Assay of Enzymes *(Continued)*
Preparation and Assay of Substrates
Special Techniques
Edited by Sidney P. Colowick and Nathan O. Kaplan

Volume VII. Cumulative Subject Index
Edited by Sidney P. Colowick and Nathan O. Kaplan

Volume VIII. Complex Carbohydrates
Edited by Elizabeth F. Neufeld and Victor Ginsburg

Volume IX. Carbohydrate Metabolism
Edited by Willis A. Wood

Volume X. Oxidation and Phosphorylation
Edited by Ronald W. Estabrook and Maynard E. Pullman

Volume XI. Enzyme Structure
Edited by C. H. W. Hirs

Volume XII. Nucleic Acids (Parts A and B)
Edited by Lawrence Grossman and Kivie Moldave

Volume XIII. Citric Acid Cycle
Edited by J. M. Lowenstein

Volume XIV. Lipids
Edited by J. M. Lowenstein

Volume XV. Steroids and Terpenoids
Edited by Raymond B. Clayton

VOLUME XVI. Fast Reactions
Edited by KENNETH KUSTIN

VOLUME XVII. Metabolism of Amino Acids and Amines (Parts A and B)
Edited by HERBERT TABOR AND CELIA WHITE TABOR

VOLUME XVIII. Vitamins and Coenzymes (Parts A, B, and C)
Edited by DONALD B. MCCORMICK AND LEMUEL D. WRIGHT

VOLUME XIX. Proteolytic Enzymes
Edited by GERTRUDE E. PERLMANN AND LASZLO LORAND

VOLUME XX. Nucleic Acids and Protein Synthesis (Part C)
Edited by KIVIE MOLDAVE AND LAWRENCE GROSSMAN

VOLUME XXI. Nucleic Acids (Part D)
Edited by LAWRENCE GROSSMAN AND KIVIE MOLDAVE

VOLUME XXII. Enzyme Purification and Related Techniques
Edited by WILLIAM B. JAKOBY

VOLUME XXIII. Photosynthesis (Part A)
Edited by ANTHONY SAN PIETRO

VOLUME XXIV. Photosynthesis and Nitrogen Fixation (Part B)
Edited by ANTHONY SAN PIETRO

VOLUME XXV. Enzyme Structure (Part B)
Edited by C. H. W. HIRS AND SERGE N. TIMASHEFF

VOLUME XXVI. Enzyme Structure (Part C)
Edited by C. H. W. HIRS AND SERGE N. TIMASHEFF

VOLUME XXVII. Enzyme Structure (Part D)
Edited by C. H. W. HIRS AND SERGE N. TIMASHEFF

VOLUME XXVIII. Complex Carbohydrates (Part B)
Edited by VICTOR GINSBURG

VOLUME XXIX. Nucleic Acids and Protein Synthesis (Part E)
Edited by LAWRENCE GROSSMAN AND KIVIE MOLDAVE

VOLUME XXX. Nucleic Acids and Protein Synthesis (Part F)
Edited by KIVIE MOLDAVE AND LAWRENCE GROSSMAN

VOLUME XXXI. Biomembranes (Part A)
Edited by SIDNEY FLEISCHER AND LESTER PACKER

VOLUME XXXII. Biomembranes (Part B)
Edited by SIDNEY FLEISCHER AND LESTER PACKER

VOLUME XXXIII. Cumulative Subject Index Volumes I-XXX
Edited by MARTHA G. DENNIS AND EDWARD A. DENNIS

VOLUME XXXIV. Affinity Techniques (Enzyme Purification: Part B)
Edited by WILLIAM B. JAKOBY AND MEIR WILCHEK

VOLUME LIV. Biomembranes (Part E: Biological Oxidations)
Edited by SIDNEY FLEISCHER AND LESTER PACKER

VOLUME LV. Biomembranes (Part F: Bioenergetics)
Edited by SIDNEY FLEISCHER AND LESTER PACKER

VOLUME LVI. Biomembranes (Part G: Bioenergetics)
Edited by SIDNEY FLEISCHER AND LESTER PACKER

VOLUME LVII. Bioluminescence and Chemiluminescence
Edited by MARLENE A. DELUCA

VOLUME LVIII. Cell Culture
Edited by WILLIAM B. JAKOBY AND IRA PASTAN

VOLUME LIX. Nucleic Acids and Protein Synthesis (Part G)
Edited by KIVIE MOLDAVE AND LAWRENCE GROSSMAN

VOLUME LX. Nucleic Acids and Protein Synthesis (Part H)
Edited by KIVIE MOLDAVE AND LAWRENCE GROSSMAN

VOLUME 61. Enzyme Structure (Part H)
Edited by C. H. W. HIRS AND SERGE N. TIMASHEFF

VOLUME 62. Vitamins and Coenzymes (Part D)
Edited by DONALD B. MCCORMICK AND LEMUEL D. WRIGHT

VOLUME 63. Enzyme Kinetics and Mechanism (Part A: Initial Rate and Inhibitor Methods)
Edited by DANIEL L. PURICH

VOLUME 64. Enzyme Kinetics and Mechanism (Part B: Isotopic Probes and Complex Enzyme Systems)
Edited by DANIEL L. PURICH

VOLUME 65. Nucleic Acids (Part I)
Edited by LAWRENCE GROSSMAN AND KIVIE MOLDAVE

VOLUME 66. Vitamins and Coenzymes (Part E)
Edited by DONALD B. MCCORMICK AND LEMUEL D. WRIGHT

VOLUME 67. Vitamins and Coenzymes (Part F)
Edited by DONALD B. MCCORMICK AND LEMUEL D. WRIGHT

VOLUME 68. Recombinant DNA
Edited by RAY WU

VOLUME 69. Photosynthesis and Nitrogen Fixation (Part C)
Edited by ANTHONY SAN PIETRO

VOLUME 70. Immunochemical Techniques (Part A)
Edited by HELEN VAN VUNAKIS AND JOHN J. LANGONE

VOLUME 71. Lipids (Part C)
Edited by JOHN M. LOWENSTEIN

VOLUME 72. Lipids (Part D)
Edited by JOHN M. LOWENSTEIN

VOLUME 73. Immunochemical Techniques (Part B)
Edited by JOHN J. LANGONE AND HELEN VAN VUNAKIS

VOLUME 74. Immunochemical Techniques (Part C)
Edited by JOHN J. LANGONE AND HELEN VAN VUNAKIS

VOLUME 75. Cumulative Subject Index Volumes XXXI, XXXII, XXXIV–LX
Edited by EDWARD A. DENNIS AND MARTHA G. DENNIS

VOLUME 76. Hemoglobins
Edited by ERALDO ANTONINI, LUIGI ROSSI-BERNARDI, AND EMILIA CHIANCONE

VOLUME 77. Detoxication and Drug Metabolism
Edited by WILLIAM B. JAKOBY

VOLUME 78. Interferons (Part A)
Edited by SIDNEY PESTKA

VOLUME 79. Interferons (Part B)
Edited by SIDNEY PESTKA

VOLUME 80. Proteolytic Enzymes (Part C)
Edited by LASZLO LORAND

VOLUME 81. Biomembranes (Part H: Visual Pigments and Purple Membranes, I)
Edited by LESTER PACKER

VOLUME 82. Structural and Contractile Proteins (Part A: Extracellular Matrix)
Edited by LEON W. CUNNINGHAM AND DIXIE W. FREDERIKSEN

VOLUME 83. Complex Carbohydrates (Part D)
Edited by VICTOR GINSBURG

VOLUME 84. Immunochemical Techniques (Part D: Selected Immunoassays)
Edited by JOHN J. LANGONE AND HELEN VAN VUNAKIS

VOLUME 85. Structural and Contractile Proteins (Part B: The Contractile Apparatus and the Cytoskeleton)
Edited by DIXIE W. FREDERIKSEN AND LEON W. CUNNINGHAM

VOLUME 86. Prostaglandins and Arachidonate Metabolites
Edited by WILLIAM E. M. LANDS AND WILLIAM L. SMITH

VOLUME 87. Enzyme Kinetics and Mechanism (Part C: Intermediates, Stereo-chemistry, and Rate Studies)
Edited by DANIEL L. PURICH

VOLUME 88. Biomembranes (Part I: Visual Pigments and Purple Membranes, II)
Edited by LESTER PACKER

VOLUME 89. Carbohydrate Metabolism (Part D)
Edited by WILLIS A. WOOD

VOLUME 107. Posttranslational Modifications (Part B)
Edited by FINN WOLD AND KIVIE MOLDAVE

VOLUME 108. Immunochemical Techniques (Part G: Separation and Characterization of Lymphoid Cells)
Edited by GIOVANNI DI SABATO, JOHN J. LANGONE, AND HELEN VAN VUNAKIS

VOLUME 109. Hormone Action (Part I: Peptide Hormones)
Edited by LUTZ BIRNBAUMER AND BERT W. O'MALLEY

VOLUME 110. Steroids and Isoprenoids (Part A)
Edited by JOHN H. LAW AND HANS C. RILLING

VOLUME 111. Steroids and Isoprenoids (Part B)
Edited by JOHN H. LAW AND HANS C. RILLING

VOLUME 112. Drug and Enzyme Targeting (Part A)
Edited by KENNETH J. WIDDER AND RALPH GREEN

VOLUME 113. Glutamate, Glutamine, Glutathione, and Related Compounds
Edited by ALTON MEISTER

VOLUME 114. Diffraction Methods for Biological Macromolecules (Part A)
Edited by HAROLD W. WYCKOFF, C. H. W. HIRS, AND SERGE N. TIMASHEFF

VOLUME 115. Diffraction Methods for Biological Macromolecules (Part B)
Edited by HAROLD W. WYCKOFF, C. H. W. HIRS, AND SERGE N. TIMASHEFF

VOLUME 116. Immunochemical Techniques (Part H: Effectors and Mediators of Lymphoid Cell Functions)
Edited by GIOVANNI DI SABATO, JOHN J. LANGONE, AND HELEN VAN VUNAKIS

VOLUME 117. Enzyme Structure (Part J)
Edited by C. H. W. HIRS AND SERGE N. TIMASHEFF

VOLUME 118. Plant Molecular Biology
Edited by ARTHUR WEISSBACH AND HERBERT WEISSBACH

VOLUME 119. Interferons (Part C)
Edited by SIDNEY PESTKA

VOLUME 120. Cumulative Subject Index Volumes 81–94, 96–101

VOLUME 121. Immunochemical Techniques (Part I: Hybridoma Technology and Monoclonal Antibodies)
Edited by JOHN J. LANGONE AND HELEN VAN VUNAKIS

VOLUME 122. Vitamins and Coenzymes (Part G)
Edited by FRANK CHYTIL AND DONALD B. MCCORMICK

VOLUME 123. Vitamins and Coenzymes (Part H)
Edited by FRANK CHYTIL AND DONALD B. MCCORMICK

VOLUME 124. Hormone Action (Part J: Neuroendocrine Peptides)
Edited by P. MICHAEL CONN

VOLUME 125. Biomembranes (Part M: Transport in Bacteria, Mitochondria, and Chloroplasts: General Approaches and Transport Systems)
Edited by SIDNEY FLEISCHER AND BECCA FLEISCHER

VOLUME 126. Biomembranes (Part N: Transport in Bacteria, Mitochondria, and Chloroplasts: Protonmotive Force)
Edited by SIDNEY FLEISCHER AND BECCA FLEISCHER

VOLUME 127. Biomembranes (Part O: Protons and Water: Structure and Translocation)
Edited by LESTER PACKER

VOLUME 128. Plasma Lipoproteins (Part A: Preparation, Structure, and Molecular Biology)
Edited by JERE P. SEGREST AND JOHN J. ALBERS

VOLUME 129. Plasma Lipoproteins (Part B: Characterization, Cell Biology, and Metabolism)
Edited by JOHN J. ALBERS AND JERE P. SEGREST

VOLUME 130. Enzyme Structure (Part K)
Edited by C. H. W. HIRS AND SERGE N. TIMASHEFF

VOLUME 131. Enzyme Structure (Part L)
Edited by C. H. W. HIRS AND SERGE N. TIMASHEFF

VOLUME 132. Immunochemical Techniques (Part J: Phagocytosis and Cell-Mediated Cytotoxicity)
Edited by GIOVANNI DI SABATO AND JOHANNES EVERSE

VOLUME 133. Bioluminescence and Chemiluminescence (Part B)
Edited by MARLENE DELUCA AND WILLIAM D. MCELROY

VOLUME 134. Structural and Contractile Proteins (Part C: The Contractile Apparatus and the Cytoskeleton)
Edited by RICHARD B. VALLEE

VOLUME 135. Immobilized Enzymes and Cells (Part B)
Edited by KLAUS MOSBACH

VOLUME 136. Immobilized Enzymes and Cells (Part C)
Edited by KLAUS MOSBACH

VOLUME 137. Immobilized Enzymes and Cells (Part D)
Edited by KLAUS MOSBACH

VOLUME 138. Complex Carbohydrates (Part E)
Edited by VICTOR GINSBURG

VOLUME 139. Cellular Regulators (Part A: Calcium- and Calmodulin-Binding Proteins)
Edited by ANTHONY R. MEANS AND P. MICHAEL CONN

VOLUME 140. Cumulative Subject Index Volumes 102–119, 121–134

VOLUME 141. Cellular Regulators (Part B: Calcium and Lipids)
Edited by P. MICHAEL CONN AND ANTHONY R. MEANS

VOLUME 142. Metabolism of Aromatic Amino Acids and Amines
Edited by SEYMOUR KAUFMAN

VOLUME 143. Sulfur and Sulfur Amino Acids
Edited by WILLIAM B. JAKOBY AND OWEN GRIFFITH

VOLUME 144. Structural and Contractile Proteins (Part D: Extracellular Matrix)
Edited by LEON W. CUNNINGHAM

VOLUME 145. Structural and Contractile Proteins (Part E: Extracellular Matrix)
Edited by LEON W. CUNNINGHAM

VOLUME 146. Peptide Growth Factors (Part A)
Edited by DAVID BARNES AND DAVID A. SIRBASKU

VOLUME 147. Peptide Growth Factors (Part B)
Edited by DAVID BARNES AND DAVID A. SIRBASKU

VOLUME 148. Plant Cell Membranes
Edited by LESTER PACKER AND ROLAND DOUCE

VOLUME 149. Drug and Enzyme Targeting (Part B)
Edited by RALPH GREEN AND KENNETH J. WIDDER

VOLUME 150. Immunochemical Techniques (Part K: *In Vitro* Models of B and T Cell Functions and Lymphoid Cell Receptors)
Edited by GIOVANNI DI SABATO

VOLUME 151. Molecular Genetics of Mammalian Cells
Edited by MICHAEL M. GOTTESMAN

VOLUME 152. Guide to Molecular Cloning Techniques
Edited by SHELBY L. BERGER AND ALAN R. KIMMEL

VOLUME 153. Recombinant DNA (Part D)
Edited by RAY WU AND LAWRENCE GROSSMAN

VOLUME 154. Recombinant DNA (Part E)
Edited by RAY WU AND LAWRENCE GROSSMAN

VOLUME 155. Recombinant DNA (Part F)
Edited by RAY WU

VOLUME 156. Biomembranes (Part P: ATP-Driven Pumps and Related Transport: The Na, K-Pump)
Edited by SIDNEY FLEISCHER AND BECCA FLEISCHER

VOLUME 157. Biomembranes (Part Q: ATP-Driven Pumps and Related Transport: Calcium, Proton, and Potassium Pumps)
Edited by SIDNEY FLEISCHER AND BECCA FLEISCHER

Volume 158. Metalloproteins (Part A)
Edited by James F. Riordan and Bert L. Vallee

Volume 159. Initiation and Termination of Cyclic Nucleotide Action
Edited by Jackie D. Corbin and Roger A. Johnson

Volume 160. Biomass (Part A: Cellulose and Hemicellulose)
Edited by Willis A. Wood and Scott T. Kellogg

Volume 161. Biomass (Part B: Lignin, Pectin, and Chitin)
Edited by Willis A. Wood and Scott T. Kellogg

Volume 162. Immunochemical Techniques (Part L: Chemotaxis and Inflammation)
Edited by Giovanni Di Sabato

Volume 163. Immunochemical Techniques (Part M: Chemotaxis and Inflammation)
Edited by Giovanni Di Sabato

Volume 164. Ribosomes
Edited by Harry F. Noller, Jr., and Kivie Moldave

Volume 165. Microbial Toxins: Tools for Enzymology
Edited by Sidney Harshman

Volume 166. Branched-Chain Amino Acids
Edited by Robert Harris and John R. Sokatch

Volume 167. Cyanobacteria
Edited by Lester Packer and Alexander N. Glazer

Volume 168. Hormone Action (Part K: Neuroendocrine Peptides)
Edited by P. Michael Conn

Volume 169. Platelets: Receptors, Adhesion, Secretion (Part A)
Edited by Jacek Hawiger

Volume 170. Nucleosomes
Edited by Paul M. Wassarman and Roger D. Kornberg

Volume 171. Biomembranes (Part R: Transport Theory: Cells and Model Membranes)
Edited by Sidney Fleischer and Becca Fleischer

Volume 172. Biomembranes (Part S: Transport: Membrane Isolation and Characterization)
Edited by Sidney Fleischer and Becca Fleischer

Volume 173. Biomembranes [Part T: Cellular and Subcellular Transport: Eukaryotic (Nonepithelial) Cells]
Edited by Sidney Fleischer and Becca Fleischer

Volume 174. Biomembranes [Part U: Cellular and Subcellular Transport: Eukaryotic (Nonepithelial) Cells]
Edited by Sidney Fleischer and Becca Fleischer

VOLUME 175. Cumulative Subject Index Volumes 135–139, 141–167

VOLUME 176. Nuclear Magnetic Resonance (Part A: Spectral Techniques and Dynamics)
Edited by NORMAN J. OPPENHEIMER AND THOMAS L. JAMES

VOLUME 177. Nuclear Magnetic Resonance (Part B: Structure and Mechanism)
Edited by NORMAN J. OPPENHEIMER AND THOMAS L. JAMES

VOLUME 178. Antibodies, Antigens, and Molecular Mimicry
Edited by JOHN J. LANGONE

VOLUME 179. Complex Carbohydrates (Part F)
Edited by VICTOR GINSBURG

VOLUME 180. RNA Processing (Part A: General Methods)
Edited by JAMES E. DAHLBERG AND JOHN N. ABELSON

VOLUME 181. RNA Processing (Part B: Specific Methods)
Edited by JAMES E. DAHLBERG AND JOHN N. ABELSON

VOLUME 182. Guide to Protein Purification
Edited by MURRAY P. DEUTSCHER

VOLUME 183. Molecular Evolution: Computer Analysis of Protein and Nucleic Acid Sequences
Edited by RUSSELL F. DOOLITTLE

VOLUME 184. Avidin-Biotin Technology
Edited by MEIR WILCHEK AND EDWARD A. BAYER

VOLUME 185. Gene Expression Technology
Edited by DAVID V. GOEDDEL

VOLUME 186. Oxygen Radicals in Biological Systems (Part B: Oxygen Radicals and Antioxidants)
Edited by LESTER PACKER AND ALEXANDER N. GLAZER

VOLUME 187. Arachidonate Related Lipid Mediators
Edited by ROBERT C. MURPHY AND FRANK A. FITZPATRICK

VOLUME 188. Hydrocarbons and Methylotrophy
Edited by MARY E. LIDSTROM

VOLUME 189. Retinoids (Part A: Molecular and Metabolic Aspects)
Edited by LESTER PACKER

VOLUME 190. Retinoids (Part B: Cell Differentiation and Clinical Applications)
Edited by LESTER PACKER

VOLUME 191. Biomembranes (Part V: Cellular and Subcellular Transport: Epithelial Cells)
Edited by SIDNEY FLEISCHER AND BECCA FLEISCHER

VOLUME 192. Biomembranes (Part W: Cellular and Subcellular Transport: Epithelial Cells)
Edited by SIDNEY FLEISCHER AND BECCA FLEISCHER

VOLUME 193. Mass Spectrometry
Edited by JAMES A. MCCLOSKEY

VOLUME 194. Guide to Yeast Genetics and Molecular Biology
Edited by CHRISTINE GUTHRIE AND GERALD R. FINK

VOLUME 195. Adenylyl Cyclase, G Proteins, and Guanylyl Cyclase
Edited by ROGER A. JOHNSON AND JACKIE D. CORBIN

VOLUME 196. Molecular Motors and the Cytoskeleton
Edited by RICHARD B. VALLEE

VOLUME 197. Phospholipases
Edited by EDWARD A. DENNIS

VOLUME 198. Peptide Growth Factors (Part C)
Edited by DAVID BARNES, J. P. MATHER, AND GORDON H. SATO

VOLUME 199. Cumulative Subject Index Volumes 168–174, 176–194

VOLUME 200. Protein Phosphorylation (Part A: Protein Kinases: Assays, Purification, Antibodies, Functional Analysis, Cloning, and Expression)
Edited by TONY HUNTER AND BARTHOLOMEW M. SEFTON

VOLUME 201. Protein Phosphorylation (Part B: Analysis of Protein Phosphorylation, Protein Kinase Inhibitors, and Protein Phosphatases)
Edited by TONY HUNTER AND BARTHOLOMEW M. SEFTON

VOLUME 202. Molecular Design and Modeling: Concepts and Applications (Part A: Proteins, Peptides, and Enzymes)
Edited by JOHN J. LANGONE

VOLUME 203. Molecular Design and Modeling: Concepts and Applications (Part B: Antibodies and Antigens, Nucleic Acids, Polysaccharides, and Drugs)
Edited by JOHN J. LANGONE

VOLUME 204. Bacterial Genetic Systems
Edited by JEFFREY H. MILLER

VOLUME 205. Metallobiochemistry (Part B: Metallothionein and Related Molecules)
Edited by JAMES F. RIORDAN AND BERT L. VALLEE

VOLUME 206. Cytochrome P450
Edited by MICHAEL R. WATERMAN AND ERIC F. JOHNSON

VOLUME 207. Ion Channels
Edited by BERNARDO RUDY AND LINDA E. IVERSON

VOLUME 208. Protein–DNA Interactions
Edited by ROBERT T. SAUER

VOLUME 209. Phospholipid Biosynthesis
Edited by EDWARD A. DENNIS AND DENNIS E. VANCE

Volume 210. Numerical Computer Methods
Edited by Ludwig Brand and Michael L. Johnson

Volume 211. DNA Structures (Part A: Synthesis and Physical Analysis of DNA)
Edited by David M. J. Lilley and James E. Dahlberg

Volume 212. DNA Structures (Part B: Chemical and Electrophoretic Analysis of DNA)
Edited by David M. J. Lilley and James E. Dahlberg

Volume 213. Carotenoids (Part A: Chemistry, Separation, Quantitation, and Antioxidation)
Edited by Lester Packer

Volume 214. Carotenoids (Part B: Metabolism, Genetics, and Biosynthesis)
Edited by Lester Packer

Volume 215. Platelets: Receptors, Adhesion, Secretion (Part B)
Edited by Jacek J. Hawiger

Volume 216. Recombinant DNA (Part G)
Edited by Ray Wu

Volume 217. Recombinant DNA (Part H)
Edited by Ray Wu

Volume 218. Recombinant DNA (Part I)
Edited by Ray Wu

Volume 219. Reconstitution of Intracellular Transport
Edited by James E. Rothman

Volume 220. Membrane Fusion Techniques (Part A)
Edited by Nejat Düzgüneş

Volume 221. Membrane Fusion Techniques (Part B)
Edited by Nejat Düzgüneş

Volume 222. Proteolytic Enzymes in Coagulation, Fibrinolysis, and Complement Activation (Part A: Mammalian Blood Coagulation Factors and Inhibitors)
Edited by Laszlo Lorand and Kenneth G. Mann

Volume 223. Proteolytic Enzymes in Coagulation, Fibrinolysis, and Complement Activation (Part B: Complement Activation, Fibrinolysis, and Nonmammalian Blood Coagulation Factors)
Edited by Laszlo Lorand and Kenneth G. Mann

Volume 224. Molecular Evolution: Producing the Biochemical Data
Edited by Elizabeth Anne Zimmer, Thomas J. White, Rebecca L. Cann, and Allan C. Wilson

Volume 225. Guide to Techniques in Mouse Development
Edited by Paul M. Wassarman and Melvin L. DePamphilis

VOLUME 226. Metallobiochemistry (Part C: Spectroscopic and Physical Methods for Probing Metal Ion Environments in Metalloenzymes and Metalloproteins)
Edited by JAMES F. RIORDAN AND BERT L. VALLEE

VOLUME 227. Metallobiochemistry (Part D: Physical and Spectroscopic Methods for Probing Metal Ion Environments in Metalloproteins)
Edited by JAMES F. RIORDAN AND BERT L. VALLEE

VOLUME 228. Aqueous Two-Phase Systems
Edited by HARRY WALTER AND GÖTE JOHANSSON

VOLUME 229. Cumulative Subject Index Volumes 195–198, 200–227

VOLUME 230. Guide to Techniques in Glycobiology
Edited by WILLIAM J. LENNARZ AND GERALD W. HART

VOLUME 231. Hemoglobins (Part B: Biochemical and Analytical Methods)
Edited by JOHANNES EVERSE, KIM D. VANDEGRIFF, AND ROBERT M. WINSLOW

VOLUME 232. Hemoglobins (Part C: Biophysical Methods)
Edited by JOHANNES EVERSE, KIM D. VANDEGRIFF, AND ROBERT M. WINSLOW

VOLUME 233. Oxygen Radicals in Biological Systems (Part C)
Edited by LESTER PACKER

VOLUME 234. Oxygen Radicals in Biological Systems (Part D)
Edited by LESTER PACKER

VOLUME 235. Bacterial Pathogenesis (Part A: Identification and Regulation of Virulence Factors)
Edited by VIRGINIA L. CLARK AND PATRIK M. BAVOIL

VOLUME 236. Bacterial Pathogenesis (Part B: Integration of Pathogenic Bacteria with Host Cells)
Edited by VIRGINIA L. CLARK AND PATRIK M. BAVOIL

VOLUME 237. Heterotrimeric G Proteins
Edited by RAVI IYENGAR

VOLUME 238. Heterotrimeric G-Protein Effectors
Edited by RAVI IYENGAR

VOLUME 239. Nuclear Magnetic Resonance (Part C)
Edited by THOMAS L. JAMES AND NORMAN J. OPPENHEIMER

VOLUME 240. Numerical Computer Methods (Part B)
Edited by MICHAEL L. JOHNSON AND LUDWIG BRAND

VOLUME 241. Retroviral Proteases
Edited by LAWRENCE C. KUO AND JULES A. SHAFER

VOLUME 242. Neoglycoconjugates (Part A)
Edited by Y. C. LEE AND REIKO T. LEE

VOLUME 243. Inorganic Microbial Sulfur Metabolism
Edited by HARRY D. PECK, JR., AND JEAN LEGALL

VOLUME 244. Proteolytic Enzymes: Serine and Cysteine Peptidases
Edited by ALAN J. BARRETT

VOLUME 245. Extracellular Matrix Components
Edited by E. RUOSLAHTI AND E. ENGVALL

VOLUME 246. Biochemical Spectroscopy
Edited by KENNETH SAUER

VOLUME 247. Neoglycoconjugates (Part B: Biomedical Applications)
Edited by Y. C. LEE AND REIKO T. LEE

VOLUME 248. Proteolytic Enzymes: Aspartic and Metallo Peptidases
Edited by ALAN J. BARRETT

VOLUME 249. Enzyme Kinetics and Mechanism (Part D: Developments in Enzyme Dynamics)
Edited by DANIEL L. PURICH

VOLUME 250. Lipid Modifications of Proteins
Edited by PATRICK J. CASEY AND JANICE E. BUSS

VOLUME 251. Biothiols (Part A: Monothiols and Dithiols, Protein Thiols, and Thiyl Radicals)
Edited by LESTER PACKER

VOLUME 252. Biothiols (Part B: Glutathione and Thioredoxin; Thiols in Signal Transduction and Gene Regulation)
Edited by LESTER PACKER

VOLUME 253. Adhesion of Microbial Pathogens
Edited by RON J. DOYLE AND ITZHAK OFEK

VOLUME 254. Oncogene Techniques
Edited by PETER K. VOGT AND INDER M. VERMA

VOLUME 255. Small GTPases and Their Regulators (Part A: Ras Family)
Edited by W. E. BALCH, CHANNING J. DER, AND ALAN HALL

VOLUME 256. Small GTPases and Their Regulators (Part B: Rho Family)
Edited by W. E. BALCH, CHANNING J. DER, AND ALAN HALL

VOLUME 257. Small GTPases and Their Regulators (Part C: Proteins Involved in Transport)
Edited by W. E. BALCH, CHANNING J. DER, AND ALAN HALL

VOLUME 258. Redox-Active Amino Acids in Biology
Edited by JUDITH P. KLINMAN

VOLUME 259. Energetics of Biological Macromolecules
Edited by MICHAEL L. JOHNSON AND GARY K. ACKERS

VOLUME 260. Mitochondrial Biogenesis and Genetics (Part A)
Edited by GIUSEPPE M. ATTARDI AND ANNE CHOMYN

Volume 261. Nuclear Magnetic Resonance and Nucleic Acids
Edited by Thomas L. James

Volume 262. DNA Replication
Edited by Judith L. Campbell

Volume 263. Plasma Lipoproteins (Part C: Quantitation)
Edited by William A. Bradley, Sandra H. Gianturco, and Jere P. Segrest

Volume 264. Mitochondrial Biogenesis and Genetics (Part B)
Edited by Giuseppe M. Attardi and Anne Chomyn

Volume 265. Cumulative Subject Index Volumes 228, 230–262

Volume 266. Computer Methods for Macromolecular Sequence Analysis
Edited by Russell F. Doolittle

Volume 267. Combinatorial Chemistry
Edited by John N. Abelson

Volume 268. Nitric Oxide (Part A: Sources and Detection of NO; NO Synthase)
Edited by Lester Packer

Volume 269. Nitric Oxide (Part B: Physiological and Pathological Processes)
Edited by Lester Packer

Volume 270. High Resolution Separation and Analysis of Biological Macromolecules (Part A: Fundamentals)
Edited by Barry L. Karger and William S. Hancock

Volume 271. High Resolution Separation and Analysis of Biological Macromolecules (Part B: Applications)
Edited by Barry L. Karger and William S. Hancock

Volume 272. Cytochrome P450 (Part B)
Edited by Eric F. Johnson and Michael R. Waterman

Volume 273. RNA Polymerase and Associated Factors (Part A)
Edited by Sankar Adhya

Volume 274. RNA Polymerase and Associated Factors (Part B)
Edited by Sankar Adhya

Volume 275. Viral Polymerases and Related Proteins
Edited by Lawrence C. Kuo, David B. Olsen, and Steven S. Carroll

Volume 276. Macromolecular Crystallography (Part A)
Edited by Charles W. Carter, Jr., and Robert M. Sweet

Volume 277. Macromolecular Crystallography (Part B)
Edited by Charles W. Carter, Jr., and Robert M. Sweet

Volume 278. Fluorescence Spectroscopy
Edited by Ludwig Brand and Michael L. Johnson

VOLUME 279. Vitamins and Coenzymes (Part I)
Edited by DONALD B. MCCORMICK, JOHN W. SUTTIE, AND CONRAD WAGNER

VOLUME 280. Vitamins and Coenzymes (Part J)
Edited by DONALD B. MCCORMICK, JOHN W. SUTTIE, AND CONRAD WAGNER

VOLUME 281. Vitamins and Coenzymes (Part K)
Edited by DONALD B. MCCORMICK, JOHN W. SUTTIE, AND CONRAD WAGNER

VOLUME 282. Vitamins and Coenzymes (Part L)
Edited by DONALD B. MCCORMICK, JOHN W. SUTTIE, AND CONRAD WAGNER

VOLUME 283. Cell Cycle Control
Edited by WILLIAM G. DUNPHY

VOLUME 284. Lipases (Part A: Biotechnology)
Edited by BYRON RUBIN AND EDWARD A. DENNIS

VOLUME 285. Cumulative Subject Index Volumes 263, 264, 266–284, 286–289

VOLUME 286. Lipases (Part B: Enzyme Characterization and Utilization)
Edited by BYRON RUBIN AND EDWARD A. DENNIS

VOLUME 287. Chemokines
Edited by RICHARD HORUK

VOLUME 288. Chemokine Receptors
Edited by RICHARD HORUK

VOLUME 289. Solid Phase Peptide Synthesis
Edited by GREGG B. FIELDS

VOLUME 290. Molecular Chaperones
Edited by GEORGE H. LORIMER AND THOMAS BALDWIN

VOLUME 291. Caged Compounds
Edited by GERARD MARRIOTT

VOLUME 292. ABC Transporters: Biochemical, Cellular, and Molecular Aspects
Edited by SURESH V. AMBUDKAR AND MICHAEL M. GOTTESMAN

VOLUME 293. Ion Channels (Part B)
Edited by P. MICHAEL CONN

VOLUME 294. Ion Channels (Part C)
Edited by P. MICHAEL CONN

VOLUME 295. Energetics of Biological Macromolecules (Part B)
Edited by GARY K. ACKERS AND MICHAEL L. JOHNSON

VOLUME 296. Neurotransmitter Transporters
Edited by SUSAN G. AMARA

VOLUME 297. Photosynthesis: Molecular Biology of Energy Capture
Edited by LEE MCINTOSH

VOLUME 298. Molecular Motors and the Cytoskeleton (Part B)
Edited by RICHARD B. VALLEE

VOLUME 299. Oxidants and Antioxidants (Part A)
Edited by LESTER PACKER

VOLUME 300. Oxidants and Antioxidants (Part B)
Edited by LESTER PACKER

VOLUME 301. Nitric Oxide: Biological and Antioxidant Activities (Part C)
Edited by LESTER PACKER

VOLUME 302. Green Fluorescent Protein
Edited by P. MICHAEL CONN

VOLUME 303. cDNA Preparation and Display
Edited by SHERMAN M. WEISSMAN

VOLUME 304. Chromatin
Edited by PAUL M. WASSARMAN AND ALAN P. WOLFFE

VOLUME 305. Bioluminescence and Chemiluminescence (Part C)
Edited by THOMAS O. BALDWIN AND MIRIAM M. ZIEGLER

VOLUME 306. Expression of Recombinant Genes in Eukaryotic Systems
Edited by JOSEPH C. GLORIOSO AND MARTIN C. SCHMIDT

VOLUME 307. Confocal Microscopy
Edited by P. MICHAEL CONN

VOLUME 308. Enzyme Kinetics and Mechanism (Part E: Energetics of Enzyme Catalysis)
Edited by DANIEL L. PURICH AND VERN L. SCHRAMM

VOLUME 309. Amyloid, Prions, and Other Protein Aggregates
Edited by RONALD WETZEL

VOLUME 310. Biofilms
Edited by RON J. DOYLE

VOLUME 311. Sphingolipid Metabolism and Cell Signaling (Part A)
Edited by ALFRED H. MERRILL, JR., AND YUSUF A. HANNUN

VOLUME 312. Sphingolipid Metabolism and Cell Signaling (Part B)
Edited by ALFRED H. MERRILL, JR., AND YUSUF A. HANNUN

VOLUME 313. Antisense Technology (Part A: General Methods, Methods of Delivery, and RNA Studies)
Edited by M. IAN PHILLIPS

VOLUME 314. Antisense Technology (Part B: Applications)
Edited by M. IAN PHILLIPS

VOLUME 315. Vertebrate Phototransduction and the Visual Cycle (Part A)
Edited by KRZYSZTOF PALCZEWSKI

VOLUME 316. Vertebrate Phototransduction and the Visual Cycle (Part B)
Edited by KRZYSZTOF PALCZEWSKI

VOLUME 317. RNA–Ligand Interactions (Part A: Structural Biology Methods)
Edited by DANIEL W. CELANDER AND JOHN N. ABELSON

VOLUME 318. RNA–Ligand Interactions (Part B: Molecular Biology Methods)
Edited by DANIEL W. CELANDER AND JOHN N. ABELSON

VOLUME 319. Singlet Oxygen, UV-A, and Ozone
Edited by LESTER PACKER AND HELMUT SIES

VOLUME 320. Cumulative Subject Index Volumes 290–319

VOLUME 321. Numerical Computer Methods (Part C)
Edited by MICHAEL L. JOHNSON AND LUDWIG BRAND

VOLUME 322. Apoptosis
Edited by JOHN C. REED

VOLUME 323. Energetics of Biological Macromolecules (Part C)
Edited by MICHAEL L. JOHNSON AND GARY K. ACKERS

VOLUME 324. Branched-Chain Amino Acids (Part B)
Edited by ROBERT A. HARRIS AND JOHN R. SOKATCH

VOLUME 325. Regulators and Effectors of Small GTPases (Part D: Rho Family)
Edited by W. E. BALCH, CHANNING J. DER, AND ALAN HALL

VOLUME 326. Applications of Chimeric Genes and Hybrid Proteins (Part A: Gene Expression and Protein Purification)
Edited by JEREMY THORNER, SCOTT D. EMR, AND JOHN N. ABELSON

VOLUME 327. Applications of Chimeric Genes and Hybrid Proteins (Part B: Cell Biology and Physiology)
Edited by JEREMY THORNER, SCOTT D. EMR, AND JOHN N. ABELSON

VOLUME 328. Applications of Chimeric Genes and Hybrid Proteins (Part C: Protein–Protein Interactions and Genomics)
Edited by JEREMY THORNER, SCOTT D. EMR, AND JOHN N. ABELSON

VOLUME 329. Regulators and Effectors of Small GTPases (Part E: GTPases Involved in Vesicular Traffic)
Edited by W. E. BALCH, CHANNING J. DER, AND ALAN HALL

VOLUME 330. Hyperthermophilic Enzymes (Part A)
Edited by MICHAEL W. W. ADAMS AND ROBERT M. KELLY

VOLUME 331. Hyperthermophilic Enzymes (Part B)
Edited by MICHAEL W. W. ADAMS AND ROBERT M. KELLY

VOLUME 332. Regulators and Effectors of Small GTPases (Part F: Ras Family I)
Edited by W. E. BALCH, CHANNING J. DER, AND ALAN HALL

VOLUME 333. Regulators and Effectors of Small GTPases (Part G: Ras Family II)
Edited by W. E. BALCH, CHANNING J. DER, AND ALAN HALL

Volume 334. Hyperthermophilic Enzymes (Part C)
Edited by Michael W. W. Adams and Robert M. Kelly

Volume 335. Flavonoids and Other Polyphenols
Edited by Lester Packer

Volume 336. Microbial Growth in Biofilms (Part A: Developmental and Molecular Biological Aspects)
Edited by Ron J. Doyle

Volume 337. Microbial Growth in Biofilms (Part B: Special Environments and Physicochemical Aspects)
Edited by Ron J. Doyle

Volume 338. Nuclear Magnetic Resonance of Biological Macromolecules (Part A)
Edited by Thomas L. James, Volker Dötsch, and Uli Schmitz

Volume 339. Nuclear Magnetic Resonance of Biological Macromolecules (Part B)
Edited by Thomas L. James, Volker Dötsch, and Uli Schmitz

Volume 340. Drug–Nucleic Acid Interactions
Edited by Jonathan B. Chaires and Michael J. Waring

Volume 341. Ribonucleases (Part A)
Edited by Allen W. Nicholson

Volume 342. Ribonucleases (Part B)
Edited by Allen W. Nicholson

Volume 343. G Protein Pathways (Part A: Receptors)
Edited by Ravi Iyengar and John D. Hildebrandt

Volume 344. G Protein Pathways (Part B: G Proteins and Their Regulators)
Edited by Ravi Iyengar and John D. Hildebrandt

Volume 345. G Protein Pathways (Part C: Effector Mechanisms)
Edited by Ravi Iyengar and John D. Hildebrandt

Volume 346. Gene Therapy Methods
Edited by M. Ian Phillips

Volume 347. Protein Sensors and Reactive Oxygen Species (Part A: Selenoproteins and Thioredoxin)
Edited by Helmut Sies and Lester Packer

Volume 348. Protein Sensors and Reactive Oxygen Species (Part B: Thiol Enzymes and Proteins)
Edited by Helmut Sies and Lester Packer

Volume 349. Superoxide Dismutase
Edited by Lester Packer

Volume 350. Guide to Yeast Genetics and Molecular and Cell Biology (Part B)
Edited by Christine Guthrie and Gerald R. Fink

VOLUME 370. RNA Polymerases and Associated Factors (Part C)
Edited by SANKAR L. ADHYA AND SUSAN GARGES

VOLUME 371. RNA Polymerases and Associated Factors (Part D)
Edited by SANKAR L. ADHYA AND SUSAN GARGES

VOLUME 372. Liposomes (Part B)
Edited by NEGAT DÜZGÜNEŞ

VOLUME 373. Liposomes (Part C)
Edited by NEGAT DÜZGÜNEŞ

VOLUME 374. Macromolecular Crystallography (Part D)
Edited by CHARLES W. CARTER, JR., AND ROBERT W. SWEET

VOLUME 375. Chromatin and Chromatin Remodeling Enzymes (Part A)
Edited by C. DAVID ALLIS AND CARL WU

VOLUME 376. Chromatin and Chromatin Remodeling Enzymes (Part B)
Edited by C. DAVID ALLIS AND CARL WU

VOLUME 377. Chromatin and Chromatin Remodeling Enzymes (Part C)
Edited by C. DAVID ALLIS AND CARL WU

VOLUME 378. Quinones and Quinone Enzymes (Part A)
Edited by HELMUT SIES AND LESTER PACKER

VOLUME 379. Energetics of Biological Macromolecules (Part D)
(in preparation)
Edited by JO M. HOLT, MICHAEL L. JOHNSON, AND GARY K. ACKERS

VOLUME 380. Energetics of Biological Macromolecules (Part E)
(in preparation)
Edited by JO M. HOLT, MICHAEL L. JOHNSON, AND GARY K. ACKERS

VOLUME 381. Oxygen Sensing (in preparation)
Edited by CHANDAN K. SEN AND GREGG L. SEMENZA

VOLUME 382. Quinones and Quinone Enzymes (Part B) (in preparation)
Edited by HELMUT SIES AND LESTER PACKER

VOLUME 383. Numerical Computer Methods, (Part D) (in preparation)
Edited by LUDWIG BRAND AND MICHAEL L. JOHNSON

VOLUME 384. Numerical Computer Methods, (Part E) (in preparation)
Edited by LUDWIG BRAND AND MICHAEL L. JOHNSON

VOLUME 385. Imaging in Biological Research, (Part A) (in preparation)
Edited by P. MICHAEL CONN

VOLUME 386. Imaging in Biological Research, (Part B) (in preparation)
Edited by P. MICHAEL CONN

Section I

Coenzyme Q: Detection and Quinone Reductases

[1] Investigation of Regulatory Mechanisms in Coenzyme Q Metabolism

By JACOB GRÜNLER and GUSTAV DALLNER

Introduction

Coenzyme Q (CoQ) is not only present in all cells and membranes of the eukaryotic cells, but it is also synthesized and broken down in all cells.[1] The lipid is a structural component, and its major part is associated with the membrane. Unlike cholesterol, CoQ itself is not serving as substrate for other products, its metabolism is strictly regulated, and variations in amount are limited. Modifications in concentration have great impact on cellular functions and normal metabolism, and for this reason it is important to study the regulatory mechanisms and establish the possibilities to interact with these processes.

Precursors for Labeling

All substrates that are intermediates in the mevalonate pathway can be used to label CoQ. However, for practical reasons, few of the radioactive precursors are used.

1. Both ^{14}C- and ^{3}H-acetate are available commercially for the lowest costs among precursors, and CoQ labeling is obtained with them in cells and tissues. However, acetate is one of the central compounds in the cellular metabolism, used in a number of processes, and it is not possible to obtain a cellular concentration that is saturating for the CoQ synthesis, particularly using animal systems. Therefore, it seems to be a less suitable precursor in any type of labeling experiments for this lipid.
2. Tritiated isopentenyl-PP (IPP) can be prepared in the laboratory even in large amounts,[2] and it is used most often in *in vitro* systems. Before using this precursor for cellular or animal systems, one should be ensured that the compound is able to enter into the cell despite the fact that it is highly negatively charged.
3. The most suitable precursor for labeling of CoQ in an *in vivo* system is radioactive mevalonate, and the ^{3}H-form can be prepared by a relatively easy synthesis in large amounts.[3] Because only the enzymes of the

[1] G. Dallner and P. J. Sindelar, *Free Radic. Biol. Med.* **29,** 285 (2000).
[2] T. Chojnacki, *Methods Enzymol.* **378,** 152 (2004).
[3] R. K. Keller, *J. Biol. Chem.* **261,** 12053 (1986).

0076-6879/04 $35.00

mevalonate pathway use this substrate, it is suitable for labeling CoQ in most biological systems. It is water soluble and penetrates cell membranes. Because of the high volatibility in ethanol solution, evaporation has to be done with care. The commercial preparations are mainly in the lactone form, but most cells and tissues effectively hydrolyze it, and therefore the precursor is present in the biologically required acid form after administration. If hydrolysis is carried out before incubation, 1 drop of 5 *M* NaOH is added to a 5-ml ethanol solution, and the pH is kept between 7 and 8. It should be evaporated with N_2 without heating and dissolved in P-buffer (~0.1 *M*, pH 7.5).

4. Farnesyl-PP is the final product of the mevalonate pathway and used as the substrate by the branch-point enzymes.[4,5] High costs and low permeability make this compound a poor choice for an isoprenoid side chain precursor of CoQ. This is true even in *in vitro* systems. The free alcohol form, ^{3}H-farnesol is applied in tissue culture system and in some cells, and it appears in mevalonate pathway end-products.[6,7] This was demonstrated to be due to the presence of two enzymes, which consecutively phosphorylate the free alcohol to farnesyl-PP.[8]

The Biological System

So far no reconstituted system is described to investigate CoQ biosynthesis and metabolism *in vitro,* and consequently, one has to apply *in vivo* approaches.

1. Isolated cells, mostly hepatocytes, are used when animal cells are studied. In the nonanimal world, *Escherichia coli* and yeast are the choice in most investigations.

2. Tissue culture cells are the most often used biological materials, because they are easy to handle and can be amplified. Of special value are the cells originating from brain (e.g., neuroblastoma cells) because in the brain the synthesis of the main mevalonate lipid, cholesterol, is very low. Although in most tissues cholesterol has a similar half-life as CoQ, in brain the $t_{1/2}$ is 4080 and 90 h, respectively.[9] In a hepatocyte, the rate of labeling

[4] J. L. Goldstein and M. S. Brown, *Nature* **343,** 425 (1990).
[5] J. Grünler, J. Ericsson, and G. Dallner, *Biochim. Biophys. Acta* **1212,** 259 (1994).
[6] D. C. Crick, D. A. Andres, and C. J. Waechter, *Biochem. Biophys. Res. Commun.* **211,** 590 (1995).
[7] S. J. Fliesler and R. K. Keller, *Biochem. Biophys. Res. Commun.* **210,** 695 (1995).
[8] M. Bentinger, J. Grünler, E. Peterson, E. Swiezewska, and G. Dallner, *Arch. Biochem. Biophys.* **353,** 191 (1998).
[9] M. Andersson, P. G. Elmberger, C. Edlund, K. Kristensson, and G. Dallner, *FEBS Lett.* **269,** 15 (1990).

of cholesterol/CoQ is 100:1 or even more, whereas in neuroblastoma cells, it is equal.

3. Slices prepared from organs (e.g., the brain) can be useful in studies in which analysis of the whole organ or animal is not applicable because of poor uptake of the precursor, or in which tissue specificity is analyzed (e.g., localization of biosynthesis).[10,11] Slices 0.5 mm-thick are prepared by use of a microtome and placed in cold Krebs-Henseleit buffer. A number of simple but also more advanced automatized systems are commercially available for this purpose. Slices of about 0.5 g are incubated in conical flasks containing 4 ml medium or buffer supplemented with 5 m*M* glucose. This system is then preincubated at 37° for 15 min before adding the appropriate radioactive precursor (2 mCi ^{3}H-acetate or 0.5 mCi ^{3}H-mevalonolactone). During incubation, the mixtures are maintained at 37°, exposed to carbogen gas (95% O_2, 5% CO_2), and shaken slowly. The incubation is terminated by rinsing the slices in cold 0.9% NaCl solution.

4. Organ perfusion, mostly the liver, is often used. Various precursors, enzyme inhibitors or activators, and drugs can be used to study the rate of synthesis and distribution of newly synthesized lipids, to analyze the transport mechanisms, the production and elimination of breakdown products, and the excretion through the circulation and bile (i.e., a large variety of questions can be answered).

The technical details to assemble the perfusion apparatus and to perform the procedure are described.[12,13] The perfusion medium is bovine or human erythrocytes washed on the day of the experiment with 0.9% NaCl several times and resuspended to a hematocrit of 20% in sterile Dulbecco's modified Eagle medium with glutamine.[14,15] The livers of the rat (150–220 g) under pentobarbital anesthesia are connected to the perfusion system by introducing the intravenous infusion cannula into the vena portae. The first 50–100 ml of perfusate is discarded. The vena cava superior is then cannulated, so that the perfusate can be collected in a beaker. From the beaker, the perfusate is recirculated to the lung for oxygenation by the artificial lung and reentry into the liver. By constructing the appropriate system, it is possible to keep the total perfusion volume as low as

[10] P. G. Elmberger, A. Kalén, E. L. Appelkvist, and G. Dallner, *Eur. J. Biochem.* **168,** 1 (1987).
[11] M. Andersson, F. Åberg, H. Teclebrhan, C. Edlund, and E. L. Appelkvist, *Mech. Ageing Dev.* **85,** 1 (1995).
[12] L. L. Miller, C. G. Bly, M. L. Watson, and W. F. Bale, *J. Exp. Med.* **94,** 431 (1951).
[13] R. L. Hamilton, M. N. Berry, M. C. Williams, and E. M. Severinghaus, *J. Lipid Res.* **15,** 182 (1974).
[14] J. B. Marsh, *Methods Enzymol.* **129,** 498 (1986).
[15] P. G. Elmberger, A. Kalén, U. T. Brunk, and G. Dallner, *Lipids* **24,** 919 (1989).

20 ml, which is advantageous in metabolic studies. An even flow during perfusion is kept up with a peristaltic pump, and the lung is continuously gassed (95% O_2, 5% CO_2). The system should also contain a pH electrode to control the pH at 7.4. It is advisable to place the whole system, including the lung and the rat, in a closed room (plastic box) and keep the temperature at 37°. Experiments can be continued for several hours, as long as the liver is functionally intact. Damage can be observed by measuring the release of some cytosolic enzymes (e.g., alanine or aspartate aminotransferases). By cannulating the common bile duct, it is possible to collect the bile continuously and analyze its content.

5. CoQ metabolism depends on and is interconnected with other metabolic pathways and influenced by factors such as hormones, age, and temperature. Consequently, the major questions in the field of the regulation of CoQ metabolism can be approached only by using the complete *in vivo* system, which involves studies on animals. Mice, rats, guinea pigs, and rabbits were used in previous investigations. Some factors should be considered in animal experimentation. The rate of CoQ synthesis is greatly increased in rat organs during the 30 days after birth, followed by a decrease.[16,17] No diurnal variations are present, and there is no dependency on the nutritional state. However when investigating the liver, it may be advantageous to use fasting rodents, because this condition decreases cholesterol synthesis to a great extent, saving precursors, like mevalonate, for CoQ synthesis. It does not seem to be significantly different between males and females. Strain differences exist, both in the level and the turnover of the lipid, and the results obtained in one strain are not necessarily valid for another strain of rat.

Depending on the substance administered, the possibilities are subcutaneous, intramuscular, intraperitoneal (ip), and intravenous injections. In most cases ip administration is applicable and usually gives the same result as intravenous injection. Water-insoluble substances may be dissolved in small amounts (0.5 ml) of ethanol, 1% Triton-X 100, or dimethyl sulfoxide (DMSO), which are tolerated by adult rats, and the uptake into the circulation from the peritoneal cavity takes only minutes. A simple way, when large amounts are administered, is to mix the lipophilic compound with a vehicle such as Intralipid (Fresenius Kabi, Uppsala, Sweden), which is a lipid emulsion used intravenously as nutrients for humans.[18] The substance in question is vortexed with Intralipid, which is suitable for ip injection.

[16] A. Kalén, E. L. Appelkvist, and G. Dallner, *Lipids* **24,** 579 (1989).
[17] M. Turunen and G. Dallner, *Chem. Biol. Interact.* **116,** 79 (1998).
[18] Y. A. Carpentier and I. E. Dupont, *World J. Surg.* **24,** 1493 (2000).

Intravenous injection into the tail vein is technically difficult, particularly in the case of larger volumes. An extremely efficient way of direct administration into the circulation is to open the abdomen after mebumal anesthesia and make the injection into the large mesenteric artery just before its entrance in the portal vein. This is technically not difficult, and in the case of radioactive compounds, it gives 8–10 times higher labeling in the liver compared with what is obtained after ip injection. When extended action is required, the abdominal wall is sutured.

The two main applications for precursor labeling of CoQ *in vivo* are the determination of the biosynthetic rate and that of the half-life. When the rate of CoQ biosynthesis is investigated in rat liver, 150 μCi ^{3}H-mevalonate/100 g is given ip, and the increase in radioactivity in CoQ is followed between 30 and 90 min.[17] The lag phase of incorporation is 30 min, followed by a linearly increasing incorporation. In mice, 150 μCi ^{3}H-mevalonate/10 g is given ip, which gives, after a 15- to 20-min lag phase, a linear rate of incorporation between 20 and 50 min.[19] In the case of the kidneys, the lag phase is 20 min, and the incorporation is linear up to 60 min. For determining the half-life of CoQ in rat, 1 mCi ^{3}H-mevalonate/180 g injected ip and the decay of radioactivity in different organs is followed between days 2 and 6.[17] In mice, 250 μCi ^{3}H-mevalonate/10 g injected ip twice at 0 and 24 h and the radioactivity is followed between 50 and 340 hours.[19] To obtain a high initial radioactivity, it may be necessary to inject the label several times initially. For labeling of prenylated proteins in the kidney of newborn rats, it is necessary to inject ^{3}H-mevalonate five times every fifth hour.[20] Mevalonate is not reused in these experiments, and the $t_{1/2}$ obtained is a valid value.

Preparation of Diet

Coenzyme Q and a large portion of drugs are not soluble in water, and the food containing the substance has to be prepared continuously to avoid breakdown processes. There is a practical procedure to perform the preparation of diet, which has proved to be efficient, time saving, inexpensive, and functional. One kilogram ordinary rat chow pellets are placed into larger glass vessels, and CoQ or the drug is dissolved in 250 ml of acetone. The mixture is added to the pellets and occasionally mixed thoroughly with a large spoon. When no acetone is left on the bottom, the pellets are placed on an aluminum foil and left in a ventilated hood overnight. In this

[19] M. Turunen, J. M. Peters, F. J. Gonzalez, S. Schedin, and G. Dallner, *J. Mol. Biol.* **297,** 607 (2000).

[20] I. Parmryd and G. Dallner, *Arch. Biochem. Biophys.* **364,** 153 (1999).

procedure, the pellets are sufficiently impregnated by the compound and consumed to the same extent as the control food.

Inhibitors of Coenzyme Q Biosynthesis

No compound is available at present for the inhibition of CoQ biosynthesis in animals. In tissue culture cells, 4-aminobenzoic and 4-chlorobenzoic acids are active by inhibition of polyprenyl-*p*-hydroxybenzoate-transferase, resulting in an extensive decrease of CoQ.[21] However, these inhibitors do not influence the biosynthetic system in rats. A decrease in liver CoQ (24%) is observed after dietary thiouracil treatment (0.3%, 20 days), which inhibits thyroid gland function.[22] Oral administration of vitamin A (5 mg, 15 days) lowers rat liver CoQ content by 32%.[22] A protein-free diet for 3 weeks decreases the CoQ amount in the liver and heart by 25–30% but not in kidney, spleen, and brain.[23] Complete protein deprivation probably has a number of other effects on the various biosynthetic systems, and the changes in CoQ biosynthesis seem to be part of an extensive modification of cellular metabolism.

The mevalonate pathway lipids are subjected to two types of regulation, with a central and a terminal target point. The central regulatory enzyme of the mevalonate pathway is 3-hydroxy-3-methylglutaryl coenzyme A (HMG-CoA) reductase, which determines the size of the common substrate pool of the branch-point enzymes, farnesyl-PP (FPP). According to the flow diversion hypothesis, the FFP pool has a much greater impact on cholesterol synthesis than on the other lipid end-products.[24] Squalene synthase has a high K_m for its substrate, and on a decrease of the FPP pool, the enzyme is not saturated, resulting in decreased cholesterol synthesis. All other branch-point enzymes, *trans*- and *cis*-prenyltransferases, and farnesyl protein transferase have higher affinity, and according to the concept, a lower FPP pool is still sufficient for their saturation. At present, however, we do not know the K_m for these enzymes in purified form. Clearly, farnesyl protein transferase has a very low K_m ($\sim$0.1 μM).[25] It seems that both *trans*- and *cis*-prenyltransferases have K_m values below that of squalene synthase; however, the differences are moderate.[26,27] This situation is of great medical interest, because HMG-CoA reductase

[21] S. S. Alam, A. M. D. Nabuiri, and H. Rudney, *Arch. Biochem. Biophys.* **171,** 183 (1975).

[22] T. Ramasarma, *J. Scient. Industr. Res.* **27,** 147 (1968).

[23] V. C. Joshi, J. Jayaraman, and T. Ramasarma, *Biochem. J.* **88,** 25 (1963).

[24] M. S. Brown and J. L. Goldstein, *J. Lipid Res.* **21,** 505 (1980).

[25] Y. Reiss, M. S. Brown, and J. L. Goldstein, *J. Biol. Chem.* **267,** 6403 (1992).

[26] J. Ericsson, A. Thelin, T. Chojnacki, and G. Dallner, *J. Biol. Chem.* **267,** 19730 (1992).

[27] H. Teclebrhan, J. Olsson, E. Swiezewska, and G. Dallner, *J. Biol. Chem.* **268,** 23081 (1993).

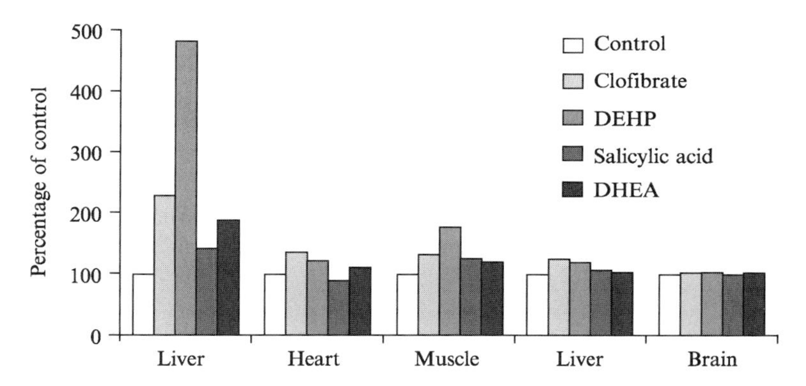

FIG. 1. Effect of a 6-week treatment with peroxisomal inducers on CoQ content in various organs.[29,30]

inhibitors, statins, are increasingly used in treatment of hypercholesterolemia. Analyses of blood in humans and the organs of rats have demonstrated that statin treatment affects not only cholesterol but also CoQ and dolichol synthesis. Dietary treatment of rats with mevinolin (500 mg/kg food, 30 days) decreases CoQ content in heart, muscle, and liver by 18%, 17%, and 11%, respectively.[28]

Activators of Coenzyme Q Biosynthesis

1. The uptake of dietery CoQ into various organs is limited and cannot compensate for the loss observed during aging and various pathological conditions. It would be of great interest to find compounds that are able to upregulate CoQ biosynthesis and increase the lipid concentration in various organs. In experimental systems, it is possible to attain this aim by administration of peroxisomal inducers to rodents.[29,30] Examples of compounds given in rats are di(2-ethylhexyl)phthalate (DEHP), 2% in the diet (w/w); clofibrate, 0.6%; 2-ethylhexanoic acid, 2%; and acetylsalicylic acid, 1% (Fig. 1).

In mice, the most often used drugs are perfluorooctanoic acid, 0.02%; WY-14643, 0.125%; and nafenopinin, 0.125%.[31] Depending on the type

[28] P. Löw, M. Andersson, C. Edlund, and G. Dallner, *Biochim. Biophys. Acta* **1165,** 102 (1992).
[29] F. Åberg, Y. Zhang, E. L. Appelkvist, and G. Dallner, *Chem. Biol. Interact.* **91,** 1 (1994).
[30] F. Åberg, Y. Zhang, H. Teclebrhan, E. L. Appelkvist, and G. Dallner, *Chem. Biol. Interact.* **99,** 205 (1996).
[31] J. K. Reddy and C. Ruiyin, *Ann. N. Y. Acad. Sci.* **804,** 176 (1996).

of experiment, the compounds are given for 1–6 weeks. A large number of compounds are now identified as inducers, and it seems that all of them are increasing the number of peroxisomes, β-oxidation of fatty acids, and the amount of CoQ. A direct relationship exists between the extent of β-oxidation induction and the increase of CoQ amount. Consequently, some of the inducers elevate the lipid concentration several-fold in the liver, whereas others give only a 10–20% induction. Most of the organs investigated are involved, but the brain is an exception. In old rats, (age of 16 months), induction of β-oxidation is 30-fold, but there is no increase of CoQ, indicating that other factors are also involved in the elevation of the lipid amount.[17] Opposed to the selective dietary uptake, the increase observed after administration of inducers occurs in all organelles and is the result of increased biosynthesis and unchanged rate of breakdown.

2. Coenzyme Q metabolism is controlled by a number of hormones, but the mechanism of this control is not yet investigated. Administration of growth hormone (0.8 IU/ip, 7 days), thyroxin (40 μg/ip, 7 days), dehydroepiandrosterone (0.6% in diet, 3 weeks), and cortisone (40 μg/ip, 7 days) increase liver CoQ amount in the rat by 15%, 55%, 75%, and 120%, respectively.[32] It is very probable that modifications in CoQ amount and synthesis described in some pathophysiological processes can be related directly to hormonal effects, and, therefore, it is of importance to know how hormones are involved (e.g., to determine hormonal levels in specific conditions).

3. Rats placed in the cold room (4°) for 10 days double the amount of CoQ in liver.[33] In mice, the increase of the lipid amount is 60% after 3 weeks in the liver, but it remains unchanged in the kidney.[34] It seems that this type of induction is a liver-specific event.

4. A diet deficient in vitamin A in mice and rat increases rat liver CoQ content.[35] Deficiency can be evoked by supplying pregnant mice with a vitamin A–deficient diet during the last 2 weeks of pregnancy and feeding the offspring with the same diet to 10–12 weeks of age. In these mice, the vitamin A level in the liver is less than 1.6% of the control value. In liver homogenate, the deficiency increases the CoQ level to 133% of the

[32] M. Turunen, E. Swiezewska, T. Chojnacki, P. Sindelar, and G. Dallner, *Free Radic. Res.* **36,** 437 (2002).

[33] H. N. Aithal, C. V. Joshi, and T. Ramasarma, *Biochim. Biophys. Acta* **162,** 66 (1968).

[34] M. Bentinger, M. Turunen, X. X. Zhang, Y. J. Y. Wan, and G. Dallner, *J. Mol. Biol.* **326,** 795 (2003).

[35] A. K. Sohlenius-Sternbeck, E. L. Appelkvist, and J. W. DePierre, *Biochem. Pharmacol.* **59,** 377 (2000).

control, and on subfractionation, this value is 141% in the mitochondrial fraction and 336% in the microsomes.

5. Squalestatin 1 is an inhibitor of squalene synthase and in Chinese hamster ovary cells causes a complete inhibition of the synthase at a 2 μM concentration. At the same time, this inhibitor results in a three-fold elevation of CoQ synthesis.[36,37] After a long-term subcutaneous infusion of rats with squalestatin 1 (15 μg/h for 14 days), the liver CoQ amount is doubled and hepatic dolichol and dolichyl-P concentrations are also increased substantially.[38] These findings agree with the conclusions obtained after administration of statins and suggest that *trans*- and *cis*-pernyltransferases are not saturated under certain conditions. Inhibition of squalene synthase increases the FPP pool to a large extent, causing an elevation of both the synthesis and amount of CoQ and dolichol.

Nuclear Receptors

Nuclear hormone receptors are ligand-dependent DNA-binding transcription factors that exert control over gene expression.[39] Most of the receptors involved in the regulation of lipid metabolism are heterodimers of the retinoid X receptor (RXR), which has three isoforms, α, β, and γ.[40] In hepatocyte-specific RXRα–deficient mice, the CoQ concentration in the liver is half that found in the control, the lipid amount is greatly increased on administration of peroxisomal inducers, and these deficient mice do not increase the CoQ amount on cold exposure.[34] Retinoid X receptor is operating as a heterodimer, and the exact nature of the receptor is not known; it is probably the liver X or farnesol X receptor. Peroxisome proliferator-activated receptor α (PPARα) is also identified as a regulator of CoQ biosynthesis.[19] In the liver of the PPARα–null mice, the CoQ amount is the same as in the control, but administration of peroxisomal inducers, such as DEHP, does not increase the lipid concentration. On the other hand, the deficiency has no effect on the induction by cold treatment. These findings are summarized in Table I.

The involvement of the previously described nuclear receptors in CoQ biosynthesis is of great interest, because their regulatory action is

[36] A. Baxter, B. J. Fitzgerald, J. L. Hutson, A. D. McCarthy, J. M. Motteram, B. C. Ross, M. Sapra, M. A. Snowden, N. S. Watson, R. J. Williams, and C. Wright, *J. Biol. Chem.* **267,** 11705 (1992).

[37] A. Thelin, E. Peterson, J. L. Hutson, A. D. McCarthy, J. Ericsson, and G. Dallner, *Biochim. Biophys. Acta* **1215,** 245 (1994).

[38] R. K. Keller, *Biochim. Biophys. Acta* **1303,** 169 (1996).

[39] S. Khorasanizadeh and F. Rastinejad, *Trends Biochem. Sci.* **26,** 384 (2001).

[40] F. Rastinejad, *Curr. Opin. Struct. Biol.* **11,** 33 (2001).

TABLE I
RXRα AND PPARα INVOLVEMENT IN COENZYME Q METABOLISM

	Biosynthesis	Induction by the peroxisomal inducer DEHP	Induction by cold treatment
RXRα	Required	Not required	Required
PPARα	Not required	Required	Not required

specifically directed to CoQ and does not influence the metabolism of cholesterol, dolichol, and dolichyl phosphate. The effects observed are the result of increased or decreased biosynthesis, and the rate of breakdown remains unchanged. Identifying the appropriate receptors will make it possible to develop new drugs acting as agonists or antagonists and as modulators of gene expression. This approach is used today for treatment of several diseases. Peroxisomal inducers, which are effective in rodents but not in humans, are an example of this method. There are nontoxic drugs, such as clofibrate, that increase the endogenous CoQ synthesis in most organs, and the newly synthesized lipid is distributed in all cellular organelles.

Breakdown of Coenzyme Q

Coenzyme Q has a high turnover, indicating an extensive breakdown of the lipid. The half-life is somewhat different in various organs of rat and, it varies between 49 and 125 h (Table II).[41]

Coenzyme Q breakdown products were analyzed in urine, feces, and some tissues of rat and rabbit after administration of CoQ7 and CoQ10, and a number of metabolites were identified by chemical synthesis and gas chromatography.[42,43] These metabolites had an intact ring, substituted as in CoQ, but the side chain was shortened to approximately one modified isoprene, which was carboxylated. The metabolites in the bile and urine of guinea pig after administration of ^{14}C–CoQ10 and deuterium-labeled CoQ10 were studied by gas chromatography–mass spectrometry after methylation and acetylation. The largest metabolite both in bile and

[41] A. Thelin, S. Schedin, and G. Dallner, *FEBS Lett.* **313,** 118 (1992).
[42] I. Imada, M. Watenabe, N. Matsumoto, and H. Morimoto, *Biochemistry* **9,** 2870 (1970).
[43] T. Nakamura, T. Ohno, K. Hamamura, and T. Sato, *Biofactors* **9,** 111 (1999).

TABLE II
Half-life of CoQ in Rat Tissues

Tissue	Half-life (h)
Liver	79
Heart	59
Kidney	125
Stomach	72
Thyroid	49
Colon	54
Muscle	50
Intestine	54
Pancreas	94
Testis	50
Thymus	104
Spleen	64

urine has the following structure: 2,3-dimethoxy-5-methyl-6-(3′-methyl-5′-carboxy-2′-pentenyl)-1,4-benzohydroquinone. The total portion of this compound in the bile was found in glucuronidated form.

The ip injection of ^{3}H-CoQ10 in rats results in the uptake of the lipid into a number of organs, and all organs also contain water-soluble metabolites.[44] In the feces, which include the constituents of the bile, water-soluble metabolites are discovered. A large part of the injected ^{3}H-CoQ is excreted by the liver without any modification and recovered from the feces. The radioactive metabolites in the urine can be isolated by sequential chromatography using aminopropyl cartridge, high-performance liquid chromatography (HPLC) with aminopropyl column, and HPLC with reversed-phase column. The two main metabolites appearing after chromatography have identical mass spectra with the main signal at *m/z* 389. On fragmentation, two signals are visible at *m/z* 79 and 80, corresponding to PO_3 and HPO_3. The main compound, previously identified in the urine, in phosphorylated form has a calculated molecular weight of 389, which is probably the compound isolated by reversed-phase HPLC (Fig. 2). The explanation for the isolation of two compounds with identical molecular weight is that the ring can be phosphorylated either at carbon 4, as it appears on Fig. 2, or at carbon 1. There are evidences that the breakdown of the endogenous CoQ follows the same pathway as the exogenous one.[43,44]

[44] M. Bentinger, G. Dallner, T. Chojnacki, and E. Swiezewska, *Free Radic. Biol. Med.* **34,** 563 (2003).

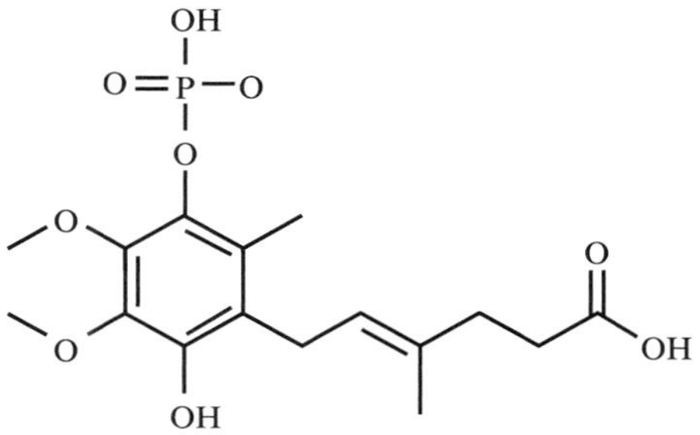

Fig. 2. Structure of the main phosphorylated metabolite of CoQ isolated from the urine.

Coenzyme Q is synthesized in all cells, and it is also broken down at all locations.[10,44] The main metabolite seems to be phosphorylated in the cell, and the water-soluble derivative is translocated to the blood, transported to the kidney, and excreted to the urine by the glomerular–tubular system. In the liver, the major derivatization is by glucuronidation, and this compound is excreted through the bile.[43] Other types of conjugations are also possible for the minor metabolites.

Metabolite identification in normal and pathological conditions is of great interest, because these products, together with nonmodified or modified intermediates of the biosynthesis, may participate in the control of the synthetic system. Metabolities that are produced in other conditions, such as ultraviolet (UV) irradiation or lipid peroxidation (if the CoQ is in the oxidized form), may also exert effects on CoQ metabolism.[45,46] Farnesol and its metabolites, like farnesoic acid and dicarboxylic acid, the hydroxylated form, esterified and glucuronidated farnesols are known effectors of various cellular processes, but their influence on CoQ metabolism has not yet been investigated.[47,48]

Effects of Exogenous Coenzyme Q

All cells are capable of synthesizing CoQ in amounts required for normal function. Consequently, the uptake of exogenous CoQ into animal organisms is limited. The uptake is important, because it may or may not

[45] H. Morimoto, I. Imada, and G. Goto, *Liebigs Ann. Chem.* **735,** 65 (1970).

[46] P. Forsmark-Andree, C. P. Lee, G. Dallner, and L. Ernster, *Free Radic. Biol. Med.* **22,** 391 (1997).

[47] J. B. Roullet, H. Xue, J. Chapman, R. Bychkov, C. M. Roullet, F. C. Luft, H. Haller, and D. A. McCarron, *J. Biol. Chem.* **272,** 32240 (1997).

[48] K. Machida, T. Tanaka, Y. Yano, S. Otani, and M. Taniguchi, *Microbiology* **145,** 293 (1999).

change cellular functions, depending on the localization of the lipid. Theoretically, the possibility exists that the uptake influences the endogenous synthesis by feedback regulation. High uptake can be deleterious, partly because of the increased mitochondrial respiration that elevates the cellular level of free radicals and partly because of the limited capacity of the CoQ reducing system, not necessarily sufficient for the complete reduction of CoQ. This situation is a possible event with isolated cells, tissue culture cells, yeast, and *Caenorhabditis elegans,* which take up all types of lipids with limited selectivity. In these systems, CoQ is incorporated into mitochondria also, because it was demonstrated that in some models a transport mechanism is operating between the plasma membrane and mitochondria.[49] Thus, when performing experiments with these biological systems, one has to control the amount CoQ taken up to avoid overloading the cellular handling capacity.

In animal systems, only a few percent of CoQ is taken up from the diet, and the plasma concentration is less than doubled.[50] The lipid increase is several-fold in the adrenal glands and ovaries, doubled or trebled in the liver and spleen, and is significant in blood mononuclear cells and aorta; the uptake is very low in heart, thymus, pancreas, islets of Langerhans, pituitary gland, and testis and absent in brain, kidney, muscle, thyroid gland, and blood polynuclear cells. Concerning the uptake into organs, it is of importance to note a few facts.

1. In blood and organs, practically all the exogenous CoQ is recovered in reduced form.[50,51]
2. The exogenous CoQ has a different localization compared with the endogenous counterpart. The exogenous lipid in liver is not found in mitochondria but mostly distributed in other organelles, mainly in lysosomes.[34] In addition, this CoQ is present to a large extent in nonmembrane-bound form.
3. The exogenous CoQ does not downregulate the biosynthesis of the endogenous lipid.[28] In some studies, even a moderate increase of endogenous CoQ amount was observed.[52,53] It is possible that the increased production of metabolites in the catabolic process has a signaling function to the biosynthetic system.

[49] C. Santos-Ocana, T. Q. Do, S. Padilla, P. Navas, and C. F. Clarke, *J. Biol. Chem.* **277,** 10973 (2002).
[50] Y. Zhang, F. Åberg, E. L. Appelkvist, G. Dallner, and L. Ernster, *J. Nutr.* **125,** 446 (1995).
[51] D. Mohr, V. W. Bowry, and R. Stocker, *Biochim. Biophys. Acta* **1126,** 247 (1992).
[52] K. Lönnrot, P. Holm, A. Lagerstedt, H. Huhtala, and H. Alho, *Biochem. Mol. Biol. Int.* **44,** 727 (1992).
[53] W. G. Ibrahim, H. N. Bhagavan, R. K. Chopra, and C. K. Chow, *J. Nutr.* **130,** 2343 (2000).

There are two main points of interest when discussing the role of CoQ uptake into the cells of various tissues. First, the main localization of CoQ in the central hydrophobic regions of the membrane does not allow additional cellular uptake under normal conditions.[54] However, in the case of deficiency — a condition not yet defined — uptake occurs, and the normal situation is re-established. An example of this situation is the genetic disturbance in children resulting in decreased CoQ biosynthesis.[55] In this case, dietary supplementation of CoQ10 dramatically improves the neuronal and muscular symptoms, indicating that this lipid is now taken up and distributed to the right localization.[56] According to this concept, membrane deficiency of CoQ will result in uptake into the appropriate organs.

Second, CoQ probably has important effects on functions that do not require a direct uptake and action as redox carrier or antioxidant. The decreased expression of β_2-integrin CD11b by monocytes found in humans on dietary administration of CoQ10 can be a powerful antiatherogenic action.[57] In a similar way, changes in the production of cytokines and interleukins in the circulation, hormonal modifications, and the use of CoQ metabolites in various signaling systems may influence functions to a great extent. Such factors can be important in cardiovascular and neurological diseases, in which CoQ exerts beneficial effects.[58,59]

Pentane Extraction

In many studies, it would be desirable to analyze the same organelle membrane but with different concentrations of the lipid and in a state when it is free-from CoQ. One way to establish this experimental system is the extraction with selected organic solvents, the most used is *n*-pentane.[60] To obtain this aim, subcellular fractions after isolation are washed and suspended in 0.15 *M* KCl and lyophilized. The lyophilized samples are vortexed in an aliquot of *n*-pentane and shaken in an ice-water bath for 5 min. The protein is sedimented by centrifugation, and the extraction is

[54] G. Lenaz, C. Bovina, M. D'Aurelio, R. Fato, G. Formiggini, M. L. Genova, G. Giuliano, M. M. Pich, U. Paolucci, G. P. Castelli, and B. Ventura, *Ann. N. Y. Acad. Sci.* **959,** 199 (2002).

[55] P. Rustin, A. Munnich, and A. Rötig, *Methods Enzymol.* **382,** 81 (2004).

[56] A. Rötig, E. L. Appelkvist, V. Geromel, D. Chretien, N. Kadhom, P. Edery, M. Lebideau, G. Dallner, A. Munnich, L. Ernster, and P. Rustin, *Lancet* **356,** 391 (2000).

[57] M. Turunen, L. Wehlin, M. Sjöberg, J. Lundahl, G. Dallner, K. Brismar, and P. J. Sindelar, *Biochem. Biophys. Res. Commun.* **296,** 255 (2002).

[58] G. P. Littarru, M. Battino, and K. Folkers, *in* "Handbook of Antioxidants" (E. Cadenas and L. Packer, eds.), p. 203. Marcel Dekker, New York, 1996.

[59] M. F. Beal, *Biofactors* **9,** 261 (1999).

[60] L. Ernster, I. Y. Lee, B. Norling, and B. Persson, *Eur. J. Biochem.* **9,** 299 (1969).

repeated four times. For reconstitution, the extracted samples are equilibrated with CoQ with or without other neutral lipids (cholesterol, dolichol). In this procedure, the lipid is dissolved in 0.25 ml *n*-pentane containing the membrane protein (8 mg). The mixture is kept in an ice-water bath for 10 min, centrifuged, and dried under nitrogen. The fractions are resuspended in 0.15 *M* KCl immediately before measurements.

n-Pentane extraction removes no proteins or phospholipids but practically all the neutral lipid CoQ, cholesterol, and dolichol.[61] After reconstitution, the complete respiratory chain and its ability for oxidative phosphorylation in mitochondria are retained at the original level. In microsomes, all electron transport enzymes, phosphatases, glycosyl transferases, and enzymes of the mevalonate pathway remain intact. During reinsertion of lipids, the available binding sites for various neutral lipids can be used for one single lipid (i.e., the membranes are enriched in the lipid in question, such as CoQ). It is also possible to prepare membranes with an increased concentration of the specific lipid and test lipid interactions by use of a mixture during reconstruction. Such experiments are of interest in studying the influence of membrane structure and composition on various functions.

[61] Å. Jakobsson-Borin, F. Åberg, and G. Dallner, *Biochim. Biophys. Acta* **1213,** 159 (1994).

[2] Methods for Characterizing TPQ-Containing Proteins

By Jennifer L. DuBois and Judith P. Klinman

Introduction

The standard 20 amino acid side chains define a repertoire of chemical function rich in potent nucleophiles but notably electrophile poor. Nature overcomes this deficiency by either supplying an exogenous cofactor to an enzyme, such as pyridoxal phosphate (PLP), pyrroloquinoline quinone (PQQ), or a metal ion, or by modifying the existing amino acid side chains in situ to supply an active site with reactive carbonyls. The latter strategy has certain advantages, because it obviates the need for transport and insertion of a reactive species across cellular space and into an enzyme. In 1990, the first of what has been shown to be a series of covalently linked, side chain–derived cofactors was discovered and definitively described as

0076-6879/04 $35.00

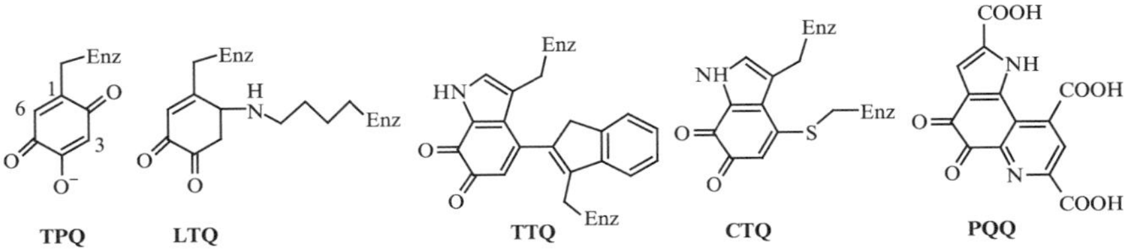

FIG. 1. Amino-acid–derived quinocofactors: topaquinone (TPQ), lysyl-tyrosylquinone (LTQ), tryptophan tryptophylquinone (TTQ), cysteine tyrosyl-quinone (CTQ), and pyrroloquinoline quinone (PQQ).

2,4,5-trihydroxyphenylalanine quinone (topaquinone, or TPQ).[1] For years, TPQ was misidentified as either PLP or PQQ, in part because its posttranslational origin was so unexpected. However, as detailed in an earlier volume in this series, TPQ was identified unequivocally through a combination of DNA sequencing, mass spectrometry, nuclear magnetic resonance (NMR), ultraviolet/visible, and resonance Raman (rR) spectroscopies.[2] Since that time, crystallography has yielded a variety of images of TPQ in the protein environment[3–8]; model chemistry has provided a means for selectively characterizing the properties of TPQ in isolation[9]; and a variety of similar, side chain–derived quinocofactors have been discovered (Fig. 1).[10–12] TPQ and the copper amine oxidases (CAOs) that house it are among the best-studied representatives of these cofactors/enzymes. Hence, work on these species serves as a useful guidepost for future work on more newly and

[1] S. M. Janes, D. Mu, D. Wemmer, A. J. Smith, S. Kaur, D. Maltby, A. L. Burlingame, and J. P. Klinman, *Science* **248,** 981 (1990).

[2] S. M. Janes and J. P. Klinman, *Methods Enzymol.* **258,** 20 (1995).

[3] R. B. Li, J. P. Klinman, and F. S. Mathews, *Structure* **6,** 293 (1998).

[4] Z. W. Chen, B. Schwartz, N. K. Williams, R. B. Li, J. P. Klinman, and F. S. Mathews, *Biochemistry* **39,** 9709 (2000).

[5] H. C. Freeman, J. M. Guss, V. Kumar, W. S. McIntire, and V. M. Zubak, *Acta Crystallogr. Sect. D-Biol. Crystallogr.* **52,** 197 (1996).

[6] E. V. S. Hogdall, G. Houen, M. Borre, J. R. Bundgaard, L. I. Larsson, and J. Vuust, *Eur. J. Biochem.* **251,** 320 (1998).

[7] V. Kumar, D. M. Dooley, H. C. Freeman, J. M. Guss, I. Harvey, M. A. McGuirl, M. C. J. Wilce, and V. M. Zubak, *Structure* **4,** 943 (1996).

[8] M. C. J. Wilce, D. M. Dooley, H. C. Freeman, J. M. Guss, H. Matsunami, W. S. McIntire, C. E. Ruggiero, K. Tanizawa, and H. Yamaguchi, *Biochemistry* **36,** 16116 (1997).

[9] M. Mure and J. P. Klinman, *J. Am. Chem. Soc.* **115,** 7117 (1993).

[10] C. Anthony, *Biochem. J.* **320,** 697 (1996).

[11] S. Datta, Y. Mori, K. Takagi, K. Kawaguchi, Z. W. Chen, T. Okajima, S. Kuroda, T. Ikeda, K. Kano, K. Tanizawa, and F. S. Mathews, *Proc. Natl. Acad. Sci. USA* **98,** 14268 (2001).

[12] I. Vandenberghe, J. K. Kim, B. Devreese, A. Hacisalihoglu, H. Iwabuki, T. Okajima, S. Kuroda, O. Adachi, J. A. Jongejan, J. A. Duine, K. Tanizawa, and J. Van Beeumen, *J. Biol. Chem.* **276,** 42923 (2001).

as yet to be discovered quinoproteins. A review of the major methods for preparing and characterizing TPQ-containing enzymes follows, with frequent reference to the CAO from *Hansenula polymorpha* (HPAO) as a specific model system.

Methods

Protein Expression and Purification

Methods for the heterologous expression of CAOs have permitted both greater protein production and a broader range of possible experiments, including site-directed mutagenesis and CAO functional studies. Three types of expression hosts are in use: yeast (or *Escherichia coli*), for the production of large amounts of mature and fully metallated enzyme[13–15]; *E. coli*, for the production of metal-free and consequently TPQ-free HPAO[16]; and drosophila S2 cells, for expression of mammalian forms of cell-surface CAOs, known as semicarbazide-sensitive amine oxidases (SSAOs).[17] The last system, which provides a means for obtaining fair amounts of highly pure mammalian protein for characterization, is expected to assist in addressing one of the more vexing and still unanswered questions about SSAOs: what is the biological role of the reaction they catalyze?

Protocols for growing and purifying apo- or holo-HPAO are summarized in Tables IA and IB. Although purification procedures for the two forms of the protein are similar, expression of metal-free HPAO requires shifting from yeast *(Saccharomyces cerevisiae)* to an *E. coli* host, with substantially different growing conditions.[16] Growth in *E. coli* provides a solution to a series of unanticipated problems. Because the yeast system was developed first, initial efforts naturally focused on producing apo-HPAO in *S. cerevisiae* grown in a metal-depleted environment. Yeast tolerate Cu-depleted conditions well; however, eliminating Zn from the growth medium is fatal. Restoring Zn to an otherwise metal-free environment results in healthy yeast, but also in HPAO purified with ~100% Zn incorporation into the Cu site. Subsequent removal of Zn from the active site is nontrivial.[18]

[13] G. A. Juda, J. A. Bollinger, and D. M. Dooley, *Protein Expr. Purif.* **22,** 455 (2001).
[14] M. A. McGuirl, C. D. McCahon, K. A. McKeown, and D. M. Dooley, *Plant Physiol.* **106,** 1205 (1994).
[15] D. Y. Cai and J. P. Klinman, *Biochemistry* **33,** 7647 (1994).
[16] D. Y. Cai, N. K. Williams, and J. P. Klinman, *J. Biol. Chem.* **272,** 19277 (1997).
[17] B. O. Elmore, J. A. Bollinger, and D. M. Dooley, *J. Biol. Inorg. Chem.* **7,** 565 (2002).
[18] S. A. Mills and J. P. Klinman, *J. Am. Chem. Soc.* **122,** 9897 (2000).

TABLE IA
CONDITIONS FOR EXPRESSION OF HPAO APOPROTEIN AND HOLOPROTEIN

	Apoprotein	Holoprotein
Vector	pET3a (Stratagene)	pDB20
Host	*E. coli* BL21DE3 (Novagen)	*S. cerevisiae* CG379 (Yeast Genetic Stock Ctr, UCB)
Growth medium	9l M9 salts plus 9l 20 g/l casein. Casein media stirred 1 h with 15 g/l Chelex-100, filtered, and combined 750 ml/750 ml with M9 medium in 4 l plastic flasks. Add 1 mg/ml sodium ampicillin and 30 ml of a 50 m*M* $MgSO_4$/20% dextrose solution per flask after autoclaving.	URA^- liquid medium 1.5 l/4 l flasks, supplemented with 10 μM $CuSO_4$
Innoculation conditions	Single colony from fresh LB/ampicillin (1 g/l) plate incubated at 37° into each of 12 flasks. Flasks at 37° on 250 RPM shaker.	Single colony from fresh URA^- plate incubated at 30° into 50 ml URA^- media overnight at 30° on 250 RPM shaker. 5 ml of this culture into 250 ml for second overnight culture. 20 ml of this culture into each of 12 flasks containing 1.5 l each.
Growth time	To OD_{600} 0.55–0.6 (~10 h); induce with 0.4 *M* IPTG; add 1/m*M* penicillamine (chelator) and grow 4 h before collecting cells by centrifugation.	To OD_{600} of 5 (~12–16 h). Collect cells by centrifugation.

LB = Luria broth; IPTG = isopropylthio-β-D-galactoside.

The HPAO gene was therefore engineered into a pET3a vector (Stratagene, La Jolla, CA) for expression in *E. coli.*[16] *E. coli* grow well in media treated extensively with chelators. To maintain as metal-free an environment as possible (with special focus on eliminating Cu(II) and Zn(II)), chelators are used in buffers and media in each step of apo-HPAO growth and purification. Plastic ware, rinsed in deionized filtered (Millipore, Bedford, MA) water, is used in place of glass, because the metals retained by glass have been found to be sufficient to remetallate the apoprotein. Specific conditions for apo-HPAO expression are as follows: selective Luria broth (LB)/ampicillin (1 mg/l) plates are streaked from frozen glycerol stocks of BL21DE3 cells (Novagen) harboring the pET3a–HPAO construct and grown for ~12 h at 37°. Single colonies are cut from plates to inoculate each of 12, 2.8-l plastic flasks, each containing 1.5 l metal-free medium.

TABLE IB

CONDITIONS FOR PURIFICATION OF HPAO APOPROTEIN AND HOLOPROTEIN

	Apoprotein	Holoprotein
Cell pellet disruption	Sonication (~70-g cell paste) followed by centrifugation.	Cell rupture by glass beads Bead-Beater (Biospec) followed by centrifugation.
Dialysis of supernatant	Against 5 m*M* potassium phosphate, 1 m*M* EDTA, 1 m*M* DDC, pH 7, three cycles of buffer changes at ≥6 h (4°).	Against 5 m*M* potassium phosphate, pH 7.2, 3 cycles of buffer changes at ≥6 h (4°). Centrifuge to remove remaining particulates.
Chromatography	(1) DEAE Sepharose ion exchange column equilibrated to pH 7, 5 m*M* potassium phosphate, 1 m*M* EDTA, 1 m*M* DDC. Eluted with 800 ml of a 5 m*M*–100 m*M* potassium gradient, followed by 200 ml of 100 m*M* potassium phosphate. Fractions selected by SDS–PAGE, combined, concentrated by ultrafiltration to less than 4 ml (Millipore XM-50 filter).	(1) DEAE Sepharose ion exchange column equilibrated to pH 7.2, 5 m*M* potassium phosphate. Eluted with 400 ml of a 5 m*M*–100 m*M* potassium phosphate gradient followed by 200 ml of a 100 m*M* to 300 m*M* gradient. Fractions selected by SDS–PAGE, combined, concentrated by ultrafiltration to <6 ml (Millipore XM-50 filter).
	(2) Sephacryl S-200 gel filtration potassium phosphate, 1 m*M* EDTA, 1 m*M* DDC. Buffer pumped through column at 0.15 ml/min, and fractions collected at 4-min intervals. ≥95% pure fractions (by SDS–PAGE) combined, concentrated (Millipore 30,000 MWCO centrifuge filters) to ≥10 mg/ml (~1 ml). Pure protein resuspended and reconcentrated 3x in 50 m*M* HEPES pH 7 or 8.	(2) Sephacryl S-300 gel filtration column equilibrated to 50 m*M* potassium phosphate. Buffer pumped through column at 0.15 ml/min, and fractions collected at 4-min intervals. ≥95% pure fractions (by SDS–PAGE) combined and concentrated (Millipore 30,000 MWCO centrifuge filters) to ≥10 mg/ml.
Typical yields	10–20 mg pure protein per 18 l culture.	150–200 mg per protein per 18 l culture.

DDC = diethyldithiocarbamic acid; MWCO = molecular weight cutoff; SDS-PAGE = sodium dodecyl sulfate-polyacrylamide gel electrophoresis.

The medium is made by combining 9 l of 2× M9 salts (recipe: 1 g/l NaCl, 6 g/l KH_2PO_4, 1 g/l NH_4Cl, 12 g/l Na_2HPO_4) with 9 l of 80 g/l casein in a 1:1 ratio (750 ml of each per flask), where the casein is first stirred for at least 1 h in the presence of 60 g/l of the general chelating material Chelex-100 (Sigma, St. Louis, MO) to eliminate metals before filtering. The 12 flasks of medium are autoclaved and cooled. Two sterile-filtered supplements are added at the time of inoculation: ampicillin (to 1 mg/l), and 30 ml per flask of 50 m*M* $MgSO_4$/20% dextrose.

Once inoculated, the flasks are put on a shaker at 37°/250 rpm for ~10 h, at which point the OD_{600} for each is measured. At an OD_{600} of 0.55–0.65, expression of HPAO is induced by adding isopropylthio-β-D-galactoside (IPTG) to 0.4 m*M*. It is critical that cells not be allowed to overgrow, because yields will be severely affected. Penicillamine is added at this point as a metal chelator (final concentration: 0.2 mg/l), to ensure that the overexpressed protein cannot acquire any remaining trace metals. Cells are grown for 4 more h and then collected by gentle centrifugation (20 min at ~4000*g*, e.g., 5 krpm in a Sorvall GS3 rotor [Kendro Laboratory Products, Newtown, CT]). The cell pellet may at this point be kept at −80° and thawed for later use or immediately sonicated to retrieve the soluble proteins.

Sonication is carried out in a rosette glass buried to the rim in ice. The cell pellet should be completely resuspended to a total volume of about 300 ml in 100 m*M* potassium phosphate buffer (pH 7) with 1 m*M* ethylene diamine tetra-acetic acid (EDTA) and 1 m*M* diethyldithiocarbamate (DDC) added as metal chelators. A sonicator from VWR Scientific (Branson Ultrasonics Corp., Danbury, CT) is used with a duty cycle of 8 and output of 30% (i.e., 90 watt power output) for three periods of 10–15 min, with 5-min breaks in between. The liquid is then loaded into 6–8 bottles and centrifuged in a Sorvall SS-34 rotor (Kendro Laboratory Products, Newtown, CT) at 18 krpm krpm (~39,000*g*) for 1 h (4°). The supernatant is retained for dialysis against 4 l of 5 m*M* potassium phosphate plus 1 m*M* EDTA and 1 m*M* DDC (pH 7) at 4°. Metal-free dialysis tubing is available (Spectrum Labs, Rancho Dominguez, CA; 50,000 molecular weight cut-off [MWCO]) and preferable for this purification. The dialysis buffer is changed at least twice at intervals of 3 h or greater. The dialyzed liquid is then ready for column purification.

Two column steps are used: anion-binding ion exchange followed by gel filtration. The anion exchange column (5 cm × 100 cm) uses diethylaminoethyl (DEAE) sepharose resin (Pfitzer, New York, NY) equilibrated to 5 m*M* potassium phosphate plus 1 m*M* EDTA and 1 m*M* DDC buffer (pH 7). The dialysate is bound and washed with 0.5 l of the equilibrating buffer, then eluted by a salt gradient of 5 m*M* to 100 m*M* potassium phosphate (800 ml), followed by 200 ml of 100 m*M* potassium phosphate. Fractions are evaluated for HPAO content by sodium dodecyl sulfate-polyacrylamide gel electrophoresis (SDS–PAGE, 10% acrylamide), with the HPAO monomer running at a molecular weight of ~75 kDa. Fractions ~60% in HPAO or greater are pooled and concentrated under N_2 gas pressure in an Amicon ultrafiltration device fitted with an XM-50 membrane (Millipore, Bedford, MA). Once ~4 ml, the protein is loaded onto a gel filration column (1 cm × 120 cm) packed with S-200 sephacryl resin (Pharmacia) equilibrated to 50 m*M* potassium phosphate plus 1 m*M* EDTA and 1 m*M* DDC buffer (pH 7). The same buffer is pumped over the column at 0.15 ml/min, and fractions are collected in ~0.5-ml volumes (4 min/fraction). Fractions are once again monitored by SDS–PAGE, and those >95% in HPAO are retained for concentration and buffer exchange. Buffer exchange is necessary, because the protein is eluted in the presence of metal chelators, which interfere with metal reconstitution experiments. A Millipore ultrafiltration centrifuge concentrator is used, with a 30,000 MWCO filter. Pure protein is concentrated to ~10–20 mg/ml in a 1-ml volume, then resuspended in 20 ml 50 m*M* HEPES, pH 7–8. Cycles of concentration and resuspension are repeated three times. Purified, concentrated protein (~10–20 mg/18 l culture) is dispensed into 50-μl aliquots, rapidly frozen in liquid N_2, and stored at -20°. Protein prepared and frozen in this way may remain stable for up to a month, although some precipitation of frozen protein has been observed. Best results have been obtained with protein kept at -4° and used within 1–2 weeks of preparation.

Purification of holoprotein from *S. cerevisiae* is largely similar to the procedure described previously,[15] with the omission of chelators and the addition of 10 μM $CuSO_4$ to the growth medium. The CG379 strain used has a URA^- genotype, so that URA^- dropout selection media is used (URA^- medium: 6.7 g/l yeast nitrogen base, 20 g/l dextrose, 5 g/l $(NH_4)_2SO_4$; autoclave). The standard URA^- medium is supplemented with a sterile-filtered nucleotide/amino acid mixture, added to 1× from a 50× stock: 2.5 g/l adenine, 2.5 g/l histidine, 3.75 g/l leucine, 2.5 g/l tryptophan. Further supplementing the medium with 10 μM $CuSO_4$ ensures the production of HPAO with stoichiometric Cu(II) incorporation. Cultures are grown from freshly streaked plates made from URA-medium plus 20 g/l agar (incubated overnight, 30°), with two scale-up stages: a single colony is inoculated into 50 ml media and grown overnight; 5 ml of this culture is used to innoculate a 250-ml overnight culture; 20 ml of this culture is used to inoculate each of 12 flasks containing 1.5 l (30°, 250 rpm shaker). These are grown for ~12–16 h until the cultures reach an optical density of ~5, at which point the cultures are centrifuged gently to collect the cell pellet. Again, the pellet may be stored for several months at -80° or immediately lysed mechanically in the presence of commercially available protease inhibitors (Sigma, St. Louis, MO), using glass or ceramic beads in a Bead Beater (Biospec Products, Bartlesville, OK). The soluble fraction is separated by centrifugation (30 min at 4° and 27,000*g*, e.g., 15,000 rpm, Sorvall SS-34 rotor) and loaded into dialysis tubes (10,000 MWCO) for dialysis against 4 l of 5 m*M* potassium phosphate buffer (pH 7). The buffer is changed three times after cycles of at least 6 h, and the dialysate is recentrifuged to eliminate any remaining particulate material. Clear dialysate liquid is loaded onto a DEAE column equilibrated to 5 m*M* potassium phosphate buffer at pH 7.2, washed with about 0.5 l of the equilibrating buffer, and then eluted in two gradients: 400 ml of a 5–100 m*M* potassium phosphate gradient, followed by 100 ml of 100 m*M* potassium phosphate buffer, and 200 ml of a 100–300 m*M* potassium phosphate gradient, followed by 200 ml of 300 m*M* potassium phosphate. Fractions are assayed by SDS–PAGE, concentrated by ultrafiltration to ~6 ml, and loaded onto a gel filtration column. Column conditions are the same for apoprotein and holoprotein, except that following the column purified fractions are merely concentrated, and the buffer is not exchanged. This procedure yields ~150–200 mg of pure protein (>95% by SDS–PAGE) per 18 l of culture.

Most recently, a method for expressing mammalian CAOs in tissue culture has been developed.[17] The CAO is overexpressed as a secreted enzyme under the control of a metallothionein promoter in *Drosophila* S2 cell culture. The cell growth medium is recovered, and its proteins are separated on heparin affinity, hydroxyapatite, and gel filtration media. These methods allow for the first time the production of large amounts of highly pure recombinant enzyme from mammalian sources.

Reactivity Assays

CAOs catalyze the oxidation of amines to aldehydes, using O_2 as the oxidant and generating H_2O_2 and NH_4^+ as by-products. Like PLP-dependent enzymes, they do so by an amino transferase mechanism involving formation of a covalent Schiff-base adduct at the C5-carbonyl of TPQ. The reaction has been shown to proceed by a ping-pong mechanism, with two kinetically separable halves:

$$E_{ox} + RCH_2NH_3^+ \rightleftharpoons E_{ox} - RCH_2NH_3^+ \rightarrow E_{red} - NH_3^+ + RCHO + H_2O \quad (1)$$

$$E_{red} - NH_3^+ + O_2 + H_2O \rightarrow E_{ox} + NH_4^+ + H_2O_2 \quad (2)$$

In the reductive half-reaction (1), a protonated amine substrate binds to the enzyme and forms a substrate Schiff base with oxidized TPQ. This undergoes active site base catalyzed proton abstraction to form a product Schiff base with concomitant reduction of TPQ. A subsequent hydrolysis step yields the corresponding aldehyde and the reduced aminoquinol form of the enzyme. In the oxidative half-reaction (2), the reduced cofactor is reoxidized by molecular oxygen to liberate NH_4^+ and recharge the cofactor to its oxidized form.[19]

Steady-state kinetic studies of the two half-reactions have been carried out under a variety of conditions (e.g., of pH, temperature, ionic strength,

[19] M. Mure, S. A. Mills, and J. P. Klinman, *Biochemistry* **41,** 9269 (2002).

D_2O, site-directed mutations, inhibitors) to test specific hypotheses. For the reductive half-reaction, k_{cat} and k_{cat}/K_M(amine) are typically measured by monitoring the absorbance change on oxidation of benzylamine to benzaldehyde (λ_{max} benzaldehyde is 250 nm, $\varepsilon = 12{,}500\ M^{-1}\ cm^{-1}$). Initial rates (first 5–10% of the total reaction) are measured by fitting a line to absorbance vs time data, and slopes are plotted as a function of [amine]. Fitting the resultant curve with the Michaelis–Menten equation yields the steady-state parameters k_{cat} and k_{cat}/K_M(amine). Air supplies a saturating amount ($\geq 10\times\ K_M$) of O_2 under most conditions ($K_M(O_2) \sim 10\ \mu M$ at pH 7.8, 25°).[20] Reaction volumes are typically 1 ml (100 mM potassium phosphate, 300 mM ionic strength, attained by adding the appropriate amount of KCl according to pH), with 1 μl of concentrated enzyme (10–15 mg/ml) added to initiate the reaction. The preferred substrate for HPAO is methylamine ($[k_{cat}]_{max} = 7.8\ s^{-1}$, $[k_{cat}/K_M(\text{amine})]_{max} = 3.2 \times 10^5\ M^{-1}\ s^{-1}$ and $[k_{cat}/K_M(O_2)]_{max} = 1.0 \times 10^6\ M^{-1}\ s^{-1}$).[20] However, use of benzylamine has the advantage of allowing for easy spectrophotometric detection of the product, as well as introduction of substitutions at the benzylic or aromatic ring positions.[21] One unit of activity is defined as the amount of active enzyme required to produce 1 μM of benzaldehyde per minute at 25°. Enyzme concentrations are determined by the standard Bradford assay and adjusted to reflect the proportion containing active TPQ, as determined by phenylhydrazine titration (see following).

For the oxidative half-reaction, k_{cat} and $k_{cat}/K_M(O_2)$ are measured by monitoring O_2 depletion with a Clark-type electrode (Yellow Springs Instruments, Yellow Springs, OH).[22] The electrode is first calibrated against the concentration of dissolved O_2 in air or O_2-saturated water (258 μM or 1272 μM, respectively, at 25°). The concentration of O_2 in a given reaction is varied by changing the flow rate of O_2 or N_2 in an N_2/O_2 mixture, which is equilibrated over the reaction while stirring. Initial rates (first 5–10% of the total reaction) are measured by fitting a line to $[O_2]$ vs time data, and slopes are plotted as a function of $[O_2]$; again, fitting the resultant curve with the Michaelis–Menten equation yields the steady-state parameters k_{cat} and $k_{cat}/K_M(O_2)$. For these measurements, N_2 is used as an inert gas. Other gases (He, Ar, CO) have been used in place of N_2 and have had no measurable effect on the kinetic parameters.[20] In all cases, saturating [amine] is used, typically ≥ 10 times K_M(amine) at the given reaction conditions (3–6 mM). Reaction volumes are typically 1 ml (100 mM potassium phosphate, adjusted to 300 mM ionic strength with KCl), with

[20] Y. Goto and J. P. Klinman, *Biochemistry* **41,** 13637 (2002).
[21] C. Hartmann, P. Brzovic, and J. P. Klinman, *Biochemistry* **32,** 2234 (1993).
[22] Q. J. Su and J. P. Klinman, *Biochemistry* **37,** 12513 (1998).

1 μl of concentrated enzyme (10–15 mg/ml) added to initiate the reaction. One unit of activity is defined as the amount of active enzyme needed to reduce 1 μM of O_2 min at 25°.

In addition, presteady-state kinetic measurements have been carried out with stopped-flow UV/visible spectroscopy for both the reductive and oxidative half-reactions.[21,23] These have aimed at identifying novel intermediates or, when possible, determining microscopic rate constants. A two-syringe mixing system has been used, with one syringe holding a concentrated enzyme solution ($\geq$40 μM enzyme in 100 mM potassium phosphate) and the other syringe carrying an equal volume of buffer and the substrate. For reductive half-reaction measurements, fully oxidized enzyme in fully deoxygenated buffer is rapidly mixed with a second syringe containing varying amounts of amine, also in buffer deoxygenated by bubbled argon gas. Spectra are then measured at small time intervals over a broad range of wavelengths using a Xe lamp as a light source and a diode array detector (e.g., Hi-Tech Scientific model SF-61 DX2, Salisbury, UK). For oxidative half-reaction measurements, fully reduced enzyme in O_2-free buffer is rapidly mixed with a second syringe containing varying amounts of dissolved O_2. The technical challenge for such experiments lies in controlling the delivery of O_2. Typically, a continuous stream of helium or argon gas is blown over separate sealed tubes containing enzyme and amine. Amine is then transferred to the enzyme-containing vessel by airtight syringe. Enzyme and amine are mixed, resulting in rapid formation of the "substrate-reduced" form of the enzyme, in which TPQ is reduced to the aminoquinol adduct. Fully substrate-reduced enzyme is then transferred to an airtight syringe for injection into the stopped flow mixer. Meanwhile, buffer containing a defined amount of dissolved O_2 is prepared by blowing N_2/O_2 mixtures over stirred buffer and measuring for dissolved O_2 using a Clark electrode. This is taken up into a second airtight syringe for injection into the stopped flow mixer. The mixer itself is prepared by scrubbing the system for dissolved O_2 using deoxygenated 5 mM sodium dithionite in water.

Finally, in addition to conventional steady-state and presteady-state kinetics measurements, the oxidative half-reaction has also been probed for O^{18}/O^{16} isotope effects.[22] These are highly sensitive to rate-limiting changes in O—O bond order. A complete technical discussion of the measurement of O_2 isotope effects can be found elsewhere.[24]

[23] S. Hirota, T. Iwamoto, S. Kishishita, T. Okajima, O. Yamauchi, and K. Tanizawa, *Biochemistry* **40,** 15789 (2001).

[24] G. Tian and J. P. Klinman, *J. Am. Chem. Soc.* **115,** 8891 (1993).

Quinone-Specific Assays

Copper amine oxidases are unique in the world of enzymes because of the TPQ they carry in proximity to Cu(II). This special combination of quinone and copper provides for a number of techniques for detecting and characterizing CAOs.

Quinone Labeling. A popular method for detecting TPQ is by what is known as redox staining, or reaction of the carbonyl-containing quinone with an electrophilic dye, nitroblue tetrazolium (NBT).[25] This is carried out by first separating the TPQ-containing protein from any contaminants by SDS–PAGE, then electroblotting the separated proteins onto a nitrocellulose membrane (0.45 μm thickness). The membrane is then soaked for 45 min in the dark in a solution containing 0.24 m*M* NBT and 2 *M* potassium glycinate (pH 10), followed by several washes with water. Quinone-containing proteins will appear as purple bands. This assay has the advantage, when used in conjunction with Western blotting and immunostaining, of indicating whether tyrosine has been processed to TPQ under a given set of conditions, making it a useful probe for studies of TPQ biogenesis. It has the disadvantage, however, of being somewhat nonspecific: a variety of quinonoid species (e.g., lysyl-tyrosyl quinone, or LTQ) will react with NBT, as will other redox cofactors, although with greatly reduced sensitivity.

Topaquinone can likewise be colorimetrically functionalized in solution. This is typically done using phenylhydrazine or its derivatives (e.g., 4-nitro-phenylhydrazine), producing brightly colored yellow or orange adducts. Although NBT staining acts as a qualitative indicator for the presence of quinones, phenylhydrazine labeling can be done quantitatively. A solution of enzyme ($\geq$10 μM) in 10–100 m*M* potassium phosphate buffer (pH 7–8) is pipetted into a microcuvette and placed into a spectrophotometer. A spectrum is taken (t_0), and five equivalents of phenylhydrazine are added from a concentrated stock (typically making a 1:100 dilution into the sample cuvette). Spectra are measured at 20- to 30-s intervals for the next 10 min. The λ_{max} for the TPQ-phenylhydrazone adduct in HPAO is 448 nm (ε = 40,500 M^{-1} cm^{-1}, pH 7–8). Work with various CAOs and TPQ model complexes has shown that different proton isomers of hydrazine adducts exist at a given pH, and different ionized forms of the adduct exist over a pH range. Note that the various forms of the adduct have different absorbance characteristics.[26,27] Plotting the absorbance at 448 nm

[25] M. A. Paz, R. Fluckiger, A. Boak, H. M. Kagan, and P. M. Gallop, *J. Biol. Chem.* **266,** 689 (1991).

[26] C. G. Saysell, J. M. Murray, C. M. Wilmot, D. E. Brown, D. M. Dooley, S. E. V. Phillips, M. J. McPherson, and P. F. Knowles, *J. Mol. Catal. B-Enzym.* **8,** 17 (2000).

[27] M. Mure, S. X. Wang, and J. P. Klinman, *J. Am. Chem. Soc.* **125,** 6113 (2003).

versus time and fitting a single exponential curve to the data accurately accounts for the change in absorbance at this wavelength. Using this and the given ε, the concentration of TPQ–phenylhydrazone is determined using Beer's law. Alternately, C-14–labeled phenylhydrazine has been used to functionalize TPQ, followed by scintillation counting to determine the uptake of radioactivity by the protein.[2] The latter has the advantage of not being dependent on the value of ε for protein-bound TPQ, which is known to vary according to protein source, buffer/pH, and mutations in the active site. It is also important to note that TPQ can assume nonproductive orientations, in which it will not be accessible by phenylhydrazine; only TPQ in its reactive conformation is labeled. Therefore phenylhydrazine labeling is a means of quantifying the reactive population of TPQ in a sample of purified enzyme. Finally, phenylhydrazine labeling may be used to distinguish TPQ from LTQ in proteins. Although both form adducts with λ_{max} $\sim$450 nm at neutral pH, this wavelength red shifts at acidic pH (1–3) for LTQ model complexes, whereas TPQ models show a strong blue shift.[27]

Spectroscopy. The simplest and most obvious spectroscopic signature of TPQ is its visible absorbance at 480 nm. Concentrated solutions of amine oxidases are consequently pink. The extinction coefficient for this absorbance is somewhat weak: $\varepsilon_{480} = 2400\ M^{-1}\ \text{cm}^{-1}$ in HPAO, pH 7. For quantitative determination of the active TPQ content, labeling with phenylhydrazine provides a much stronger UV/visible signal.[28]

A second method that has proved especially useful for studies of quinones is resonance Raman spectroscopy. The various C—C, C=O, and C—H stretching modes of TPQ have been well characterized.[29,30] Experiments in D_2O and $H_2{}^{18}O$ have shown, respectively, that the hydrogen at the C3 position and the oxygen at the C5 position are solvent-exchangeable.[31] The rates of solvent exchange have been measured using rR. These correlate with the degree of solvent accessibility at the two ring positions. TPQ oriented in its productive conformation has its C5 oxygen facing the active site base and a solvent-filled substrate channel, with its C3 hydrogen facing away from the channel. Thus, isotopic exchange is fast at C5 and slow at C3. The opposite is observed when the cofactor is in its flipped, nonreactive conformation. Thus, rR can be used as a general tool to determine ring conformation as a function of a selected condition. For example, using

[28] M. M. Palcic and S. M. Janes, *Methods Enzymol.* **258,** 34 (1995).

[29] P. Moenne-Loccoz, N. Nakamura, V. Steinebach, J. A. Duine, M. Mure, J. P. Klinman, and J. Sanders-Loehr, *Biochemistry* **34,** 7020 (1995).

[30] D. E. Brown, M. A. McGuirl, D. M. Dooley, S. M. Janes, D. Mu, and J. P. Klinman, *J. Biol. Chem.* **266,** 4049 (1991).

[31] D. Y. Cai, J. Dove, N. Nakamura, J. Sanders-Loehr, and J. P. Klinman, *Biochemistry* **36,** 11472 (1997).

rR in conjunction with site-directed mutagenesis, several conserved residues in the active site have been shown to have roles in maintaining TPQ in its productive orientation.[32,33]

Finally, electron paramagnetic resonance (EPR) has a unique application in CAOs. Paramagnetic Cu(II) is EPR-detectable, whereas Cu(I) is EPR-silent. Thus, EPR can be used to determine the oxidation state of Cu under a given set of conditions and can further describe the electronic structure of Cu(II). Cu(I) appears in the mechanism for amine oxidation only as an equilibrium partner with the substrate-reduced, Cu(II)-aminoquinol form of the enzyme. Therefore, EPR has been used to specifically probe this Cu(I)/Cu(II) equilibrium, notably in temperature-jump experiments that measure the rate of return to equilibrium after a rapid temperature perturbation.[34]

Quinone Biogenesis. TPQ is generated post-translationally in a spontaneous process that depends only on the active site Cu and O_2. Two observations first indicated that this was true. First, HPAO heterologously expressed in *S. cerevisiae,* a host with no endogenous CAOs, has fully processed and functional TPQ, suggesting that special host-produced factors are not necessary for TPQ formation.[35] Second, CAO expressed in hosts without adequate Cu(II) supplementation in the growth medium is purified with less than an equivalent of Cu(II) per subunit. The reduced Cu titer results in correspondingly reduced TPQ levels. Readdition of Cu(II) to apoprotein in air results in the spontaneous formation of TPQ, which can be monitored by UV/visible spectroscopy by means of the characteristic 480-nm TPQ absorbance.[36]

Experiments in which TPQ formation is monitored are carried out in one of two ways.[37] First, an equivalent of Cu(II) can be added aerobically to a concentrated solution of enzyme ($\geq$20 μM in dimeric HPAO or $\geq$40 μM in HPAO monomer). Typically $CuCl_2$ is used, and reactions are carried out in buffers such as HEPES that do not chelate Cu(II). Cu is added in as small a volume as possible (e.g., 4 μl of a 1 mM solution is added to 40 μM HPAO monomer in 96 μl) and rapidly mixed to offset protein precipitation. Addition of Cu(II) to concentrated enzyme often results in precipitation of protein and is a greater problem at higher

[32] E. L. Green, N. Nakamura, D. M. Dooley, J. P. Klinman, and J. Sanders-Loehr, *Biochemistry* **41,** 687 (2002).

[33] J. Plastino, E. L. Green, J. Sanders-Loehr, and J. P. Klinman, *Biochemistry* **38,** 8204 (1999).

[34] P. N. Turowski, M. A. McGuirl, and D. M. Dooley, *J. Biol. Chem.* **268,** 17680 (1993).

[35] D. Y. Cai and J. P. Klinman, *J. Biol. Chem.* **269,** 32039 (1994).

[36] R. Matsuzaki, S. Suzuki, K. Yamaguchi, T. Fukui, and K. Tanizawa, *Biochemistry* **34,** 4524 (1995).

[37] J. E. Dove, B. Schwartz, N. K. Williams, and J. P. Klinman, *Biochemistry* **39,** 3690 (2000).

[enzyme] (e.g., $\geq$100 μM HPAO). Spectra are then measured at 30- or 60-s intervals for 45 min to 1 h ($t_{1/2} \sim$ 10 min at 50 mM HEPES, pH 7.0) until TPQ formation is complete.

Alternately, the same concentrated solution of apoprotein can be prepared in a gas-tight, septum-sealed cuvette (custom made by fusing a small round-bottom flask to the top of a quartz microcuvette). The protein is then made anaerobic, either by purging the sample on ice with a slow flow of argon gas for $\geq$30 min (depending on cuvette size) or by several cycles of evacuation followed by argon back filling. Spreading the sample (e.g., $\sim$100–250 μl volume) out in the bulbous part of the cuvette increases its surface area, allowing for better gas exchange. A stock solution of 1 mM $CuCl_2$ is likewise made anaerobic by continuous bubbling of argon. An equivalent of Cu(II) is added to the sample by gas-tight syringe, and the sample is shaken to mix. If spectra are taken at this point, a band at 380 nm is immediately observed. This is a Cu-dependent absorbance that disappears over time, presumably as Cu enters the active site. When the 380 nm absorbance has returned to baseline ($\leq$1 h, 25°), the septum is removed from the cuvette, and the sample is flushed with air or O_2 gas from a line to initiate TPQ formation. As before, spectra are measured at 30- or 60s-intervals for 45 min to 1 h, until TPQ formation is complete.

Because TPQ has a low extinction coefficient and because sample concentrations are limited practically and by precipitation concerns, the overall change in absorbance at 480 nm is typically small. Spectra are therefore often visualized as subtractions of the spectrum at time t from the spectrum at time t_0. The t_0 spectrum is generally taken immediately before addition of the reaction-initiating reagent (i.e., either Cu(II) or O_2). Once data are measured, the absorbance at 480 nm may be plotted versus time and fitted to a single exponential to obtain a first-order rate constant, k_{TPQ}.

Formation of one equivalent of TPQ requires consumption of two equivalents of O_2. Consumption of O_2 is the second major observable that can be measured during biogenesis.[38] This has typically been done with a temperature-controlled Clark-type oxygen-sensitive electrode, from Yellow Springs Instruments. Samples are at least 10 μM in HPAO monomer, in 999 μl of 50 mM HEPES, 25°. The sample is stirred rapidly, and TPQ formation is initiated by injection of 1 μl of 10 mM $CuCl_2$ (one equivalent of Cu(II)). The concentration of dissolved O_2 can be varied by equilibrating solutions of enzyme with various mixtures of N_2/O_2 gas. In general, for O_2 concentrations much greater or less than that in air, leakage of O_2 out of or into the reaction chamber can contribute to significant background in the measurements. In those cases, use of a sealed, air-tight

[38] B. Schwartz, J. E. Dove, and J. P. Klinman, *Biochemistry* **39,** 3699 (2000).

reaction chamber is necessary. Such a chamber can be custom-made using vacuum-tight glassware equipped with sealable needle ports.

Conclusions

Work with CAOs has progressed to the stage where these enzymes can be used as virtual laboratories for exploring detailed mechanistic questions. Such a detailed understanding will complement efforts to understand broader questions about CAOs, their roles in nature, and their place in the wider family of quinoproteins.

[3] Aldo-Keto Reductases and Formation of Polycyclic Aromatic Hydrocarbon *o*-Quinones

By Trevor M. Penning

Introduction

*A*ldo-*k*eto *r*eductases (AKRs) are a rapidly growing protein superfamily that is highly conserved across prokaryotes and eukaroytes.[1,2] These are generally monomeric reduced nicotinamide adenine dinucleotide phosphate (NAD[P]H)–linked oxidoreductases that are soluble and consist of approximately 320 amino acids and have molecular weights of 34–37 kDa. These enzymes often convert carbonyl-containing substrates to alcohols; aldehydes are converted to primary alcohols, and ketones are converted to secondary alcohols. When the carbonyl functionality is found on a natural substrate (steroid hormone or prostaglandin) or a drug or xenobiotic, conversion to the corresponding alcohol functionalizes the product for conjugation and elimination. These enzymes thus play a central role in the metabolism of endogenous substrates, drugs, xenobiotics, and carcinogens and are likely to be as important as the cytochrome P450 (CYP) superfamily in dealing with toxic insults.

Currently, there are 114 members in the AKR superfamily distributed across 14 families. For a complete listing and nomenclature visit www.med.upenn.edu/akr. Proteins that belong to separate families have less than 40% sequence identity, and proteins that have greater than 60%

[1] J. M. Jez, T. G. Flynn, and T. M. Penning, *Biochem. Pharmacol.* **54,** 639 (1997).

[2] J. M. Jez, M. J. Bennett, B. P. Schlegel, M. Lewis, and T. M. Penning, *Biochem. J.* **326,** 625 (1997).

0076-6879/04 $35.00

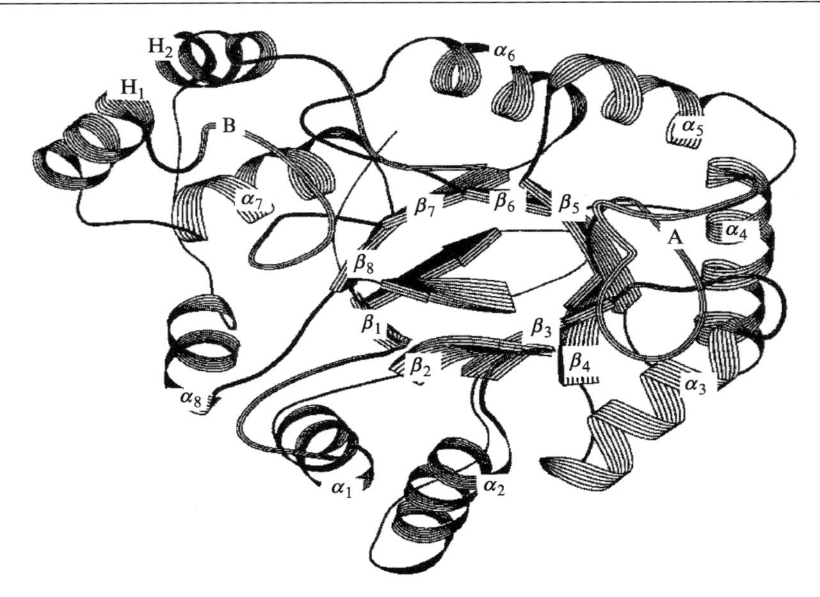

FIG. 1. Structural motif of AKRs, showing the characteristic $(\alpha/\beta)_8$-barrel of the AKRs. Reproduced with permission from Hoog *et al.*, *Proc. Natl. Acad. Sci. USA* **91**, 2517 (1994), copyright (1994), National Academy of Sciences, USA.

similarity belong to the same subfamily. Each of the AKRs has common properties. They catalyze an ordered bi–bi reaction in which pyridine nucleotide binds first and leaves last. They are A-face–specific dehydrogenases and catalyze the transfer of the 4-pro-*R*-hydride ion from the nicotinamide cofactor to the acceptor carbonyl.[3] NADP(H) binding is accompanied by the formation of a loose complex that isomerizes to a tight complex.[4,5] The rate of release of $NADP^+$ is a slow event and is controlled by the slow isomerization of the tight complex to the loose complex before release of cofactor.

There are 13 crystal structures of AKRs in the Protein Data Bank. Each crystal structure has a common $(\alpha/\beta)_8$-barrel motif, in which an α-helix and a β-strand alternate eight times.[6–8] The β-strands coalesce at the core of the structure to comprise the staves of a barrel (Fig. 1). In the available ternary

[3] L. J. Askonas, J. W. Ricigliano, and T. M. Penning, *Biochem. J.* **278,** 835 (1991).
[4] C. E. Grimshaw, K. M. Bohren, C. J. Lai, and K. H. Gabbay, *Biochemistry* **34,** 14356 (1995).
[5] K. Ratnam, H. Ma, and T. M. Penning, *Biochemistry* **38,** 7856 (1999).
[6] S. S. Hoog, J. E. Pawlowski, P. M. Alzari, T. M. Penning, and M. Lewis, *Proc. Natl. Acad. Sci. USA* **91,** 2517 (1994).
[7] M. J. Bennett, B. P. Schlegel, J. M. Jez, T. M. Penning, and M. Lewis, *Biochemistry* **35,** 10702 (1996).

complexes, the cofactor lies across the lip of the barrel, and the substrate is orientated perpendicular to the cofactor. Three large loops at the back of the barrel define substrate specificity. At the base of the barrel, a catalytic tetrad exists that is almost entirely conserved across the superfamily and consists of Tyr, Lys, His, and Asp, in which the catalytic tyrosine functions as the general acid–base in the proton relay to the substrate.[9]

Several AKRs have been implicated in carcinogen metabolism; these include the dihydrodiol dehydrogenases that oxidize polycyclic aromatic hydrocarbon *trans*-dihydrodiols to reactive and redox-active *o*-quinones (AKR1C1–AKR1C4 and AKR1A1);[10–14] the NNK (4-*N*-methyl-*N*-nitrosamino)-1-(3-pyridyl)-1-butanone) carbonyl reductases that reduce tobacco-specific nitrosamino-ketones to their corresponding alcohols (AKR1C1–AKR1C4);[15] and the aflatoxin dialdehyde reductases (AKR7A subfamily).[16,17] This chapter will focus on the AKRs that display dihydrodiol dehydrogenase activity and their role in polycyclic aromatic hydrocarbon (PAH) activation.

Routes of Polycyclic Aromatic Hydrocarbon Activation

Polycyclic aromatic hydrocarbons are major environment pollutants; they are prevalent in tobacco smoke and are suspected human lung carcinogens. PAHs such as benzo[*a*]pyrene (BP) are metabolically activated to mediate their deleterious effects. Three major routes of PAH activation have been described (Fig. 2). In the first route, CYP peroxidase, in the presence of a peroxide substrate, will catalyze peroxide bond cleavage and generate higher oxidation states of iron (compound I and compound II), which can be reduced in the presence of BP to yield a highly reactive BP radical

[8] M. J. Bennett, R. H. Albert, J. M. Jez, H. Ma, T. M. Penning, and M. Lewis, *Structure* **5,** 799 (1997).
[9] B. P. Schlegel, J. M. Jez, and T. M. Penning, *Biochemistry* **37,** 3538 (1998).
[10] A. Hara, H. Taniguchi, T. Nakayama, and H. Sawada, *J. Biochem.* **108,** 250 (1990).
[11] A. Hara, K. Matsurra, Y. Tamada, K. Sato, Y. Miyabe, Y. Deyashiki, and T. Ishida, *Biochem. J.* **313,** 373 (1996).
[12] Y. Deyashiki, H. Taniguchi, T. Amano, T. Nakayama, A. Hara, and H. Sawada, *Biochem. J.* **282,** 741 (1992).
[13] M. E. Burczynski, R. G. Harvey, and T. M. Penning, *Biochemistry* **37,** 6781 (1998).
[14] N. T. Palackal, S-H. Lee, R. G. Harvey, I. A. Blair, and T. M. Penning, *J. Biol. Chem.* **277,** 24799 (2002).
[15] A. Atalla and E. Maser, *Chem. Biol. Inter.* **130,** 737 (2001).
[16] E. M. Ellis, D. J. Judah, G. E. Neal, and J. D. Hayes, *Proc. Natl. Acad. Sci. USA* **90,** 10350 (1993).
[17] L. P. Knight, T. Primiano, J. D. Groopman, T. W. Kensler, and T. R. Sutter, *Carcinogenesis* **20,** 1215 (1999).

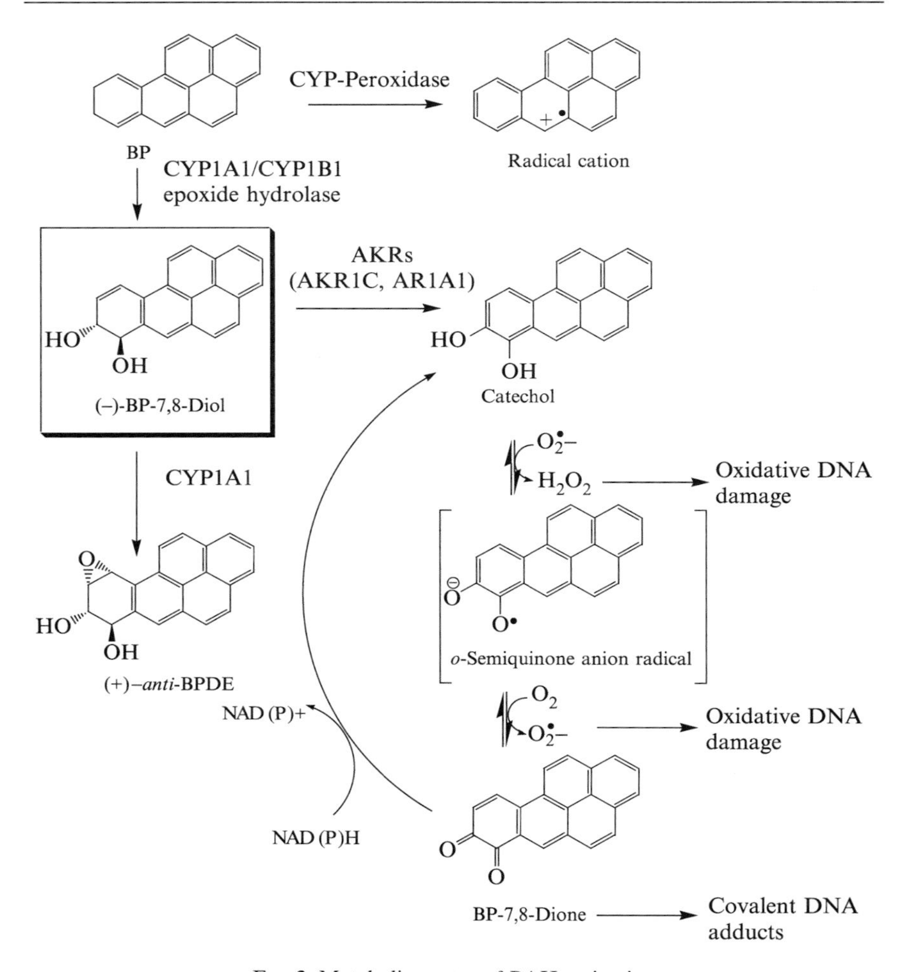

FIG. 2. Metabolic routes of PAH activation.

cation that will alkylate DNA.[18–20] Alternately, BP is converted to BP-7,8-oxide (an arene oxide) by CYP1A1/CYP1B1, which is then hydrolyzed by epoxide hydrolase to yield a *trans*-dihydrodiol (7,8-dihydroxy-,7,8-dihydro-benzo[*a*]pyrene or (–)-BP-7,8-diol). This route of metabolism is

[18] E. L. Cavalieri and E. G. Rogan, *Xenobiotica* **25,** 677 (1995).

[19] P. D. Devanesan, N. V. S. RamaKrishna, R. Todorovic, E. G. Rogan, E. L. Cavalieri, H. Jeong, R. Jankowiak, and G. J. Small, *Chem. Res. Toxicol.* **5,** 302 (1992).

[20] L. Chen, P. D. Devanesan, S. Higginbotham, F. Ariese, R. Jankowiak, G. J. Small, E. G. Rogan, and E. Cavalieri, *Chem. Res. Toxicol.* **9,** 897 (1996).

stereoselective, because only the (−)-BP-7,8-diol is formed *in vivo*.[21] The (−)-BP-7,8-diol is then a substrate for the two remaining pathways of PAH activation. It can either undergo a second monoxygenation catalyzed by CYP1A1/CYP1B1 to yield (+)-*anti*-BPDE (7,8-dihydroxy-9,10-epoxy-7,8,9,10-tetrahydrobenzo[*a*]pyrene)[21,22] or it can undergo $NADP^+$-dependent oxidation catalyzed by AKRs to yield the corresponding reactive and redox-active *o*-quinone (benzo[*a*]pyrene-7,8-dione; BP-7,8-dione).[23,24] Compelling evidence exists that (+)-*anti*-BPDE is an ultimate carcinogen derived from BP.

Rat and Human Aldo-Keto-Reductases with Dihydrodiol Dehydrogenase Activity

Aldo-keto reductases that convert PAH *trans*-dihydrodiols to PAH *o*-quinones do so because they have dihydrodiol dehydrogenase activity.[25] In this reaction, the *trans*-dihydrodiol is oxidized in the presence of $NADP^+$ to yield a ketol that spontaneously rearranges to a catechol. The catechol is air sensitive and undergoes two one-electron autoxidations to produce an *o*-semiquinone anion radical, which is then converted to the *o*-quinone (Fig. 2). The *o*-quinones are reactive electrophiles that can undergo Michael addition reactions with macromolecules (protein, RNA, and DNA) or react with cellular nucleophiles (e.g., GSH). Once formed, the *o*-quinones can also be reduced back by two electrons to the catechol using cellular reducing equivalents (e.g., NAD[P]H).[25] These *o*-quinones are not substrates for quinone reductase.[26] Alternately, the *o*-quinones can undergo successive two one-electron reduction reactions back to the catechol using NADPH cytochrome P450 reductases.[26] Each time the catechol is reformed, it is reoxidized by molecular oxygen to form reactive oxygen species (ROS) (superoxide anion ($O_2^{\bullet -}$), hydroxyl radical ($OH^\bullet$), and hydrogen peroxide [H_2O_2]) and the fully oxidized *o*-quinone. This establishes a futile redox cycle that generates ROS multiple times. The ROS generated may then contribute to the initiation and promotional phases of PAH carcinogenesis.

[21] H. V. Gelboin, *Physiol. Rev.* **60,** 1107 (1980).
[22] A. H. Conney, *Cancer Res.* **42,** 4875 (1982).
[23] T. E. Smithgall, R. G. Harvey, and T. M. Penning, *J. Biol. Chem.* **261,** 6184 (1986).
[24] T. E. Smithgall, R. G. Harvey, and T. M. Penning, *J. Biol. Chem.* **263,** 1814 (1988).
[25] T. M. Penning, M. E. Burczynski, C-F Hung, K. D. McCoull, N. T. Palackal, and L. S. Tsuruda, *Chem. Res. Toxicol.* **12,** 1 (1999).
[26] L. Flowers-Geary, R. G. Harvey, and T. M. Penning, *Biochemistry (Life-Sci. Adv.)* **11,** 49 (1992).

The most extensively studied AKRs that have dihydrodiol dehydrogenase (DD) activity are as follows: AKR1C9 (rat 3α-hydroxysteroid/dihydrodiol dehydrogenase); AKR1C1 (human 3α(20α)-hydroxysteroid dehydrogenase (HSD) or DD1); AKR1C2 (human type 3 3α-HSD and bile-acid–binding protein or DD2); AKR1C3 (human type 2 3α-HSD/type 5 17β-HSD and DDx); AKR1C4 (human type 1 3α-HSD/DD4); and human aldehyde reductase (AKR1A1) (Table I).

Experimental Strategies

To study the role of AKRs in PAH activation requires a source of the recombinant enzymes, reliable assay methods, a source of PAH *trans*-dihydrodiol substrates, and protocols for trapping the highly reactive PAH *o*-quinones. A source of PAH *o*-quinones is also required to study their chemical and biological properties. Biological properties of interest include their cytotoxicity and genotoxicity (DNA lesions and mutations).

CAUTION: All PAH are potentially hazardous and should be handled in accordance with the "NIH Guidelines for the Laboratory Use of Chemical Carcinogens."

Expression and Purification of Aldo-Keto Reductases

Cloning, Expression, and Purification of Recombinant Aldo-Keto Reductases

Cloning. Historically, AKR1C9 cDNA was cloned from a λgt11 expression library using a rabbit anti-rat AKR1C9 polyclonal antibody to yield a construct in pBluescriptSK$^+$.[27] Subsequent subcloning led to the construction of prokaryotic expression vectors (pKK-3α-HSD (AKR1C9) and pET16b-3α-HSD (AKR1C9), in which expression of the cDNA is driven by the hybrid *trc* promoter.[28,29]

By contrast, the human enzymes (AKR1C1–AKR1C4 and AKR1A1) can be cloned by isoform-specific reverse transcriptase–polymerase chain reaction (RT–PCR).[13] In this approach, first-strand cDNA synthesis is conducted using human hepatoma (HepG2) cell polyA$^+$-RNA as template, an oligo-dT primer, and Superscript II RNAse H$^-$RT (Gibco BRL, Gaithersburg, MD). Aliquots of the first-strand cDNA library are then mixed with the 5′ and 3′ isoform-specific primers (Table II), dNTPs, and Vent-DNA polymerase

[27] J. E. Pawlowski, M. Huizinga, and T. M. Penning, *J. Biol. Chem.* **266,** 8820 (1991).
[28] J. E. Pawlowski and T. M. Penning, *J. Biol. Chem.* **269,** 13502 (1994).
[29] H. Ma and T. M. Penning, *Proc. Natl. Acad. Sci. USA* **96,** 11161 (1999).

TABLE I
RELATIONSHIP BETWEEN RAT AND HUMAN AKR1C FAMILY MEMBERS

Nomenclature	Enzyme	Other names	Percent sequence identity	Tissue distribution
AKR1C9	Rat dihydrodiol dehydrogenase	3α-HSD	67	Liver
AKR1C1	Human dihydrodiol dehydrogenase 1	$20\alpha(3\alpha)$HSD	100	Lung > liver > testis > mammary gland
AKR1C2	Human dihydrodiol dehydrogenase 2	Type 3 3α-HSD/bile-acid–binding protein	98	Lung & liver > prostate, testis & mammary gland
AKR1C3	Human dihydrodiol dehydrogenase X	Type 2 3α-HSD/type 5 17β-HSD	87	Mammary gland > prostate > liver >lung
AKR1C4	Human dihydrodiol dehydrogenase 4	Type 1 3α-HSD chlordecone reductase	83	Liver specific
AKR1A1	Human dihydrodiol dehydrogenase 3	Aldehyde reductase	46	Ubiquitous

TABLE II
OLIGONUCLEOTIDE PRIMERS USED FOR ISOFORM-SELECTIVE RT-PCR OF HUMAN AKRI cDNAs

Isoform	Primer	Primer sequence
AKR1C1	5′-primer	5′-CCAG**CCATGG**ATTCGAAATAT-3′
	3′-primer	3′-CCATG<u>TTA</u>ATATTCATCAGAG-5′
AKR1C2	5′-primer	5′-ACAG**CCATGG**ATTCGAAATAC-3′
	3′-primer	3′-TCCATG<u>TTA</u>ATATTCATCAGAA-5′
AKR1C3	5′-primer	5′-CAGG**CCATGG**ATTCCAAACAG-3′
	3′-primer	3′-TCCATG<u>TTA</u>ATATTCATCTGAAT-5′
AKR1C4	5′-primer	5′-**CCATGG**ATCCCAAATATCAGC-3′
	3′-primer	3′-TCTATGC<u>TAA</u>TATTCATCTGAA-5′
AKR1A1	5′-primer	5′-GGGGG**CCATGG**CGGTTCCTG-3′
	3′-primer	3′-GGGAAA<u>TTA</u>CTGGGCARGACTCTG-5′

*Nco*I-engineered start codons are in bold; stop-codons are underlined.

(New England Biolabs, Beverly, MA). PCR is conducted for 30 cycles, and the PCR products (1.0 kB) are purified by agarose gel electrophoresis. PCR-amplified cDNAs are then ligated into the TA cloning vector pCRII (Invitrogen, Carlsbad, CA). The fidelity of positive inserts is established by dideoxysequencing. Coding regions are then excised using *Nco*I and *BamH*1 partial digestion and inserted into the compatible *Nco*I and *BamH*I sites of the prokaryotic pET16b vector (Novagen, Madison, WI). Use of the *Nco*I site in pET16b intentionally results in the loss of the histidine tag, so that the expressed AKRs contain only their deduced amino acid sequence. His-tagged fusion proteins are avoided, because they generally require purification by Ni^{2+} chromatography, and AKR1 family members are metal sensitive.

Expression and Purification of Recombinant Aldo-Keto Reductases. pET-16b-AKR constructs are transformed into competent *Escherichia coli* host expression strain C41 (DE3).[13,14,30] This host strain is provided by Dr. J. E. Walker of the Medical Research Council Laboratory of Molecular Biology, Cambridge, U. K. Cultures are streaked on Luria-Bertania plates containing 100 μg/ml ampicillin for positive selection of single colonies. Single colonies are grown overnight in 5-ml starter cultures. Aliquots of each starter culture (2.5 ml) are used to inoculate 2 × 100-ml cultures, which after 5 h are used to inoculate 2 × 2 liters of Luria-Bertania/ampicillin media. Once the absorbance at 600 nm has reached 0.6–0.8 OD units, each 2-liter culture is induced with 2 ml 1 *M* IPTG (isopropyl-β-D-thiogalactosidase) for overnight protein production.

[30] N. T. Palackal, M. E. Burczynski, R. G. Harvey, and T. M. Penning, *Biochemistry* **40,** 10901 (2001).

Cells are harvested by centrifugation (8 × 250-ml bottles), washed twice in 10 m*M* TRIS-HCl, pH 8.6, containing 1 m*M* ethylene diamine tetraacetic acid (EDTA) and 1 m*M* 2-mercaptoethanol, and each pellet is suspended in 5 ml of the same buffer. AKR proteins are released from the cell pellets by sonication, and following centrifugation, the sonicates are dialyzed into 10 m*M* TRIS-HCl, pH 8.6, containing 1 m*M* EDTA, 1 m*M* 2-mercaptoethanol, and 20% glycerol. The dialyzed protein is then applied to DE-52 cellulose column equilibrated in the dialysis buffer, and the AKR protein is eluted with a linear salt gradient of 10–250 m*M* KCl, where the proteins elute at the mid-point in the gradient (pI = 6.1). Following dialysis into 10 m*M* potassium phosphate, pH 7.0, containing 1 m*M* EDTA, 1 m*M* 2-mercaptoethanol, and 20% glycerol, the sample is purified by Sepharose-blue affinity column chromatography. After application, the sample is batch-eluted in a purified concentrated form in column buffer containing 1.0 *M* KCl. The purified AKRs are then dialyzed into protein storage buffer (20 m*M* potassium phosphate, pH 7.0, containing 1 m*M* EDTA, 1 m*M* 2-mercaptoethanol, and 30% glycerol).

Specific activities are monitored throughout the purification by measuring the NAD^+-dependent oxidation of 75 μM androsterone in 100 m*M* potassium phosphate, pH 7.0, at 25° (AKR1C9 or AKR1C4), by measuring the NAD^+-dependent oxidation of 1 m*M* 1-acenaphthenol in 100 m*M* potassium phosphate, pH 7.0, at 25° (AKR1C1-AKR1C3), or by measuring the NADPH-dependent reduction of 4-nitrobenzaldehyde in 100 m*M* potassium phosphate buffer, pH 7.0, at 25° (AKR1A1).[13,30] The purity of peak chromatographic fractions is concurrently monitored by sodium dodecyl sulfate-polyacrylamide gel electrophoresis (SDS–PAGE), in which expression of a 37-kDa protein is detected. Specific activities for the homogeneous recombinant proteins are 1.6 (AKR1C9), 2.1 (AKR1C1), 2.5 (AKR1C2), 2.8 (AKR1C3), 0.21 (AKR1C4), and 6.0 (AKR1A1) μmol substrate oxidized per minute per milligram, where protein concentration is measured on the basis of a Lowry determination relative to bovine serum albumin (BSA). Typically, 40 mg of homogeneous recombinant protein can be obtained from a 4-liter culture. Representative purification tables are given for AKR1C1–AKR1C4 and AKR1A1 (Table III).[13,30]

Assay Methods for Dihydrodiol Dehydrogenase Activity

Continuous spectrophotometric assays based either on the NAD^+-dependent oxidation of androsterone or 1-acenaphthenol can be used to monitor activity throughout the purification of the recombinant AKRs. However, these assays do not monitor the oxidation of PAH-*trans*-dihydrodiols directly; the following assays are recommended.

TABLE III

Purification Schemes for Five Recombinant Human Dihydrodiol Dehydrogenase (AKR) Isoforms

Protein	Purification step	Volume (ml)	Total protein (mg)	Total activity (μmol min^{-1} mg^{-1})	Specific activity (μmol min^{-1} mg^{-1})	Purification factor (-fold)	Yield (%)
AKR1C1	Sonicate	61	460	59[a]	0.13		
	DE52 cellulose	16	23	46	2.0	15	78
	Blue Sepharose	8	20	42	2.1	16	71
AKR1C2	Sonicate	77	510	42[a]	0.08		
	DE52 cellulose	18	19	39	2.1	25	93
	Blue-Sepharose	11	15	37	2.5	31	88
AKR1C3	Sonicate	51	310	65[a]	0.21		
	DE52 cellulose	56	25	51	2.0	10	78
	Blue-Sepharose	5	12	34	2.8	13	52
AKR1C4	Sonicate	44	270	4.6[a]	0.02		
	DE52 cellulose	12	23	2.8	0.12	6	61
	Blue-Sepharose	8	13	2.7	0.21	11	59
AKR1A1	Sonicate	70	470	66[b]	1.42	1	100
	DE52 cellulose	12.5	88.5	233	2.63	1.85	34.7
	Blue-Sepharose	7.1	30.5	183	6.0	4.23	27.4

[a] Specific activities throughout the purifications were measured using 75 μM androsterone (AKR1C_4) or 1 mM 1-acenaphthenol (AKR1C1-AKR1C3) as the substrate in reaction mixtures containing 100 mM KPO_4 (pH 7.0) and 2.3 mM NAD^+ at 25°.

[b] Specific activities throughout the purification were measured with 1 mM nitrobenzadehyde as substrate in reaction mixtures containing 100 mM KPO_4 (pH 7.0) and 200 μM NADHPH at 25°.

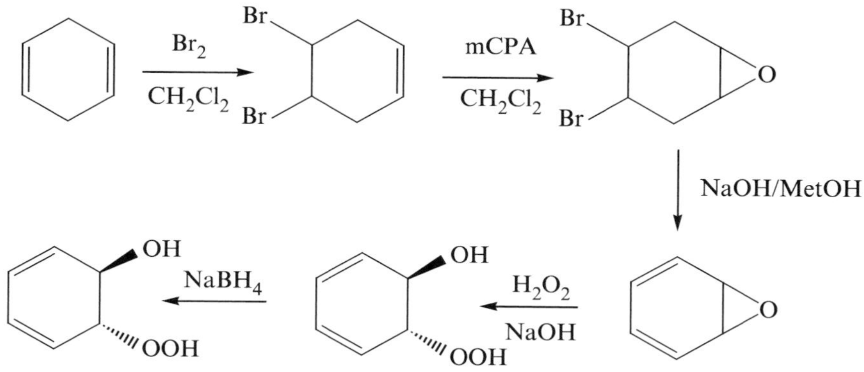

FIG. 3. Synthesis of benzenedihydrodiol from cylcohexadiene. mCPA = metachloroperoxybenzoic acid.

NADP⁺-Dependent Oxidation of Benzene Dihydrodiol

Both continuous and discontinuous assays have been developed to monitor the $NADP^+$-dependent oxidation of the model *trans*-dihydrodiol (benzene dihydrodiol; 1,2-*trans*-dihydroxy-1,2-dihydro-cyclohexa-3,5-diene). Both assays require synthesis of the benzenedihydrodiol substrate.[31,32] This is accomplished by the stoichiometric bromination of cyclohexadiene to yield 1,2-dibromo-cyclohexene, followed by epoxidation to yield 1,2-dibromo-4,5-epoxy-cyclohexane, and subsequent dehydrohalogenation to yield benzene oxide. Benzene oxide is then subjected to alkali peroxidation to yield the peroxy intermediate, which is reduced with $NaBH_4$ to yield the final product that is purified by thin-layer chromatography (TLC) and continuous extraction (Fig. 3). The synthesis is low yielding because of the propensity of benzene oxide to revert to phenol. Typically, 100 mg of the product can be obtained if the synthesis is conducted on a 5-g scale.

The continuous assay measures the $NADP^+$-dependent oxidation of benzene dihydrodiol in mixtures containing 1 m*M* benzene dihydrodiol, 2.3 m*M* $NADP^+$, and 50 m*M* glycine, pH 9.0, buffer at 25°. The discontinuous assay exploits the observation that benzene dihydrodiol is the only *trans*-dihydrodiol that is oxidized to a stable hydroquinone (catechol).[33]

[31] E. Vogel and H. Gunther, *Angew. Chem. Int. Ed. Eng.* **6,** 385 (1967).

[32] A. M. Jeffrey, H. J. Yeh, D. M. Jerina, R. M. Demarinis, C. H. Foster, D. E. Piccolo, and G. A. Berchtold, *J. Am. Chem. Soc.* **96,** 6929 (1974).

[33] J. K. Ivins and T. M. Penning, *Cancer Res.* **47,** 680 (1987).

The assay links catechol production to catechol-*O*-methyl transferase (COMT) using [^{3}H]-*S*-adenosyl-L-methionine (SAM) as methyl donor to yield [^{3}H]-guaiacol, which is isolated by extraction and counted. Assays are conducted in 0.1-ml systems containing 50 m*M* potassium phosphate, pH 7.8, 1.0 m*M* DTT, 1.0 m*M* $MgCl_2$, 2.3 m*M* $NADP^+$, 50 m*M* [^{3}H]-SAM, 10 units COMT, and 1.0 m*M* benzene dihydrodiol. Reactions are initiated by enzyme and quenched with 200 μl 1 *M* HCl containing 2 μl/ml guaiacol as carrier. The reaction mixture is then extracted into 2 × 1.0 ml toluene. By counting the radioactive extracts and knowing the specific radioactivity of SAM, it is possible to compute specific activities. Because the assay is radiometric, it provides a 1000–5000-fold increase in sensitivity over the spectrophotometric assay (Fig. 4). Its advantage is that it can be used to measure dihydrodiol dehydrogenase activity in cell culture and in tissue extracts.

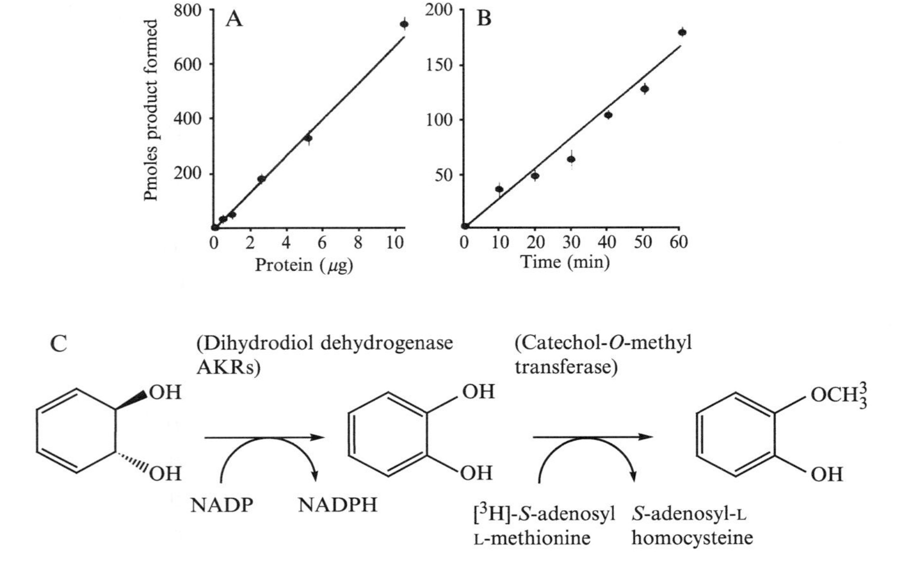

FIG. 4. Coupled radiochemical assay for the detection of dihydrodiol dehydrogenase. (A) Linearity with protein where varying amounts of the 40–75% ammonium sulfate fraction of rat liver cytosol were incubated with 1 m*M* benezene dihydrodiol, 2.3 m*M* $NADP^+$, 50 μM [^{3}H]-*S*-adenosyl-L-methionine (0.10 μCi/nmol), and 10 units of COMT. (B) Linearity with time in which the assay was initiated with 2.6 μg of protein. (C) The basis of the assay in which the formation of catechol is linked to catechol-*O*-methyl transferase to yield [^{3}H]-guaiacol. After Ivins and Penning.[33]

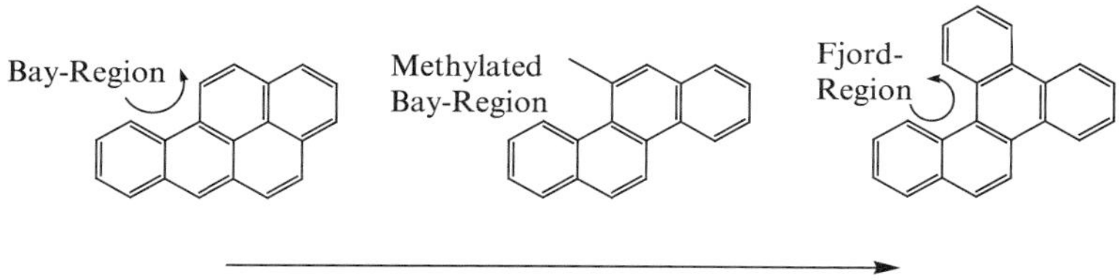

Bay-Region

Methylated Bay-Region

Fjord-Region

Increased carcinogenicity based on the corresponding diol-epoxides

FIG. 5. Structural features that increase the carcinogencity of PAH.

PAH Trans-*Dihydrodiol Oxidation*

Monitoring the oxidation of PAH *trans*-dihydrodiols requires a source of substrates. The substrate of most relevance is (±)-*trans*-7,8-dihydroxy-7,8-dihydrobenzo[*a*]pyrene and is available in limited amounts from the NCI, Chemical Carcinogen Reference Standard Repository (MidWest Research Institute, Kansas City, MO). Examination of a structural series of PAH *trans*-dihydrodiols is important from a structure-activity perspective, because issues of regio-, stereo-, and conformational chemistry determine whether the *trans*-dihydrodiol is a proximate carcinogen or not. For example, non-K region (−) *R,R trans*-dihydrodiols are potent proximate carcinogens (e.g., (−) *trans*-7,8-dihydroxy-7,8-dihydrobenzo[*a*]pyrene), but other stereo and regio-isomers are not (e.g., (+) *trans*-7,8-dihydroxy-7,8-dihydrobenzo[*a*]pyrene and (±) *trans*-4,5-dihydroxy-4,5-dihydrobenzo[*a*]pyrene).[34] In systematic studies on *trans*-dihydrodiol oxidation, substrates containing increasing ring-number, bay-region methylation, and *fjord* regions need to be examined, because the presence of these structural features often determines the carcinogenicity of the corresponding diol-epoxide. The structural features of interest are shown in Fig. 5, and PAH *trans*-dihydrodiols often examined as substrates are shown in Fig. 6. In this article, there is not room to describe the synthesis of all the *trans*-dihydrodiol substrates.

The synthesis of (±) *trans*-1,2-dihydroxy-1,2-dihydro-phenanthrene is fairly typical. This is prepared from 1-keto-2,3,4-trihydro-phenanthrene (Aldrich, Milwaukee, WI), as follows: reduction of 1-keto-2,3,4-trihydro-phenantrene with $NaBH_4$ in methanol affords the corresponding alcohol, which is dehydrated with catalytic quantities of HCl in acetic acid, to give the 3,4-dihydroarene.[35] Prevost reaction of the dihydroarene using silver

[34] A. Dipple and R. G. Harvey, eds. American Chemical Society: Washington, D.C. p. 1–17, 1984.

[35] H. Yagi and D. M. Jerina, *J. Am. Chem. Soc.* **97,** 3185 (1975).

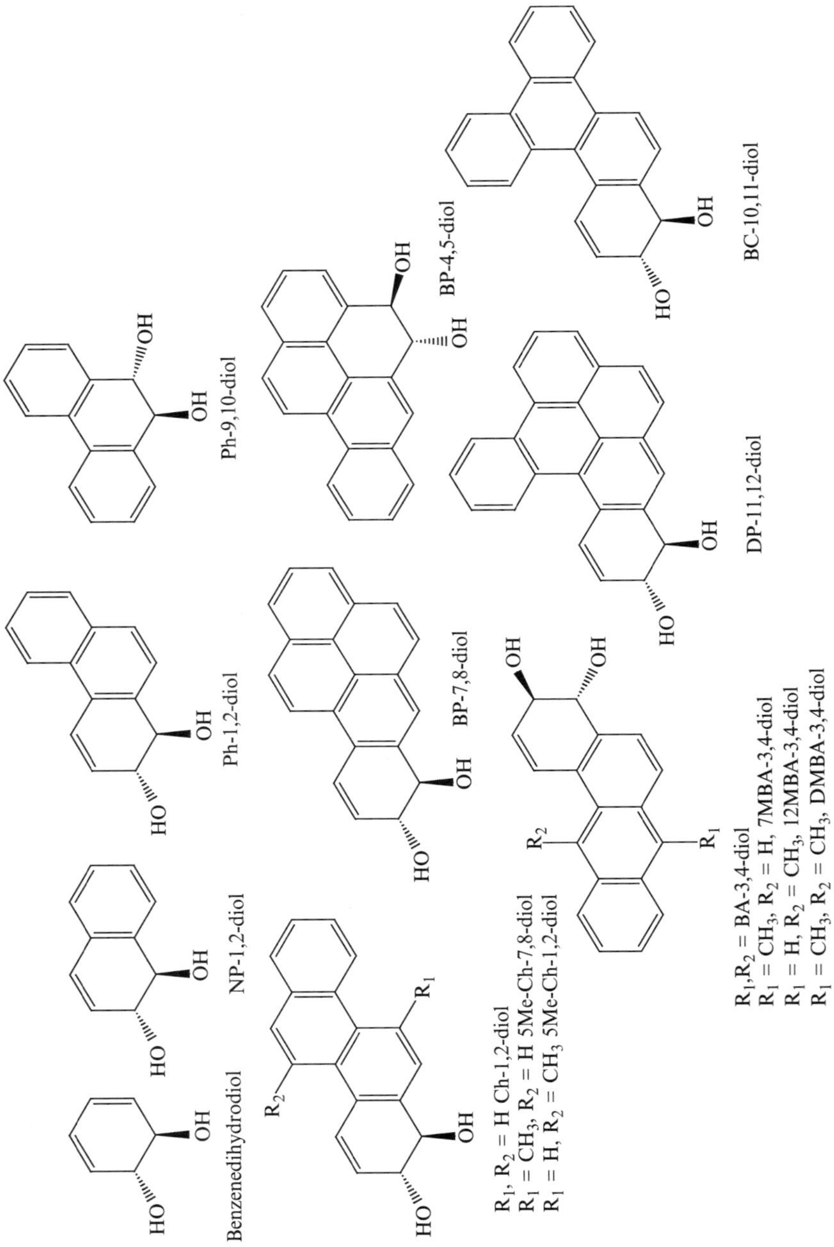

Benzenedihydrodiol

NP-1,2-diol

Ph-1,2-diol

Ph-9,10-diol

R_1, R_2 = H Ch-1,2-diol
R_1 = CH_3, R_2 = H 5Me-Ch-7,8-diol
R_1 = H, R_2 = CH_3 5Me-Ch-1,2-diol

BP-7,8-diol

BP-4,5-diol

R_1,R_2 = BA-3,4-diol
R_1 = CH_3, R_2 = H, 7MBA-3,4-diol
R_1 = H, R_2 = CH_3, 12MBA-3,4-diol
R_1 = CH_3, R_2 = CH_3, DMBA-3,4-diol

DP-11,12-diol

BC-10,11-diol

benzoate and iodine yields the *trans*-dibenoxytetrahydroarene,[36] which is reduced with 2,3-dichloro-5,6-dicyano-1,4-benzo quinone (DDQ) to yield *trans*-dibenzoxydihydrophenanthrene.[37] Hydrolysis of the resultant dibenzoate with sodium methoxide yields the desired product, (±) *trans*-1,2-dihydroxy-1,2-dihydro-phenanthrene.

Note that (±) *trans*-9,10-dihydroxy-9,10-dihydrophenanthrene is prepared by reducing phenanthrene-9,10-dione with lithium aluminium hydride.[38,39] The remaining compounds can be synthesized by the routes cited (±) *trans*-1,2-dihydroxy-1,2-dihydro-chrysene;[40] (±) *trans*-1,2-dihydroxy-1,2-dihydro-5-methyl-chrysene; (±) *trans*-7,8-dihydroxy-7,8-dihydro-5-methyl-chrysene;[41] (±) *trans*-3,4-dihydroxy-3,4-dihydro-benz[*a*]anthracene;[42] (±) *trans*-3,4-dihydroxy-3,4-dihydro-7-methyl-benz[*a*]anthracene;[43] (±) *trans*-3,4-dihydroxy-3,4-dihydro-12-methyl-benz[*a*]anthracene; (±) *trans*-3,4-dihydroxy-3,4-dihydro-7,12-dimethyl-benz[*a*]anthracene;[44] (±) *trans*-10, 11-dihydroxy-10,11-dihydro-benz[*g*]chrysene;[45] and (±) *trans*-11,12-dihydroxy-11,12-dihydro-dibenzo[*a,l*]pyrene.[46]

To determine whether these compounds are substrates for AKRs, we have relied on two assays. First, we have monitored the oxidation of 20 μM (±)-*trans*-dihydrodiol in the presence of 2.3 mM $NADP^+$ in 3[(1,1-dimethyl-2-hydroxyethyl)amino]-2-hydroxypropanesulfonic acid (AMPSO) buffer at pH 9.0 by measuring changes in absorbance at 340 nm.[14,23,30] Under these conditions, the reaction is pseudo-first-order, where $K_m \gg [S]$, so that the Michaelis–Mention equation simplifies to $v/[S] = V_{max}/K_m$.

[36] R. E. Lehr, M. Schaefer-Ridder, and D. M. Jerina, *J. Org. Chem.* **42,** 736 (1977).
[37] P. P. Fu and R. G. Harvey, *Tetrahedron Lett.* **24,** 2059 (1977).
[38] J. Booth, J. E. Boyland, and E. E. Turner, *J. Chem. Soc.* 1188 (1950).
[39] C. Cortez and R. G. Harvey, *Org. Synth.* **58,** 12 (1978).
[40] P. P. Fu and R. G. Harvey, *J. Org. Chem.* **44,** 3778 (1979).
[41] J. Pataki, H. Lee, and R. G. Harvey, *Carcinogenesis* **4,** 399 (1983).
[42] R. G. Harvey and K. B. Sukumaran, *Tetrahedron Lett.* **24,** 2387 (1977).
[43] H. H. Lee and R. G. Harvey, *J. Org. Chem.* **44,** 4948 (1979).
[44] H. H. Lee and R. G. Harvey, *J. Org. Chem.* **51,** 3502 (1986).
[45] R. G. Harvey, "Polycyclic Aromatic Hydrocarbons," p. 129–188. Academic Press, New York, 1997.
[46] H. S. Gill, P. L. Kole, J. C. Wiley, K-M. Li, S. Higginbotham, E. G. Rogan, and E. L. Cavalieri, *Carcinogenesis* **15,** 2455 (1994).

FIG. 6. Structures of PAH *trans*-dihydrodiol substrates for AKRs. All *trans*-dihydrodiols shown are racemic; NP = naphthalene, Ph = phenanthrene; Ch = chrysene; 5Me-Ch = 5-methylchrysene; BP = benzo[*a*]pyrene; BA = benz[*a*]anthracene; 7MBA = 7-methylbenz[*a*]anthracene; 12MBA = 12-methylbenz[*a*]anthracene; DMBA = 7,12-dimethylbenz[*a*]anthracene, DP = dibenz[*a,l*]pyrene; and BC = benzo[*g*]chrysene.

This has allowed us to report the utilization ratios for a large number of *trans*-dihydrodiol substrates for various AKRs.[13,23,30,47,48] In assigning utilization ratios, it is important to consider that the initial velocities are inhibited by the 8% dimethyl sulfoxide (DMSO) cosolvent and require correction. Because these absorbance assays give relatively small changes, the assay is usually validated by monitoring the disappearance of the *trans*-dihydrodiol substrate by reverse-phase–high-performance liquid chromatography (RP–HPLC).[13,14]

Reverse-phase–high performance liquid chromotography assays are conducted in 0.1 ml 100 m*M* potassium phosphate buffer, pH 7.8, or 50 m*M* glycine (pH 9.0) containing 2.3 m*M* $NADP^+$ and 20 μM *trans*-dihydrodiol. The substrate is dissolved in DMSO, and the final organic solvent concentration is 8%. Reactions are initiated by the addition of homogeneous enzyme (2–10 μg) and incubated for 2 h at 37°. Control incubations are performed in the absence of $NADP^+$ or purified enzyme. Reaction mixtures are extracted into 0.2 ml ethyl acetate, pooled, and evaporated to dryness, the residues are redissolved in methanol (50 μl) for RP-HPLC analysis.

Reverse-phase–high-performance liquid chromatography analysis is conducted by injecting aliquots (20 μl) onto a Zorbax Ultrasphere-ODS (10 μm; 4.6 $\times$ 25 mm; Dupont, Wilmington, DE) reverse-phase column. Compounds are separated by use of a 70-min linear gradient of 60–80% methanol-water (v/v). The solvent flow rate is 0.5 ml/min, and the chromatographic system is run at ambient temperature. The absorbance is monitored at 254 nm.[13,48] Chromatography has been performed on a Perkin-Elmer (Albany, Boston, MA) LC480 diode-array HPLC, a Beckman-System Gold HPLC (Fullerton, CA) and a Waters Alliance system (Milford, MA). Comparison of specific activities for AKR1C9 yields a value of 5–6 nmoles of ($\pm$)BP-7,8-diol oxidized/min/mg for the spectrophotometric assay, and 2.0 nmoles of ($\pm$)BP-7,8-diol oxidized/min/mg based on the RP-HPLC assay. Typical progress curves for the consumption of ($\pm$)BP-7,8-diol by AKR1C1–AKR1C4 based on RP-HPLC assays are shown (Fig. 7).[13]

To determine the sterochemical course of *trans*-dihydrodiol oxidation, RP-HPLC assays are performed until the reaction reaches 50% completion. In some instances, this is the end-point of the reaction.[13,23] At this juncture, the unreacted isomer(s) is extracted into ethyl acetate, the remaining substrate is purified by TLC, and the CD spectra are recorded. Sufficient material for these assays requires an absorbance of 1.0 OD unit at 254 nm. The sign of the Cotton effect can be related to the stereochemical assignment for the unreacted isomer, which is determined

[47] T. E. Smithgall, R. G. Harvey, and T. M. Penning, *Cancer Res.* **48,** 1227 (1988).

[48] L. Flowers-Geary, R. G. Harvey, and T. M. Penning, *Chem. Res. Toxicol.* **5,** 576 (1992).

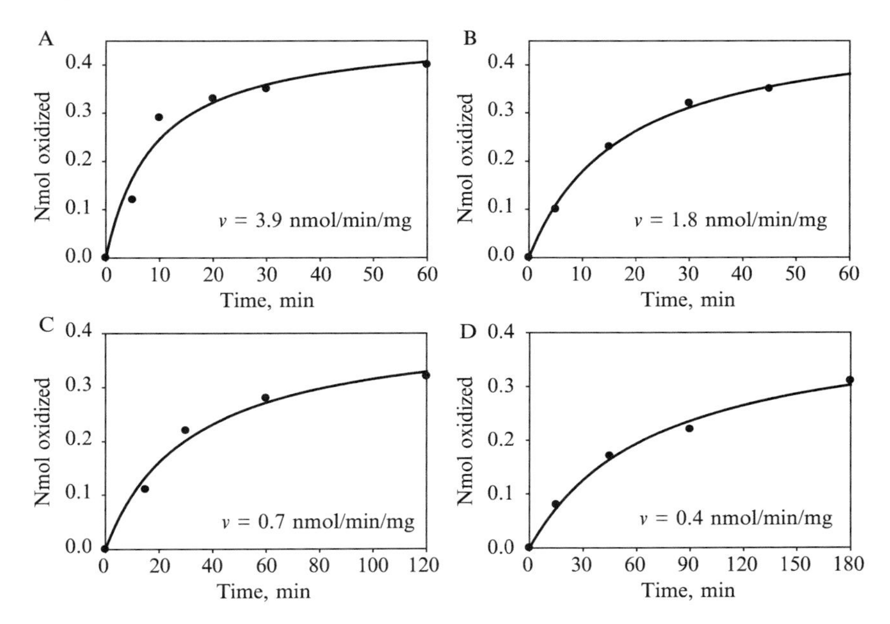

FIG. 7. Progress curves for the oxidation of BP-7,8-diol by AKR1C1–AKR1C4. Initial progress curves for the oxidation of the proximate carcinogen (±)-[1,3-^{3}H]-BP-diol catalyzed by recombinant AKR1Cs. Following RP-HPLC analysis, the total counts per minute in the (±)-[1,3-^{3}H]-BP-diol peak remaining at each time point was used to calculate the nmoles of BP-diol oxidized and was plotted versus time. The initial velocities (insets) were calculated from the linear portion of the data as shown. (A) AKR1C2; (B) AKR1C1; (C) AKR1C4; and (D) AKR1C3. Reprinted with permission from Burczynski *et al.*, *Biochemistry* **37,** 6781 (1998), copyright 1998, American Chemical Society.

independently by nuclear magnetic resonance (NMR) spectroscopy. AKR1C1–AKR1C4 oxidize both stereoisomers of racemic *trans*-dihydrodiols; however, pseudo-first-order analysis of progress curves for (±)BP-7,8diol oxidation show that one isomer is preferentially oxidized. CD spectra showed that for AKR1C1 and AKR1C2 the (+)-*S,S*-isomer was oxidized before the (−)-*R,R*-isomer (Fig. 8).

PAH-Trans-*Dihydrodiol Specificity of Aldo-Keto Reductases*

By use of these assays, a picture of the *trans*-dihydrodiol specificity of AKRs has emerged: (1) PAH *trans*-dihydrodiols of increasing ring size are substrates; (2) K-region *trans*-dihydrodiols, e.g., (±)-*trans*-4,5-dihydroxy-4,5-dihydrobenzo[*a*]pyrene, are not substrates; (3) AKR1C1–AKR1C4 do not exhibit stereospecificity in that they will oxidize both

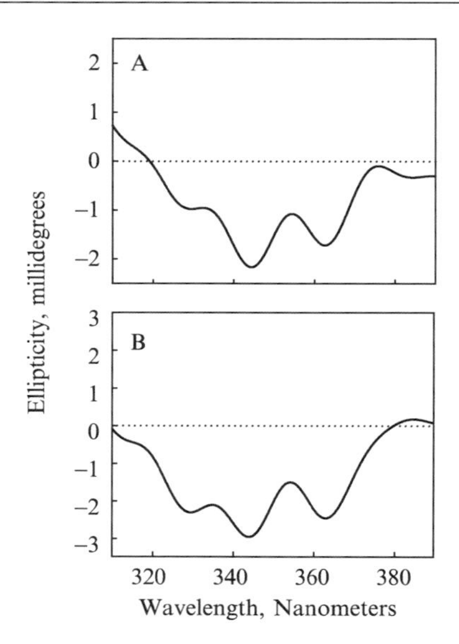

FIG. 8. CD spectroscopy to determine the stereochemical preference of BP-7,8-diol oxidation. AKR1C1–AKR1C4 oxidize both stereoisomers of (±)-BP-7,8-diol. Pseudo-first-order plots of BP-7,8-diol oxidation catalyzed by AKR1C1 and AKR1C2 indicated that decay curves for BP-7,8-diol were biphasic, indicating a preference for one isomer. RP-HPLC assays were used to monitor the progress curves of 10-ml reactions containing 50 μM BP-7,8-diol, 2.3 mM $NADP^+$ in 50 mM glycine buffer, pH 9.0, and 8% DMSO. When 50% of the diol was consumed, the reaction was quenched with ethyl acetate, and unoxidized dihydrodiol substrates remaining were isolated by TLC and their CD-spectra recorded. The CD spectrum of the racemic (±)-BP-7,8-diol was subtracted from the enantiomerically enriched substrates to determine the sign of the Cotton effect. (A) CD spectra of the unreacted BP-7,8-diol recovered from the AKR1C2 reaction. (B) CD spectra of the unreacted BP-7,8-diol recovered from the AKR1C1 reaction. Reprinted with permission from Burczynski *et al.*, *Biochemistry* **37,** 6781 (1998), copyright 1998, American Chemical Society.

the (+) and the (−) stereoisomers of racemic BP-7,8-diol and DMBA-3,4-diol; (4) AKR1A1 is stereoselective in that it will only oxidize the (−)-BP-7,8-diol; and (5) depending on the ring arrangement, different AKRs can be involved in the metabolism of the PAH *trans*-dihydrodiol.[13,14,23,30,49]

[49] M. E. Burczynski, N. T. Palackal, R. G. Harvey, and T. M. Penning, *Polycyclic Aromatic Compounds* **215,** 205 (1999).

Characterization of PAH *o*-Quinones as Products of PAH *trans*-Dihydrodiol Oxidation

PAH *o*-quinones produced by the AKR-dependent oxidation of PAH *trans*-dihydrodiols are highly reactive and readily form buffer conjugates, making their isolation difficult. They can be isolated from *in vitro* reactions if these are performed in the presence of a potent nucleophile as a trapping agent (e.g., 2-mercaptoethanol). The trapped product is a thio-ether conjugate that arises from 1,4-Michael addition to the PAH-*o*-quinone. Validation of the structure of the product requires liquid chromatography/mass spectrometry (LC/MS) characterization of the thio-ether conjugate and its comparison to an authentic synthetic standard.[24]

Preparation of the 2-Mercaptoethanol Conjugates of PAH o-*Quinones*

Reactions (10 ml) containing 50 m*M* potassium phosphate buffer, pH 7.0, 5 m*M* 2-mercaptoethanol, and 20 μM PAH *o*-quinone in 8% DMSO are incubated for 18 h at 37°. The reaction is terminated by extraction with ethyl acetate (2.0 ml). Organic extracts are combined, dried over sodium sulfate, and the organic solvent is removed under reduced pressure. The resulting residue is chromatographed on a 20- × 20-cm semipreparative silica TLC plate using chloroform methanol (4:1) as mobile phase. For the mercaptoethanol BP-7,8-dione conjugate, a single purple spot Rf = 0.77 is observed, which can be eluted from the plate and analyzed by electron ionization mass spectometry (EIMS), LC/MS, and [^{1}H]-NMR.[24]

Characterization of the Product of Aldo-Keto Reductase Oxidation of BP-7,8-diol by Trapping with 2-Mercaptoethanol

Incubations (50 ml) containing 20 μM BP-7,8-diol, 2.3 m*M* $NADP^+$, 5 m*M* 2-mercaptoethanol, and 8% DMSO are reacted with purified recombinant AKR (1 mg) for 20 h at 37°. Reactions are terminated by extraction with ethyl acetate, and the product is isolated and characterized as before.

EIMS spectra for BP-7,8-dione thio-ether conjugate yield a m/z = 358 M^+; and fragment ions that correspond to the successive loss of 2 carbonyls m/z 358 → 330, and 298 → 270, which is diagnostic of the structure.[24] LC/MS gives m/z = 359 MH^+; and m/z = 331 (MH^+-CO).[30] [^{1}H]-NMR shows the diagnostic loss of the C9-vinyl proton δ = 6.6 ppm and its replacement by a singlet because of substitution at C10.[24] LC/MS data were acquired on a Finnigan LCQ ion trap mass spectrometer (ThermoQuest, San Jose, CA) equipped with a Finnigan atmospheric pressure chemical (APCI) source. The on-line chromatography was performed using a Waters Alliance 2690 HPLC system (Waters Corp., Milford,

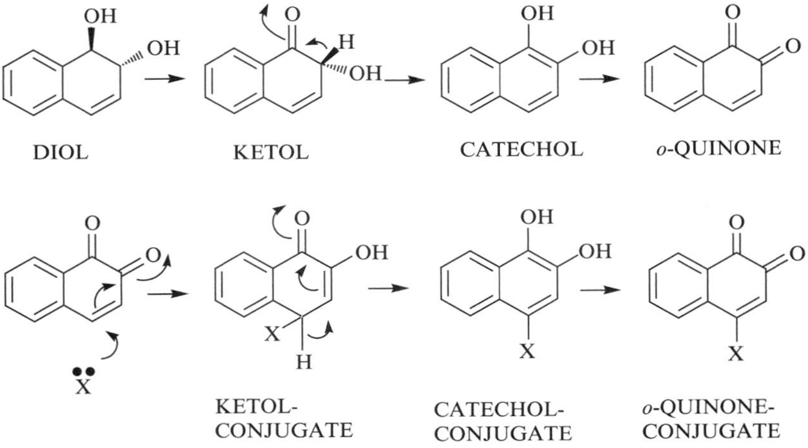

DIOL KETOL CATECHOL *o*-QUINONE

KETOL-CONJUGATE CATECHOL-CONJUGATE *o*-QUINONE-CONJUGATE

FIG. 9. Conversion of PAH *trans*-dihydrodiols to PAH *o*-quinone thio-ether conjugates. Reprinted with permission from Smithgall *et al.*, *J. Biol. Chem.* **263,** 1814 (1988), copyright 1988. American Society for Biochemistry and Molecular Biology.

MA). A YMC C18 ODS-AQ column was used at a flow rate of 0.9 ml/min. Solvent A was 5 m*M* ammonium acetate in water containing 0.01% trifluoroacetic acid, and solvent B was 5 m*M* ammonium acetate in methanol containing 0.01% trifluoroacetic acid with the gradient run as follows: 30% solvent B at 0 min, 30% solvent B at 5 min, 100% solvent B at 16 min, 100% solvent B at 24 min, and 30% solvent B at 26 min.[30]

The identification of PAH-*o*-quinone-thio ether conjugates in enzymatic reactions with AKRs suggests the following sequence of events: oxidation of the *trans*-dihydrodiol would yield a ketol that tautomerizes to yield the air-sensitive catechol, the catechol undergoes autoxidation to yield the PAH *o*-quinone, reaction with 2-mercaptoethanol yields a ketol, which rearranges to the catechol, which air oxidizes to the *o*-quinone-conjugate (Fig. 9).

Oxygen Metabolism During Aldo-Keto Reductase–Mediated *Trans*-Dihydrodiol Oxidation

The conversion of *trans*-dihydrodiols to fully oxidized PAH *o*-quinones catalyzed by AKRs suggests that this reaction sequence will be accompanied by concomitant changes in ROS ($O_2^{\bullet}$, $OH^{\bullet}$, and H_2O_2). Enzymatic *trans*-dihydrodiol oxidation occurs concomitantly with the depletion of molecular oxygen, the formation of superoxide anion, and the generation of H_2O_2.[50]

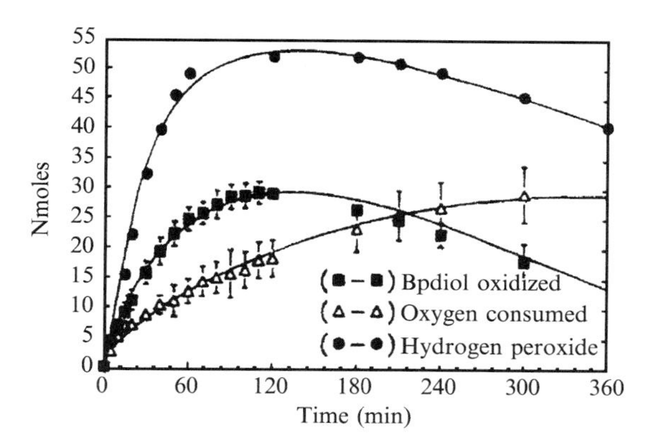

FIG. 10. Oxygen uptake and hydrogen peroxide formation during the enzymatic oxidation of (±)BP-7,8-diol. Reactions contained 20 μM (±)BP-7,8-diol in 8% DMSO, 2.3 mM $NADP^+$, and 50 mM glycine buffer, pH 9.0. Oxygen uptake was measured in 600-μl chambers with a micro-oxygen electrode. Nanomoles of BP-diol oxidized were monitored spectrophotometrically at 340 nm and nanomoles of hydrogen peroxide formed were measured discontinuously with horseradish peroxidase and tetramethylbenzidine as coreductant. Reprinted with permission from Penning *et al.*, *Chem. Res. Toxicol.* **9,** 84 (1996), copyright 1996, American Chemical Society.

The consumption of molecular oxygen during the enzymatic oxidation of *trans*-dihydrodiols can be monitored using Clark-style oxygen electrodes in 600-μl chambers containing the reaction components for *trans*-dihydrodiol oxidation (Fig. 10). The reaction is initiated by the addition of enzyme. The output from the electrode is connected through an amplifier (Instech, Plymouth Mtg., PA), which allows the simultaneous display of oxygen concentration and the rate of oxygen uptake as a function of time. The data acquisition software allows the recording of 1–1000 data points/min. All oxygen uptake experiments are performed at 25°, assuming that the amount of dissolved oxygen is 0.137 μmol in 0.6 ml at 25° at 760 mmHg.

The concurrent formation of H_2O_2 during the enzymatic oxidation of PAH–*trans*-dihydrodiols is measured by linking H_2O_2 production to the horseradish peroxidase–catalyzed oxidation of 3,3,5,5-tetramethylbenzidine $\varepsilon = 30{,}680\ M^{-1}\ \mathrm{cm}^{-1}$.[51]

The concurrent formation of superoxide anion during the enzymatic oxidation of PAH–*trans*-dihydrodiol is measured by monitoring the rate

[50] T. M. Penning, S. T. Ohnishi, T. Ohnishi, and R. G. Harvey, *Chem. Res. Toxicol.* **9,** 84 (1996).

[51] H. H. Liem, F. Cradenas, M. Tavassoli, M. B. Poh-Fitzpatrick, and U. Muller-Eberhard, *Anal. Biochem.* **98,** 388 (1979).

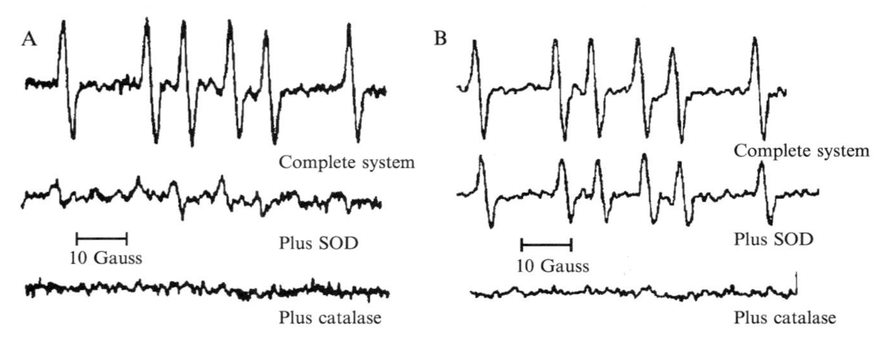

FIG. 11. EPR spectra obtained during the enzymatic oxidation of (±)-BP-7,8-diol using DMPO as the spin-trapping agent. (A) Complete system for the enzymatic oxidation of (±)BP-7,8 diol plus 50 m*M* DMPO taken at 78 min (upper spectrum); complete system for the enzymatic oxidation of (±)BP-7,8 diol plus 50 m*M* DMPO and SOD taken at 60 min (middle spectrum); complete system for the enzymatic oxidation of BP-diol plus 50 m*M* DMPO and catalase taken at 71 min (lower spectrum). The complete system contained 45 μM (±)BP-7,8 diol, 2.3 m*M* $NADP^+$, 8% DMSO 50 m*M* glycine buffer, pH 9.0, and 50 m*M* DMPO in 200 μl. (B) Superoxide-generating system (hypoxanthine/xanthine oxidase) plus DMSO as cosolvent and DMPO as spinning trapping agent, complete system at 57 min (upper spectrum); complete system plus SOD at 38 min; and complete system plus catalase at 56 min (lower spectrum). The complete system contained 800 μM hypoxanthine, 8% DMSO, 100 m*M* potassium phosphate buffer, pH 8.0, plus 50 m*M* DMPO and 30 mU xanthine oxidase. Reprinted with permission from Penning *et al.*, *Chem. Res. Toxicol.* **9,** 84 (1996), copyright 1996, American Chemical Society.

of acetylated cytochrome *c* reduction that is inhibited by superoxide dismutase at 550 nm ($\varepsilon = 19{,}600\ M^{-1}\ \text{cm}^{-1}$).[52] Typical traces of oxygen metabolism during the enzymatic oxidation of BP-7,8-diol are shown in Fig. 10. Inspection of these traces shows that during the autoxidation of the intermediate catechol, H_2O_2 is produced before O_2 consumption.

Direct measurement of ROS produced in the reaction can be monitored with spin-trapping agents such as DMPO in the presence of DMSO cosolvent. Reaction systems (200 μl) contain 45 μM (±)BP-7,8-diol, 2.3 m*M* $NADP^+$, 8% DMSO, and 50 m*M* glycine buffer, pH 9.0, plus 50 m*M* DMPO. Reactions are initiated by the addition of 50 μg enzyme. EPR spectra are recorded on a Varian E109 EPR (Palo Alto, CA) spectrometer operating in the X-band (9.25 GHz) with 0.1 *M* modulation amplitude and 200 mW microwave power with a receiver gain set at 5×10^4.[50]

A six-line hyperfine splitting pattern is observed during the AKR mediated oxidation of (±)BP-7,8-diol (Fig. 11). This splitting pattern can be

[52] A. Azzi, C. Montecucco, and C. Richter, *Biochem. Biophys. Res. Commun.* **65,** 597 (1975).

assigned to DMPO-CH_3 and is abolished by the presence of superoxide dismutase (SOD) and catalase, suggesting that it is both $O_2^{\bullet}$ and H_2O_2 dependent. An identical DMPO spin-adduct is observed with DMPO and DMSO, using hypoxanthine and xanthine oxidase to produce superoxide anion. In this sequence, the spin-adduct is only destroyed by catalase but is unaffected by SOD, which will produce H_2O_2. Thus, the difference between the two spin-adducts is that the one formed in the enzymatic reaction has a dependency on $O_2^{\bullet}$, suggesting that this radical is the initiating radical, whereas the spin-adduct produced in the model system is entirely dependent on H_2O_2.[50]

Formation of the DMPO-CH_3 spin-adduct can be best explained by the following scheme:

$$O_2^{\bullet-} + O_2^{\bullet-} + 2H^+ = H_2O_2 + O_2$$
$$H_2O_2 + Fe^{2+}(\text{trace}) = Fe^{3+} + OH^- + \dot{O}H^{\bullet}$$
$$OH^{\bullet} + DMSO = MSOH + CH_3^{\bullet}$$
$$CH_3^{\bullet} + DMPO = DMPO - CH_3$$

(where *MSOH* is methanesulfinic acid)

On the basis of the temporal relationship that H_2O_2 is produced before O_2 is consumed and that superoxide anion is the initiating radical, a mechanism for auto-oxidation is proposed. Superoxide anion first acts as base to remove a proton from the catechol to form a hydroperoxy radical. The hydroperoxy radical then acts as oxidant on the catecholate anion to yield an *o*-semiquinone anion radical and hydrogen peroxide. The *o*-semiquinone radical is then oxidized by molecular oxygen to produce the fully oxidized PAH *o*-quinone and superoxide anion. Trace metal ions are believed to cause the initiating Fenton chemistry (Fig. 12).

Detection of the Aldo-Keto Reductose Pathway in Cell Culture

Detection of AKR expression in human cells can be achieved using isoform specific RT-PCR (for transcripts); immunoblot analysis using rabbit anti-rat AKR1C9 antisera (which cross-reacts with AKR1C1–AKR1C4), and by direct functional enzymatic assays (see earlier). By use of this approach, a number of human cells lines that represent a null environment for AKR expression have been identified and are available for stable transfection: MCF-7 (mammary carcinoma) and H358 and H-441 (bronchoalveolar cells).[53] By contrast, the same assays show that HepG2 and the human lung adenocarcinoma cell line A549 are rich sources of AKRs.[13,14]

[53] L. Tsuruda, Y-t. Hou, and T. M. Penning, *Chem. Res. Toxicol.* **14,** 856 (2001).

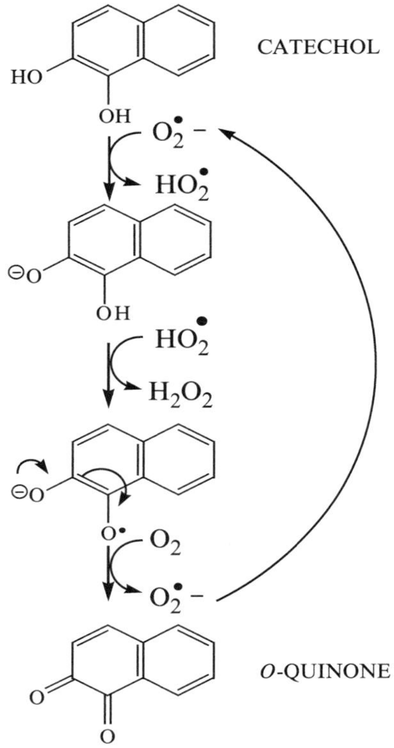

CATECHOL

O-QUINONE

FIG. 12. Mechanism of catechol autooxidation. Reprinted with permission from Penning *et al.*, *Chem. Res. Toxicol.* **12,** 1 (1999), copyright 1999, American Chemical Society.

The AKR-dependent production of PAH *o*-quinones has been detected in rat hepatocytes, MCF7-cells stably overexpressing AKR1C9, and human lung cell lysates A549.[14,53,54] In rat hepatocytes, BP-7,8-dione formation was characterized by EIMS and abolished by the AKR inhibitor indomethacin.[54] In the MCF-7 AKR1C9 transfectants, BP-7,8-dione was detected by cochromatography versus synthetic standards and by diode-array spectrometry.[53] The quinone was not formed in the mock-transfected controls. In A549 cells, the conversion of DMBA-3,4-diol to DMBA-3,4-dione was monitored by trapping the quinone as its *mono-* and *bis*-thioether conjugates, which can be characterized by LC/MS.[14]

[54] L. Flowers-Geary, R. G. Harvey, and T. M. Penning, *Carcinogenesis* **16,** 2707 (1996).

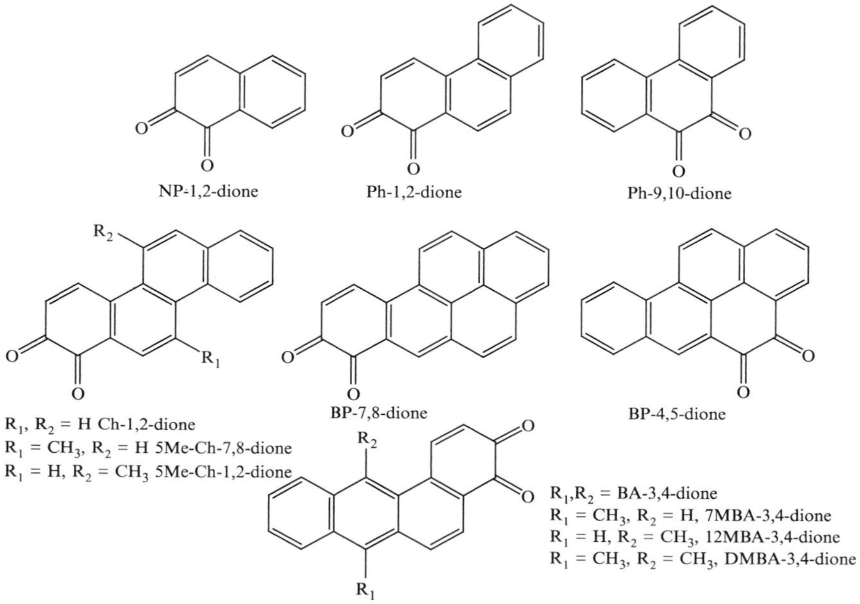

NP-1,2-dione Ph-1,2-dione Ph-9,10-dione

R_1, R_2 = H Ch-1,2-dione
R_1 = CH_3, R_2 = H 5Me-Ch-7,8-dione
R_1 = H, R_2 = CH_3 5Me-Ch-1,2-dione

BP-7,8-dione BP-4,5-dione

R_1,R_2 = BA-3,4-dione
R_1 = CH_3, R_2 = H, 7MBA-3,4-dione
R_1 = H, R_2 = CH_3, 12MBA-3,4-dione
R_1 = CH_3, R_2 = CH_3, DMBA-3,4-dione

FIG. 13. Structures of PAH *o*-quinones: products of AKR reactions. NP = naphthalene, Ph = phenanthrene; Ch = chrysene; 5Me-Ch = 5-methylchrysene; BP = benzo[*a*]pyrene; BA = benz[*a*]anthracene; 7MBA = 7-methylbenz[*a*]anthracene; 12MBA = 12-methylbenz[*a*]anthracene; and DMBA = 7,12-dimethylbenz[*a*]anthracene, after Sukumaran and Harvey.[58]

Properties of PAH *o*-Quinones Produced by Aldo-Keto Reductases

To study the properties of PAH *o*-quinones produced by AKRs, it is necessary to synthesize these quinones in sufficient quantities. For each *trans*-dihydrodiol substrate for AKRs, the corresponding *o*-quinone product is desired (Fig. 13). The quinones synthesized include naphthalene-1,2-dione (NP-1,2-dione)[55] and phenanthrene-1,2-dione.[56] The remainder: chrysene-1,2-dione, 5-methylchrysene-1,2-dione, 5-methylchrysene-7,8-dione, benzo[*a*]pyrene-7,8-dione (BP-7,8-dione), benz[*a*]anthracene-3,4-dione, 7-methyl-benz[*a*]anthracene-3,4-dione (BA-3,4-dione), 7,12-dimethylbenz[*a*]-anthracene-3,4-dione (DMBA-3,4-dione), and benzo[*g*]chrysene-10,11-dione can be synthesized by somewhat similar routes.[57,58] The synthetic

[55] L. F. Fieser and A. H. Blatt, "Organic Synthesis," p. 430. Wiley, New York, 1943.
[56] L. F. Fieser, *J. Am. Chem. Soc.* **51,** 1896 (1929).
[57] P. P. Fu, C. Cortez, K. B. Sukumaran, and R. G. Harvey, *J. Org. Chem.* **44,** 4265 (1979).
[58] K. B. Sukamaran and R. G. Harvey, *J. Org. Chem.* **45,** 4407 (1980).

sequence to BP-7,8-dione, involves reduction of 7-keto-8,9,10-trihydrobenzo[*a*]pyrene to yield the 7-hydroxy-7,8,9,10-tetrahydrobenzo[*a*]pyrene. Following dehydration, epoxidation to the 7,8-oxide provides a route to the enol-acetate, which is then oxidized in the presence of Fremy's salt with a phase-transfer catalyst to yield the BP-7,8-dione (Fig. 14).[59,60]

Three Classes of PAH o-*Quinones Produced by Aldo-Keto Reductases*

PAH *o*-quinones produced by AKRs differ in their reactivity and redox activity. Reactivity is determined by measuring their bimolecular rate constants for the addition of thiols (2-mercaptoethanol, *N*-acetyl-L-cysteine, and glutathione) on the basis of RP-HPLC analyses.[59] Bimolecular rate constants for the addition of *N*-acetyl-*L*-cysteine with either (+)-*anti*-BPDE or BP-7,8-dione show that the quinones are considerably more reactive.[59,60] Redox activity is determined by measuring the rate of NADPH oxidation and the rate of consumption of molecular oxygen during redox-cycling in the absence and presence of subcellular fractions (e.g., mitochondria and microsomes).[61,62] These properties predict the cytotoxicity of the *o*-quinones and the mechanisms responsible for cell death.

Class I *o*-quinones (e.g., NP-1,2-dione and DMBA-3,4-dione) are highly electophilic and redox active. They are potent cytotoxins affecting hepatoma cell viability (acute effect) and cell survival (chronic effect). They produce the most superoxide anion and semiquinone radical. Cell death is associated with a change in redox state decreased $NADPH^+/NADP^+$ and NAD^+, oxidized:reduced glutathione ratios.[62,63]

Class II *o*-quinones (e.g., 5-methlychrysene-1,2-dione and benz[a]anthracene-3,4-dione) are less electrophilic but are not redox active. They are potent cytoxins affecting cell survival only. They produce mainly *o*-semiquinone anion radicals, and cell death is likely due to macromolecule damage by the *o*-semiquinone radical.[61,62]

Class III *o*-quinones (e.g., BP-7,8-dione) are as electophilic as class II *o*-quinones and are redox active but produce mainly superoxide anion, suggesting that the *o*-semiquinone radical may be transient. They are potent cytotoxins affecting cell viability only. Cell death is likely due to GSH depletion, which occurs without a change in redox state.[61,62]

[59] V. S. Murty and T. M. Penning, *Chem-Biol. Inter.* **84,** 169 (1992).

[60] N. B. Islam, S. C. Gupta, H. Yagi, D. M. Jerina, and D. L. Whalen, *J. Am. Chem. Soc.* **112,** 6363 (1990).

[61] L. Flowers-Geary, R. G. Harvey, and T. M. Penning, *Chem. Res. Toxicol.* **6,** 252 (1993).

[62] L. Flowers-Geary, W. Bleczinski, R. G. Harvey, and T. M. Penning, *Chemico-Biol. Inter.* **99,** 55 (1996).

[63] L. Flower, S. T. Ohnisni, and T. M. Penning, *Biochemistry* **36,** 8640 (1997).

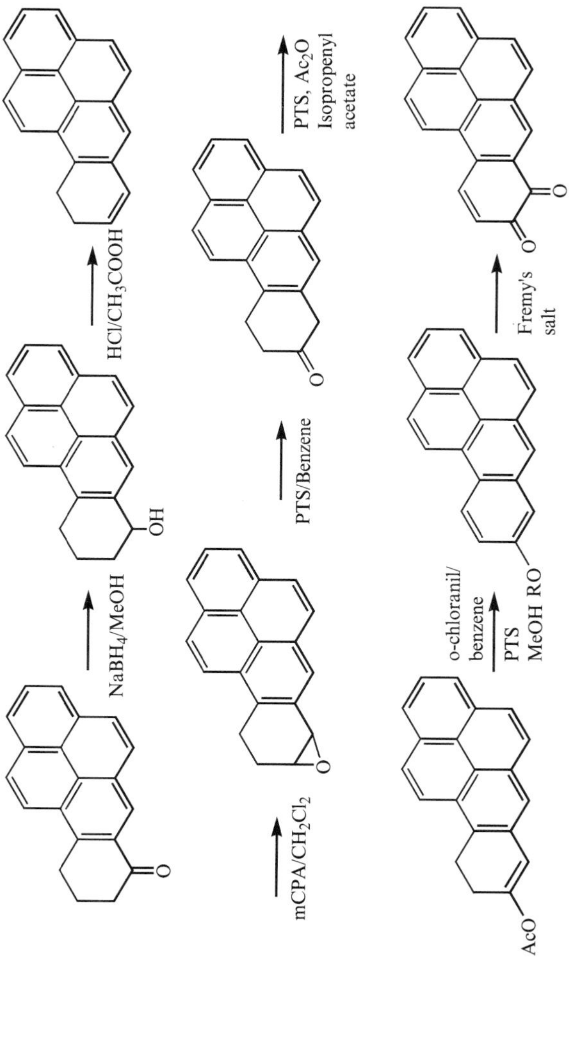

Fig. 14. Synthesis of BP-7,8-dione. mCPA = meta-chloroperoxybenzoic acid; PTS = p-tolvenesulfonic acid.

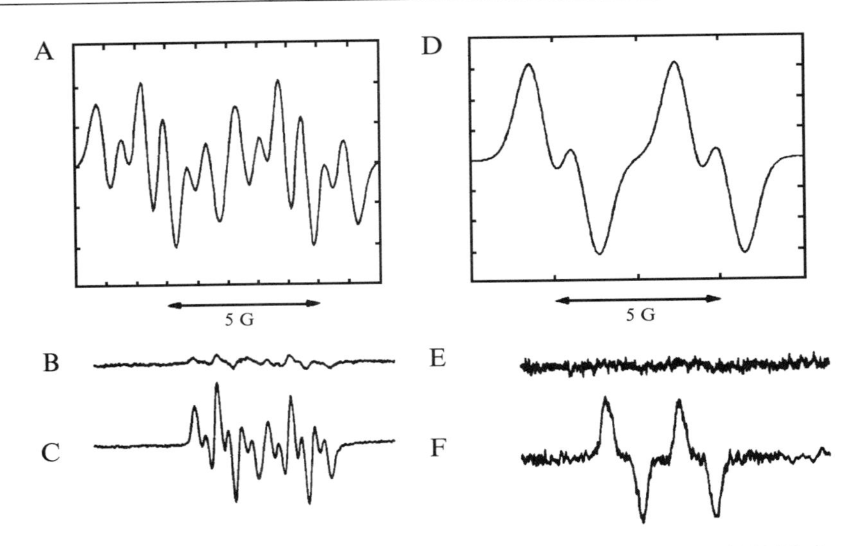

FIG. 15. EPR spectra of *o*-semiquinone (SQ) radicals of NP-1,2-dione and BP-7,8-dione. (A) Simulated spectrum of the SQ radical of NP-1,2-dione. (B) Spectrum of 0.78 m*M* NP-1,2-dione taken in 8% DMSO. (C) Spectrum of 0.78 m*M* NP-,1,2-dione plus 10 m*M* NADPH taken anaerobically. (D) Simulated spectrum of the SQ radical of BP-7,8-dione. (E) Spectrum of 50 μM BP-7,8-dione taken in 8% DMSO. (F) Spectrum of 50 μM BP-7,8-dione plus 10 m*M* NADPH taken anaerobically.

Detection of o-*Semiquinone Anion Radicals*

Reduction of representative PAH *o*-quinones (naphthalene-1,2-dione and benzo[a]pyrene-7,8-dione) in the presence of NADPH coupled with EPR leads to the detection of the *o*-semiquinone radicals. *o*-Semiquinone anion radicals may differ in their stability. Naphthalene-1,2-*o*-semiquinone radical gives a 12-line hyperfine splitting pattern, indicating that the free radical can be stabilized throughout the ring system. By contrast, benzo[*a*]pyrene-7,8-*o*-semiquinone radical gives a 4-line hyperfine splitting pattern, showing that the free radical is stabilized only over the terminal benzo ring and may be shorter lived; this would be consistent with the existing classification of the *o*-quinones (Fig. 15).[63]

PAH o-*Quinone Reactivity with Amino Acids and Proteins*

PAH *o*-quinones can also undergo 1,4-Michael addition with the side chains of reactive amino acids to form conjugates. These reactions are likely to modify cellular proteins leading to a change in protein function.

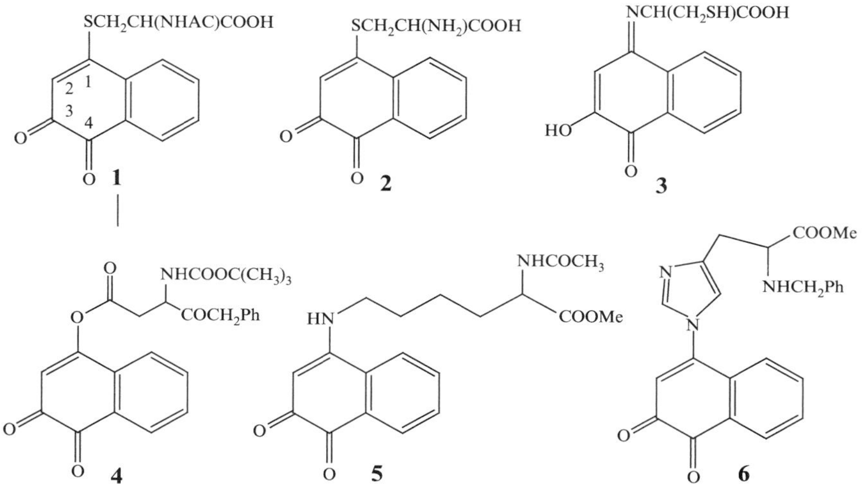

FIG. 16. PAH *o*-quinone amino acid conjugates. *N*-acetyl-*L*-cysteine (*N*-acetyl-*S*-(3,4-dihydro-3,4-dioxo-1-naphthyl)-*L*-cysteine) conjugate **1;** *L*-cysteine (*L*-cysteine (1-*S*-(3,4-dihydro-3,4-dioxo-1-naphthyl)-*L*-cysteine) conjugate **2;** *p*-iminoquinone-*L*-cysteine (*N*-3-hydroxy-4-oxo-1-naphthyl)imino-*L*-cysteine) conjugate **3;** protected *L*-aspartyl (*O*-(3,4-dihydroxy-1-naphthyl)-*N*-t-Boc-*L*-aspartyl benzyl ester) conjugate **4;** protected *L*-lysine (1-*N*-(3,4-dihydro-3,4-dioxo-1-naphthyl)-*N*α-acetyl-*L*-lysine methyl ester) conjugate **5;** and protected *L*-histidyl (1-*N*-(3,4-dihydro-3,4-dioxo-1-naphthyl)-*N*α-benzyl-*L*-histidyl methyl ester **6**. Please note that the ring numbering changes in the conjugates versus the unreacted quinone, naphthalene-1,2-dione.

Many transcription factors (e.g., AP-1, Nf-κB) contain reactive thiols, and it is speculated that their modification by PAH *o*-quinones may cause a change in cell signaling. Model reactions have been performed with NP-1,2-dione and amino acids. Reactions with *N*-acetyl-*L*-cysteine and cysteine are informative, because they show that after the initial addition reaction, subsequent rearrangements can occur, leading to more complex products (e.g., *p*-iminoquinones) (Fig. 16). Michael addition products have also been characterized for the addition of *N*-*t*-boc-*L*-aspartic acid α-benzoyl ester, *N*-α-acetyl-*L*-lysine methyl ester, and *N*-α-benzyl-*L*-histidine methyl ester to show that reactive amino acid side chains can form discrete amino acid conjugates.[64]

[64] G. Sridhar, V. S. Murty, S-H. Lee, I. A. Blair, and T. M. Penning, *Tetrahedron Lett.* **57,** 407 (2001).

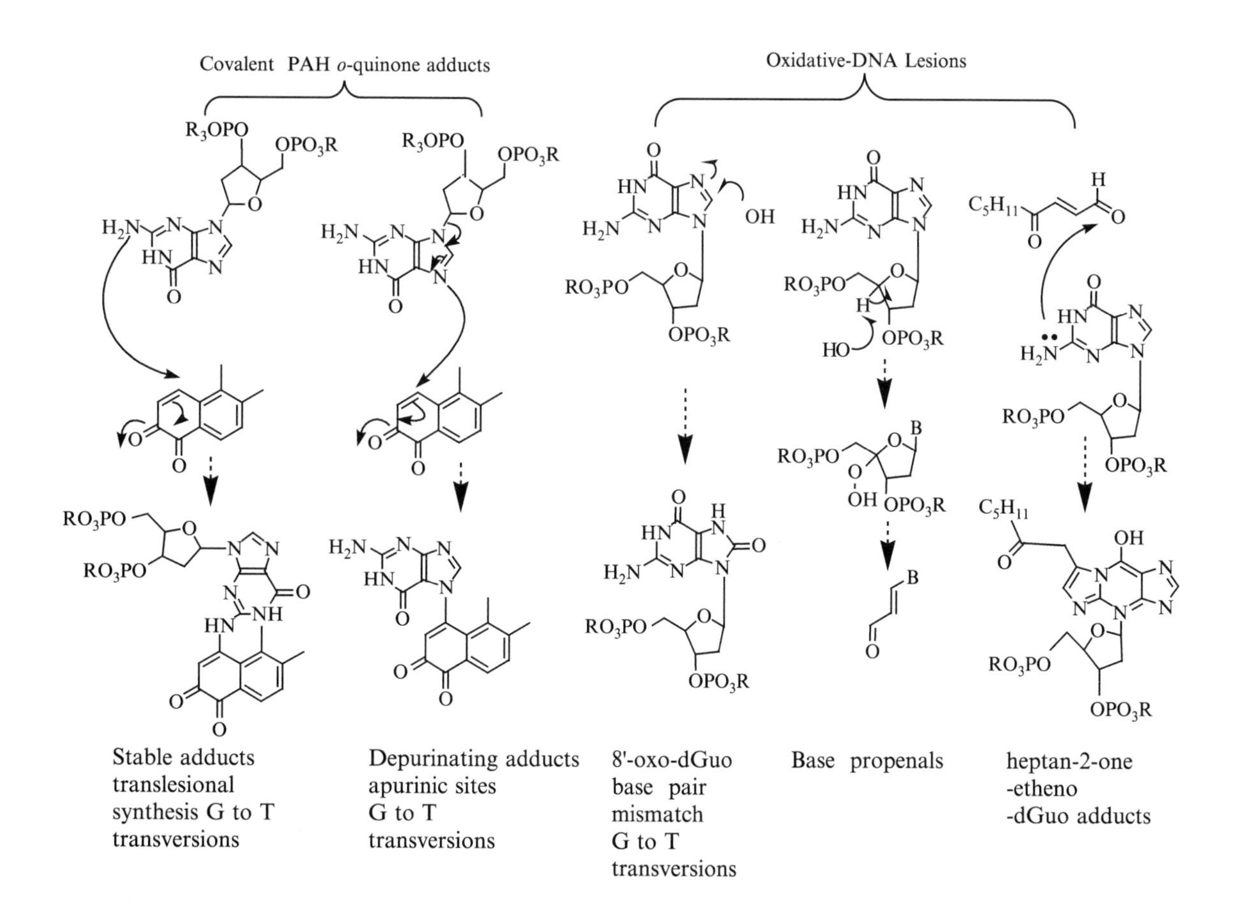
Covalent PAH o-quinone adducts
Oxidative-DNA Lesions
Stable adducts translesional synthesis G to T transversions
Depurinating adducts apurinic sites G to T transversions
8'-oxo-dGuo base pair mismatch G to T transversions
Base propenals
heptan-2-one -etheno -dGuo adducts

DNA Lesions Caused by PAH o-*Quinones*

PAH *o*-quinones will cause both covalent and oxidative modification of DNA (Fig. 17). Interest exists in characterizing these reactions, because they may lead to mutation of critical proto-oncogenes and tumor suppressor genes, increase the mutational load, and contribute to the causation of human lung cancer. The most common mutations seen in human lung cancer are G to T transversions in *K-ras* (12th codon) and G to T transversions in *p53* (codons, 245, 248, and 249).[65–67] DNA lesions observed with PAH *o*-quinones should provide routes to these mutations if they are to a play role in lung carcinogenesis.

Covalent Adducts with DNA. PAH *o*-quinones have the capacity to form stable N^2-deoxyguanosine and N^6-deoxyadenosine adducts (Fig. 17). Detection of stable BP-7,8-dione DNA adducts can be achieved by reacting [^{3}H]-BP-7,8-dione (Chemsyn, Lenexa, KS) with calf-thymus DNA, followed by enzymatic digestion to the constituent deoxyribonucleosides. This leads to the isolation of a single major adduct that coelutes on RP-HPLC, with the adduct formed by addition of the *o*-quinone to oligo-p$(dG)_{10}$, following enzymatic digestion.[68] By contrast, no adducts were detected when [^{3}H]-BP-7,8-dione was reacted with poly-dC, poly-dT, or poly-dA. No formal structure exists for the stable N^2-deoxyguanosine adduct because of the difficulty in synthesizing authenticated standards.

Current routes to these standards include two approaches. In the first route, there is activation of the *o*-quinone by bromination followed by attack with the *O*-TBDMS–protected deoxyribonucleoside. In the second approach, the amino-quinone is formed followed by attack from the halogenated deoxyribonucleoside.[69] Bulky stable deoxyribonucleoside adducts could give rise to G to T transversions, depending on whether there is

[65] S. Rodenhius, *Cancer Biol.* **3,** 241 (1992).

[66] T. Soussi, *The p53 Database.* http://p53.curie.fr/p53%20site%20version%202.0/database/p53 database.html, 2001.

[67] P. Hainaut and G. P. Pfeifer, *Carcinogenesis* **22,** 367 (2001).

[68] M. Shou, R. G. Harvey, and T. M. Penning, *Carcinogenesis* **14,** 475 (1993).

[69] S. R. Gopishetty, R. G. Harvey, S-H. Lee, I. A. Blair, and T. M. Penning, Aldo-Keto reductases and toxicant metabolism. ACS Symposium Series (T. M. Penning and J. Mark Petrash, ed.), Vol. 865, pp. 127–138. Oxford University Press, 2003.

FIG. 17. Covalent and oxidative DNA lesions caused by PAH *o*-quinones. Reprinted with permission from Penning *et al.*, *Chem. Res. Toxicol.* **12,** 1 (1999), copyright 1999, American Chemical Society.

error-free or error-prone translesional synthesis by replicative DNA polymerases.

PAH *o*-quinones can also form N7-guanine and N7-adenine depurinating adducts, leaving behind an abasic site (Fig. 17). N7-guanyl depurinating adducts of NP-1,2-dione, phenanthrene-1,2-dione, and BP-7,8-dione can be synthesized by conducting reactions with the *o*-quinone and deoxyribonucloside in the presence of acetic acid.[70]

Typical reactions involve reacting a PAH *o*-quinone (10 mg) in acetonitrile/DMSO (2 ml) to a stirred solution of a five-fold molar excess of dGuo in a 1:1 mixture of acetic acid/water (4 ml). The reaction mixture is purified by solid-phase extraction on a C18-Sep-Pak cartridge, which is washed successfully with 1% acetic acid/water (30 ml) and water (3 ml). The product is then eluted with 75% methanol/water, and following removal of the organic solvent, the final product is obtained by lyophilization and analyzed by RP-HPLC.

LC/ESI/MS analysis of the purified 7-(benzo[*a*]pyrene-7,8-dion-10-yl) guanine adduct gave a protonated molecular ion at $m/z = 432$, consistent with the N7 depurinating adduct. Collision-induced dissociation (CID) of the $m/z = 432$ ion generated adduct derived fragment ions at $m/z = 281$ (MH^+-guanine) and 253 (MH^+-guanine-CO). Additional product ions consistent with the proposed structure were observed at $m/z = 415$ (MH^+-NH_3), 404 (MH^+-CO), 389 (MH^+-HNCO), 362 (MH^+-CO, NH_2CN), and 308 (MH^+-CO, NH_2CN, 2HCN) (Fig. 18). These adducts have also been detected in *o*-quinone reactions with calf-thymus DNA. In terms of mutagenesis, abasic sites created by depurination provide a route to G to T transversions. DNA polymerases often introduce an A opposite the abasic site so that on replication a G to T transversion occurs.[71]

Oxidative Damage of DNA. Different forms of oxidative DNA damage are possible as a result of PAH *o*-quinone redox-cycling (Fig. 17). The ROS produced could cause DNA strand breaks (single and double strand breaks); formation of oxidatively damaged bases (e.g., 8-oxo-2-deoxyguanosine (8-oxo-dGuo) and thymine glycol); the formation of base-propenals from 4′OH radical attack of deoxyribose, leading to malondialdehyde-dG adducts; and the indirect formation of adducts that may arise from lipid peroxidation breakdown products. Many of these lesions can be seen with PAH *o*-quinones.[63]

DNA strand scission can be detected using RF ϕX174DNA, and DNA fragmentation can be detected using poly dG•poly dC by agarose

[70] K. D. McCoull, D. Rindgen, I. A. Blair, and T. M. Penning, *Chem. Res. Toxicol.* **12,** 237 (1999).

[71] D. Sagher and B. Strauss, *Biochemistry* **22,** 4518 (1983).

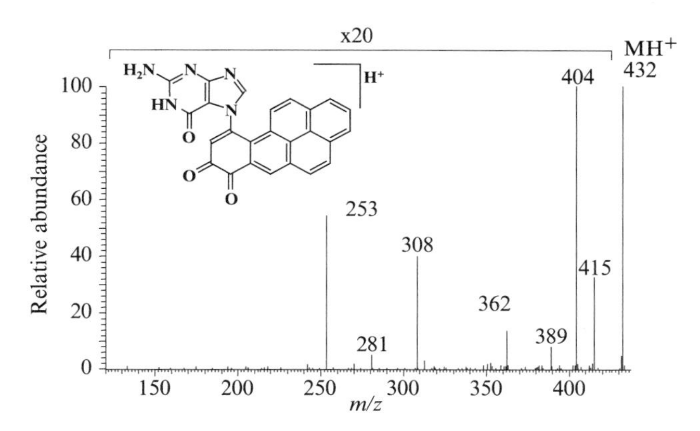

FIG. 18. Collision-induced dissociation spectrum of the *m/z* 432 (MH^+) ion of the benzo[*a*]pyrene-7,8-dione-guanine adduct, obtained on a Finnigan LCQ ion trap. Reprinted with permission from McCoull *et al.*, *Chem. Res. Toxicol.* **12,** 237 (1999), copyright 1999, American Chemical Society.

gel electrophoresis following incubation with submicromolar concentrations of PAH *o*-quinones under redox-cycling conditions. Redox-cycling conditions are maximized by the presence of NADPH and $CuCl_2$, whereas DNA strand scission or fragmentation is abolished by enzymatic scavengers of ROS (SOD and catalase) and chemical scavengers of ROS (Tiron and mannitol). Under anaerobic conditions in which only the *o*-semiquinone anion radical is formed, no DNA strand scission is observed.[64] Thus, DNA strand scisson is entirely dependent on ROS. The DNA strand scission that is seen *in vitro* may lead to cytotoxic rather than mutagenic consequences.

PAH *o*-quinones in the presence of NADPH produce sufficient ROS to form 8-oxo-deoxyguanosine (8-oxo-dGuo). Interest in this lesion exists, because the C8 oxygen causes base mispairing with adenine if the lesion is unrepaired.[72,73] Like depurination, these lesions would also provide a straightforward route to G to T transversions. Thus, both depurination and 8-oxo-dGuo formation could provide routes to the G to T transversions that dominate in *K-ras* and *p53* in lung cancer.

Sensitive detection of 8-oxo-dGuo is required to detect this lesion, and the method of choice has been Electrochemical (EC)-HPLC, which provides sensitivity in the fmole range, where the amount of 8-oxo-dGuo is

[72] A. P. Breen and J. P. Murphy, *Free Radical Biol. Med.* **18,** 1033 (1995).
[73] D. Wang, D. A. Kreutzer, and J. M. Essigmann, *Mutation Res.* **400,** 99 (1998).

TABLE IV
DETECTION OF 8-OXO-DGUO IN SALMON TESTIS DNA EXPOSED TO NP-1,2-DIONE UNDER REDOX CYCLING CONDITIONS

Treatment	Peak area		Ratio
	dGuo	8-oxo-dGuo	8-oxo-dGuo/dGuo
DNA w/Dig Enz	235 μC	8.2 nC	$2.4/10^5$
DNA w/DMSO, S9, NADPH, Dig Enz	230 μC	8.3 nC	$2.5/10^5$
DNA w/NP-1,2-dione, S9, NADPH, Dig Enz	228 μC	38 nC	$23.0/10^5$
DNA w/NP-1,2-dione, NADPH, Dig Enz	229 μC	525 nC	$239/10^5$
DNA w/NP-1,2-dione, NADPH, SOD, Dig Enz	253 μC	2230 nC	$892/10^5$

NP-1,2-dione = naphthalene-1,2-dione.

normalized to dGuo. Pronounced increases in the amount of 8-oxo-dGuo are seen when salmon testis DNA is treated with NP-1,2-dione under redox cycling conditions (20 μM NP-1,2-dione, 200 μM NADPH) (Table IV). Levels of 8-oxo-dGuo are decreased by the addition of the S9 fraction from rat liver because of protein sequestration but elevated with SOD because of the increased production of H_2O_2. Measurements of 8-oxo-dGuo and the precautions necessary to prevent adventitious oxidation of dG are described elsewhere in this series.[74]

PAH o-*Quinones and p53 Mutagenesis*

We have adapted a yeast reporter gene assay that allows us to detect mutation of p53 mediated by PAH *o*-quinones produced by AKRs.[75,76] In this assay, p53 cDNA is treated with PAH *o*-quinones in the absence and presence of redox-cycling conditions (at concentrations of *o*-quinone that are insufficient to cause significant strand scission). The resulting mixture of wt p53 and lesioned DNA is mixed with a gap-repair plasmid and used to transform a host yeast strain containing a p53 reporter gene. The gapped-repair plasmid contains sites for homologous recombination. The transformed host produces wt and mutated p53 from an ADH promoter 5′ to the sites of homologous recombination. The host also has integrated into its chromosome an adenine reporter gene driven by the p21 promoter.

[74] M. K. Shigenaga, J-W. Park, K. C. Cundy, C. J. Gimeno, and B. N. Ames, *Methods Enzymol.* **186,** 521 (1990).
[75] D. Yu, T. M. Penning, and J. M. Field, *Polycyclic Aromatic Compounds* **3–4,** 881 (2002).
[76] D. Yu, J. A. Berlin, T. M. Penning, and J. M. Field, *Chem. Res. Toxicol.* **15,** 832 (2002).

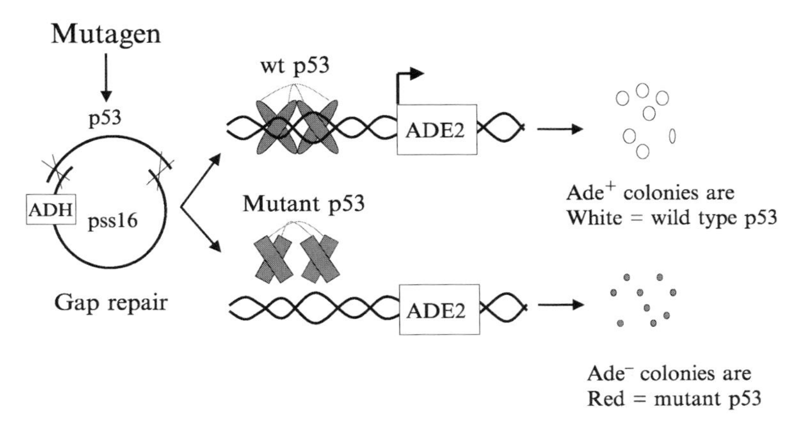

FIG. 19. Yeast reporter assay for p53 mutation. Reprinted with permission from Yu *et al.*, *Chem. Res. Toxicol.* **15,** 832 (2002), copyright 2002, American Chemical Society.

Wt p53 will bind to the p21 promoter, and in the presence of limiting adenine, the yeast colonies will synthesize adenine and turn white. By contrast, mutated p53 will be unable to bind to the p21 promoter, and in the presence of limiting adenine, the yeast colonies will not synthesize adenine, and colonies will turn red. This positive selection allows us to score change-in-function mutations in p53 that eliminate its transcriptional competency (Fig. 19).

By use of this assay, we have shown that BP-7,8-dione is mutagenic at submicromolar concentrations, provided a redox-cycling system is present (Table V), that the mutation rate (red colonies/total colonies $\times$ 100) is abolished by ROS scavengers, and that a significant number of G to T transversions were observed, implicating 8-oxo-dGuo in their formation.

Aldo-Keto Reductase Stable Transfection Systems to Measure DNA Lesions and p53 Mutations

Strong evidence exists for the formation of stable and depurinating PAH *o*-quinone DNA adducts and for the formation of oxidative lesions based on *in vitro* measurements; however, these DNA lesions have not been detected within a cellular context. Stable overexpression of human AKRs into heterologous systems will address this concern. Several systems currently exist; these include the stable overexpression of AKR1A1 (aldehyde reductase) in human lung bronchoalevolar cells (H-358 and H-441) and the observation that A549 cells endogenously overexpress

TABLE V
NP-1,2-DIONE AND BP-7,8-DIONE MUTATE p53 UNDER REDOX CYCLING CONDITIONS

	NP-1,2-dione Colonies			BP-7,8-dione Colonies		
Treatment	Total	Red	% red	Total	Red	% red
Quinone alone[a]	922	11	1.2	3380	32	0.9
Quinone + NADPH	896	13	1.5	3160	31	1.0
μM quinone + NADPH + $CuCl_2$ [b]						
0 μM	798	16	2.0	3290	34	1.0
0.031 μM	812	31	3.8	3284	41	1.2
0.0625 μM	1232	54	4.4	3080	51	1.7
0.125 μM	756	46	6.1	2536	69	2.1
0.25 μM	484	45	9.3	2200	113	5.1

[a] p53 mutagenesis was determined in the presence of 20 μM PAH *o*-quinone alone.
[b] p53 mutagenesis was determined in the presence of increasing PAH *o*-quinone concentration in the presence of 1 mM NADPH and 100 μM $CuCl_2$.

AKR1C1–AKR1C3. Treatment of cells with PAH *trans*-dihydrodiol substrates would be predicted to cause the DNA lesions described and should be attenuated by AKR1A1 inhibitors (alrestat and tolerstat) or AKR1C inhibitors (indomethacin and 6-medroxyprogesterone acetate).

MCF7-AKR1A1 and A549-AKR1C1–AKR1C3 overexpressing cells will also allow us to determine whether exposure of these cells to PAH *trans*-dihydrodiols will cause change in function mutations in p53. MCF7 cells and A549 cells contain wt p53; following treatment of the cells with PAH *trans*-dihydrodiols, p53 can be obtained by RT-PCR and substituted into the yeast reporter gene assay to detect p53 mutations. Increased p53 mutation should be attenuated by AKR inhibitors. Such an approach will provide a proof-of-principle concept that AKRs generate reactive and redox-active *o*-quinones and cause change-in-function mutations in p53.

Acknowledgments

The methods and techniques described could not have been developed without the contribution of many talented colleagues, including Drs. Thomas E. Smithgall, Lynn Flowers, Varanasi Murty, Kirsten McCoull, Elizabeth Glaze, Michael E. Burczynski, Nisha Palackal, and Sridhar Gopishetty. Synthetic approaches were developed in collaboration with Dr. Ronald G. Harvey at the Ben May Institute for Cancer Research, and mass-spectrometry was performed in the laboratory of Dr. Ian A. Blair at the Center for Cancer Pharmacology, University of Pennsylvania School of Medicine. Many PAH metabolites were made available from the National Cancer Institute Chemical Carcinogen Standard Reference Repository, and the work was supported by grants R01 CA39505 and P01 CA092537 awarded to TMP.

[4] Redox Cycling of β-Lapachone and Structural Analogues in Microsomal and Cytosol Liver Preparations

By SILVIA FERNÁNDEZ VILLAMIL, ANDRÉS O. M. STOPPANI, and MARTA DUBIN

Introduction

Quinones are widely distributed in nature and make up an important group of substrates for flavoenzymes. Lipophilic *o*-naphthoquinones possess antibacterial, antifungal, trypanocidal, and cytostatic effects. Among those quinones, β-lapachone[1,2] (3,4-dihydro-2,2-dimethyl-2*H*-naphtho [1,2*b*]pyr-an-5,6-dione) isolated from the lapacho tree *(Tabebuia avellanedae)* has proved to be an effective cytostatic agent in different human tumor cells, such as murine leukemia, melanoma, hepatoma, human leukemia, colon carcinoma, lymphoma, and glioma, as well as epidermoid laryngeal, ovarian, breast, lung, and prostate cancer.[3–14] On these grounds, β-lapachone has

[1] I. L. D'Albuquerque, M. C. N. Maciel, A. R. Schuler, M. do C. de Araujo, G. Medeiros Maciel, M. da S. B. Cavalcanti, D. Gimino Martins, and A. Lins Lacerda, *Revta. Inst. Antibiot. Univ. Recife* **12,** 31 (1972).

[2] A. O. M. Stoppani, S. Goijman, M. Dubin, S. H. Fernández Villamil, M. P. Molina Portela, A. M. Biscardi, and M. Paulino, *Trends Comp. Biochem. Physiol.* **7,** 1 (2000).

[3] M. Dubin, S. H. Fernández Villamil, and A. O. M. Stoppani, *Medicina-Buenos Aires* **61,** 343 (2001).

[4] A. B. Pardee, Y. Z. Li, and C. J. Li, *Curr. Cancer Drug Targets* **2,** 227 (2002).

[5] Y. Z. Li, C. J. Li, A. Ventura Pinto, and A. B. Pardee, *Mol. Med.* **5,** 232 (1999).

[6] A. Samali, H. Nordgren, B. Zhivotovsky, E. Peterson, and S. Orrenius, *Biochem. Biophys. Res. Commun.* **255,** 6 (1999).

[7] M. E. Dolan, B. Frydman, C. B. Thompson, A. M. Diamond, B. J. Garbiras, A. R. Safa, W. T. Beck, and L. J. Marton, *Anti-Cancer Drugs* **9,** 437 (1998).

[8] C. J. Li, L. Averboukh, and A. B. Pardee, *J. Biol. Chem.* **268,** 22463 (1993).

[9] C. J. Li, C. Wang, and A. B. Pardee, *Cancer Res.* **55,** 1512 (1995).

[10] B. Frydman, L. J. Marton, J. S. Sun, K. Neder, D. T. Witiak, A. A. Liu, H.-M Wang, Y. Mao, H.-Y.Wu, M. M. Sanders, and L. F. Liu, *Cancer Res.* **57,** 620 (1997).

[11] A. Vanni, M. Fiore, A. De Salvia, E. Cundari, R. Ricordy, R. Ceccarelli, and F. Degrassi, *Mutat. Res.* **401,** 55 (1998).

[12] Y. -P. Chau, S.-G. Shiah, M.-J. Con, and M.-L. Kuo, *Free Radic. Biol. Med.* **24,** 660 (1998).

0076-6879/04 $35.00

been suggested for clinical use; its effects have often been described as apoptosis or necrosis, depending on target cells, time, and drug dose.

β-Lapachone redox cycling in the presence of reductants and oxygen yields[15] reactive oxygen species (ROS), including superoxide anion radical O_2^-, hydroxyl radical $^\bullet OH$, hydrogen peroxide H_2O_2, and singlet oxygen $^1O_2$, whose cytoxicity explains β-lapachone activity in cells.

Quinones can undergo enzymatic one-electron reduction catalyzed by microsomal reduced nicotinamide adenine dinucleotide phosphate (NADPH) cytochrome P450 reductase to the semiquinone.[16] In the presence of molecular oxygen, the semiquinone radical can transfer an electron and generate the O_2^-. This reaction results in shunting electrons toward oxygen, as a futile pathway for reduction equivalents otherwise used for cytochrome P450 reductase-dependent reactions.[15] Superoxide can dismutate, by a superoxide dismutase (SOD)-catalyzed reaction to H_2O_2, and $^\bullet OH$ would then be formed by the iron-catalyzed reduction of peroxide by means of the Fenton reaction. All these highly reactive species may react directly with DNA or other cellular macromolecules, such as lipids and proteins, causing cell damage.

Unlike most other cellular reductases, two-electron reduction of quinone can also be catalyzed by cytosolic and mitochondrial DT-diaphorase (DTD), quinone oxidoreductase, E.C. 1.6.99.2 (NQO1), directly to the hydroquinone.[17–19] DTD-mediated production of the hydroquinone, which can be readily conjugated and excreted from the cell, constitutes a protective mechanism against these types of damage. It has been suggested that the reducing activity of DTD protects cells from the toxicity of naturally occurring xenobiotics containing quinone moieties. In addition to its protective effects, DTD can also reduce certain quinones to more reactive forms.[20,21] Interestingly, DTD is overexpressed in a number of tumors, including breast, colon, and lung cancers, compared with surrounding normal tissue.[22] This observation, more than any other, suggests that drugs that are

[13] S. M. Planchon, S. M. Wuerzberger-Davis, J. J. Pink, K. A. Robertson, W. G. Bornmann, and D. A. Boothman, *Oncol. Rep.* **6,** 485 (1999).

[14] J. J. Pink, S. M. Planchon, C. Tagliarino, M. E. Varnes, D. Siegel, and D. A. Boothman, *J. Biol. Chem.* **275,** 5416 (2000).

[15] M. Dubin, S. H. Fernandez Villamil, and A. O. M. Stoppani, *Biochem. Pharmacol.* **39,** 1151 (1990).

[16] S. Fernandez Villamil, M. Dubin, M. P. Molina Portela, L. J. Perissinotti, M. A. Brusa, and A. O. M. Stoppani, *Redox Report* **3,** 245 (1997).

[17] L. Ernster, *Methods Enzymol.* **10,** 309 (1967).

[18] L. Ernster, R. W. Estabrook, P. Hochstein, and S. Orrenius, *Chem. Scr.* **27A,** 1 (1987).

[19] C. Lind, E. Cadenas, P. Hochstein, and L. Ernster, *Methods Enzymol.* **186,** 287 (1990).

[20] E. Cadenas, P. Hochstein, and L. Ernster, *Adv. Enzymol.* **65,** 97 (1992).

[21] E. Cadenas, *Biochem. Pharmacol.* **49,** 127 (1995).

[22] P. Joseph, T. Xie, Y. Xu, and A. K. Jaiswal, *Oncol. Res.* **6,** 525 (1994).

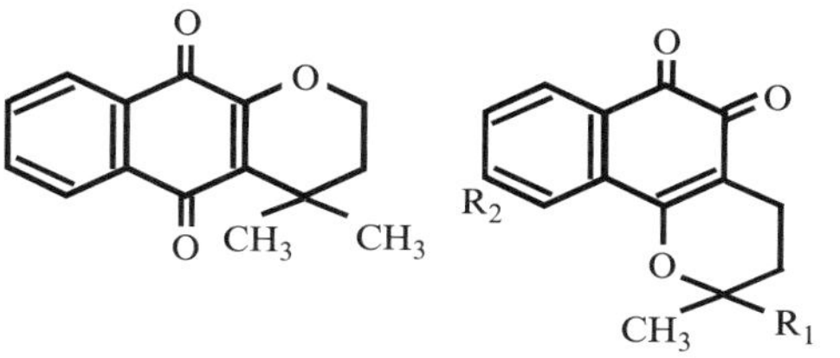

α-lapachone β-lapachone

Quinone	R_1	R_2
CG 8-935	C_2H_5	H
CG 9-442	C_6H_5	H
CG 10-248	CH_3	Cl

FIG. 1. Structures of α- and β-lapachone and β-lapachone analogues.

activated by DTD (e.g., mitomycin C, streptonigrin) should show significant tumor-specific activity. DTD expression has been proposed as a major determinant of β-lapachone–mediated apoptosis and lethality. On the other hand, redox cycling of β-lapachone catalyzed by DTD could also be a futile cycle and a possible component of the cytotoxic mechanism.[14] Because DTD can use either NADPH or reduced nicotinamide adenine dinucleotide (NADH) as electron donors, this cycle would lead to a substantial loss of NADH and NADPH, with a concomitant rise in NAD^+ and $NADP^+$ levels. Such depletion of reduced enzyme cofactors may be a critical stage for the activation of the apoptotic pathway after β-lapachone treatment. One novel aspect of β-lapachone toxicity is the apparent activation of calpain followed by NAD(P)H depletion, showing an atypical cleavage pattern of poly(ADP-ribose)polymerase (PARP).[14]

Recently, it was reported that β-lapachone selectively induces apoptosis in transformed cells but not in proliferating normal cells, which is an unusual property not shared by conventional chemotherapeutic agents. It activates checkpoints in the absence of DNA damage. This selective induction of apoptosis is preceded by the rapid and sustained increase in the E2F1 level and activity in cancer cells. Taken jointly, the preceding results suggest direct checkpoint activators as selective agents against transformed cells.[23]

β-Lapachone cytotoxicity prompted the synthesis of a number of *o*-naphthoquinones, to establish the structural requirements for optimal therapeutic use. Several of those quinones (CG 10–248, 3,4-dihydro-2,2-dimethyl-9-chloro-2*H*-naphtho[1,2*b*]pyran-5,6-dione; CG 9–442, 3,4-dihydro-2-methyl-2-phenyl-2*H*-naphtho[1,2*b*]pyran-5,6-dione; and CG 8–935, 3,4-dihydro-2-methyl-2-ethyl-2*H*-naphtho[1,2*b*]pyran-5,6-dione) (Fig. 1)

[23] Y. Li, X. Sun, J. T. LaMont, A. B. Pardee, and Ch. J. Li, *Proc. Natl. Acad. Sci. USA* **100,** 2674 (2003).

proved in some assay systems to be more effective than β-lapachone itself.[24]

Materials and Methods

Microsomal and Cytosol Preparations

Microsomal and cytosol preparations were obtained from the livers of 20-h fasted, male Wistar rats, 240–280 g fed a Purine-like chow (A. C. A., Buenos Aires, Argentina), whose protein content was 23.4% and included all the essential amino acids. After the rats were rapidly decapitated, the liver was removed, weighed, washed, and homogenized in a Potter tissue grinder with a Teflon pestle, using 4 ml of homogenization medium per 2.5 g of tissue. All these steps were carried out in the cold. Following homogenization of the livers in TRIS-KCl buffer (50 m*M* TRIS-HCl, 150 m*M* KCl, pH 7.4) and centrifugation at 11,000*g* (15 min), the supernatant was centrifuged for 60 min at 105,000*g*. The microsomal pellet was washed twice with 150 m*M* KCl by centrifugation for 1 h at 105,000*g*, resuspended in 150 m*M* KCl, and either used immediately or stored in liquid nitrogen up to 3 months. The final supernatant named "cytosol" was used for DTD assays or stored at -70° for 15 days. No superoxide dismutase or catalase activities were found in the microsomal suspension.[15]

DTD activity was measured spectrophotometrically at 340 nm and 30° using NADPH as the immediate electron donor and menadione as the intermediate electron acceptor. The reaction mixture contained 100 μM NADPH, 100 μM menadione, 0.1 *M* K^{+}-phosphate, pH 7.4, and 10–30 μl cytosol for a total volume of 3.0 ml.

Each assay was repeated in the presence of 10 μM dicoumarol, and activity attributed to DTD was the inhibited by dicoumarol.[17]

The protein concentration was measured by the Biuret method.

Quinone Preparations

CG-quinones were supplied by Ciba Geigy-Novartis (Basel, Switzerland), and α-lapachone (3,4-dihydro-2,2-dimethyl-2*H*-naphthol[2,3*b*]-pyran-5,10 dione) and β-lapachone were obtained from a program for the synthesis of antiparasitic drugs at the Universidade Federal de Rio de Janeiro, Brazil. *o*-Naphthoquinones are unstable and degrade over time, so the compounds assayed were examined by HPLC, indicating more than

[24] K. Schaffner-Sabba, K. H. Schmidt-Ruppin, W. Wehrli, A. R. Schuerch, and J. W. F. Wasley, *J. Med. Chem.* **27,** 990 (1984).

98% purity. Quinones were dissolved in dimethylformamide (DMFA); the corresponding volume of DMFA was added to control samples.

Electron Spin Resonance (ESR) Measurements

ESR measurements for microsomal preparations were performed at room temperature using a Bruker ER 200 tt x-band ESR spectrometer (Bruker Analytische Messtechnic GMBH, Rheistellen, Karlsruhe, Germany) equipped with a TE 102 cavity. General instrumental conditions were microwave power, 21 mw; microwave frequency, 9.90 GHz; modulation frequency, 100 KHz; and time constant, 0.5 s. The field was centered at 3500*G*.

ESR measurements for cytosol preparations were performed using a Bruker ER 1100 ESR spectrometer. General instrumental conditions were microwave power, 20 mw; microwave frequency, 10 GHz; modulation frequency, 50 KHz; time constant, 0.2 s. The field was centered at 3490*G*. Modulation amplitude, gain, and scan rates were as indicated in legends to Figs. 2, 3, and 4.

Redox Cycle of β-Lapachone and Related *o*-Naphthoquinones Catalyzed by Microsomal NADPH Cytochrome P450 Reductase

ESR Study of Semiquinone Radical Formation by NADPH Cytochrome P450 Reductase

Principle. In the hepatic microsomal mono-oxygenase system, electrons donated from NADPH are transferred to the quinone (Q) through NADPH cytochrome P450 reductase. The semiquinone radical ($Q^{\bullet}$) generation can be detected by ESR.

$$Q + NADPH \xrightarrow{\textit{Cyt}\ \text{P450}\ \textit{reductase}} Q^{\bullet} + NADP^{+} + H^{+} \quad (1)$$

Procedure. The NADPH cytochrome P450 reductase assay medium contained liver microsomes (6 mg · ml^{-1} protein), 50 m*M* TRIS-HCl, pH 7.4, 10 m*M* NADPH and NADPH regenerating system (10 m*M* G_6P, 12 U · ml^{-1} G_6P dehydrogenase, and 3.0 m*M* $MgCl_2$). To obtain anaerobic conditions for the ESR measurements, the assay media (less quinone) were flushed with a nitrogen stream. At the same time, the quinone solution was also flushed, with nitrogen, as earlier. A sample of the deoxygenated, concentrated quinone solution was added to the deoxygenated assay medium, and the mixture was further flushed with nitrogen as earlier for 1 min. Last, the reaction medium was transferred to the spectrometer cell, flushed with nitrogen, and the spectrum was recorded. The time elapsing from the reaction mixture preparation to ESR spectrum recording was roughly 10 min, a period considered as the incubation time.[16]

ESR Spectrum of β-Lapachone–Related Semiquinones

The ESR spectrum of CG 10–248 semiquinone, after quinone reduction by the microsomal NADPH cytochrome P450 reductase system, is shown in Fig. 2A. Hyperfine splittings (HFSC) indicate spin couplings at protons at C7, C8, and C10 of the naphthalene ring. The observed quadruplet indicates three nuclei of 0.5 spin. HFSC values varied between 0.140 and 0.150 mT. The chlorine atom has a 1.5 nuclear spin, and no splitting is observed, probably because of its very low HFSC value.

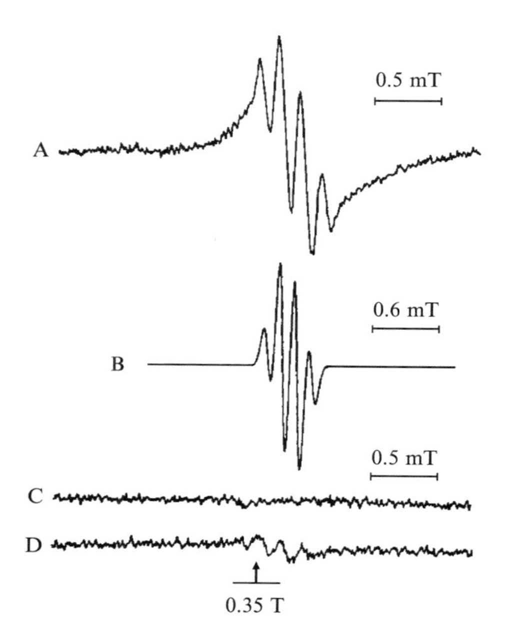

FIG. 2. ESR spectrum of CG 10–248 semiquinone after quinone reduction by the liver microsomal NADPH cytochrome P450 system. (A) The reaction mixture contained 5 m*M* CG, 10 m*M* NADPH, NADPH regenerating system, liver microsomes, 50 m*M* TRIS-HCl, pH 7.4, under nitrogen. (B) Computed simulation of the semiquinone spectrum using a spin-coupling constant value of 0.145 mT. (C) and (D) same as (A), except for quinone (C) or microsomes (D) omission. Instrumental conditions: modulation amplitude, 0.08 mTpp; gain 1.0×10^6, scan rate, 0.36 mT/min. Other experimental conditions were as described in "Materials and Methods" and in the text. From S. Fernandez Villamil, M. Dubin, M. P. Molina Portela, L. J. Perissinotti, M. A. Brusa, and A. O. M. Stoppani, *Redox Report* **3,** 245 (1997).

The kinetics of semiquinone formation shows that maximum values were reached after a 10-min incubation, scarcely varying thereafter (data not shown). The computed spectrum, using the average HFSC value (0.145 mT), fits well with the experimental results obtained (Fig. 2B). Omission of quinone or reductant system prevented the appearance of the semiquinone signal (Fig. 2C and D).

Figure 3A shows the CG 8–935 semiquinone spectrum after quinone reduction by the microsomal NADPH cytochrome P450 reductase system. With CG 8–935, a quintuplet signal indicates four spin couplings at protons at C7–C10 of the naphthalene ring. HFSC values were similar to those calculated for CG 10–248. Computational analysis reproduced the experimental spectrum (Fig. 3B). Similar results were obtained with CG 9–442 (data not shown), demonstrating that protons at the methyl, ethyl, and phenyl group or pyran ring failed to contribute to the ESR signal.

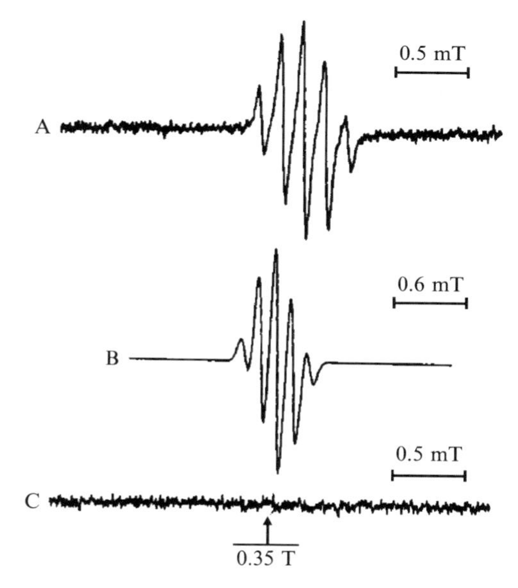

Fig. 3. ESR spectrum of CG 8–935 semiquinone after quinone reduction by the liver microsomal NADPH cytochrome P450 system. (A) The reaction mixture contained 5 m*M* CG, 10 m*M* NADPH, NADPH regenerating system, liver microsomes, 50 m*M* TRIS-HCl, pH 7.4, under nitrogen. (B) Computed simulation of the semiquinone spectrum. (C) Same as (A) except for microsomes omission. Instrumental conditions: modulation amplitude, 0.08 mTpp; gain 0.8×10^6; scan rate, 0.9 mT/min. Other experimental conditions were as described in "Materials and Methods" and in the text. From S. Fernandez Villamil, M. Dubin, M. P. Molina Portela, L. J. Perissinotti, M. A. Brusa, and A. O. M. Stoppani, *Redox Report* **3,** 245 (1997).

TABLE I
EFFECTS OF NAPHTHOQUINONES ON MICROSOMAL SUPEROXIDE ANION PRODUCTION

Quinone (5 μM)	Superoxide anion production (nmol min^{-1} mg^{-1} protein)
CG 10–248	27.4
CG 9–442	17.3
CG 8–935	14.7
β-Lapachone	17.9
α-Lapachone	3.4
None	1.9

Superoxide anion production by microsomes was measured by the adrenochrome method at 30°. Other experimental conditions are described in the text. Values are the means of triplicate measurements.

Effects of β-*Lapachone and Related* o-*Naphthoquinones on Superoxide Anion Production by Liver Microsomes*

Principle. Semiquinones in aerobic conditions are reoxydized, generating O_2^{-}.

$$Q^{\bullet} + O_2 \rightarrow Q + Q_2^{-} \tag{2}$$

$$\text{Adrenaline} + O_2^{-} \rightarrow O_2 + \text{Adrenochrome} \tag{3}$$

Production of O_2^{-} can be determined by the adrenochrome assay by measuring the absorption change at 485–575 nm ($\varepsilon = 2.96\ mM^{-1}\ cm^{-1}$).[25]

Procedure. The reaction mixture consisted of microsomes (0.20 mg·ml^{-1} protein), 23 mM Na^{+}/K^{+} phosphate buffer, pH 7.4, 130 mM KCl, 1 mM epinephrine, 5 μM quinone, and the NADPH generating system. Adrenochrome production was measured at 30°, using an Amino $DW2_a$ spectrophotometer (American Instrument Company, Silver Spring, MD) at 485–575 nm. Control samples were supplemented with DMFA.

The *o*-naphthoquinones redox cycling, in microsomal preparations, can readily be demonstrated by O_2^{-} generation, as shown in Table I, because *o*-quinones were about five-fold more effective O_2^{-} generators than the *p*-quinone, as exemplified by the β-lapachone/α-lapachone pair. Addition of SOD (6 U · ml^{-1}) confirmed that the absorbance increase was due to O_2^{-}.

[25] H. P. Misra and I. Fridovich, *J. Biol. Chem.* **247,** 188 (1972).

Effects of β-Lapachone and Related *o*-Naphthoquinones on Cytochrome P450-Catalyzed Reactions

In cytochrome P450-catalyzed reactions, electrons donated from NADPH are transferred to cytochrome P450 through NADPH cytochrome P450 reductase. Considering that the cytochrome P450 reductase catalyzes quinone redox-cycling, it was assumed that this reaction may lead to diversion of electrons from cytochrome P450, thereby inhibiting cytochrome P450-dependent reactions. To test this hypothesis, β-lapachone and related *o*-naphthoquinones were assayed on aniline 4-hydroxylase and aminopyrine *N*-demethylase activities, at fixed quinone concentrations.[15]

Aminopyrine N-*Demethylase Determination*

Principle. Many drugs are dealkylated by hepatic microsomal enzymes, and the method described here can be applied to such substrates.

$$\text{Aminopyrine} \xrightarrow[\searrow\ \text{Formaldehyde}]{\text{Microsomes, NADPH}} \text{Monomethyl-4-aminoantipyrine} \xrightarrow[\searrow\ \text{Formaldehyde}]{\text{Microsomes, NADPH}} \text{4-Aminoantipyrine} \quad (4)$$

The preceding reaction indicates that dealkylation activity may be determined by measuring the formation of either formaldehyde or 4-aminoantipyrine[26]; in the following experiments we measured formaldehyde formation.

Formaldehyde generated during incubation is trapped as the semicarbazone (by semicarbazide in the incubation mixture) and measured by the spectrophotometric procedure of Nash,[27] based on the Hantzsch reaction. The Nash reaction has been widely used, because it is simple, fast, and accurate. The Hantzsch reaction requires a β-diketone (acetylcetone), an aldehyde (formaldehyde), and an amine (NH_2 from ammonium acetate) as shown in Eq. 5. With the indicated reactants, the product formed is 3,5-diacetyl-1,4-dihydrolutidine (DDL), which can be monitored by its absorption at 415 nm.

$$\text{Acetylacetone} + \text{Ammonia} + \text{Formaldehyde} \rightarrow \text{DDL} + H_2O \quad (5)$$

[26] P. Mazel, *in* "Fundamentals of Drug Metabolism and Drug Disposition" (B. N. LaDu, H. Mandel, and E. Way, eds.), p. 546. Williams & Wilkins, Baltimore, 1971.

[27] T. Nash, *J. Biol. Chem.* **55,** 416 (1953).

Procedure. Into a tube immersed in ice add the following solutions in order: 1.7 m*M* aminopyrine, 4.2 m*M* $MgCl_2$, 7.5 m*M* semicarbazide, NADPH generating system containing 0.10 m*M* $NADP^+$, 1.7 m*M* G_6P, and 0.4 $U \cdot ml^{-1}$ G_6P dehydrogenase in 0.05 *M* Na^+/K^+ phosphate buffer, pH 7.4, and 10 or 5 μM β-lapachone or related *o*-naphthoquinones preflushed with oxygen. The reaction was started by adding microsomal suspension equivalent to 0.7 $mg \cdot ml^{-1}$ protein. Total volume was 6.0 ml. Blanks carried incubates through the procedure without substrate, whereas controls received the same volume of quinone solvent (DMFA). DMFA fails to interfere with the assay up to 5%.

Incubate with shaking for 30 min at 37°. At the end of the incubation period, remove the tubes and add 2.0 ml of 15% zinc sulfate to each one to stop the reaction. Mix well and wait 5 min. Add 2.0 ml of saturated barium hydroxide to each tube. Again, mix well and wait 5 min. Centrifuge for 10 min using a high speed to completely settle the precipitate. Transfer 5.0 ml of the supernatant to a test tube. Add 2.0 ml of Nash reagent (30% ammonium acetate, 0.4% v/v acetylacetone), mix well, and place in a water bath (60°) for 20 min. Filter if the solutions are cloudy. Measure the absorbance spectrophotometrically at 415 nm. A small amount of endogenous formaldehyde must be subtracted from the total amount of formaldehyde formed by setting the blanks to zero absorbance. Determine the quantity of formaldehyde formed from the standard curve.

Preparation of Standard Formaldehyde Curve. A standard formaldehyde curve may be prepared using formaldehyde solution (40%). Although formaldehyde solution contains 12% methanol, the final dilutions are such that methanol does not interfere in the reaction. Dilute the formaldehyde solution in water so as to obtain the following concentrations: 4, 2, 1, and 0.5 $\mu g \cdot ml^{-1}$.

Comments

It is essential to use proper concentrations of barium and zinc. Barium hydroxide is much more soluble in boiling water.

The Nash reagent is quite stable and may be kept in the refrigerator for many weeks.

The optimal pH for the Hantzsch reaction ranges from 5.5–6.5.

Large amounts of acetaldehyde interfere with the measurements of formaldehyde, but at molar formaldehyde concentration acetaldehyde produces only 1% interference. Drug de-ethylation to acetaldehyde could be a source of the latter aldehyde under certain experimental conditions. Acetone, chloral, and glucose do not interfere, but microsomes and soluble fractions obtained in sucrose solutions yield lower values of formaldehyde. Amines compete with ammonia in the reaction.

The reaction product is relatively stable but is affected by prolonged exposure to light and oxidizing agents.

Color development is faster at 60° (10–20 min) than at 37° (60 min).

Aniline Hydroxylase Determination

Principle. Hydroxylation of aniline by cytochrome P450 can be investigated in the presence of liver microsomes in NADPH- or cumene hydroperoxide (CHP)–dependent systems.[28]

The rate of aniline metabolism *in vitro* may be determined by measuring the quantity of *p*-aminophenol formed according to the following reaction:

$$\text{Aniline} \xrightarrow[\text{}]{\text{Microsomes} \atop \text{NADPH or CHP}} p\text{-aminophenol} \qquad (6)$$

In this procedure, trichloroacetic acid (TCA) is used to precipitate the protein, and the quantity of *p*-aminophenol produced by aniline hydroxylation is determined in an aliquot of the TCA supernatant fraction. Phenol is then added to form the blue phenol–indophenol complex, which is measured at 640 nm.[29]

Procedure

Method 1 (NADPH-Dependent System). Into a series of flasks, immersed in ice, add the following components: 1.25 m*M* aniline hydrochloride, NADPH generating system (1.2 m*M* $NADP^+$, 2.5 m*M* G_6P, 0.5 U ml^{-1} G_6P dehydrogenase, 6.0 m*M* $MgCl_2$, dissolved in 0.1 *M* Na^+/K^+ phosphate buffer, pH 7.4), and β-lapachone or related quinones preflushed with O_2. The reaction was started by adding 1.0-ml microsomes suspension equivalent to 1.7 $mg \cdot ml^{-1}$ protein. Total incubation volume was 4.0 ml. Blanks carried incubates through the procedure without substrate. Controls received the same volume of quinone solvent (DMFA). DMFA interferes with the reaction, so that, depending on the volume to obtain the appropriate quinone concentration to be tested, the corresponding control with DMFA must be included. Incubate the flasks with rapid shaking for 20 min at 37°. At the end of the incubation period, add 2.0 ml of 20% TCA to the incubation flasks. Mix well and centrifuge. To a 2.0-ml aliquot

[28] I. I. Karuzina, A. I. Varenitsa, and A. I. Archakov, *Biokhimiya* **48,** 1788 (1983).

[29] B. G. Lake, *in* "Biochemical Toxicology" (K. Snell and B. Mullock, eds.), p. 206. IRL Press, Oxford, 1987.

of the TCA supernatant add 1.0 ml of 10% Na_2CO_3 and mix well. Add 2.0 ml of 2% phenol in 0.2 *N* NaOH and allow the color to develop for 30 min at 37°. Read the absorbance in a spectrophotometer at 640 nm. Determine the micrograms of *p*-aminophenol formed from a previously plotted standard curve. Multiply this number by 3 to obtain the total amount of *p*-aminophenol in the incubation flasks.

Preparation of Standard p-*Aminophenol Curve*

Prepare a series of tubes containing *p*-aminophenol HCl (1–4 $\mu g \cdot ml^{-1}$) in 6.67% TCA (2.0 ml), because this is the final concentration of TCA after adding 20% TCA to the incubate. To 2.0 ml of each concentration in TCA (in duplicate) add 1.0 ml of 10% Na_2CO_3 followed by 2.0 ml of 2% phenol in 0.2 *N* NaOH. Allow color to develop and read the absorbance in a spectrophotometer at 640 nm.

Method 2 (Cumene Hydroperoxide-Dependent System)

It is known that cytochrome P450 can catalyze oxidation reactions of many substrates, using organic hydroperoxides as active oxygen donors. In this method hydroxylation of aniline in the microsomes was measured in the presence of CHP.[28]

Procedure. For the assay using CHP, the reaction mixture contained 3.0 m*M* aniline, 0.25 m*M* CHP, 80 m*M* TRIS-HCl buffer, pH 7.6, and microsomes (2.0 $mg \cdot ml^{-1}$ protein). Incubation was for 30 min at 37°. Total incubation volume was 4.0 ml. Blanks carried incubates through the procedure without substrate, whereas controls received the same volume of quinone solvent (DMFA). DMFA interferes with the reaction, so that controls with the same volume of DMFA must be included. At the end of the incubation period, add 7.7% TCA to the incubation flasks. Mix well and centrifuge. To a 2.0-ml aliquot of the TCA supernatant add 1.0 ml of 10% Na_2CO_3 and mix well. Add 2.0 ml of 2% phenol in 0.2 *N* NaOH and allow the color to develop for 30 min at 37°. Read the absorbance in a spectrophotometer at 640 nm.

Comments

The procedures described for the determination of microsomal aniline formation by hydroxylase activity are both simple and sensitive.

The cofactor mixture must be preincubated briefly before adding enzyme.

Hydrogen peroxide (100 m*M*) can be used instead of CHP, but the latter is better as a cosubstrate than the former. Because the presence of residual hydrogen peroxide may prevent the development of color in the following reaction, it must be previously decomposed by adding 8000 $U \cdot ml^{-1}$ catalase.

The incubation period should be as short as possible, because microsomes incubation with *p*-aminophenol for about 30 min leads to its loss.

Inhibition of Cytochrome P450-Catalyzed Reaction by β-Lapachone and Related o-*Naphthoquinones*

A comparative effect of *o*-naphthoquinones on microsomal aminopyrine *N*-demethylase and aniline hydroxylase activities is shown in Table II.

With 10 μM quinone, β-lapachone and CG quinones inhibited the demethylase activity by 70–77%, lesser effects being obtained with the 5 μM concentration. The naphthoquinones also inhibited the aniline 4-hydroxylase activity using NADPH as electron donor by 66–77% (Table II). However, for aniline hydroxylase activity with CHP, which also supports cytochrome P450-catalyzed reactions by a different mechanism (Table II), no significant inhibition was observed. Inhibition of cytochrome P450

TABLE II
EFFECTS OF *o*-NAPHTHOQUINONES ON MICROSOMAL AMINOPYRINE *N*-DEMETHYLASE (A) AND ANILINE 4-HYDROXYLASE (B) ACTIVITIES

		Inhibition of enzyme activities (%)		
			B	
Quinone	Quinone concentration (μM)	A	NADPH	CHP
CG 10–248	5	52	10	1.4
	10	76	77	2.8
CG 9–442	5	46	19	ND
	10	77	66	3.6
CG 8–935	5	21	10	5.6
	10	74	71	7.0
β-Lapachone	5	51	26	4.2
	10	70	68	2.8

Experiment A: the reaction mixture contained microsomes (0.7 $mg \cdot ml^{-1}$ protein), 1.7 m*M* aminopyrine, 7.5 m*M* semicarbazide, 4.2 m*M* $MgCl_2$, 0.10 m*M* $NADP^+$, 1.7 m*M* G_6P, 0.4 $U \cdot ml^{-1}$ G_6P dehydrogenase, 50 m*M* KH_2PO_4-Na_2HPO_4, pH 7.4, flushed with O_2 before use. Incubation was for 30 min at 37°. Experiment B: the reaction mixture contained microsomes (~2 $mg \cdot ml^{-1}$ protein), aniline, and NADPH generating system (NADPH) or cumene hydroperoxide (CHP) as indicated earlier. Control activity (nmol substrate min^{-1} mg $protein^{-1}$): 0.99 (A), 0.31 (B, NADPH), and 0.71 (B, CHP). Values are the means of triplicate measurements. Other experimental conditions are described in the text. ND: not done.

reactions may imply inhibitor binding, which can be monitored by spectral changes in the Soret absorption of cytochrome P450. Under standard experimental conditions and using 5–25 μM quinone, however, no spectral changes could be observed (data omitted).

These observations support the hypothesis that, in the 0–10 μM concentration range, β-lapachone and related CG quinones inhibit cytochrome P450-catalyzed reactions by diverting reduction equivalents from NADPH to dioxygen.[15]

Redox Cycle of β*-Lapachone and Related* o*-Naphthoquinones Catalyzed by Cytosol DT-Diaphorase*

Principle. In the NAD(P)H/*o*-naphthoquinone/DTD/oxygen system, the redox cycle involves two phases, namely: (1) the reductive phase and (2) the oxidative phase. The former is represented by Eq. 7 involving Q and QH_2, which are the *o*-naphthoquinone and the corresponding hydroquinone, respectively:

$$2NAD(P)H + Q \rightarrow 2NAD(P)^+ + QH_2 \tag{7}$$

Concerning the hydroquinone formed in Eq. 7, it autoxidizes with formation of ROS in a process inhibited by SOD. Ionization of the QH_2 is a necessary prerequisite for the reaction, and oxidation is initiated by reaction of anion QH^- with molecular oxygen forming superoxide and semiquinone ($Q^\bullet$) (Eq. 8).

$$H^+ + QH^- + O_2 \leftrightarrow Q^\bullet + O_2^- + 2H^+ \tag{8}$$

The oxidative phase of the NAD(P)H/*o*-naphthoquinone/DTD/oxygen system redox cycle involves three reactions termed initiation, propagation, and termination.[30–32] The initiation reaction (Eq. 8) can be demonstrated by semiquinone production by ESR, oxygen uptake, and O_2^- production.

The propagation reaction of hydroquinone oxidation proceeds by way of O_2^- in agreement with Eq. 9 and Eq. 10.

$$H^+ + QH^- + O_2^- \leftrightarrow Q^\bullet + H_2O_2 \tag{9}$$

$$Q^\bullet + O_2 \leftrightarrow Q + O_2^- \tag{10}$$

[30] T. Ishii and I. Fridovich, *Free Radic. Biol. Med.* **8,** 21 (1990).
[31] R. Munday, *Free Radic. Biol. Med.* **26,** 1475 (1999).
[32] R. Munday, *Free Radic. Res.* **35,** 145 (2001).

Finally, SOD-catalyzed O_2^- disproportionation (Eq. 11) and semiquinone disproportionation (Eq. 12) terminate the naphthoquinone redox cycle in cytosol.

$$O_2^- + O_2^- + 2H^+ \rightarrow H_2O_2 + O_2 \tag{11}$$

$$Q^\bullet + Q^\bullet + H^+ \rightarrow Q + QH^- \tag{12}$$

ESR Study of Semiquinone Radical Formation after CG Reduction by the NADPH/DT-Diaphorase System

Procedure. The reaction mixture contained 21 m*M* NADPH, 8.4 m*M* CG 10–248, 360 μl cytosol (1.6 $U \cdot ml^{-1}$ DTD), 0.1 *M* K^+ phosphate, pH 7.4, under air, total volume, 0.47 ml. Spectra were recorded immediately after completing the reaction mixture.

Semiquinone is an immediate product of hydroquinone oxidation by one-electron transfer to dioxygen. Figure 4A shows the ESR signal of CG semiquinone after CG reduction by the NADPH/DTD system. Hyperfine splittings indicate spin couplings at protons at C7, C8, and C10 of the naphthalene ring, as described previously. No signals were observed when cytosol or CG was omitted (Figs. 4B, 4C).

Effect of CG 10–248 Addition on Oxygen Uptake by Liver Cytosol Fraction

Principle. o-Naphthoquinones redox cycling in the presence of NAD(P)H and DTD consumed oxygen, a reaction used for monitoring the redox-cycle operation. Oxygen uptake results from hydroquinone oxidation according to Eq. 8.

Procedure. Oxygen uptake rate was polarographically measured with a Model 5/6 Oxygraph (Gilson Medical Electronics, Madison, WI), fitted with a Clark-type oxygen electrode.[33] The reaction mixture is placed in the electrode chamber, and the solution is stirred by a Teflon-coated magnetic bar driven from a stirrer located under the electrode. The chamber should be maintained at 30° by circulating water. *o*-Naphthoquinone solution was injected through the small opening in the top of the chamber, using a Hamilton-type syringe. Continuous measurements were taken on a chart recorder.

[33] I. Stadler, *in* "Free Radical and Antioxidant Protocols" (D. Armstrong, ed.), p. 3. Humana Press, Totowa, New Jersey, 1988.

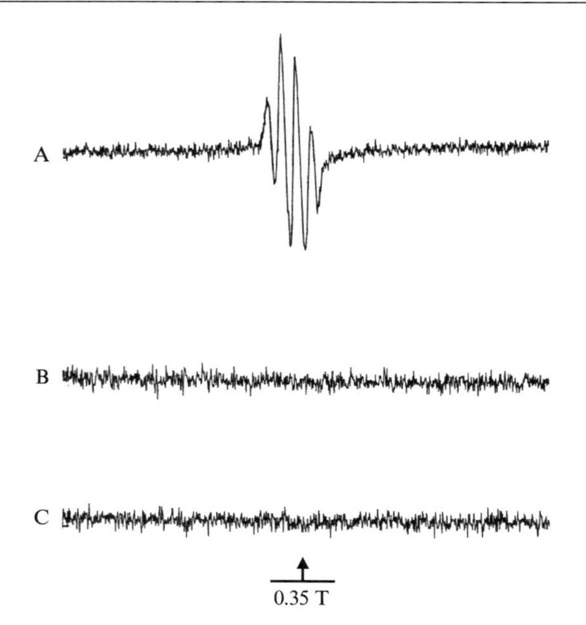

FIG. 4. ESR spectrum of CG 10–248 semiquinone after quinone reduction by the NADPH/DTD system. The reaction mixture contained 21 mM NADPH, 8.4 mM CG (A) or 1% DMFA (B), 360 μl cytosol (1.6 U · ml^{-1} DTD), 0.1 M K^+ phosphate buffer, pH 7.4. (C) Same condition as in (A) except for cytosol omission. Instrumental conditions: modulation amplitude, 0.946 G; gain 1.0 × 10^5. Other experimental conditions were as described in "Materials and Methods" and in the text.

Comment

The Teflon electrode membrane should be replaced daily.

The effect of CG 10–248 (CG) on oxygen uptake by cytosol fraction is shown in Fig. 5. In experiment A, the reaction mixture contained 300 μM NADPH, 100 μM CG 10–248 and cytosol (153 mU · ml^{-1} DTD), 0.1 M K^+ phosphate buffer, pH 7.4, total volume 1.8 ml. In experiment B, the same experimental conditions were maintained, except that 25 μM dicoumarol (DC) (a DTD inhibitor) was added as indicated in Fig. 5.

Figure 5A shows that CG addition to the NADPH/DTD system increased several fold the rate of oxygen consumption, compared with the one depending on endogenous substrate oxidation (150/6.7 $\mu M \cdot \text{min}^{-1}$). Oxygen uptake ceased when approximately 50% of the reaction mixture oxygen had been consumed. A second addition of NADPH re-established the oxidation rate although at a lower value (89 $\mu M \cdot \text{min}^{-1}$) than during the first NADPH oxidation. Finally, the reaction stopped when all the reaction mixture oxygen was consumed. The addition of dicoumarol (Fig. 5B) produced significant inhibition of oxygen uptake, in the same system, after the first and the second NADPH addition compared with experiment A. The dicoumarol effect was not immediate as shown by the kinetics of the oxidation inhibition.

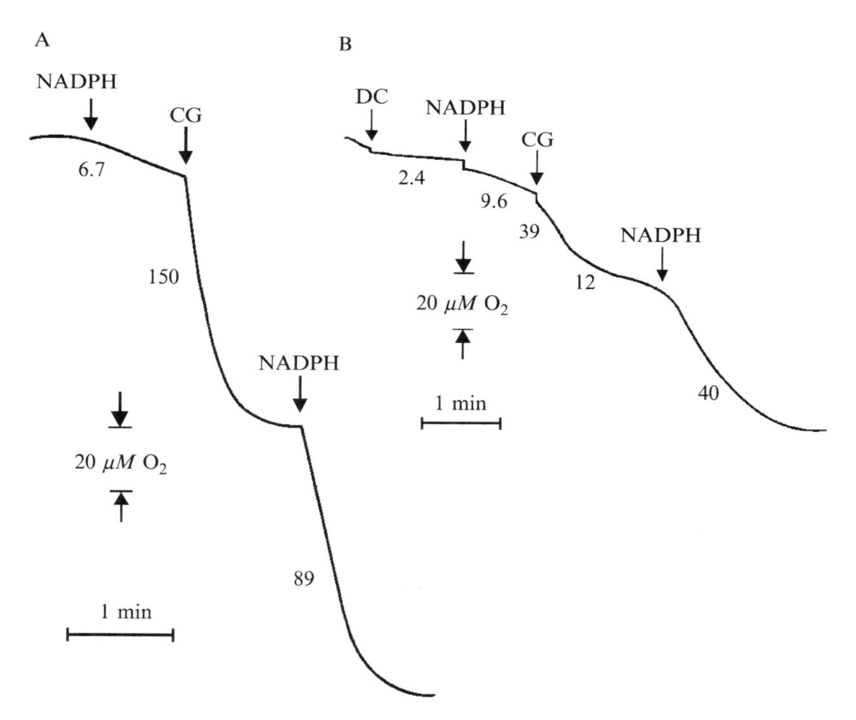

FIG. 5. Oxygen uptake by the NADPH/*o*-naphthoquinone/DTD system. (A) The reaction mixture contained 300 μM NADPH, 100 μM CG 10–248, cytosol (153 $\text{mU} \cdot \text{ml}^{-1}$ DTD), 0.1 M K^+ phosphate buffer, pH 7.4, total volume 1.8 ml. (B) Same conditions as in (A), except that 25 μM dicoumarol (DC) was added. Oxygen uptake was measured polarographically. The number near each tracing indicates oxygen uptake rate ($\mu M \cdot \text{min}^{-1}$).

Effect of o-*Naphthoquinones Addition on Superoxide Anion Formation by Cytosol Fraction*

Principle. In liver cytosol fraction, superoxide anion production was measured by the acetyl–cytochrome *c* reduction according to Eq. 13.[34]

$$O_2^- + \text{Ac cyt } c(3+) \rightarrow O_2 + \text{Ac cyt } c(2+) \qquad (13)$$

Procedure. The reaction mixture contained quinone, 30 μM acetyl-cytochrome *c*, 0.1 mM K^+ phosphate buffer, pH 7.4, and cytosol (19 $mU \cdot ml^{-1}$ DTD), total volume 3.0 ml. The reaction was started by adding 300 μM NADPH, prepared immediately before use, and the rate of reaction in the linear phase was calculated. Reduced acetyl-cytochrome *c* produced during the reaction was monitored spectrophotometrically at 550–540 nm at a constant temperature (30°) in an Aminco Chance $DW2_a$ spectrophotometer (American Instrument Company, Silver Spring, MD) ($\varepsilon = 19.1\ mM^{-1}\ cm^{-1}$). The addition of SOD (150 $U \cdot ml^{-1}$) confirmed that absorbance increase was due to O_2^-.

β-Lapachone and related quinones may be included in the group of quinones generating redox labile hydroquinones that autoxidize to produce O_2^-.

Table III shows the effect of several naphthoquinones on O_2^- production by the NADPH/*o*-naphthoquinone/DTD system. At 100 μM concentration, *o*-naphthoquinones were more effective than α-lapachone (*p*-naphthoquinone). Assayed *o*-naphthoquinones, including β-lapachone, showed similar activities, despite structural differences. β-Lapachone was somewhat more effective than the other *o*-naphthoquinones, but differences were not significant. With the 10 μM quinone concentration, O_2^- production was still effective. Dicoumarol inhibited 86–94% O_2^- production.

Figure 6 shows the effect of increasing concentrations of CG 10–248 on O_2^- production by the NADPH/*o*-naphthoquinone/DTD system. In the 0–10 μM range, O_2^- production increased almost linearly as a function of CG concentration, but in the 10–50 μM range, O_2^- production failed to vary to a significant degree, apparently as a result of substrate saturation of DTD active site. SOD inhibition confirmed the O_2^- production by the quinone redox cycling.

[34] A. Azzi, C. Montecucco, and C. Richter, *Biochem. Biophys. Res. Commun.* **65,** 597 (1975).

TABLE III
SUPEROXIDE PRODUCTION BY THE NADPH/NAPHTHOQUINONE/DTD SYSTEM: EFFECT OF NAPHTHOQUINONE STRUCTURE AND DICOUMAROL

	Superoxide anion production ($\mu M \cdot min^{-1}$)	
	Quinone: 100 μM	Quinone: 10 μM
CG 10–248	39.89 ± 0.36 (92)	26.05 ± 0.78 (87)
CG 9–442	41.05 ± 0.20 (88)	25.26 ± 0.0 (86)
CG 8–935	38.68 ± 0.78 (87)	28.00 ± 0.3 (92)
β-Lapachone	42.00 ± 0.36 (94)	25.52 ± 0.78 (91)
α-Lapachone	5.84 ± 0.78	ND
None	1.10 ± 0	1.10 ± 0

The reaction mixture contained 300 μM NADPH, 19 $mU \cdot ml^{-1}$ DTD, 30 μM acetyl-cytochrome *c*, 0.1 *M* K^+ phosphate, pH 7.4, naphthoquinone as stated previously. Values in parentheses indicate inhibition of O_2^- production by 15 μM dicoumarol. Values represent means ± SD (n = 3). ND: not done.

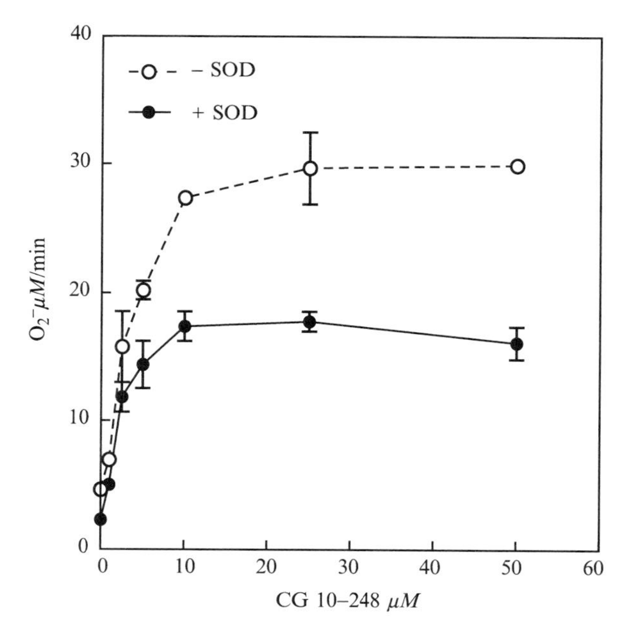

FIG. 6. Effect of CG 10–248 (CG) concentration and SOD on O_2^- production by the NADPH/*o*-naphthoquinone/DTD system. The reaction mixture contained 300 μM NADPH, cytosol (19 $mU \cdot ml^{-1}$ DTD), 30 μM acetyl-cytochrome *c*, 0.1 *M* K^+ phosphate buffer, pH 7.4, and CG as indicated on the abscissa, total volume, 3.0 ml. – SOD: sample without SOD; + SOD: sample containing 150 $U \cdot ml^{-1}$ SOD. Values are means ± SD. ($n = 3$) and represent O_2^- production μM min^{-1}.

CG 10–248 Redox Cycling by Cytosol Fraction

Reduction of naphthoquinones can also be monitored by absorption measurements at characteristic wavelengths.

Procedure. The reaction mixture contained 0.1 M K^+ phosphate buffer, cytosol (95 mU ml^{-1} DTD), and 100 μM CG 10–248 (CG); final volume 3.0 ml. The effect of NADPH on CG redox cycling was monitored by absorbance variation at 448 nm, a wavelength at which the hydroquinone does not interfere.

Figure 7A illustrates typical results obtained with CG 10–248 and rat liver cytosol. The initial rate of CG reduction by the NADPH/DTD system, which generates CG hydroquinone, was relatively fast (180 μM min^{-1}) but decreased to such an extent that CG reduction ceased when the level of CG concentration was about 35% of its initial value. Such minimum value was observed for a few seconds, and then absorbance increased at a rate of 99 $\mu M \cdot min^{-1}$, with semiquinone and/or original quinone generation, at approximately half the rate of CG reduction. Finally, the

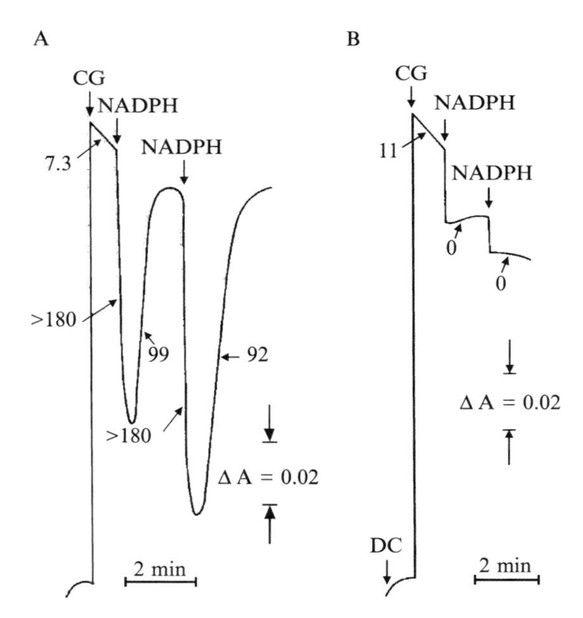

FIG. 7. CG 10–248 redox cycling in the presence of the NADPH/DTD/oxygen system. Cytosol was incubated with 100 μM CG 10–248 (CG) in the absence (A) or presence (B) of 15 μM dicoumarol (DC). Arrows indicate NADPH addition. The number near each tracing indicates the rate of quinone reduction or quinol oxidation.

CG redox level stabilized at approximately 80% of the initial value. It should be noted that the spectroscopic method failed to distinguish semiquinone from quinone, but, nevertheless, for kinetic reasons the quinone was assumed to be the main contributor to absorbance.

Figure 7B shows the effect of dicoumarol on CG 10–248 redox cycling in cytosol fraction. Experimental conditions were as indicated in Fig. 7A, except for dicoumarol addition as indicated in Fig. 7B. No redox cycling was observed under these experimental conditions.

Conclusions

The lipophilic *o*-naphthoquinones β-lapachone and structural analogues quinones (CG quinones) are proposed as cytostatic, trypanocidal, and antiviral agents. With rat liver microsomal NAD(P)H cytochrome P450 reductase or cytosol flavoenzyme DTD, these quinones constitute redox systems, which in the presence of oxygen generate ROS. *o*-Naphthoquinones redox cycling, catalyzed by the NADPH cytochrome P450 reductase, generate in microsomal liver preparations: (1) semiquinone free radicals, (2) ROS, and (3) inhibition of cytochrome P450-dependent reactions, exerting cytotoxic effects. Hydroquinones are the immediate products of quinones reduction by the DTD-dependent systems (Eq. 7). Three types of hydroquinones formed by that reaction have been proposed by Cadenas[21]: (1) redox-stable hydroquinones; (2) redox-labile hydroquinones that subsequently reoxidize, with formation of semiquinone and ROS; and (3) redox-labile semiquinones that immediately rearrange to potent electrophils undergoing biological alkylating reactions.[21] Our observations with β-lapachone and related *o*-naphthoquinones indicate that the corresponding hydroquinones must be included in the second group in agreement with (1) the semiquinone spectrum, demonstrated by ESR spectroscopy; (2) semiquinone (or quinone) production, demonstrated by optical spectroscopy; and (3) the effect of dicoumarol on the quinone redox cycling and oxygen consumption by the NADPH/*o*-naphthoquinone/DTD system. These reactions associated with DTD activity seem to rule out the contention proposing DTD as an antioxidant enzyme protecting against quinone toxicity.

Acknowledgments

We thank Ms. Alejandra Veron, Cristina Lincon, and Sara del Valle for technical assistance and Ms. Dolores Podestá for helpful figure design. This work was supported by grants from CONICET (National Research Council, Argentina), UBA (University of Buenos Aires), and CEDIQUIFA (Buenos Aires).

[5] Quinone Chemistry and Melanogenesis

By EDWARD J. LAND, CHRISTOPHER A. RAMSDEN, and PATRICK A. RILEY

Introduction

Melanin is a polymeric dark pigment that is widely distributed in nature. The detailed structure of melanins has not been explained, but there are several classes of pigments comprising linked indole moieties that are recognized, eumelanins (brown-black pigments) and pheomelanins (reddish brown pigments) comprise the major categories.[1] Most melanins are closely linked to protein matrices in the form of melanoproteins and may comprise a mixture of eumelanic and pheomelanic elements. The major difference between eumelanins and pheomelanins is the extent to which their structure incorporates sulfur. Pheomelanin formation involves a thiol addition reaction and gives rise to benzothiazine residues that are incorporated into the pigment and alter the conjugation profile and hence the absorbance properties of the chromophore.

The divergence of the melanogenic pathway occurs after the initial oxidation step that yields dopaquinone (Fig. 1). The current analytical approach to the classification of melanins depends on the assessment of the comparative levels of degradation products that are considered characteristic of indoles and benzothiazine residues. The method, based on the analytical data of Nicolaus and Prota and colleagues, has been developed principally by Ito and his collaborators.[1]

Several physical methods for the analysis of melanins are available, including solid-state nuclear magnetic resonance (NMR) spectroscopy using ^{13}C and ^{15}N as probes,[2] and electron paramagnetic resonance (EPR) spectroscopy.[3] The total amount can be estimated from the dry weight of the material resistant to acid hydrolysis or from the optical absorbance of tissue dissolved in aqueous Soluene-350 (Perkin Elmer, Buckinghamshire, UK).[4] Generally, the absorbance at 500 nm gives a reasonably good estimate of the total melanin content, and some degree of differentiation of types of melanin may be possible by comparisons of absorbance at 400, 500, or 650 nm. The most reliable method of quantitative analysis of eumelanins and pheomelanins relies on chemical degradation and high-performance liquid chromatography (HPLC) separation of the degradation products.

The basis of this method (Fig. 2) is the yield of (1) pyrrole-2,3,5-tricarboxylic acid (PTCA) from 5,6-dihydroxyindole-2-carboxylic acid (DHICA)

[1] G. Prota, "Melanins and Melanogenesis," Academic Press, San Diego, CA, 1992.

0076-6879/04 $35.00

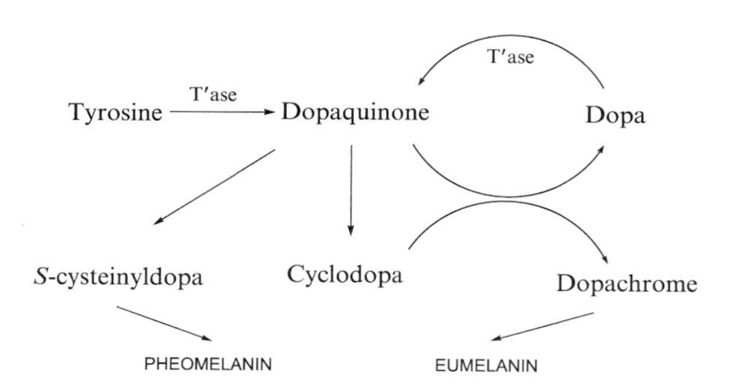

FIG. 1. Divergence of melanogenic pathways.

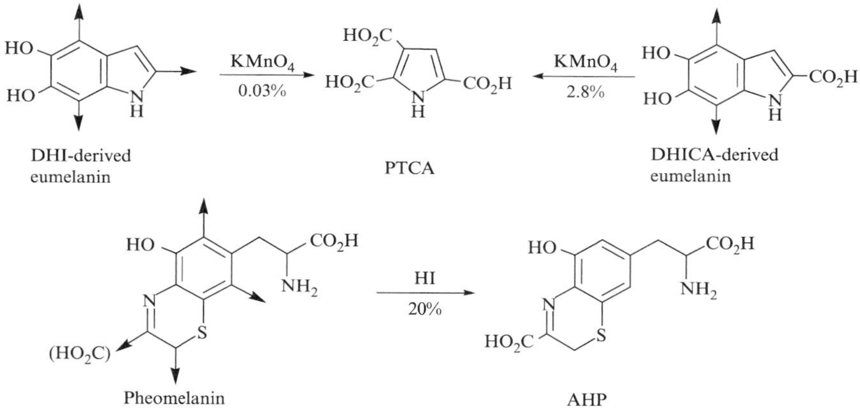

HO HO N H
$KMnO_4$ 0.03%
HO_2C HO_2C N H CO_2H
$KMnO_4$ 2.8%
HO HO N H CO_2H

DHI-derived eumelanin

PTCA

DHICA-derived eumelanin

HO CO_2H NH_2 N S (HO_2C)
HI 20%
HO CO_2H NH_2 N S HO_2C

Pheomelanin

AHP

FIG. 2. Chemical degradation of melanins.

(2.8%) and 5,6-dihydroxyindole (DHI) (0.03%) after permanganate oxidation as an indication of eumelanin and (2) aminohydroxyphenylalanine (AHP) from benzothiazine derivatives by hydriodic acid hydrolysis as an indicator of pheomelanin content. These products are separated by HPLC and quantified by ultraviolet (UV) detection.[5]

[2] G. A. Duff, J. E. Roberts, and N. Foster, *Biochemistry* **27,** 7112 (1988).
[3] W. S. Enochs, M. J. Nilges, and H. M. Swartz, *Pigment Cell Res.* **6,** 91 (1993).
[4] J. Borovansky, *Mikrochim. Acta* **2,** 423 (1978).
[5] S. Ito and K. Fujita, *Anal. Biochem.* **144,** 527 (1985).

The biogenesis of melanin (termed melanogenesis) seems to have evolved from a process that generates *ortho*-quinones from monohydric and dihydric phenols. This oxidation is accomplished by a set of enzymes collectively termed phenoloxidases and includes tyrosinases (enzymes capable of converting monophenols to *ortho*-quinones) and catechol oxidases (which oxidize catechols to the corresponding *ortho*-quinones). The phenoloxidase enzymes seem to have arisen early in evolution and are found in prokaryotes, plants, and animals. This class of proteins depends for their action on the presence of two copper atoms at the catalytic site. These are coordinated to histidine residues and situated close to each other, enabling binding of molecular oxygen. The catalytic function is modified by access of the substrate molecules. In one form, found in hemocyanin, the active site is screened, and the metalloprotein acts as the principal oxygen carrier of the circulation. In recent years, a number of excellent studies of the active site have been made.[6,7]

Although in mammals the melanogenic pathway seems to involve several enzymatic steps, the nature and importance of which are unclear, the requisite and necessary process of melanogenesis is the generation of *ortho*-quinones, and normal melanin formation does not occur in the absence of tyrosinase. Inactive tyrosinase mutants lead to albinism in which melanin is absent. Partial variants of albinism are the outcome of partially active enzymes. Much is known about the tyrosinase mutants in human albinism.[8]

In vertebrates, the initial substrate of the melanogenic pathway is tyrosine, and conversion to the corresponding *ortho*-quinone gives rise to dopaquinone. The *ortho*-quinone possesses an electronic configuration that promotes nucleophilic addition to the ring, and the dopaquinone side chain includes an amino group that is able to cyclize by a Michael addition reaction, thus generating a dihydroxyindoline derivative from which, by subsequent reactions, the indole moieties of eumelanin are derived. Clearly, other nucleophilic additions may take place, and exoreactions are of importance in some of the other processes with biological significance associated with quinone generation such as cross-linking reactions as in insect sclerotization, feather strengthening, and balanid adhesion. Nucleophilic addition of cysteine to dopaquinone gives rise to cysteinyldopa, which is the starting molecule of the pheomelanin pathway.

[6] H. Decker, R. Dillinger, and F. Tuczek, *Angew. Chem. Int. Ed.* **39,** 1591 (2000).

[7] J. C. García-Borrón and F. Solano, *Pigment Cell Res.* **15,** 162 (2002).

[8] W. S. Oetting and R. A. King, *Pigment Cell Res.* **7,** 285 (1994).

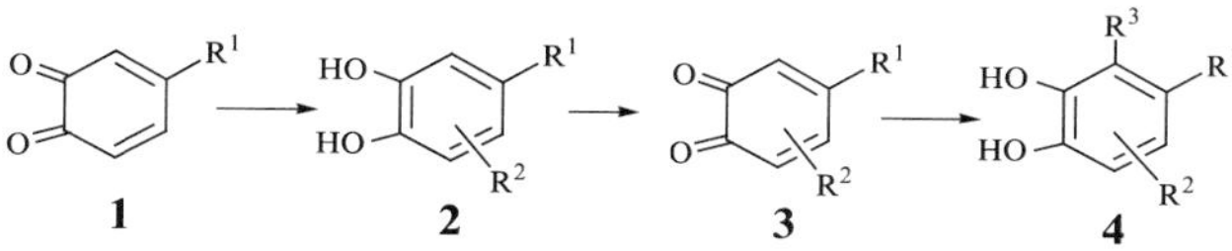

FIG. 3. Catechol functionalization by means of *ortho*-quinones.

In Vivo ortho-Quinone Chemistry

Ortho-quinones **1** (Fig. 3) are reactive, electron-deficient species. When generated in a biological environment, they rapidly react, often by way of a nonenzymatic pathway, to give more stable products.[9] The driving force for this reactivity is conversion to aromatic catechols **2** (*ortho*-diphenols). Catechol formation occurs by reduction, by addition of a nucleophile, or by rearrangement (tautomerism) followed by nucleophilic addition. The addition reactions result in functionalization of the ring by the nucleophile. Because the catechol products **2** are readily reoxidized *in vivo* to new *ortho*-quinones **3**, these pathways rapidly lead to multifunctionalized macrostructures **4** such as melanin.

Three important alternative modes of reaction of *ortho*-quinones in biological systems can be identified: (1) addition, (2) reduction, and (3) tautomerism. These pathways are illustrated in the following sections with selected examples.

Addition Reactions

Electron-deficient *ortho*-quinones react with electron-rich nucleophilic species such as amines (RNH_2) and thiols (RSH). Addition can be intermolecular or intramolecular.

Thiols such as cysteine and glutathione (GSH) readily add to *ortho*-quinones. Typically, cysteine has been shown to react with dopaquinone **6**, formed by enzymatic oxidation of dopa **5**, to give predominately 5-*S*-cysteinyldopa **7** (74%) (Fig. 4).[1,10] This arises from 1,6-addition of the thiol to the quinone. The reaction is not, however, site-specific, and smaller amounts of the alternative 1,6-adduct **8** (14%) (2-*S*-cysteinyldopa) together with traces of the 1,4-adduct **9** (1%) (6-*S*-cysteinyldopa) are also formed. 1,6-Addition is clearly favored. A small amount of 2,5-*S,S*-dicysteinyldopa (5%) is also formed, probably by oxidation of the major product **7** followed by 1,6-addition to the new *ortho*-quinone (cf. **2** → **3** → **4**). These thiol

[9] M. G. Peter, *Angew. Chem. Int. Ed.* **28,** 555 (1989).
[10] S. Ito and G. Prota, *Experientia* **33,** 1118 (1977).

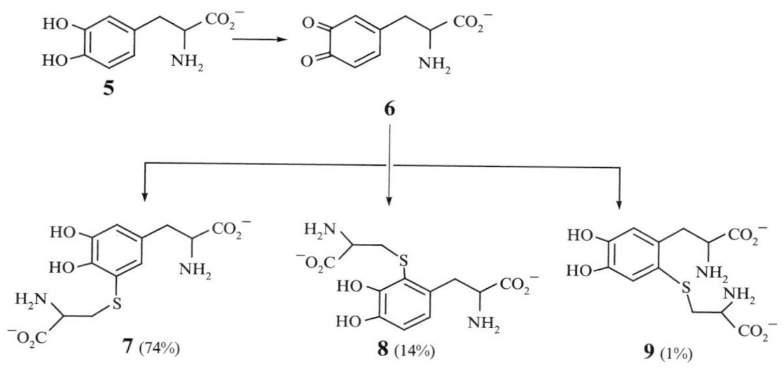

FIG. 4. Intermolecular addition.

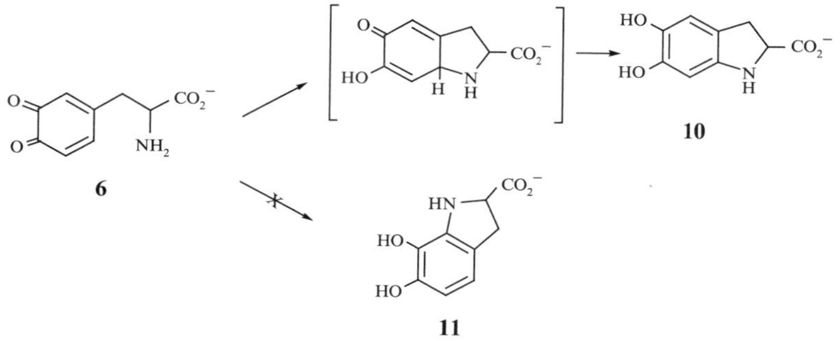

FIG. 5. Intramolecular addition.

additions are important in pheomelanin formation. Reaction with glutathione gives a similar product profile,[11] and *ortho*-quinones are known to undergo similar addition reactions with oxygen nucleophiles (e.g., H_2O) and carbon nucleophiles (e.g., indoles).

The intramolecular addition of amines to *ortho*-quinones is an important biosynthetic pathway and leads to indole derivatives (Fig. 5). Dopaquinone **6** cyclizes by the mechanism shown to give cyclodopa (leukodopachrome) **10**. This reaction, which competes with thiol addition, is formally a 1,4-addition and is sometimes referred to as a Michael addition. The alternative cyclization (**6** → **11**) by way of 1,6-addition, which is the preferred

[11] S. Ito, A. Palumbo, and G. Prota, *Experientia* **41,** 960 (1985).

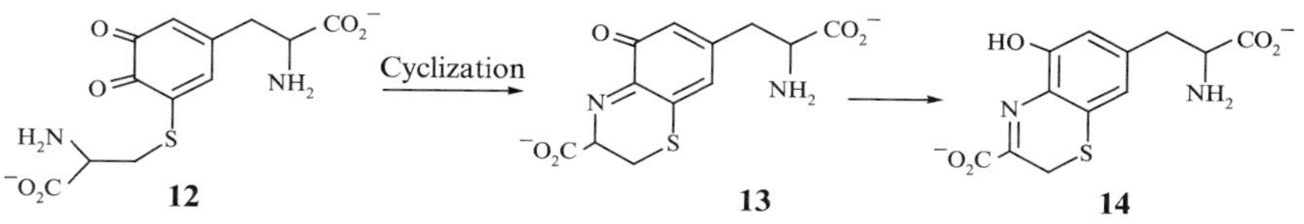

FIG. 6. Intramolecular addition–elimination.

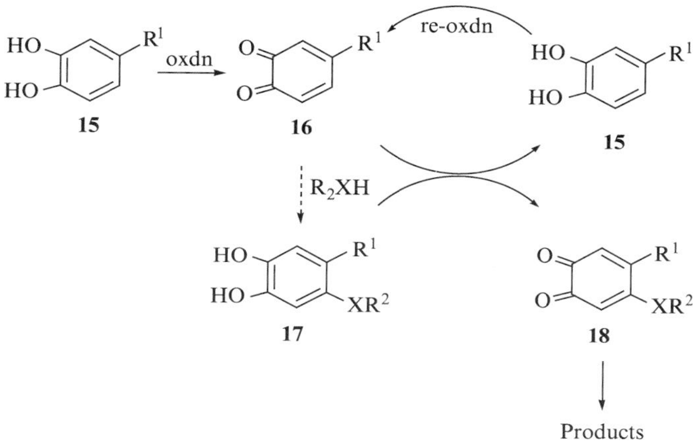

FIG. 7. Redox exchange.

position of reaction for intermolecular nucleophilic addition, is not observed, possibly because of steric constraints.

An alternative mode of intramolecular reaction of *ortho*-quinone amines is addition-elimination (Fig. 6). In these reactions the amine attacks one of the carbonyl groups with the subsequent elimination of water. This is exemplified by the cyclization of 5-*S*-cysteinyldopaquinone **12** to form a 1,4-benzothiazine **13** in the pheomelanin pathway.[1] The relatively unstable quinone imine **13** then undergoes further transformation to aromatic products (e.g., **14**).

Reduction (Redox Exchange)

Ortho-quinones and catechols form a redox system that leads to equilibration with a new quinone-catechol pair (**16** + **17** → **15** + **18**) (Fig. 7). The direction and extent of this redox exchange depends on the ring substituents and reactant concentrations. This nonenzymatic route to new

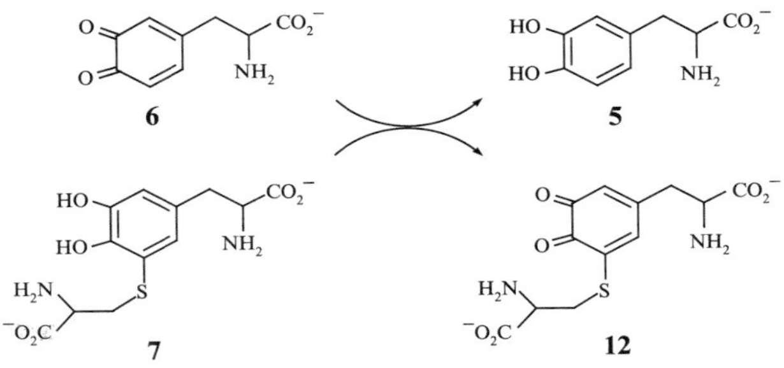

FIG. 8. Formation of 5-*S*-cysteinyldopaquinone by redox exchange.

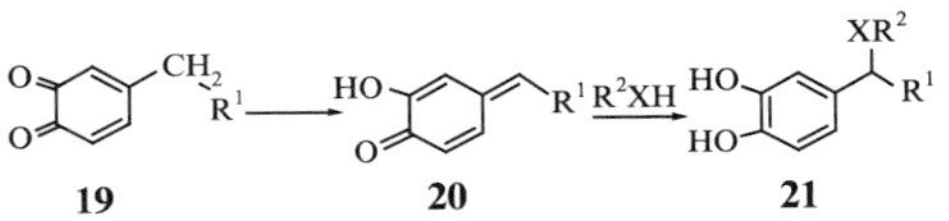

FIG. 9. Tautomerization.

ortho-quinones **18** and catechols **15** plays an important role in both eumelanin and pheomelanin formation. It has been shown, for example, that dopaquinone **6** reacts spontaneously with 5-*S*-cysteinyldopa **7** to give dopa **5** and 5-*S*-cysteinyldopaquinone **12** (Fig. 8).[12] The latter may then cyclize to give pheomelanin precursors (Fig. 6) or react with cysteine to give 2,5-*S,S*-dicysteinyldopa.

In melanin formation, an important feature of the redox system shown in Fig. 7 is that the initially formed *ortho*-quinone **16** can act as both precursor (by means of nucleophilic addition) and oxidant of the catechol **17**. The regenerated catechol **15** can then be reoxidized enzymatically, leading to further product formation. The role and significance of this redox mechanism in melanin formation is discussed in the next section.

Rearrangement (Tautomerism)

If an *ortho*-quinone derivative carries a suitable functional group an alternative reaction pathway to addition is rearrangement (isomerization) to a *para*-quinomethane (**19** → **20**).[9] Like *ortho*-quinones, *para*-quinomethanes are reactive electron-deficient species, and their ultimate fate is reaction with nucleophiles to form catechols **21** (Fig. 9). This

[12] E. J. Land and P. A. Riley, *Pigment Cell Res.* **13,** 273 (2000).

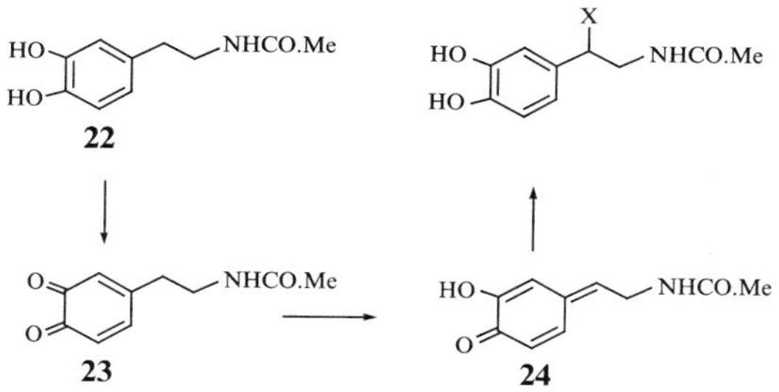

FIG. 10. Quinomethane involvement in sclerotization.

tautomerisation pathway is increasingly favored as the protons of the alkyl substituent (e.g., $-CH_2R^1$) become more acidic and may be enzyme assisted (isomerases).

This mode of reaction is important in the sclerotization mechanism in insect cuticles in which *N*-acetyldopamine **22** is oxidized to the *ortho*-quinone **23**.[13,14] Quinone isomerase–mediated formation of the *para*-quinomethane **24** then occurs, followed by rapid reaction with a nucleophile such as the imidazole group of a histidine residue, thus incorporating catechols in the cuticular region (Fig. 10).

Ortho-Quinone Amines and Melanin Pigment Formation

The Raper–Mason Pathway

Ortho-quinone chemistry plays an important role in melanin biosynthesis in both mammals and invertebrates. The early stages of melanin generation starting from tyrosine **25** were established by Raper and Mason.[15,16] Figure 11 shows an updated version of the "Raper–Mason" scheme for formation of eumelanin and pheomelanin precursors in mammals. A similar scheme operates in insects and other arthropods.[14] However, there are some fundamental differences, including initiation by phenoloxidase (PO) oxidation of tyrosine.

The mammalian melanogenic pathway is initiated by oxidation of tyrosine **25** by the copper-containing enzyme tyrosinase,[17] which converts the

[13] S. O. Anderson, M. G. Peter, and P. Roepstorff, *Comp. Biochem. Physiol.* **113B,** 689 (1996).
[14] M. Sugumaran, *Pigment Cell Res.* **15,** 2 (2002).
[15] H. S. Raper, *Physiol. Rev.* **8,** 245 (1928).
[16] H. S. Mason, *J. Biol. Chem.* **172,** 83 (1948).
[17] C. Olivares, J. C. García-Borrón, and F. Solano, *Biochemistry* **41,** 679 (2002).

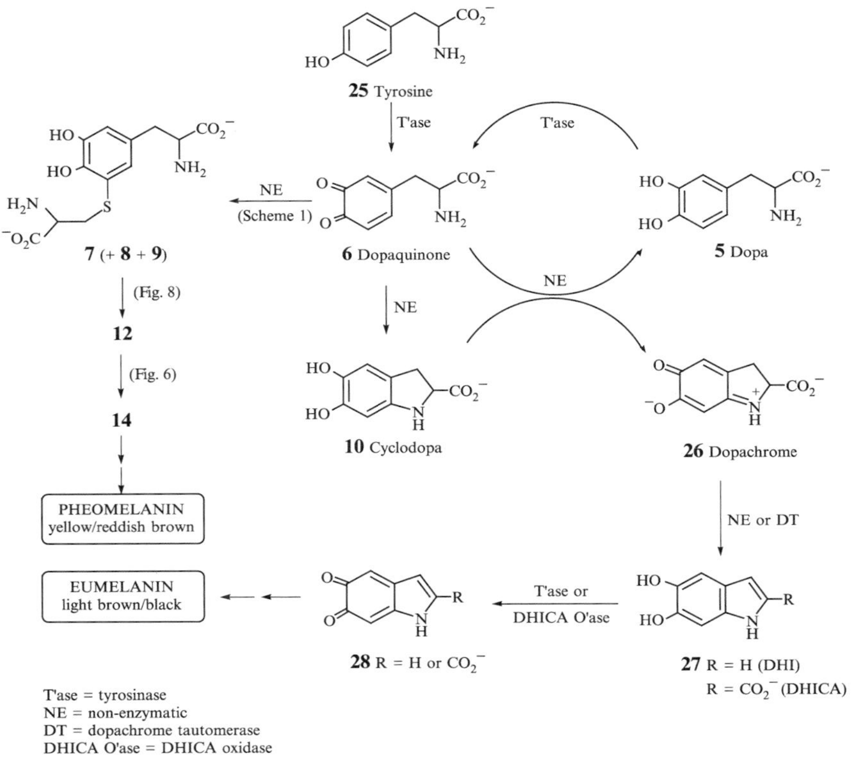

T'ase = tyrosinase
NE = non-enzymatic
DT = dopachrome tautomerase
DHICA O'ase = DHICA oxidase

FIG. 11. Updated Raper–Mason scheme for melanogenesis.

phenolic function into an *ortho*-quinone **6** in one step. Tyrosinase will also oxidize the catecholic function of dopa **5** to give dopaquinone **6** but by a mechanism that has distinct differences that are discussed in the next section.

The dopaquinone **6**, formed by tyrosinase oxidation of tyrosine **25**, then reacts by means of one of two competing nonenzymatic pathways.[12] Intramolecular amine addition gives cyclodopa **10**, and subsequent redox exchange of this catechol **10** with its precursor **6** gives dopachrome **26** and dopa **5**. Thus, although dopa **5** is formed and accumulates in the pathway, it is primarily formed by an indirect nonenzymatic route and not by direct oxidation of tyrosine **25**. Further reaction of dopachrome **26** leads to eumelanin pigment by means of the indoles **27** and the indolequinones **28**. Alternatively, the dopaquinone **6** reacts with cysteine to give

S-cysteinylcatechols (e.g., **7**) and *ortho*-quinone **12** (Fig. 8), leading to pheomelanin. Eumelanin is therefore characterized by indole fragments derived by means of cyclodopa formation, whereas pheomelanin is characterized by 1,4-benzothiazine fragments derived by means of the intermediate **14**.[1]

Tyrosinase Oxidation of Phenolamines and Catecholamines

Tyrosinase belongs to the class-3 copper proteins that include hemocyanins, which are oxygen carriers in arthropods and mollusks.[18] On the basis of x-ray structures,[19,20] there is strong evidence to suggest that the oxygen binding sites of hemocyanins and the active sites of tyrosinases are very similar.[21] In hemocyanins, access of phenols to the copper binding site is prevented by an amino acid residue.[22,23] Sequence similarity studies suggest that tyrosinase (mammals) is related to molluscan hemocyanins and phenoloxidase (PO) (insects) is related to arthropod hemocyanins.[14]

The active site of tyrosinase contains two copper atoms, and this results in the enzyme existing in three discrete oxidation states (Fig. 12).[24] Native tyrosinase occurs primarily as *met*-tyrosinase, which cannot bind oxygen and in which both copper atoms have the Cu^{II} oxidation state. Two-electron reduction of *met*-tyrosinase gives *deoxy*-tyrosinase in which the copper atoms are in the Cu^{I} oxidation state. *Deoxy*-tyrosinase binds dioxygen to give *oxy*-tyrosinase. The interrelationship of the different forms of tyrosinase is shown in Fig. 12.

The relationship between the oxidation states of tyrosinase (Fig. 12) is such that the mechanisms and stoichiometries of tyrosinase oxidation of

$Cu^{II}(A)$ –O(H)– $Cu^{II}(B)$ $\xrightarrow{2e^- + H^+}$ $Cu^{I}(A)$ ------- $Cu^{I}(B)$ + H_2O $\xrightarrow{O_2}$ $Cu^{II}(A)$ <O–O> $Cu^{II}(B)$

Met-tyrosinase *Deoxy*-tyrosinase *Oxy*-tyrosinase

FIG. 12. Tyrosinase oxidation states.

[18] K. E. Van Holde, K. I. Miller, and H. Decker, *J. Biol. Chem.* **276,** 15563 (2001).
[19] C. Gerdemann, C. Eicken, and B. Krebs, *Acc. Chem. Res.* **35,** 18 (2002).
[20] H. Decker and F. Tuczek, *TIBS* **25,** 392 (2000).
[21] H. Decker and N. Terwilliger, *J. Exp. Biol.* **203,** 1777 (2000).
[22] H. Decker and T. Rimke, *J. Biol. Chem.* **273,** 25889 (1998).
[23] H. Decker, M. Ryan, E. Jaenicke, and N. Terwilliger, *J. Biol. Chem.* **276,** 17796 (2001).
[24] E. I. Solomon and M. D. Lowery, *Science* **259,** 1575 (1993).

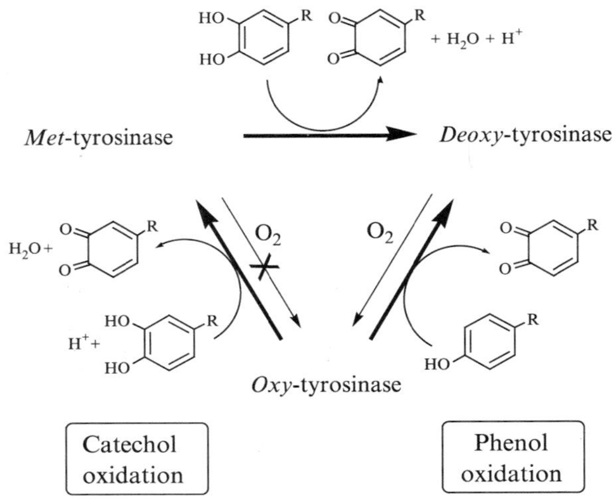

FIG. 13. *Oxy*-tyrosinase oxidation of catechols and phenols.

phenols and catechols are different. The transformations leading to *ortho*-quinone formation are summarized in Fig. 13. It can be seen that *oxy*-tyrosinase oxidizes both phenols and catechols to *ortho*-quinones. Phenol oxidation gives *deoxy*-tyrosinase, which can bind more dioxygen, and the oxidation cycle can continue. Catechol oxidation by *oxy*-tyrosinase gives *met*-tyrosinase, which cannot bind dioxygen. This *met*-tyrosinase must be reduced by a second catechol molecule to give *deoxy*-tyrosinase and subsequent regeneration of *oxy*-tyrosinase.

A consequence of the transformations shown in Fig. 13 is that native tyrosinase, which occurs mainly in the more stable oxidized Cu^{II} form (*met*-tyrosinase), will not oxidize phenols without prior reduction to *deoxy*-tyrosinase (Cu^{I}). In practice, when native tyrosinase is used to oxidize a phenol, there is a characteristic lag period ($\sim$ 30 min) in the enzyme kinetics.[17,25,26] During this period, slow nonenzymatic catechol formation takes place, leading to *deoxy*-tyrosinase formation (Fig. 13), and the oxidation rate accelerates until the maximum velocity is achieved. The lag period can be abolished by addition of a small amount of a catechol, which immediately reduces the *met*-tyrosinase, and the reaction then proceeds at maximum rate.

[25] P. A. Riley, *Cell. Mol. Biol.* **45,** 951 (1999).

[26] E. J. Land, C. A. Ramsden, and P. A. Riley, *Acc. Chem. Res.* **36,** 300 (2003).

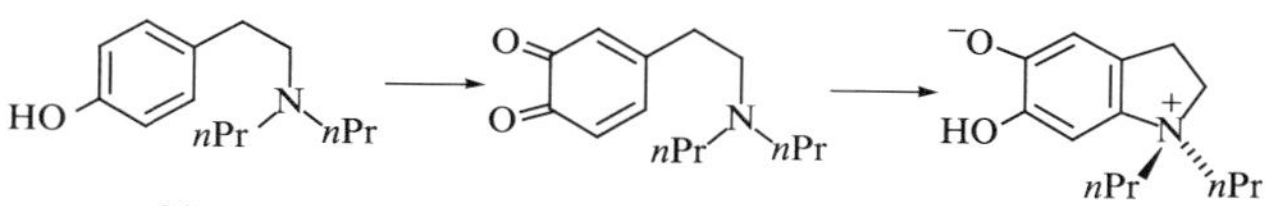

HO nPr N nPr **29** → O O nPr N nPr **30** → $^{-}$O HO N+ nPr nPr **31**

FIG. 14. Betaine formation.

For phenolamine substrates, such as tyrosine **25,** the slow activation of the tyrosinase during the lag period occurs by means of catechol formation by nonenzymatic redox exchange. This can be appreciated by considering the fate of tyrosine **25** in Fig. 11. The presence of a small amount of *oxy*-tyrosinase in the native enzyme leads to dopaquinone **6** and cyclization to cyclodopa **10** followed by redox exchange between cyclodopa and dopaquinone to give dopa and dopachrome (**6** + **10** → **5** + **26**) (Fig. 11). The dopa **5** is a substrate for *met*-tyrosinase but is in competition with tyrosine **25**, which is able to bind to the active site but cannot be oxidized. Therefore, the initial activation of the enzyme by conversion to *oxy*-tyrosinase (Fig. 13) is slow but accelerates as more dopa is formed. This accelerating activation of the enzyme by indirect catechol formation continues until the maximum rate of oxidation is achieved.

Evidence for the nonenzymatic formation of catechols by means of redox exchange during melanogenesis has been provided by the study of structurally modified phenolamines. *N,N*-Dialkyltyramines (e.g., **29**) remain unoxidized by native tyrosinase even after prolonged exposure.[27] In this case, oxidation and cyclization of the resulting *ortho*-quinone **30** leads to formation of a betaine (e.g., **31**) (Fig. 14). These betaines are not catechols, redox exchange does not occur, and the enzyme remains unactivated. If catechol formation and activation did not rely on nonenzymatic addition followed by redox exchange, the nature of the side chain would not prevent activation. In practice, betaine formation, (e.g., **31**) blocks the progress of the activation mechanism and also prevents other nucleophiles (e.g., H_2O) from generating a catechol. It is conceivable that some catechol may be released from the enzyme during *oxy*-tyrosinase oxidation of a phenol to an *ortho*-quinone.[28] However, this is deactivating, because in the process the *oxy*-tyrosinase would be converted to *met*-tyrosinase. Direct formation of catechol by means of monophenol oxidase activity

[27] C. J. Cooksey, P. J. Garratt, E. J. Land, S. Pavel, C. A. Ramsden, P. A. Riley, and N. P. M. Smit, *J. Biol. Chem.* **272,** 26226 (1997).

[28] J. N. Rodríguez-López, L. G. Fenoll, M. J. Peñalver, P. A. García-Ruiz, R. Varón, F. Martínez-Ortíz, F. García-Cánovas, and J. Tudela, *Biochim. Biophys. Acta* **1548,** 238 (2001).

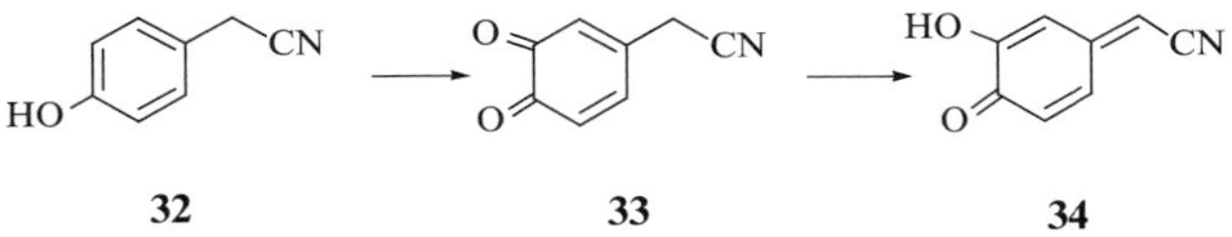

32 33 34

FIG. 15. Quinomethane formation.

would therefore prolong the lag period and is not associated with dopachrome formation.

In contrast to many other 4-substituted phenols, 4-hydroxybenzylcyanide **32** also fails to achieve autoactivation of tyrosinase after a lag period but is readily oxidized by preactivated enzyme.[29] In this case, it has been shown that the initially formed *ortho*-quinone **33**, which would normally lead to catechol formation by intermolecular addition, rapidly isomerizes to the *para*-quinomethane **34** (Fig. 15). This reactive species then binds to protein by intermolecular addition (cf. Fig. 9) and cannot participate in an activation mechanism. The rapid isomerization **33** → **34** is attributable to the acidity of the methylene protons (CH_2CN).

Measurement of Tyrosinase Activity

Many methods exist for the measurement of melanogenic activity, which include techniques that are designed to estimate the rate of melanogenesis (i.e., the entire pathway) such as the radiometric method of Chen and Chavin,[30] which used the incorporation of ^{14}C-tyrosine into acid-insoluble material, or to infer the contribution of enzymes such as dopachrome tautomerase (tyrosinase-like protein-2, TRP-2) and DHICA oxidase (TRP-1). These methods have been critically reviewed by Solano and García-Borrón.[31]

The reactions catalyzed by tyrosinase (i.e., the conversion of monophenols [e.g., tyrosine] or catechols [e.g., dopa] to the corresponding *ortho*-quinone [dopaquinone]) involve oxygen uptake, removal of the substrate, and generation of products that are readily detectable by spectrophotometric means. Thus, there are several methods that may be used to measure

[29] C. J. Cooksey, P. J. Garratt, E. J. Land, C. A. Ramsden, and P. A. Riley, *Biochem. J.* **333,** 685 (1998).

[30] Y. Chen and W. Chavin, *Anal. Biochem.* **13,** 234 (1965).

[31] F. Solano and J. C. García-Borrón, *in* "The Pigmentary System" (J. J. Nordlund, R. E. Boissy, V. J. Hearing, R. A. King, and J.-P. Ortonne, eds.), p. 461. Oxford University Press, Oxford, 1998.

activity of the enzyme. A much used method depends on the formation of dopachrome,[16] which has a characteristic absorbance in the 480-nm region, and a definition of a unit of tyrosinase is based on the rise in absorbance at this wavelength[32] under prescribed conditions using L-3,4-dihydroxyphenylalanine (dopa) as substrate. A variant of this method[33] uses HPLC separation of the products of oxidation, but this method is technically more demanding and is not very sensitive. In the presence of plenty of enzyme, oximetry is a well-established method. Polarimetric recordings using Clark-type electrodes are reasonably sensitive, and recent developments with oxygen-sensitive fluorescent probes have added another option. A highly sensitive technique dependent on oxygen-sensitive probes for EPR spectroscopy has been used, but this method is subject to a number of technical difficulties connected with oxygen diffusion rates in the sample chamber. We have developed a combined oximetric and spectrophotometric method that allows accurate estimation of stoichiometric data, because it permits separate identification of dopaquinone and subsequent products such as dopachrome.[27] This technique uses a silica spectrophotometry cuvette modified to accept the tip of an oxygen probe orthogonal to and at the same height as the path of the light beam. The reaction mixture is stirred by a corrugated disk-shaped magnetic stirrer, which is rotated on the vertical wall of the cuvette by a small rotary motor. The substrate is introduced into the cuvette through a capillary inlet in a stopper. Temperature control is by a thermostatically regulated air stream. Timed scans of the spectrum are made with a Hewlett-Packard diode array spectrophotometer (Wokingham, UK), and the oxygen uptake is recorded with a calibrated oxygen electrode.

Other methods include the detection of adducts of dopaquinone, which include cysteinyldopa.[34] The simplest of these methods uses the Besthorn's reagent (3-methyl-2-benzothiazolinone, MBTH), which forms an adduct with dopaquinone that absorbs strongly at 500 nm.[35]

Undoubtedly, the most sensitive method for the measurement of monohydric phenol oxidation by tyrosinase is the method of Pomerantz,[36] which uses the displacement of the hydrogen atom *ortho* to the 4-hydroxyl group of tyrosine by the oxygen insertion reaction. Thus, by use of 3,5-^{3}H-tyrosine, the catalytic reaction can be followed by measurement

[32] H. W. Duckworth and J. E. Coleman, *J. Biol. Chem.* **245,** 1613 (1970).
[33] B. Jergil, C. Lindbladh, H. Rorsman, and E. Rosengren, *Acta Derm. Venereol. (Suppl.)* **63,** 468 (1983).
[34] G. Agrup, L. E. Edholm, H. Rorsman, and E. Rosengren, *Acta Derm. Venereol.* **63,** 59 (1983).
[35] A. J. Winder and H. Harris, *Eur. J. Biochem.* **198,** 317 (1991).
[36] S. H. Pomerantz, *Biochem. Biophys. Res. Commun.* **16,** 188 (1964).

of the tritiated water released. There are some problems that arise from long incubation times in which hydrogen exchange reactions lead to high background counts, but the procedure is widely used and, in capable hands, extremely sensitive.

Histochemical methods of demonstrating tyrosinase have been used in tissue sections and separated epidermal sheets. The first method was described by Bruno Bloch[37] using dopa as a substrate to demonstrate pigment cells in the skin. Although tyrosine and other monophenols have been used, most histochemical staining methods rely on the generation of insoluble melanin from dopa (generally a 0.1% solution in phosphate buffer at pH 6.8). Histochemical methods have also been used to stain tyrosinase separated by gel electrophoresis under nondenaturing conditions as have fluorographic techniques to detect incorporation of radioactive precursors into melanin.

Pulse Radiolysis and the Study of Unstable Melanogenic *ortho*-Quinones

Many of the *ortho*-quinones involved in melanogenesis are highly unstable, and so the quantification of their reactions requires the use of some type of fast reaction technique. The time-resolved method that has been found to be by far the most useful in this field is pulse radiolysis. This technique has been used to generate, within a few milliseconds, readily detectable amounts of many unstable *ortho*-quinones formed from the corresponding stable catechols by disproportionation of the initially formed semiquinone radicals.[38]

In pulse radiolysis,[39] the absorption change induced in a sample solution by a pulse of ionizing radiation, typically of duration 10 ns to 5 μs, is monitored by an analyzing light beam passing through the sample and reaching a detector (photomultiplier or photodiode, depending on the wavelength of interest) by means of a monochromator. The detector converts changes in the analyzing light intensity into electrical signals, which are digitized, displayed, then stored and treated by microcomputer. The transient absorbances thus obtained are calculated as a function of time and wavelength. Conventional absorption spectroscopy remains the most widely used technique for monitoring the spectra and kinetics of formation

[37] B. Bloch, *Z. Physiol. Chem.* **9,** 226 (1916).

[38] E. J. Land, *J. Chem. Soc. Faraday Trans.* **89,** 803 (1993).

[39] R. V. Bensasson, E. J. Land, and T. G. Truscott, "Excited States and Free Radicals in Biology and Medicine: Contributions from Flash Photolysis and Pulse Radiolysis." Oxford University Press, Oxford, 1993.

and decay of transient intermediates. However, there are other monitoring techniques such as luminescence, photoacoustic spectroscopy, diffuse reflectance in solids, Rayleigh and Raman scattering, conductivity, polarography, and EPR.

In any dilute solution, it is the solvent that absorbs most of the high-energy radiation. In water, within a nanosecond of radiation deposition, the main primary radicals formed are reducing hydrated electrons, e^-_{aq}, and oxidizing $OH^\bullet$. The usual way of generating exclusively oxidizing radical species is to irradiate the aqueous solution saturated with nitrous oxide, which converts e^-_{aq} into extra $OH^\bullet$ [Eq. (1)]:

$$e^-_{aq} + N_2O + H_2O \rightarrow OH^\bullet + N_2 + OH^- \tag{1}$$

Because $OH^\bullet$ is such a vigorous oxidizing agent and tends to add to solutes, as well as abstracting electrons or hydrogen atoms, milder oxidizing agents are the radicals $Br_2^{\bullet -}$ and $N_3^\bullet$, which can be produced from $OH^\bullet$ by adding a high concentration of the corresponding halide or pseudo-halide ion [Eqs. (2)–(4)]:

$$OH^\bullet + Br^- \rightarrow OH^- + Br^\bullet \tag{2}$$

$$Br^- + Br^\bullet \rightarrow Br_2^{\bullet -} \tag{3}$$

$$OH^\bullet + N_3^- \rightarrow OH^- + N_3^\bullet \tag{4}$$

A disadvantage of the azide system over the bromide system is that nucleophilic N_3^-/HN_3 can add to *ortho*-quinones in some circumstances, unless the conditions are carefully controlled.[40]

A typical solution that might be used to generate an *ortho*-quinone is N_2O-saturated aqueous 5×10^{-3} M catechol + 10^{-1} M KBr. The rationale behind this choice of concentrations is now described using, as an example, the important catechol 3,4-dihydroxyphenylalanine (dopa). Because the concentration of N_2O when saturated in water is $\approx 10^{-2}$ M, and the rate constant for reaction of e^-_{aq} + N_2O is $\approx 10^{10}$ M^{-1} s^{-1},[41] the half-life of e^-_{aq} under such conditions is a few nanoseconds. Because neither dopa nor

[40] C. Lambert, T. G. Truscott, E. J. Land, and P. A. Riley, *J. Chem. Soc. Faraday Trans.* **87,** 2939 (1991).

[41] G. V. Buxton, C. L. Greenstock, W. P. Helman, and A. B. Ross, *J. Phys. Chem. Ref. Data* **17,** 513 (1988).

Br^- react quickly with e^-_{aq}, this half-life is short enough to ensure that all the e^-_{aq} are scavenged by N_2O, giving further $OH^•$. Regarding the fate of the $OH^•$ radicals, the rate constants for its reactions with Br^- and dopa are both approximately 10^{10} M^{-1} s^{-1}. Thus, use of a concentration of Br^- of 20 times that of dopa ensures that more than 90% of the $OH^•$ radicals are scavenged by Br^- to give $Br_2^{•-}$. The latter reacts with dopa ($k \approx 10^8$ M^{-1} s^{-1})[12] within a few microseconds to yield dopasemiquinone, which goes on to disproportionate over approximately a millisecond to dopaquinone (wavelength max, 400 nm, extinction coefficient $\approx$2000 M^{-1} cm^{-1}). By following the subsequent changes in absorption at 400 nm or by looking at products appearing at other wavelengths, one can measure quantitatively the rate constants for various reactions of dopaquinone. In the absence of further additives, the rate of cyclization to cyclodopa, a Michael addition, can be followed. On the other hand, a comparison of the rate of decay at 400 nm in the presence and absence of additional 10^{-4} M thiol can lead to a rate constant for this further nucleophilic reaction.

The Balance between Eumelanogenesis and Pheomelanogenesis

As mentioned in the "Introduction," differences in surface pigmentation in vertebrates, although partly quantitative in terms of the rate of pigment turnover (which is largely dependent on the rate of synthesis and transfer of melanin), are predominantly influenced by the qualitative nature of the pigment formed. In general, the darker melanins (termed eumelanin) are composed of indolic elements. Lighter (reddish brown) melanins (pheomelanin) contain significant amounts of sulfur incorporated into benzothiazine residues that form part of the polymeric structure of the pigment. The precise synthetic mechanisms contributing to the formation of these pigments are not known, but the major point at which divergence between the eumelanogenic and pheomelanogenic pathways takes place occurs in the initial phase of melanin biosynthesis (Fig. 11).

The initial product of tyrosinase activity is dopaquinone. This *ortho*-quinone can undergo a number of reactions as detailed previously. In general terms, there is competition between: (1) endocyclization of dopaquinone to form cyclodopa, which is an indole precursor; and (2) exogenous Michael addition reactions (Fig. 11). The endocyclization involves the amino group on the side chain which, in dopaquinone, is able to form a five-membered ring, a favored option for cyclization (see preceding discussion). Nevertheless, the measured rate constant for this reaction is relatively slow, especially under the acidic conditions thought to exist in the melanosomes. Even at neutral pH, the rate constant of dopaquinone cyclization is no more than 7.6 s^{-1}.[42] Nucleophilic addition reactions with exogenous amino groups are unlikely to compete successfully with the cyclization through reaction with the vicinal amino group on the side chain.

But other, more powerful nucleophiles, such as thiols, are able to react with dopaquinone to give sulfur-containing precursors of pheomelanin. The pheomelanogenic pathway involves the reaction of dopaquinone and cysteine to form cysteinyldopa, predominantly as the 5-cysteinyldopa isomer (and a smaller amount of the 2-cysteinyldopa product). The rate constant of the reaction between dopaquinone and cysteine has been measured and at neutral pH is 3×10^7 $M^{-1}s^{-1}$.[43] Thus, other things being equal, the cyclization and thiol addition reactions would be balanced in the presence of approximately 10^{-7} molar cysteine.

An interesting facet of the pathway of divergence between eumelanin and pheomelanin generation is that, *in vitro,* dopaquinone is able to act as the oxidizing agent in redox exchange reactions both with cyclodopa (the endocyclization product of dopaquinone) and cysteinyldopa (the addition product with cysteine), giving rise to dopachrome (a eumelanin precursor) and cysteinyldopaquinone (a pheomelanin precursor). The redox exchange rate constants have been measured and are, respectively, 5.3×10^6 M^{-1} s^{-1} [44] and 8.8×10^5 M^{-1} s^{-1}.[12] Taking dopachrome and cysteinyldopaquinone as the starting points of eumelanogenesis and pheomelanogenesis, respectively (Fig. 11), the index of divergence[26] is unity when the cysteine concentration is in the micromolar range. Although the intramelanosomal cysteine levels have not been determined, control of the extent of pheomelanogenesis by regulation of melanosomal access by cysteine has been proposed.[45]

Biological Significance of the Melanogenic Pathway

The first step in the formation of melanin is the generation of *ortho*-quinones from phenolic precursors. Because this step has been strongly conserved in evolution, the production of *ortho*-quinones must be endowed with properties beneficial to the survival of the organism and, as noted earlier, one of the most notable features of *ortho*-quinones is their ability

[42] M. R. Chedekel, E. J. Land, A. Thompson, and T. G. Truscott, *J. Chem. Soc. Chem. Commun.* 1170 (1984).

[43] A. Thompson, E. J. Land, M. R. Chedekel, K. V. Subbarao, and T. G. Truscott, *Biochim. Biophys. Acta* **843,** 49 (1985).

[44] E. J. Land, S. Ito, K. Wakamatsu, and P. A. Riley, *Pigment Cell Res.* **16,** 487 (2003).

[45] S. B. Potterf, V. Virodor, K. Wakamatsu, M. Furumura, C. Santis, S. Ito, and V. J. Hearing, *Pigment Cell Res.* **12,** 4 (1999).

to undergo facile addition reactions with nucleophiles such as thiol and amino groups. This property has been used in plants both as a means of antibiosis by damaging invasive organisms ("cytotoxic" action) and as a way of modifying and hardening the protective exterior layers such as are found in seed envelopes ("cross-linking" action). The participation of quinone formation in the "stress responses" of plants is exemplified by the browning reactions in potatoes and bananas, which takes place when any surface damage permits access of the phenol oxidase to the separately segregated dopamine substrate. Similar processes occur in the fruiting bodies of fungi.

In insects, the cytotoxic potential of *ortho*-quinone generation is the basis of their immune system, and quinones form an important component of defensive sprays. It is likely that the ink of cephalopods exerts its action not so much by obscuring vision (which is of limited significance in guiding predators in deep waters) as by the deterrent action on sensitive chemoreceptor organelles of reactive *ortho*-quinones in the ink, which contains a mixture of tyrosinase and substrate molecules.

The cross-linking action of quinones is involved in insect cuticular hardening. The process has been studied quite extensively and relies on the ability of certain side-chain–substituted substrate molecules to be successively oxidized so as to form cross-links between adjacent proteins.[14]

Another example of cross-linking is the participation of tyrosinase in balanid adhesion. The protein glues of the blue mussel have been characterized and are members of a dopa-rich family of polypeptides that cross-link by an oxidative tanning process. Mussels produce protein "byssus" threads secreted by the mussel foot, which anchor the mussel to solid substrata in the intertidal region. These threads consist of collagen fibers embedded in a polyphenolic protein matrix in which there is a high level of post-translational hydroxylation of tyrosine and proline residues to give dopa and hydroxyproline, respectively. On oxidation of the dopa residues to the *ortho*-quinone, these proteins form cross-links that stabilize the adhesion. Similar substrate peptides have been isolated in several species of balanids, and the cross-linking reactions are tyrosinase-catalyzed.[46]

These examples are what may be termed initial benefits of melanogenesis in that they derive from the primary oxidation that leads to the production of pigment. The phases of the evolution of melanin pigmentation have been summarized[47] as involving the following steps:

[46] H. Yamamoto and H. Tatehata, *J. Marine Biotechnol.* **2,** 95 (1995).

[47] P. A. Riley, *in* "Melanin: Its Role in Human Photoprotection" (L. Zeise, M. R. Chedekel, and T. B. Fitzpatrick, eds.), p. 1. Valdemar Publishing Co., Overland Park, 1995.

1. Secretion of an *ortho*-quinone–generating enzyme with defensive and cross-linking potential
2. Intracellular retention of the enzyme with accumulation of a polymerized product of *ortho*-quinones and their derivatives, permitting photoreceptor screening
3. Limitation of pigment production to specific cells, leading to organismal patterning with roles in camouflage and display
4. Transference of pigment to acceptor cells, enabling regional effects on durability, photoprotection, and excretion

Some of these actions are dependent on the structure and properties of melanin. For example, the increased durability of pigmented tissues, demonstrated by the greater resistance to mechanical damage and greater strength of black compared with white feathers, derives from the polymeric structure of melanin.[48]

The detailed structure of melanin is unknown, but there is overwhelming evidence that the basic structure entails a backbone of indole moieties that contain both quinone and hydroquinone residues.[1] A high degree of conjugation exists in the polymer, which imbues it with strong photon absorption in the visible and UV spectrum. The bathochromic characteristic of melanin results in significant photon absorption even in the infrared portion of the spectrum, and this may permit it to act as a means of thermal absorption in poikilotherms and animals in cold climates, where solar radiation constitutes a significant factor in energy conservation. Photoprotection is widely considered to be the most significant action of melanin in humans, and there is good evidence that increased epidermal melanization is associated with a significantly decreased risk of skin cancer related to sun exposure. The mechanism may involve the absorption of energetic photons that might damage important cellular structures such as DNA and therefore constitutes an aspect of screening. Alternatively, the protective action may involve the generation of free radicals that are toxic to cells that have been exposed to potentially genotoxic doses of light.[49] Because of the conjugated structure of melanin, there is an equilibrium between the quinone and hydroquinone moieties in the polymer, which renders the pigment a facile electron exchanger. It can, thus, act as a free radical generator or scavenger.[50] In addition, the relatively high density of negative surface

[48] A. H. Robins, "Biological Perspectives on Human Pigmentation" Cambridge University Press, Cambridge, 1991.

[49] P. A. Riley, *in* "The Physiology and Pathophysiology of the Skin" (A. Jarrett, ed.), Vol. 3, p. 1104. Academic Press, London, 1974.

[50] G. Sichel, C. Corsaro, M. Scalia, S. Sciuto, and E. Geremia, *Cell Biochem. Function* **5,** 123 (1987).

charge on melanin renders it a cation trap, and it has been suggested that this constitutes one of the biological benefits conferred by surface pigment, because by desquamation in keratocytes melanin provides an excretory pathway for metals and other cationic materials.[51]

Possible Therapeutic Implications of Quinone Chemistry

The reactivity of the *ortho*-quinone products of melanogenesis, which proves of value when the process is exteriorized, necessitates the containment of the process within membrane-delimited organelles (melanosomes) when pigment production takes place intracellularly. Even with this biochemical segregation, melanogenic cells (as opposed to acceptor cells) require cytosolic safeguards to prevent cytotoxic effects of leaking quinone intermediates.[52] Thus, the notion of the use of the cytotoxicity inherent in the process of melanogenesis to target melanocyte tumors would seem to be a viable therapeutic strategy for malignant melanoma. Melanoma is an aggressive malignancy that is increasing in incidence. The disseminated form of this cancer exhibits marked resistance to conventional chemotherapeutic regimens. Among the abnormalities of melanoma cells that have been described is overexpression of tyrosinase and the incompleteness of the melanosomal membranes.[53] Hence, the use of the melanogenic pathway as a means of delivering targeted cytotoxicity to melanoma cells has been advocated.[54]

As has been discussed previously, the possession of a potentially reactive side chain by the natural melanogenic substrate, tyrosine, ensures that the lifetime of any free dopaquinone is limited by rapid endocyclization. Therefore, to maximize the chance of quinone reactions with vital cellular components by exo-reactions, analogue substrates lacking reactive side chains have been investigated.[55–57] This approach was initiated by the depigmenting action of certain phenolic antioxidants in man and

[51] P. A. Riley, *Biologist* **44,** 408 (1997).

[52] N. P. M. Smit, S. Pavel, and P. A. Riley, *in* "Role of Catechol Quinone Species in Cellular Toxicity" (C. R. Creveling, ed.), p. 191. Graham Publishing, Johnson City, TX, 2000.

[53] J. Borovansky, P. Mirejovsky, and P. A. Riley, *Neoplasma* **38,** 393 (1991).

[54] P. A. Riley, *Eur. J. Canc.* **27,** 1172 (1991).

[55] S. Naish-Byfield, C. J. Cooksey, A. J. M. Latter, C. I. Johnson, and P. A. Riley, *Melanoma Res.* **1,** 273 (1991).

[56] C. J. Cooksey, E. J. Land, C. A. Ramsden, and P. A. Riley, *Anti-Cancer Drug Design* **10,** 119 (1995).

[57] C. J. Cooksey, E. J. Land, F. A. P. Rushton, C. A. Ramsden, and P. A. Riley, *Quant. Struct. Act. Relat.* **15,** 498 (1996).

animal models, which was shown to result from the selective toxicity to melanocytes.

Several pathways of toxic action have been proposed, and *ortho*-quinones are known to react with nucleophilic groups of proteins and nucleic acids. In cellular systems, one of the major detoxification pathways is by reaction with glutathione to form an adduct, and it is probable that the intracellular glutathione has to be depleted before cells are irreversibly damaged.[58] Because the cellular glutathione concentration is in the millimolar range, this imposes considerable limitations on the efficacy of chemotherapy based on quinone generation by the introduction of analogue melanogenic substrates, as has been shown for 4-hydroxyanisole.

More recently, an alternative strategy has been investigated, which is based on the use of the cyclization reaction to trigger the release of a cytotoxic compound with an established mode of action at relatively low concentrations, such as nitrogen mustard.[59] This strategy depends on labilization of the bond between the cytotoxic agent and the "carrier" molecule, which acts as a substrate for tyrosinase. Conversion of the substrate to the corresponding *ortho*-quinone prompts the nucleophilic cyclization reaction which, by electronic redistribution, permits facile hydrolytic cleavage of the prodrug linkage and release of the toxic agent. A number of such compounds are currently under investigation.[60]

[58] P. A. Riley, C. J. Cooksey, C. I. Johnson, E. J. Land, A. J. M. Latter, and C. A. Ramsden, *Eur. J. Canc.* **33,** 135 (1997).

[59] A. M. Jordan, T. H. Khan, H. M. I. Osborn, A. Photiou, and P. A. Riley, *Bioorg. Med. Chem.* **7,** 1775 (1999).

[60] A. M. Jordan, T. H. Khan, H. Malkin, H. M. I. Osborn, A. Photiou, and P. A. Riley, *Bioorg. Med. Chem.* **9,** 1549 (2001).

[6] Quinoids Formed from Estrogens and Antiestrogens

By Judy L. Bolton, Linning Yu, and Gregory R. J. Thatcher

Introduction

Estrogens and Antiestrogens and Cancer Risk

A firm link between female reproductive variables and increased risk of cancer developing in the breast and endometrium has been established from epidemiological studies.[1–4] The longer women are exposed to estrogens, either through early menarche and late menopause and/or through estrogen replacement therapy, the higher is the risk of certain hormone-dependent cancer developing. It used to be thought that the numerous benefits of estrogen replacement therapy, including the relief of menopausal symptoms, decrease in coronary heart disease, osteoporosis, stroke, and Alzheimer's disease, justified the use of long-term estrogen replacement therapy. However, the release of the Women's Health Initiative Study in July 2002 cast serious doubt on this paradigm for the treatment of postmenopausal women.[5] This study was halted 3 years early because of significant increases in breast cancer, coronary heart disease, stroke, and cardiovascular disease, with the most recent data suggesting an increase in vascular dementia in women older than 65 taking estrogen replacement therapy.[6] These troubling findings have emphasized the crucial need for developing alternative estrogen replacement formulations that maintain the beneficial properties of estrogens without generating adverse side effects. To develop such "wonder" drugs, it is first necessary to understand the cytotoxic mechanisms of estrogens and in particular their ability to cause DNA damage.

Selective estrogen receptor modulators (SERMs) have been developed for treatment of menopause-associated osteoporosis, with a primary

[1] B. E. Henderson, R. Ross, and L. Bernstein, *Cancer Res.* **48,** 246 (1988).

[2] J. G. Liehr, *Mutat. Res.* **238,** 269 (1990).

[3] H. S. Feigelson and B. E. Henderson, *Carcinogenesis* **17,** 2279 (1996).

[4] G. A. Colditz, *J. Natl. Cancer Inst.* **90,** 814 (1998).

[5] J. E. Rossouw, G. L. Anderson, R. L. Prentice, A. Z. LaCroix, C. Kooperberg, M. L. Stefanick, R. D. Jackson, S. A. Beresford, B. V. Howard, K. C. Johnson, J. M. Kotchen, and J. Ockene, *JAMA* **288,** 321 (2002).

[6] S. A. Shumaker, C. Legault, S. R. Rapp, L. Thal, R. B. Wallace, J. K. Ockene, S. L. Hendrix, B. N. Jones, A. R. Assaf, R. D. Jackson, J. M. Kotchen, S. Wassertheil-Smoller, and J. Wactawski-Wende, *JAMA* **289,** 2651 (2003).

0076-6879/04 $35.00

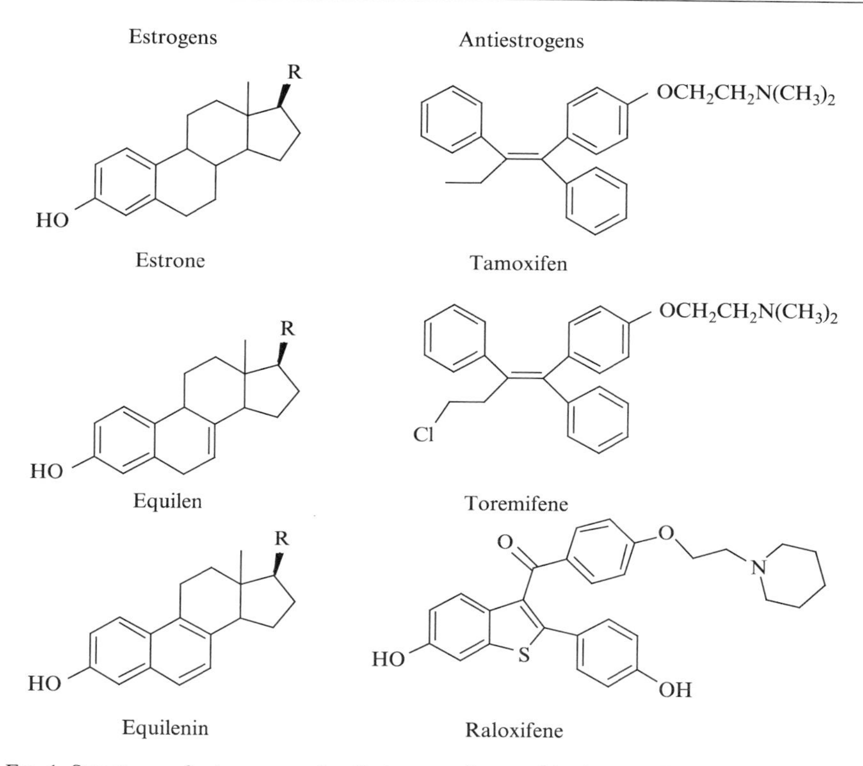

FIG. 1. Structures of estrogens and antiestrogens discussed in the text. R = OH or carbonyl.

application being in the treatment and prevention of breast cancer in postmenopausal women.[7] The most widely used SERM is tamoxifen (Fig. 1), which has been extensively used for the treatment of hormone-dependent breast cancer and more recently as a chemopreventive agent in women at risk for breast cancer; however, like the estrogens associated with estrogen replacement therapy, tamoxifen also can lead to an increased endometrial cancer risk.[8,9] The molecular mechanism(s) involved in the carcinogenic action of estrogens and antiestrogens still remains both controversial and elusive.[10] They can both be metabolized to reactive intermediates such as *o*-quinones and quinone methides (Fig. 2), which could cause DNA

[7] T. A. Grese and J. A. Dodge, *Current Pharm. Design.* **4,** 71 (1998).
[8] M. A. Seoud, J. Johnson, and J. C. Weed, Jr., *Obstet. Gynecol.* **82,** 165 (1993).
[9] D. H. Phillips, *Carcinogenesis* **22,** 839 (2001).
[10] R. F. Service, *Science* **279,** 1631 (1998).

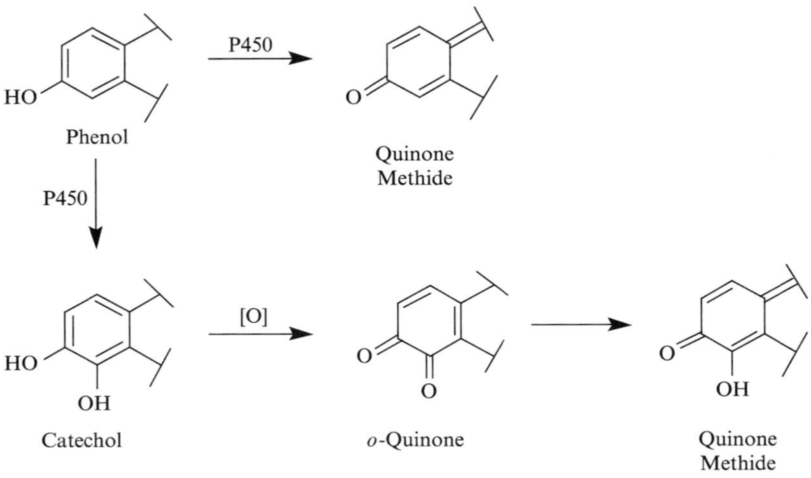

FIG. 2. General bioactivation scheme showing the formation of quinone methides and *o*-quinones.

damage directly through the formation of DNA adducts or indirectly through the generation of reactive oxygen species that oxidize DNA.[11] The focus of this report is the role of quinoids in the DNA damage induced by estrogens and antiestrogen metabolites.

Experimental Procedures

CAUTION: *Estrogen and antiestrogen metabolites were handled in accordance with the NIH Guidelines for the Laboratory Use of Chemical Carcinogens.*[12] All solvents and chemicals were purchased from either Aldrich Chemical Company (Milwaukee, WI) or Fisher Scientific (Itasca, IL) unless stated otherwise. 4-Hydroxytamoxifen,[13] 4-hydroxytoremifene,[13] 3,4-dihydroxytamoxifen,[14] and 3,4-dihydroxytoremifene[15] were synthesized as described previously. Droloxifene[16] and raloxifene[17] were

[11] J. L. Bolton, E. Pisha, F. Zhang, and S. Qiu, *Chem. Res. Toxicol.* **11,** 1113 (1998).

[12] NIH Guidelines for the Laboratory Use of Chemical Carcinogens. Washington, DC: U.S. Government Printing Office, 1981.

[13] S. Gauthier, J. Mailhot, and F. Labrie, *J. Org. Chem.* **61,** 3890 (1996).

[14] F. Zhang, P. Fan, X. Liu, L. Shen, R. B. van Breemen, and J. L. Bolton, *Chem. Res. Toxicol.* **13,** 53 (2000).

[15] D. Yao, F. Zhang, L. Yu, Y. Yang, R. B. van Breemen, and J. L. Bolton, *Chem. Res. Toxicol.* **14,** 1643 (2001).

synthesized according to the literature procedures. 4-Hydroxyequilenin[18] and 4-hydroxyequilin[19] were also synthesized as previously described.

Chemical Oxidation: Preparation of Antiestrogen and Estrogen Quinoids

o-Quinones. A mixture of 3,4-dihydroxytoremifene (4.4 mg), fresh silver oxide (440 mg), and acetonitrile (4 ml) was stirred for 15 min at 60°. After filtration, the solution was concentrated to a final volume of 1 ml. Similar procedures have been used to generate the *o*-quinones of 2-OHE and 4-OHE.[20] The equine *o*-quinones are formed spontaneously under physiological conditions by autoxidation of the corresponding catechol.

Quinone Methides. 4-Hydroxytamoxifen or 4-hydroxytoremifene was oxidized to quinone methides using activated manganese dioxide in acetone.[21] After filtration of the manganese dioxide, the resulting solution was concentrated fourfold and used for product and kinetic studies. Ultraviolet (UV) spectra were measured on a Hewlett-Packard (Palo Alto, CA) 8452A photodiode array UV/vis spectrophotometer.

Microsomal Oxidation: Oxidation of Antiestrogens by Rat Liver Microsomes

Female Sprague–Dawley rats (200–220 g) were obtained from Sasco Inc. (Omaha, NE). The rats were pretreated with dexamethasone to induce P450 3A isozymes. They were given intraperitoneal injections of dexamethasone in corn oil (100 mg/kg) daily for 3 consecutive days and were killed on day 4. Rat liver microsomes were prepared, and protein and P450 concentrations were determined as described previously.[22] A solution containing the substrate (0.05 m*M*), rat liver microsomes (1 n*M* P450/ml), GSH (0.5 m*M*), and a NADPH-generating system (including 1.0 m*M* $NADP^+$, 5.0 m*M* $MgCl_2$, 5.0 m*M* isocitric acid, and 0.2 unit/ml isocitric

[16] R. McCage, G. Leclercq, N. Legros, J. Goodman, M. Blackburn, M. Jarman, and A. B. Foster, *J. Med. Chem.* **32,** 2527 (1989).
[17] C. D. Jones, M. G. Jevnikar, A. J. Pike, M. K. Peters, L. J. Black, A. R. Thompson, J. F. Falcone, and J. A. Clemens, *J. Med. Chem.* **27,** 1057 (1984).
[18] L. Shen, E. Pisha, Z. Huang, J. M. Pezzuto, E. Krol, Z. Alam, R. B. van Breemen, and J. L. Bolton, *Carcinogenesis* **18,** 1093 (1997).
[19] M. Chang, F. Zhang, L. Shen, N. Pauss, I. Alam, R. B. van Breemen, S. Blond-Elguindi, and J. L. Bolton, *Chem. Res. Toxicol.* **11,** 758 (1998).
[20] S. L. Iverson, L. Shen, N. Anlar, and J. L. Bolton, *Chem. Res. Toxicol.* **9,** 492 (1996).
[21] M. M. Marques and F. A. Beland, *Carcinogenesis* **18,** 1949 (1997).
[22] J. A. Thompson, A. M. Malkinson, M. D. Wand, S. L. Mastovich, E. W. Mead, K. M. Schullek, and W. G. Laudenschlager, *Drug Metab. Dispos.* **15,** 833 (1987).

dehydrogenase) in 50 m*M* phosphate buffer (pH 7.4, 1.0 ml total volume) was incubated for 30 min at 37°.[20] For control incubations, $NADP^+$ was omitted. The reactions were terminated by chilling in an ice bath followed by the addition of perchloric acid (50 μl). An aliquot (200 μl) was analyzed by high-performance liquid chromatography (HPLC) using a modification of the method of Chen *et al.*[23] HPLC analysis was performed using a 4.6 × 250 mm, 5 μM Agilent (Palo Alto, CA) Zorbax Rx-C8 column with a Shimadzu (Columbia, MD) HPLC system with UV detection at 284 nm. The Shimadzu HPLC system consisted of LC-10A gradient HPLC system equipped with an SIL-10A auto injector, an SPD-M10AV UV/vis photodiode array detector, and an SPD-10AV detector. The mobile phase consisted of acetonitrile-water containing 10% methanol and 0.4% formic acid (pH 2.5) at a flow rate of 1.0 ml/min and was programmed for a linear increase from 10–30% acetonitrile during a 60-min period and increased to 90% acetonitrile over the remaining 10 min. Liquid chromatography–mass spectrometry–mass spectrometry (LC-MS-MS) spectra were obtained using a Micromass (Manchester, UK) Quattro II triple quadrupole mass spectrometer equipped with a Waters (Milford, MA) 2487 UV detector and a Waters 2690 HPLC system. Collision-induced dissociation (CID) was carried out using a range of collision energies from 25–70 eV and an argon collision gas pressure of 2.7 μbar. Representative positive ion electrospray LC-MS-MS data for one of the *o*-quinone mono- and di-GSH conjugates, quinone methide GSH conjugates, and the raloxifene catechols are as follows. *o*-Quinone-di-SG: MS/MS of *m/z* 551 $[M + 2H]^{2+}$, 513 (45%) $[M + 2H\text{-}Gly]^{2+}$, 486 (100%) $[M + 2H\text{-}Glu]^{2+}$, 448 (30%) $[M + 2H\text{-}Gly\text{-}Glu]^{2+}$, 422 (20%) $[M + 2H\text{-}2Glu]^{2+}$, 112 (10%) $[(CH_2)_2N(CH_2)_5]^+$. *o*-Quinone-SG: MS-MS of *m/z* 795 $[M + H]^+$, 777 (35%) $[MH\text{-}H_2O]^+$, 720 (45%) $[MH\text{-}Gly]^+$, 666 (100%) $[MH\text{-}Glu]^+$, 720 (45%) $[MH\text{-}Gly]^+$, 112 (52%) $[(CH_2)_2N(CH_2)_5]^+$. Quinone methide-SG: MS-MS of *m/z* 779 $[M + H]^+$, 761 (40%) $[MH\text{-}H_2O]^+$, 650 (100%) $[MH\text{-}Glu]^+$, 112 (52%) $[(CH_2)_2N(CH_2)_5]^+$. Catechol: MS-MS of *m/z* 490 $[M + H]^+$, 128 (40%) $[O(CH_2)_2N(CH_2)_5]^+$, 112 (100%) $[(CH_2)_2N(CH_2)_5]^+$.

Reaction of Antiestrogen and Estrogen o-*Quinones with Deoxynucleosides*

A solution of toremifene-*o*-quinone (0.5 m*M*) in acetonitrile was combined with deoxynucleosides (25 m*M*) in phosphate buffer (pH 7.4) in a total volume of 5.0 ml. The solution was incubated for 5 h at 37°,

[23] Q. Chen, J. S. Ngui, G. A. Doss, R. W. Wang, X. Cai, F. P. DiNinno, T. A. Blizzard, M. L. Hammond, R. A. Stearns, D. C. Evans, T. A. Baillie, and W. Tang, *Chem. Res. Toxicol.* **15,** 907 (2002).

extracted using Oasis solid-phase extraction cartridges (Waters, Miford, MA), eluted with methanol, and concentrated to a final volume of 250 μl. Aliquots (50 μl) were analyzed directly by LC-MS-MS. LC-MS-MS analysis was performed using a 4.6 $\times$ 150 mm Ultrasphere C-18 column (Beckman, San Ramon, CA) with on-line UV absorbance detection at 280 nm and electrospray MS-MS detection. A Micromass (Manchester, UK) Q-TOF 2 hybrid tandem mass spectrometer equipped with electrospray and a Waters (Milford, MA) 2690 HPLC was used for LC-MS-MS. In all experiments, the electrospray capillary voltage was held at +3.1 kV, and the cone voltage was at 30 V. Source block and drying gas temperatures were maintained at 110 and 330°, respectively. Nitrogen was used as both nebulizing gas and drying gas. Following the mass selection for $[M + H]^+$ and $[M + Na]^+$ ions of the possible adducts by the quadrupole mass analyzer, CID was carried out with argon at the collision energy of 30 eV. The acceleration voltage of Q-TOF was 9.1 kV, and the end plate of the reflectron was 11.2 kV. The TOF cycle time was 43 μs for the mass range of $m/z < 1000$ Dalton. The data acquisition time for one product ion spectrum of each selected precursor was set at 1 s. The voltage applied on the MCP detector was 2200 V. The resolution of the reflectron TOF was approximately 9000 (FWHH). The mobile phase consisted of 25% methanol in 0.5% ammonium acetate (pH 3.5) at a flow rate of 1.0 ml/min for 5 min, which was increased to 35% methanol over 1 min, then to 50% methanol over the next 39 min, and finally increased to 95% methanol over the remaining 10 min. Toremifene-*o*-quinone was shown to react with all four deoxynucleosides. Three adducts were obtained with thymidine, five adducts were detected with deoxyguanosine, two adducts were observed with deoxyadenosine, and eight adducts were obtained with deoxycytosine. The 3,4-dihydroxytoremifene thymidine adducts all gave similar positive ion electrospray mass spectra with protonated molecules at m/z 678 (Fig. 3).

4-OHEN (20 mg, 0.07 m*M*, dissolved in 1 ml methanol) was incubated with dA (0.14 m*M* in 200 μl dimethyl sulfoxide [DMSO]) in pH 7.4 potassium phosphate buffer (25 m*M*, 20 ml) at 37° for 5 h. The solution was incubated for 5 h at 37°, extracted using Oasis solid-phase extraction cartridges, eluted with methanol, and concentrated to a final volume of 250 μl. Aliquots (50 μl) were analyzed directly by LC-MS-MS using the afore mentioned Q-TOF LC-MS-MS and a 2.0 $\times$ 250 mm C_{18} Beckman column. The mobile phase consisted of 5% methanol containing 0.5% ammonium acetate (pH 3.5) for 5 min, a linear gradient to 27% methanol over 15 min, isocratic for 25 min, then a 15-min gradient to 60% methanol, then a 5-min gradient to 90% methanol. The flow rate was 0.2 ml/min.

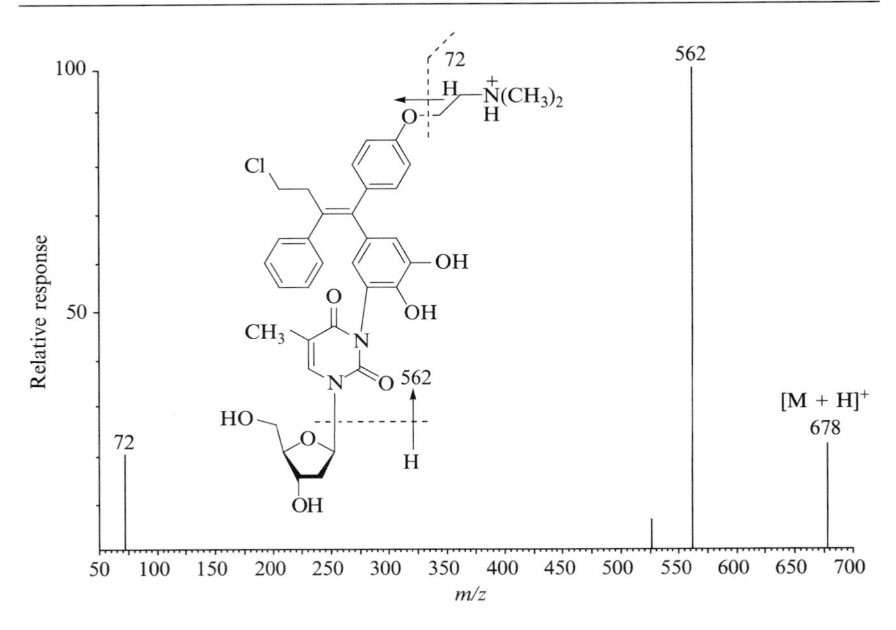

FIG. 3. Positive ion electrospray MS-MS with CID of the protonated molecule of 3,4-di-hydroxytoremifene thymidine adduct at *m/z* 678.

Results and Discussion

Tamoxifen, and its active metabolite, 4-hydroxytamoxifen, are antiestrogens that can be viewed as the prototypical SERMs. However, the search to optimize the mix of estrogenic and antiestrogenic activity continues to generate new SERMs, many of which may be categorized into families, which one might expect to show similar chemical and biological reactivity. As we collect data on the chemical and biological pathways leading to formation of reactive intermediates and the reactions of these intermediates in biological systems, it may be possible to anticipate, predict, and circumvent the formation of cytotoxic and genotoxic metabolites. The experimental methods that we have generally applied to estrogens and antiestrogens are described previously in detail for the specific cases of tamoxifen, toremifene, raloxifene, and 4-hydroxyequilenin.

DNA Adducts Formed from Antiestrogen Quinoids

Oxidation of tamoxifen may generate: (1) a resonance-stabilized carbocation from α-hydroxytamoxifen; (2) quinone methides; and (3) *o*-quinones.[24,25] At least 12 DNA adducts have been detected by ^{32}P-postlabeling in the livers of rats given tamoxifen.[26] One family of DNA adducts is produced by initial P450-catalyzed aromatic hydroxylation of tamoxifen producing 4-hydroxytamoxifen as the proximate carcinogen.[27] This phenol can form an electrophilic quinone methide by a 2-electron oxidation mechanism (Fig. 2). The quinone methides formed from tamoxifen seem to be unusually long lived under physiological conditions likely because of stabilization imposed by the two aryl substituents and the π system of the additional vinyl group.[28] Most quinone methides react instantaneously with GSH, whereas it was found that the rate of reaction of the tamoxifen quinone methide could be measured using a conventional spectrophotometer (Fig. 4).[29] ^{32}P-postlabeling experiments of chemical or microsomal oxidation of 4-hydroxytamoxifen in the presence of DNA gave adducts that cochromatographed with *in vivo* hepatic DNA adducts isolated from mice treated with tamoxifen.[30] Coadministration of tamoxifen with the sulfotransferase inhibitor, pentachlorophenol, enhanced formation of these adducts 11-fold,[31] presumably the result of the greater availability of 4-hydroxytamoxifen, which had not been depleted through sulfate conjugation. The DNA adducts have been fully characterized as those resulting from the reaction of the tamoxifen quinone methide with the exocyclic amino groups of guanine and adenine similar to what was observed with the quinone methide formed from 2-OHE.[21] Unlike the DNA adducts formed from the tamoxifen carbocation, which have been detected in the endometrial tissue of women taking tamoxifen,[32] the tamoxifen quinone methide DNA adducts have not been detected in patients to date. However, a recent study comparing the mutagenic potential of the DNA adducts formed from the tamoxifen quinone methide with those obtained through the carbocation pathway concluded that the former were two-fold to seven-fold more mutagenic despite having a lower level of incidence.[33] Finally, it should be noted that another triphenyl antiestrogen, toremifene,

[24] D. H. Phillips, *Carcinogenesis* **22,** 839 (2001).
[25] J. L. Bolton, *Toxicology* **177,** 55 (2002).
[26] P. Carthew, K. J. Rich, E. A. Martin, F. De Matteis, C. K. Lim, M. M. Manson, M. Festing, N. H. White, and L. L. Smith, *Carcinogenesis* **16,** 1299 (1995).
[27] G. A. Potter, R. McCague, and M. Jarman, *Carcinogenesis* **5,** 439 (1994).
[28] J. L. Bolton, E. Comeau, and V. Vukomanovic, *Chem.-Biol. Interact.* **95,** 279 (1995).
[29] P. Fan, F. Zhang, and J. L. Bolton, *Chem. Res. Toxicol.* **13,** 45 (2000).
[30] B. Moorthy, P. Sriram, D. N. Pathak, W. J. Bodell, and K. Randerath, *Cancer Res.* **56,** 53 (1996).
[31] K. Randerath, B. Moorthy, N. Mabon, and P. Sriram, *Carcinogenesis* **15,** 2087 (1994).
[32] S. Shibutani, N. Suzuki, I. Terashima, S. M. Sugarman, A. P. Grollman, and M. L. Pearl, *Chem. Res. Toxicol.* **12,** 646 (1999).
[33] K. I. McLuckie, M. N. Routledge, K. Brown, M. Gaskell, P. B. Farmer, G. C. Roberts, and E. A. Martin, *Biochemistry* **41,** 8899 (2002).

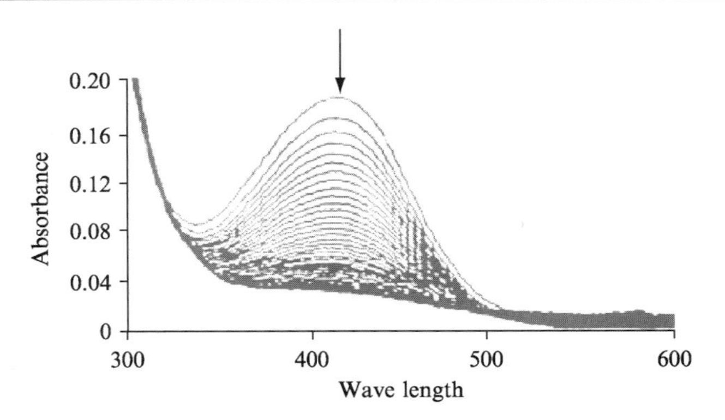

FIG. 4. Kinetic studies on the reaction of the tamoxifen quinone methides with GSH. Incubations contained 4-hydroxytamoxifen quinone methide (0.15 m*M*), GSH (50 m*M*), pH 7.4, 37°. Scans every 10 s.

undergoes similar oxidative metabolism to give 4-hydroxytoremifene, which is further oxidized to the corresponding quinone methide.[29] It is not known whether quinone methide formation from toremifene has any biological relevance *in vivo;* however, the presence of the β-chlorine group on the ethyl side chain does destabilize this quinone methide, which may explain why there have been no reports of endometrial cancers with this SERM.[34]

It has been shown that 4-hydroxytamoxifen and 4-hydroxytoremifene can be further hydroxylated at the 3-position producing the corresponding catechols (Fig. 2).[35] Similarly, another SERM, droloxifene (3-hydroxytamoxifen) can form the same catechol as 4-hydroxytamoxifen. Similar to the catechol estrogens, these SERM catechols are readily oxidized to *o*-quinones by a variety of oxidative enzymes and metal ions, and it is possible that alkylation/oxidation of cellular macromolecules by the corresponding *o*-quinones could contribute to the carcinogenic effects of these SERMs. It has been shown that 4-hydroxytamoxifen–mediated alkylation of microsomal proteins could be inhibited by 17–23% in incubations containing *S*-adenosyl-L-methionine and endogenous catechol *O*-methyltransferase.[36] These data suggest that the *o*-quinone formed from 4-hydroxytamoxifen is at least partially responsible for tamoxifen protein binding. In support of this, the ocular toxicity reported in some patients on high doses of tamoxifen may be caused by tyrosinase-mediated *o*-quinone formation in the cornea.[37] The *o*-quinones formed from 4-hydroxytamoxifen,

[34] A. U. Buzdar and G. N. Hortobagyi, *J. Clin. Oncol.* **16,** 348 (1998).
[35] V. C. Jordan, *Breast Cancer Res. Treatment* **2,** 123 (1982).
[36] S. S. Dehal and D. Kupfer, *Cancer Res.* **56,** 1283 (1996).
[37] S. G. Nayfield and M. B. Gorin, *J. Clin. Oncol.* **14,** 1018 (1996).

droloxifene, and 4-hydroxytoremifene reacted with deoxynucleosides to give corresponding adducts.[14,15] However, the toremifene-*o*-quinone was shown to be considerably more reactive than the tamoxifen-*o*-quinone in terms of both kinetic data as well as the yield and type of deoxynucleoside adducts formed. Because thymidine formed the most abundant adducts with the toremifene-*o*-quinone, sufficient material was obtained for characterization by nuclear magnetic resonance (NMR) and tandem mass spectrometry. The spectral data showed that adduct formation took place between the N^3 ring nitrogen of thymidine and the 5-position on the catechol ring (Fig. 3).

There has been a recent report in the literature of oxidation of the archetypical benzothiophene SERM, raloxifene, mediated by cytochrome P450 3A4.[23] Although the mechanistic details are not certain, the formation and characterization of glutathione conjugates were rationalized by two alternate pathways; one by way of an arene oxide reactive intermediate and the other by way of an extended quinone methide. Given the structural similarity of raloxifene to the SERMs discussed previously, it is far more likely that the mechanism involves quinone methide formation. Interestingly, the enzyme was observed to be irreversibly inhibited during raloxifene bioactivation. The possibility of covalent modification of P450 3A4 by a reactive intermediate emphasizes the potential for oxidative metabolites to modify and interfere with the function of biopolymers, especially DNA. We have recently confirmed quinone methide and *o*-quinone formation from raloxifene in our own laboratories (Fig. 5).[38] Incubation of raloxifene with rat liver microsomes and GSH gave two di- and three mono *o*-quinone GSH conjugates in addition to the previously reported three mono GSH quinone methide conjugates. Three raloxifene catechols were also detected and characterized by LC-MS-MS. These data raise the possibility that this could be a general bioactivation mechanism for other SERMs of similar structure.

Endogenous Estrogen Quinoid DNA Adducts. The endogenous estrogens, estrone and estradiol, are oxidized by cytochrome P450 producing two catechols. Both catechols are further oxidized by virtually any oxidative enzyme or metal ion to give *o*-quinones. The *o*-quinone formed from 2-hydroxyestrone (2-OHE) has a half-life of 47 s, whereas the 4-hydroxyestrone (4-OHE) *o*-quinone is considerably longer lived ($t_{1/2} = 12$ min).[20] In the absence of thiol nucleophiles, both *o*-quinones isomerize to hydroxylated quinone methides, which are much more potent electrophiles than

[38] L. Yu, F. Zhang, D. Nikolic, W. Li, R. B. van Breemen, and J. L. Bolton, *American Chemical Society*, Boston (2002).

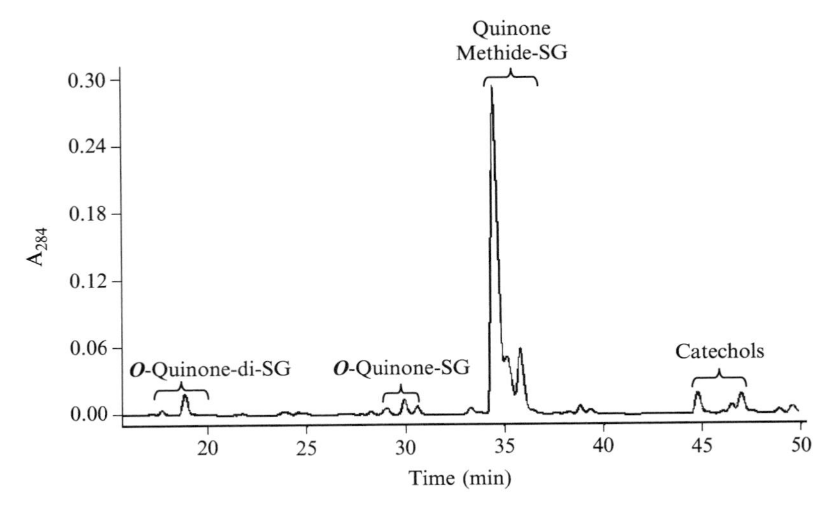

FIG. 5. HPLC chromatogram of raloxifene GSH conjugates. Incubations were conducted for 30 min with 0.05 m*M* substrate and rat liver microsomes (1.0 nmol P450/ml) in the presence of an NADPH-generating system and 0.5 m*M* GSH at 37°. Under these conditions raloxifene elutes at 53.7 min.

o-quinones.[39] The *o*-quinones and/or quinone methides have been shown to cause a variety of DNA damage *in vitro* and in animal models, including alkylation of DNA. Early work by Dwivedy *et al.*[40] showed that the synthetic *o*-quinones or the catechol estrogens in the presence of peroxidase reacted with DNA to give the same DNA adducts as analyzed by ^{32}P-postlabeling. These adducts were later identified as the N^7-guanine adduct from 4-OHE as expected from reaction of guanine with the C^1 position on 4-OHE-*o*-quinone, giving an unstable adduct that readily depurinates.[41] Interestingly, the more unstable 2-OHE-*o*-quinone did not react directly with DNA but rather isomerized to a quinone methide, which alkylated guanine and adenine at the exocyclic amino groups. DNA adducts have been isolated from rats treated with 4-OHE-*o*-quinone. This *o*-quinone was directly injected into the rat mammary gland, the mammary tissue was subjected to Soxhlet extraction, and the extracts were analyzed by HPLC to determine the extent of depurinating adducts formed *in vivo*.[42] The adduct detected was 4-OHE-N^7-guanine, identical to that obtained

[39] J. L. Bolton and L. Shen, *Carcinogenesis* **17,** 925 (1996).

[40] I. Dwivedy, P. Devanesan, P. Cremonesi, E. Rogan, and E. Cavalieri, *Chem. Res. Toxicol.* **5,** 828 (1992).

[41] D. E. Stack, J. Byun, M. L. Gross, E. G. Rogan, and E. L. Cavalieri, *Chem. Res. Toxicol.* **9,** 851 (1996).

from the model studies with 4-OHE-*o*-quinone and dG. DNA also was isolated from the mammary tissue, hydrolyzed to deoxynucleosides, and analyzed by HPLC. No stable adducts were detected under these conditions, which suggests that only apurinic sites would be formed from reaction of 4-OHE-*o*-quinone with DNA *in vivo*. However, more recent work from two separate laboratories has provided evidence for both stable adducts and deglycosylated adducts formed from catechol estrogen *o*-quinones, and therefore formation of stable adducts on DNA, as well as the loss of purines from the DNA strands, could be possible.[43,44] In support of this, 4-hydroxyestradiol (4-OHE_2) and 4-hydroxyestrone (4-OHE) dG adducts have recently been detected in human breast tumor tissue using combined nano LC-nano ES tandem mass spectrometry.[45] Although the adducts were not unambiguously identified, these data suggest that stable adducts and deglycosylated adducts could be formed from the catechol estrogens, either from reaction with *o*-quinones or quinone methides.

DNA strand breaks have also been reported as the result of a synergistic effect between catechol estrogens and NO, possibly by means of generation of the oxidant and nitrating agent, peroxynitrite.[46,47] Because elevation of cellular NO is an estrogenic response and quinoids are among the products of peroxynitrite action on phenols, the interactions between NO, estrogens, and antiestrogens warrant further study.

DNA Adducts Formed from Equine Estrogen Quinoids

As far as equine estrogens present in estrogen replacement formulations such as Premarin are concerned, equilin is metabolized to equal amounts of 2- and 4-hydroxylated catechols, as well as 4-hydroxyequilenin (4-OHEN), in liver microsomes.[48–50] In contrast, equilenin forms

[42] E. L. Cavalieri, D. E. Stack, P. D. Devanesan, R. Todorovic, I. Dwivedy, S. Higginbotham, S. L. Johansson, K. D. Patil, M. L. Gross, J. K. Gooden, R. Ramanathan, R. L. Cerny, and E. G. Rogan, *Proc. Natl. Acad. Sci. USA* **94,** 10937 (1997).

[43] I. Terashima, N. Suzuki, S. Itoh, I. Yoshizawa, and S. Shibutani, *Biochemistry* **37,** 8803 (1998).

[44] O. Convert, C. Van Aerden, L. Debrauwer, E. Rathahao, H. Molines, F. Fournier, J. C. Tabet, and A. Paris, *Chem. Res. Toxicol.* **15,** 754 (2002).

[45] J. Embrechts, F. Lemiere, W. V. Dongen, E. L. Esmans, P. Buytaert, E. van Marck, M. Kockx, and A. Makar, *J. Am. Soc. Mass Spectrom.* **14,** 482 (2003).

[46] Y. Yoshie and H. Ohshima, *Free Radic. Biol. Med.* **24,** 341 (1998).

[47] B. Paquette, A. M. Cantin, S. Kocsis-Bedard, S. Barry, R. Lemay, and J. P. Jay-Gerin, *Chem. Res. Toxicol.* **14,** 547 (2001).

[48] R. H. Purdy, P. H. Moore, M. C. Williams, H. W. Goldzheher, and S. M. Paul, *FEBS Lett.* **138,** 40 (1982).

[49] F. Zhang, Y. Chen, E. Pisha, L. Shen, Y. Xiong, R. B. van Breemen, and J. L. Bolton, *Chem. Res. Toxicol.* **12,** 204 (1999).

[50] J. L. Bolton, M. A. Trush, T. M. Penning, G. Dryhurst, and T. J. Monks, *Chem. Res. Toxicol.* **13,** 135 (2000).

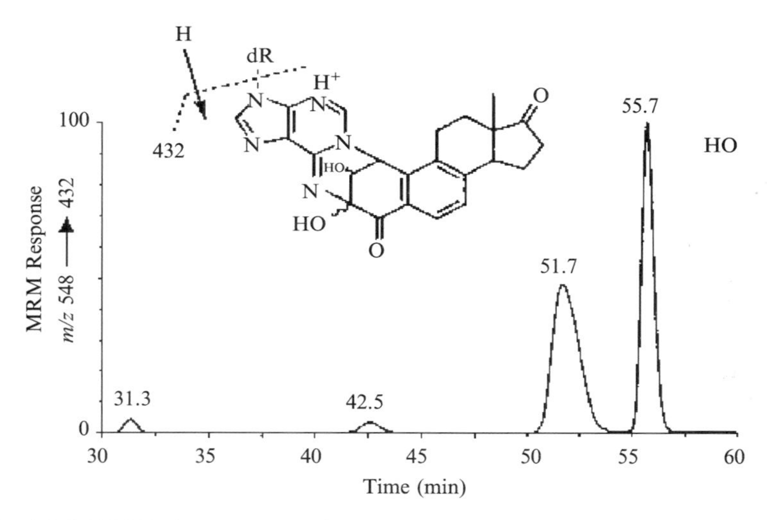

FIG. 6. LC-MS-MS chromatograms showing the elution of 4-OHEN-dA adducts. Product ions of *m/z* 432 are shown that were formed from the protonated molecules of *m/z* 548 during CID with the Q-TOF tandem mass spectrometer.

exclusively 4-hydroxylated metabolites, which could be problematic, because 2-hydroxylation of endogenous estrogens is regarded as a benign metabolic pathway, whereas 4-hydroxylation could lead to carcinogenic metabolites. Interestingly, the 4-hydroxylated equine catechol estrogens [4-OHEN, 4-hydroxyequilin (4-OHEQ)] both autoxidize to *o*-quinones without the need for enzymatic or metal ion catalysis.[18,19] The 4-OHEQ-*o*-quinone readily isomerizes to 4-OHEN-*o*-quinone, and as a result, most of the biological effects caused by catechol metabolites of equilin could be due to 4-OHEN-*o*-quinone formation.[49]

The quinoids of the equilenin metabolite 4-OHEN reacted with 2′-deoxynucleosides generating very unusual cyclic adducts.[11,51–53] For example, the HPLC chromatogram and structure of a 4-OHEN dA adduct is shown in Fig. 6. Because 4-OHEQ is converted to 4-OHEN and 4-OHEN-*o*-quinone, the same 4-OHEN adducts were observed during incubations with 4-OHEQ and deoxynucleosides or DNA.[21] Deoxyguanosine (dG), deoxyadenosine (dA), or deoxycytosine (dC) all gave four

[51] L. Shen, S. Qiu, R. B. van Breemen, F. Zhang, Y. Chen, and J. L. Bolton, *J. Am. Chem. Soc.* **119,** 11126 (1997).

[52] L. Shen, S. Qiu, Y. Chen, F. Zhang, R. B. van Breemen, D. Nikolic, and J. L. Bolton, *Chem. Res. Toxicol.* **11,** 94 (1998).

[53] J. Embrechts, F. Lemiere, W. V. Dongen, and E. L. Esmans, *J. Mass Spectrom.* **36,** 317 (2001).

isomers; however, no products were observed for thymidine under similar physiological conditions, which emphasizes the importance of the exocyclic amino group in adduct formation. An *in vivo* study has recently been completed in which 4-OHEN was injected into the mammary fat pads of Sprague–Dawley rats.[54] Analysis of cells isolated from the mammary tissue for DNA single strand breaks and oxidized bases using the comet assay showed a dose-dependent increase in both types of lesions. In addition, LC-MS-MS analysis of extracted mammary tissue showed the formation of an alkylated depurinating guanine adduct. Finally, extraction of mammary tissue DNA, hydrolysis to deoxynucleosides, and analysis by LC-MS-MS showed the formation of stable cyclic dG and dA adducts as well as oxidized bases. These data showed that 4-OHEN induced four different types of DNA damage that must be repaired by different mechanisms. Finally, Embrecht *et al.*[45] have recently shown that these 4-OHEN DNA adducts could be detected in human breast tissue obtained from patients taking Premarin. These results suggest that 4-OHEN has the potential to be a potent carcinogen through the formation of a variety of DNA lesions *in vivo,* which could produce mutations leading to initiation of the carcinogenic process in the endometrium or the breast.

Concluding Remarks

In conclusion, it has been shown that quinones and quinone methides can be formed from both estrogens and antiestrogens. These reactive species can cause damage in cells either through alkylation of cellular macromolecules and/or through generation of reactive oxygen species. The type of DNA adducts formed vary considerably with the structure and reactivity of the quinoid with some generating unstable adducts, which would readily depurinate, and others forming stable lesions, which would have to be repaired by different mechanisms. Given the link between exposure to estrogens and antiestrogens, metabolism of these compounds, and increased risk of some hormone-dependent cancers, it is crucial that factors that affect the formation, reactivity, and cellular targets of quinoids be thoroughly explored.

Acknowledgments

This work was supported by NIH Grants CA73638 and CA79870.

[54] F. Zhang, S. M. Swanson, R. B. van Breemen, X. Liu, Y. Yang, C. Gu, and J. L. Bolton, *Chem. Res. Toxicol.* **14,** 1654 (2001).

[7] Ubiquinone and Plastoquinone Metabolism in Plants

By EWA SWIEZEWSKA

Introduction

Ubiquinone (UQ) and plastoquinone (PQ) function as carriers of electrons in the mitochondrial inner membrane and chloroplast thylacoids, respectively. In the reduced form, both UQ and PQ have been shown to act as antioxidants[1,2] in plant cells. Plastoquinone is also involved in the chlororespiratory pathway[3] and serves as the effector of carotenoid biosynthesis.[4] In addition, activation of specific kinases was keyed to the redox status of the PQ pool.[5] On the other hand, UQ is the obligatory cofactor of uncoupling proteins[6] (responsible for the dissipation of the mitochondrial chemiosmotic gradient), producing heat rather than ATP; this observation has been made for bovine heart mitochondria, nevertheless, could be possibly generalized for plant uncoupling proteins described recently.[7] Finally, UQ and PQ as the constituents of biological membranes modulate their physicochemical properties. Molecular dynamics simulations of UQ inside a lipid bilayer suggests two preferred positions of the molecule; one close to the phospholipid headgroups, the other in the membrane midplane.[8] It has also been shown that the isoprenoid side chain length rather than the redox state of prenylquinones determines their effectiveness in perturbation of thermotropic properties of the lipid bilayer.[9] As postulated recently, cation leak inhibition resulting in saving of metabolic energy might be attributed to certain isoprenoid compounds (e.g., UQ for mitochondria and PQ for the chloroplasts).[10]

[1] T. Hundall, P. Forsmark-Andree, L. Ernster, and B. Andersson, *Arch. Biochem. Biophys.* **1271,** 195 (1995).
[2] V. N. Popov, A. C. Purvis, V. P. Skulachev, and A. M. Wagner, *Biosci. Rep.* **21,** 369 (2001).
[3] T. S. Feild, L. Nedbal, and D. R. Ort, *Plant Physiol.* **116,** 1209 (1980).
[4] S. R. Norris, T. R. Barrette, and D. DellaPenna, *Plant Cell* **7,** 2139 (1995).
[5] V. Vener, P. J. van Kann, P. R. Rich, I. I. Ohad, and B. Andersson, *Proc. Natl. Acad. Sci. USA* **94,** 1585 (1997).
[6] K. S. Echtay, E. Winkler, and M. Klingenberg, *Nature* **408,** 609 (2000).
[7] M. Laloi, M. Klein, J. W. Riesmeier, B. Muller-Rober, C. Fleury, F. Bouillaud, and D. Ricquier, *Nature* **389,** 135 (1997).
[8] J. A. Soderhall and A. Laaksonen, *J. Phys. Chem. B* **105,** 9308 (2001).
[9] M. Jemiola-Rzeminska, B. Mysliwa-Kurdziel, and K. Strzalka, *Chem. Phys. Lipids* **114,** 169 (2002).
[10] T. H. Haines, *Prog. Lipid. Res.* **40,** 299 (2001).

0076-6879/04 $35.00

Biosynthesis of PQ and UQ has been studied in detail, and sequences of reactions leading from the isoprenoid precursors, namely prenyl diphosphate (solanesyl diphosphate for nine-isoprene-unit long side chain) and aromatic "head" (4-hydroxybenzoate and homogentizate for UQ and PQ, respectively) have been described. Fundamental studies are summarized in this series by J. Soll for PQ[11] and by F. Lutke-Brinkhaus and H. Kleining for UQ.[12]

A few enzymes of both pathways have been characterized at the molecular level. Recently published cloning of solanesyl diphosphate synthase[13] of *Arabidopsis thaliana* provides a good platform for future studies. According to the computer prediction, this enzyme can be transported into chloroplasts and possibly mitochondria, whereas kinetic analysis indicates that geranylgeranyl diphosphate and farnesyl diphosphate are its preferred substrates. Examination of the total spinach microsomal fraction performed earlier[14] reveals that farnesyl diphosphate in contrast to geranyl and geranylgeranyl diphosphates is not accepted as the precursor of the prenylated quinone intermediates. Whether these differences in substrate specificity indicate the existence of solanesyl diphosphate synthase isoenzymes in endoplasmic reticulum (ER) and chloroplasts remains a task for future investigations. The only eukaryotic hydroxybenzoate prenyltransferase purified[15] and cloned[16] so far is geranyl diphosphate:4-hydroxybenzoate geranyltransferase from *Lithospermum erythrorhizon* involved in shikonin biosynthesis; this enzyme is localized at the ER. Corresponding homogentisate phytyltransferase responsible for the biosynthesis of tocopherols (structurally related to PQ) in *A. thaliana* has been cloned[17] recently. Of other enzymes catalyzing the biosynthetic reactions downstream the prenyl group transfer only the product of the *A. thaliana* gene homologous to yeast *COq3* encoding a methyltransferase involved in UQ biosynthesis is characterized[18] and found to be localized within mitochondrial membranes.

Although the structures of intermediates of the UQ and PQ biosynthetic pathways are generally accepted, questions concerning the intracellular localization of the biosynthetic reactions have been raised when UQ was proved to be formed in the ER in addition to mitochondria of the rat

[11] J. Soll, *Meth. Enzymol.* **148,** 383 (1987).
[12] F. Lutke-Brinkhaus and H. Kleining, *Meth. Enzymol.* **148,** 486 (1987).
[13] K. Hirooka, T. Bamba, E. I. Fukusaki, and A. Kobayashi, *Biochem. J.* **370,** 679 (2003).
[14] E. Swiezewska, G. Dallner, B. Andersson, and L. Ernster, *J. Biol. Chem.* **268,** 1494 (1993).
[15] A. Muhlenweg, M. Melzer, S. M. Li, and L. Heide, *Planta* **205,** 407 (1998).
[16] K. Yazaki, M. Kunihisa, T. Fujisaki, and F. Sato, *J. Biol. Chem.* **277,** 6240 (2002).
[17] E. Collakova and D. DellaPenna, *Plant Physiol.* **127,** 1113 (2001).
[18] M. H. Avelange-Macherel and J. Joyard, *Plant J.* **14,** 203 (1998).

liver.[19] In addition, involvement of the same precursor, all-*trans*-prenyl diphosphate, might give additional support to this localization. Assuming the shared localization of both prenyltransferases in the ER, the existence of the efficient and selective transport of the end products, UQ and PQ (or possibly intermediates), from the site of biosynthesis to the site of biological activity has to be accepted. Spinach leaf cells seem to be a good experimental model for these studies, because they accumulate both lipids of interest. Estimation of the content of PQ and UQ in subcellular fractions isolated from spinach leaf cells[14] indicates a high level of prenylated quinones not only, as expected, in chloroplasts (solely PQ) and mitochondria (UQ only) but also in the membranes of the Golgi apparatus. For PQ the ratio chloroplasts/Golgi is 1:1, whereas for UQ it is mitochondria/Golgi equals 10:1, further supporting the concept of the export machinery.

Chemicals

All chemicals are commercially available; however, preparation of tritium-labeled precursors of high specific activity is of crucial importance. Synthesis of (R,S)-[5-^{3}H]mevalonate lactone has been described earlier,[20] and [^{3}H]isopentenyl diphosphate is described in this series.[21] Solanesyl diphosphate is prepared by two-step synthesis performed as described previously with some modifications. After reduction[22] of solanesal with [^{3}H]$NaBH_4$ (60 Ci/mmol or higher, Amersham Biosciences UK Limited, Little Chalfont, UK), the product all-*trans*-[1-^{3}H]solanesol should be carefully purified on a Silica gel column (240–400 mesh, Merck, Darmstadt, Germany), and this is achieved with a narrow long column (0.8×10 cm for 20–50 μmoles of solanesol) and a mild gradient of diethyl ether in hexane (100×100 ml, from 0–20%). To visualize separation of α-*cis* (eluting slightly earlier from the column) and α-*trans* isomers, thin-layer chromatogram (Silica gel plate 60, Merck, Darmstadt, Germany) should be developed in benzene/ethyl acetate 95:5, by volume [1-^{3}H]solanesol thus prepared is phosphorylated according to the method described earlier[23] with minor changes, namely, chloroform is avoided and instead separation of [1-^{3}H]solanesyl monophosphate and diphosphate is performed on DEAE-Sephadex (acetate form, Pharmacia Biotech, Uppsala, Sweden) in methanol. Fractions containing phosphates are pooled, methanol is evaporated,

[19] A. Kalen, B. Norling, E. L. Appelkvist, and G. Dallner, *Biochem. Biophys. Acta* **926,** 70 (1987).
[20] R. K. Keller, *J. Biol. Chem.* **261,** 12053 (1986).
[21] T. Chojnacki, *Meth. Enzymol.* **378,** 152 (2004).
[22] R. W. Keenan and M. Kruczek, *Anal. Bioch.* **69,** 504 (1975).
[23] L. L. Danilov, T. N. Druzhinina, N. A. Kalinchuk, S. S. Maltsev, and V. N. Shibayev, *Chem. Phys. Lipids* **51,** 191 (1989).

and the remaining ammonium acetate is removed after addition of water by means of freeze-drying. The residual white powders of [1-^{3}H]solanesyl diphosphate triammonium salt and [1-^{3}H]solanesyl monophosphate diammonium salt are dissolved in benzene/ethanol 7:3, by volume or *n*-propanol/ammonia/water 6:3:1, by volume. If lower volumes of final solutions are required, the mixture of chloroform/methanol 2:1, by volume, alkalized with ammonia (2%) could also be used; however, prolonged storage might result in the decomposition of the substrate. The high rate of decomposition of [^{3}H]solanesol and especially [^{3}H]solanesyl diphosphate of high specific activity should be taken into consideration, and it is recommended that substrates be used as soon as possible.

Chromatography

Incubation products after extraction are purified on Silica gel column (240–400 mesh, Merck, in Pasteur pipette) eluted with 25% diethyl ether in hexane or chloroform, evaporated with nitrogen, and dissolved in 30 μl of chloroform/methanol (2:1, by volume) before analysis by reversed-phase high-performance liquid chromatography (HPLC)[14] using a C-18 column (Knauer or Hewlett-Packard, ODS-Hypersil 3 μm, 4.5 $\times$ 60 mm). A linear gradient from the initial methanol/water (9:1, by volume) in pump A to methanol/2-propanol (4:1, by volume) in pump B at flow rate 1.5 ml/min, with a program time of 30 or 45 min is applied. The absorbance at 210 nm and radioactivity of the eluate are monitored using an ultraviolet (UV) detector connected to a radioactivity flow detector (Radiomatic Instruments, Tampa, FL) with a cell volume of 0.5 ml. To confirm the identity of labeled prenylated quinones, aliquots of incubation products purified on a Silica column and supplemented with standard of PQ-9 and UQ-9 are subjected to reduction with $NaBH_4$ (saturated ethanolic solution) just before the injection to the HPLC. Thin-layer chromatography (TLC) is performed as described[14] on Silica gel 60 plates (Merck) developed in hexane/diethyl ether (7:3, by volume); lipids are visualized with iodine vapor or autoradiography (in case of labeled compounds).

Enzyme Assay for the Synthesis of Ubiquinone and Plastoquinone

Chloroplasts, mitochondria, total microsomes, and Golgi membranes of spinach leaves are prepared according to the procedures described earlier;[14] chloroplast envelope membranes are prepared as described.[24]

[24] S. Osowska-Rogers, E. Swiezewska, B. Andersson, and G. Dallner, *Biochem. Biophys. Res. Commun.* **205,** 714 (1994).

Formation of prenylated quinones is followed from the step of transfer of the solanesyl group from solanesyl diphosphate to the aromatic precursors. Conversion of thus obtained 3-nonaprenyl-4-hydroxybenzoate (NPHB) toward UQ-9 requires the involvement of three hydroxylations, three methylations (*S*-adenosyl methionine serves as the donor of the methyl group), and decarboxylation. The sequence of reactions leading from prenylated homogentizate to PQ-9 is much shorter, engaging decarboxylation proceeding simultaneously with prenylation[11] and resulting in 3-nonaprenyl-5-methylquinol (NPMQ), and one *S*-adenosylmethionine–dependent methylation. Two types of reaction mixtures are used. When the assay is aimed at testing the activities of prenyltransferases, the incubation mixture described earlier[14,23] contains 200–250 nCi of [^{3}H]solanesyl diphosphate sp. act. 17 Ci/mmol, aromatic precursors (10–50 μM), $MgCl_2$ (10 m*M*), and subfraction equivalent to 0.5–50 μg of protein in 50 m*M* imidazole buffer, pH 7.5. Tested detergents, other divalent cations, cysteine, and KF[11,12,18] are not stimulatory if not inhibitory. Substitution of [^{3}H]solanesyl diphosphate with [^{3}H]isopentenyl diphosphate is also possible. When synthesis of UQ and PQ is monitored, the incubation mixture is supplemented with 10 m*M* adenosine triphosphate (ATP), 10 m*M* phosphoenolpyruvate, 75 units pyruvate kinase, 5 m*M* reduced nicotinamide adenine dinucleotide (NADH), 5 m*M* reduced nicotinamide adenine dinucleotide phosphate (NADPH), and 1 m*M* *S*-adenosylmethionine. Incubation is carried on routinely at 37° for 1 h and stopped by the addition of chloroform/methanol/water to final ratio 1:1:0.3, by volume. After 15–30 min, the extract is adjusted to a final C/M/W ratio of 3:2:1, and the lower phase is evaporated under nitrogen.

Identification of In Vitro*–Labeled Products*

Products observed during PQ and UQ *in vitro* biosynthesis are purified and analyzed as described in the "Chromatography" section. Two major labeled products observed in the applied HPLC system as peak 1 eluting at 3–4 min and peak 2 at 19–20 min are collected (splitting of the eluate is applied) and further characterized on the thin-layer plate (see "Chromatography"). Mobility of compounds confirms the identity of products 1 and 2 as NPHB and NPMQ, respectively. As expected from the biosynthetic route, no other PQ intermediate is observed; surprisingly, the same is found for UQ under the applied experimental conditions. However, when methyl donor, ATP-generating system, required nucleotides, and much higher amounts of protein equivalents are included in the incubation mixture, formation of low amounts of final products is also observed (HPLC and TLC). Reduction of putative labeled quinones with $NaBH_4$ (see

"Chromatography") results in shift of [^{3}H]UQ and [^{3}H]PQ peaks toward the retention time of reduced ubiquinol-9 and plastoquinol-9 standards. It should be underlined that the rate of formation NPHB highly exceeds that of NPMQ, and the final products PQ and UQ are detected only in trace amounts.

Assay for *In Vivo* Localization of Plastoguinone and Ubiquinone Synthesis in Spinach Cells

Leaves of etiolated spinach seedlings are labeled as described[25] with 0.5 mCi [^{3}H]mevalonate (sp. act. 3.5 Ci/mmol, see preceding) in a medium described earlier[25] (5 m*M* KNO_3, 1.5 m*M* $Ca(NO_3)_2$, 1 m*M* $MgSO_4$, 1 m*M* KH_2PO_4, 1 m*M* NH_4Cl, 156 μM EDTA, 72 μM $FeSO_4$, 46 μM H_3BO_3, 6.2 μM $MnSO_4$, 0.8 μM $ZnSO_4$, 0.3 μM $CuSO_4$, 0.7 μM MoO_3, 0.2 μM NH_4VO_3) supplemented with 0.05% Tween 80 in the presence of drugs known to interfere with lipid and protein intercellular transport (i.e, brefeldin A [7.5 μ/ml], colchicine [0.2%], *m*-chlorophenylhydrazone [CCCP, 20 μM], monensine [50 μM]. After labeling, leaves are washed with ice-cold culture medium and homogenized using a mortar and pestle. Obtained homogenate is centrifuged, and two main subfractions, plastid/mitochondrial (10 min, 6000 rpm) and microsomal (15 min, 80,000 rpm), are extracted in chloroform/methanol/water 1:1:0.3, by volume. After 15–30 min, the extract is adjusted to a final CMW ratio of 3:2:1 and lower phase is evaporated under nitrogen.

Identification of In Vivo*–Labeled Products*

Neutral lipid fraction analyzed as described in the "Chromatography" section shows efficient formation of [^{3}H]PQ and [^{3}H]UQ from [^{3}H]mevalonate as judged from comigration of labeled products with standards (HPLC and TLC). Chemical reduction of labeled PQ and UQ (see preceding) is an important part of the identification procedure. Subcellular distribution of *de novo*–labeled PQ and UQ suggests the efficient transport of prenylated quinones from the site of biosynthesis toward their sites of deposition and biological activity. The use of drugs results in partial inhibition of this translocation and allows preliminary characterization of the intracellular transfer mechanism. It seems of importance to use etiolated seedlings to obtain measurable incorporation of the precursor. Nevertheless, extensive biosynthesis of labeled UQ-9 and UQ-10 in contrast to four-times lower rate of formation of labeled PQ is noted. This could be explained by the

[25] M. Wanke, G. Dallner, and E. Swiezewska, *Biochim. Biophys. Acta* **1463,** 188 (2000).

higher content PQ than UQ in the tissue. Yet this observation should also be discussed in the context of the newly described mevalonate-independent pathway (MEP or Rohmer's pathway) of isoprenoid biosynthesis in plants.[26,27] The MEP pathway was found to be operating during the biosynthesis of the side chain of PQ in tobacco cells, whereas UQ was of mevalonate (MEV pathway) origin.[28] On experimental conditions described earlier, the rate of labeling of PQ from mevalonate is lower than that of ubiquinone, however, high enough to consider the mevalonate pathway as significant for PQ biosynthesis, suggesting that this lipid might be derived from both MEV and MEP.

Estimation of Plastoquinone and Ubiquinone Half-Life

Roots are excised from etiolated spinach seedlings grown on vermiculite, and the experimental solution[29] supplemented with 0.5 mCi of [^{3}H]mevalonate, sp. act. app. 3.5 Ci/m*M* is fed in through the cut stems as described earlier.[30] After 24 h of labeling under continuous light, plants are rinsed and placed in a medium devoid of [^{3}H]mevalonate, and incubation is continued for 60 h under the same temperature and light conditions. Plants are harvested at various time points, homogenized, and extracted (see "Chromatography"). The purified neutral lipid fraction is analyzed as described earlier. The linear semilogarithmic plot of the decay of specific radioactivity in [^{3}H]PQ and [^{3}H]UQs observed for approximately 60 h makes the calculation of the half-life possible. Again, the use of etiolated seedlings is important to increase the incorporation of labeled precursor. Possible cross-talk between the MEV and MEP pathway should be taken into consideration as discussed briefly earlier.

Conclusions

Several experimental approaches have been used to study the biosynthesis of UQ and PQ in plants. Comparison of the specific activity of the prenyltransferases responsible for the coupling of prenyl side chain with the aromatic acceptors reveals the highest values for the microsomal/Golgi fraction, hence formation of NPHB is also detected in mitochondria[14] and even chloroplast envelopes.[24] Parallel experiments with metabolic labeling

[26] M. Rohmer, *in* "Comprehensive Natural Product Chemistry" (D. Cane, ed.), Vol. 2, p. 45. Pergamon, 1999.

[27] H. K. Lichtenthaler, *Annu. Rev. Plant Physiol. Plant Mol. Biol.* **50,** 47 (1999).

[28] Disch, A. Hemmerlin, T. J. Bach, and M. Rohmer, *Biochem. J.* **331,** 615 (1998).

[29] P. A. Siegenthaler and F. Depesy, *Eur. J. Biochem.* **61,** 575 (1976).

[30] M. Wanke, E. Swiezewska, and G. Dallner, *Plant Sci.* **154,** 183 (2000).

of spinach seedlings followed by fractionation point out to the same direction, suggesting microsomes/Golgi membranes as the primary site of UQ and PQ biosynthesis, as well as the existence of the machinery responsible for the translocation of both quinones from microsomes to mitochondria and plastids, respectively. Diverse mechanisms of PQ and UQ sorting in a reconstituted cell-free system of membrane fractions isolated from dark-grown spinach seedlings is also suggested in the preliminary studies.[31] Rapid turnover of UQ and PQ[30] explains the requirement for the efficient translocation of both lipids to compensate the catabolic processes, although neither the products nor the mechanisms have been identified yet.

Future experiments are required to propose a complete model of the biosynthesis of prenylated quinones in plants. Methods described here provide the tool for explanation of the localization of the PQ and UQ biosynthetic machinery in the spinach cell. Identification of the ER/Golgi apparatus as the important locus of this process does not exclude involvement of other cellular compartments. The organization of the isoprenoid biosynthesis in plants might be far more complex[32] (e.g., several enzymes have been recently assigned to the liverwort oil bodies[33]).

Acknowledgments

This work was partly supported by the Polish State Committee for Scientific Research (KBN), grants No. 6P04A 077 21 and No. 6P04A025 19.

[31] M. Wanke, E. Swiezewska, and G. Dallner, *Plant Physiol. Biochem.* **39,** 467 (2001).
[32] I. Parmryd and G. Dallner, *Biochem. Soc. Trans.* **24,** 677 (1996).
[33] C. Suire, F. Bouvier, R. A. Backhaus, D. Begu, M. Bonneu, and B. Camara, *Plant Physiol.* **124,** 971 (2000).

[8] Extramitochondrial Reduction of Ubiquinone by Flavoenzymes

By Mikael Björnstedt, Tomas Nordman, and Jerker M. Olsson

Introduction

This chapter summarizes current methods for extramitochondrial regeneration of the reduced antioxidant active form of ubiquinone, ubiquinol, by flavoenzymes belonging to a family of pyridine nucleotide oxidoreductases and studies of these reductions in homogenates and cytosols isolated from cell lines.

0076-6879/04 $35.00

Since the discovery that extramitochondrial ubiquinol possess antioxidant functions, efforts have been made to characterize the physiological enzyme reduction system(s).[1–3] Ubiquinol and/or vitamin E are suggested to inhibit lipid peroxidation by scavenging lipid peroxyl radicals (Fig. 1). Furthermore, the protective effect of ubiquinone has also been proposed to be achieved by scavenge of the perferryl radical, thereby preventing the initiation of lipid peroxidation.[2] These important functions require continuous regeneration of ubiquinol from the oxidized ubiquinone.

For the antioxidant function of ubiquinol and because it is widely spread in all membranes, it is of great importance that the reduced form can be regenerated at all these locations. Many investigators have so far studied the regeneration of ubiquinone, and different quinone reductases have been proposed as reduction enzymes.[4–9]

It was published that cytosolic enzymes could reduce ubiquinone extramitochondrially.[5,6,10] One group described that cytosolic enzymes involved in the reduction are reduced nicotinamide adenine dinucleotide phosphate (NADPH) and NADH dependent and have flavin adeninedinucleotide (FAD) as prosthetic group.[7,8] A decade ago, a microsomal ubiquinone reductase was also proposed but not fully investigated.[4] Other studies suggested a cytosolic NADPH-dependent ubiquinone reductase not inhibited by dicumarol or rotenone, but this reductase has so far not been isolated.[10] Other investigators have proposed that the main extramitochondrial ubiquinone reductase is DT-diaphorase,[5,6,11] which is a homodimeric enzyme with a molecular weight of 55 kDa.[12,13] However, the quinone reductase activity of DT-diaphorase decreases with increasing numbers of

[1] R. E. Beyer, *Free Radic. Biol. Med.* **8,** 545 (1990).

[2] L. Ernster, *in* "Active Oxygens, Lipid Peroxides, and Antioxidants" (K. Yagi, ed.), pp. 1–38. CRC Press, New York, 1993.

[3] B. Frei, M. C. Kim, and B. N. Ames, *Proc. Natl. Acad. Sci. USA* **87,** 4879 (1990).

[4] T. Shigemura, D. Kang, K. Nagata-Kuno, K. Takeshige, and N. Hamasaki, *Biochim. Biophys. Acta* **1141,** 213 (1993).

[5] R. E. Beyer *et al., Proc. Natl. Acad. Sci. USA* **93,** 2528 (1996).

[6] R. E. Beyer *et al., Mol. Aspects Med.* **18,** (1997).

[7] T. Takahashi, T. Yamaguchi, M. Shitashige, T. Okamoto, and T. Kishi, *Biochem. J.* **309,** 883 (1995).

[8] T. Takahashi, T. Okamoto, and T. Kishi, *J. Biochem.* (*Tokyo*) **119,** 256 (1996).

[9] T. Kishi, T. Takahashi, A. Usui, and T. Okamoto, *Biofactors* **10,** 131 (1999).

[10] T. Kishi, T. Takahashi, A. Usui, N. Hashizume, and T. Okamoto, *Biofactors* **9,** 189 (1999).

[11] L. Landi, D. Fiorentini, M. C. Galli, J. Segura-Aguilar, and R. E. Beyer, *Free Radic. Biol. Med.* **22,** 329 (1997).

[12] L. Ernster, *in* "Pathophysiology of Lipid Peroxides and Related Free Radicals" (K. Yagi, ed.), pp. 149–168. Japan Scientific Societies Press, Tokyo, 1998.

[13] C. Lind, E. Cadenas, P. Hochstein, and L. Ernster, *Meth. Enzymol.* **186,** 287 (1990).

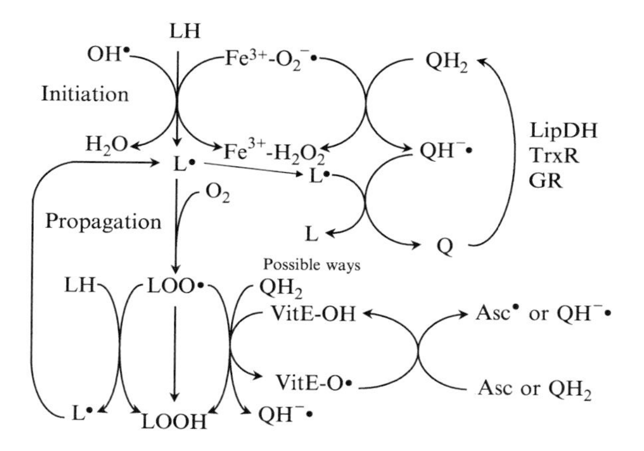

FIG. 1. Schematic illustration of lipidperoxidation. Fe^{3+}-O_2^{-}•, perferryl radical; OH•, hydroxyl radical; L•, carbon centered lipid radical; LH, unsaturated fatty acid; LOO•, lipid peroxyl radical; LOOH, lipid hydroperoxide; QH_2, ubiquinol; QH•, semiquinones; Q, ubiquinone; VitE-OH, vitamin E or α-tocopherol; VitE-O•, α-tocopheryl radical; Asc, ascorbate; Asc•, ascorbate radical; LipDH, lipoamide dehydrogenase; TrxR, thioredoxin reductase; GR, glutathione reductase.

isoprene units in the side chain of quinones and the enzyme has a very low reductase activity, with the mammalian ubiquinones containing the longest side chains (i.e., ubiquinone-9 and 10).[13] The highest rate of extramitochondrial ubiquinone reduction so far described is accomplished by flavoenzymes belonging to a unique family of pyridine nucleotide oxidoreductases.[14–17] These enzymes, lipoamide dehydrogenase (LipDH), mammalian thioredoxin reductase (TrxR-1), and glutathione reductase (GR), share structural and functional similarities.[18] All three enzymes are homodimers with subunits of approximately 55 kDa, including a functional important FAD and a disulfide/dithiol in each subunit. Lipoamide dehydrogenase is mainly known as part of several multienzyme complexes in the inner mitochondrial membrane[19,20] but is also present extramitochondrially.[21] Lipoamide

[14] J. M. Olsson, L. Xia, L. C. Eriksson, and M. Björnstedt, *FEBS Lett.* **448,** 190 (1999).

[15] L. Xia, M. Björnstedt, T. Nordman, L. C. Eriksson, and J. M. Olsson, *Eur. J. Biochem.* **268,** 1486 (2001).

[16] L. Xia *et al., J. Biol. Chem.* **278,** 2141 (2003).

[17] T. Nordman, M. Björnstedt, and J. M. Olsson, submitted 2003.

[18] C. H. J. Williams, *in* "Chemistry and Biochemistry of Flavoenzyme" (F. Müller, ed.), Vol. 3, pp. 121–211. CRC Press, Boca Raton, 1992.

[19] T. Hayakawa, T. Kanzaki, T. Kitamura, Y. Fukuyoshi, Y. Sakurai, K. Koike, T. Suematsu, and M. Koike, *J. Biol. Chem.* **244,** 3660 (1969).

[20] M. Hirashima, T. Hayakawa, and M. Koike, *J. Biol. Chem.* **242,** 902 (1967).

dehydrogenase reduces ubiquinone with either NADH or NADPH as cofactors at similar rates,[14] and the rate of reduction is stimulated by the presence of zinc and at acidic pH.[14,15] The fact that FAD alone chemically reduces ubiquinone in the presence of NADPH, but at a considerably lower rate compared with enzymatic reduction, indicates that the FAD moiety of these enzymes is involved in the reduction.

The mammalian cytosolic TrxR-1 differs from the other members of the flavoenzyme family of pyridine nucleotide oxidoreductases, because this enzyme is a selenoenzyme with very broad substrate specificity.[22] In the penultimate C-terminal part of each subunit, a selenocysteine residue is located, making a functional part with the *N*-terminal cysteines of the other subunit. This selenocysteine residue, forming a selenylsulfide with the adjacent cysteine residues, is essential for activity.[18,23] Thioredoxin reductase-1 reduces ubiquinone with an apparent K_m of 22 μM and a maximal rate of 12 nmol reduced Q10/min and mg protein.[16] The reduction of ubiquinone by TrxR-1 has a sharp physiological pH optimum at pH 7.5, and the rates are similar with either NADPH or, remarkably for this enzyme, NADH as cofactors. The ubiquinone reductase activity of this enzyme is, like all other tested reactions, selenium dependent, because enzyme lacking the penultimate C-terminal selenocysteine residue failed to function as an ubiquinone reductase.[16] The efficacy of TrxR-1 has also been proven in a cellular context using transfected TrxR-1 overexpressing stable cell lines; homogenates and cytosols from the overexpressing cells more efficiently reduced ubiquinone compared with homogenates and cytosols from control cells.[16] Furthermore, the specific activity (i.e., the selenium saturation of the TrxR-1) is most important for the ubiquinone reductase activity. The addition of selenium, in increasing concentrations, to the transfected cells resulted in increasing activity of TrxR-1 accompanied by increasing ubiquinone reductase activity.[16] Taken together, all these data clearly demonstrate mammalian cytosolic TrxR-1 to be the most efficient ubiquinone reductase so far tested, and there is a close connection between the antioxidant ubiquinol and the essential trace element selenium. Previously selenium was suggested to be an important factor for ubiquinone and vitamin E, because several publications show that selenium deficiency is associated with lower concentrations of both vitamin E and ubiquinol.[24–27]

[21] M. J. Danson, *Biochem. Soc. Trans.* **16,** 87 (1988).

[22] E. S. J. Arnér and A. Holmgren, *Eur. J. Biochem.* **267,** 6102 (2000).

[23] L. Zhong, E. S. J. Arnér, and A. Holmgren, *Proc. Natl. Acad. Sci. USA* **97,** 585 (2000).

[24] H. Chen and A. L. Tappel, *Free Radic. Biol. Med.* **18,** 949 (1995).

[25] R. W. Scholz, L. A. Minicucci, and C. C. Reddy, *Biochem. Mol. Biol. Int.* **42,** 997 (1997).

[26] S. Vadhanavikit and H. E. Ganther, *Biochem. Biophys. Res. Commun.* **190,** 921 (1993).

[27] S. Vadhanavikit and H. E. Ganther, *Mol. Aspects Med.* **15,** 103 (1994).

Current knowledge regarding the important role of TrxR-1 in the reduction of ubiquinone and the essential role of selenium offers a mechanistic explanation to these well-known connections.

The third member of the flavoenzyme family of pyridine nucleotide oxidoreductases, GR, also reduces ubiquinone.[17] This reaction is as for lipoamide dehydrogenase also stimulated by zinc and has an acidic pH optimum.

Although current data suggest TrxR-1 to be the predominant physiological ubiquinone reductase, a complex and vital function like the regeneration of ubiquinone surely requires several parallel enzyme systems. Differences in pH optima, subcellular localization, cofactor preference, and preference in the length of the isoprenoid side chain may, on the contrary, offer a multifunctional system to ensure efficient regeneration at many different situations, thereby inhibiting lipid peroxidation, membrane damage, and cell death.

Enzymatic Reduction of Ubiquinone by Flavoenzymes and Cell Extracts

Principle

Commercially available flavoenzymes of the pyridine nucleotide oxidoreductases (i.e., lipoamide dehydrogenase (EC 1.8.1.4), TrxR-1 (EC 1.6.4.5), and GR (EC 1.6.4.2)) are used to characterize the reduction of ubiquinone. Specific thioredoxin reductases lacking the selenocysteine moiety are also used to study the selenium dependency of the reaction by this enzyme (i.e., *Escherichia coli,* mutant human TrxR-1, and truncated rat TrxR-1).[16]

Reagents

Lipoamide Dehydrogenase. 50 m*M* TRIS-HCl, pH 7.45, or TRIS-malate, 0.1%, Triton X-100, 2 m*M* NADH or NADPH, and 50 μM ubiquinone-10 (added in ethanol not exceeding a final concentration of 2% in the reaction mixture).

Thioredoxin Reductase. 50 m*M* TRIS-malate, pH 7.45, 0.4% sodium deoxycholate, 1 m*M* EDTA, 2 m*M* NADH or NADPH, and 50 μM ubiquinone-10 (added in ethanol not exceeding a final concentration of 2% in the reaction mixture).

Glutathione Reductase. 50 m*M* TRIS-HCl, pH 7.45, or citrate, pH 4.5, 0.1% Triton X-100, 2 m*M* NADH or NADPH, and 50 μM ubiquinone-10 (added in ethanol not exceeding a final concentration of 2% in the reaction mixture).

Extracts from Cell Lines. 50 mM TRIS-HCl, pH 7.45, 0.4% sodium deoxycholate, 1 mM EDTA, 2 mM NADPH, and 50 μM ubiquinone-10 (added in ethanol not exceeding a final concentration of 2% in the reaction mixture).

Procedure

Lipoamide Dehydrogenase. To the reaction mixture (100 μl) amounts from 1–90 μg of the enzyme are added. The reaction mixtures are kept dark under nitrogen at 37° during the incubation time continuing from 2.5–30 min. To enhance the activity at physiological pH with NADPH as cofactor, $ZnCl_2$ may be added in concentrations up to 500 μM. The pH could be changed in the interval from 4.5–8.5 using the different buffers. The reaction is terminated by addition of methanol.

Thioredoxin Reductase. To the reaction mixture (100 μl) amounts from 1–10 μg of the enzyme are added. The reaction mixtures are kept dark under nitrogen at 37° during the incubation time continuing from 5–30 min. The optimal reaction rate is at the physiological pH. The reaction is terminated by the addition of methanol.

Glutathione Reductase. To the reaction mixture (100 μl) concentrations from 5–20 μg of the enzyme are added. The reaction mixtures are kept dark under nitrogen at 37° during the incubation time continuing from 2.5–30 min. To enhance the activity at physiological pH with NADPH as cofactor, $ZnCl_2$ is added in concentrations up to 25 μM. The pH could be changed in the interval from 4.5–8.5 using the different buffers. The reaction is terminated by the addition of methanol.

Extracts from Cell Lines. The cells are harvested, and both homogenate and cytosol are prepared according to the following procedures: The homogenates are obtained after repeating freezing/thawing or homogenization in 50 mM TRIS-HCl, pH 7.5, and centrifugation at 10,000g for 10 min at +4°. The cytosol is isolated by sonication in 50 mM phosphate buffer, pH 7.0, and centrifuged at 105,000g for 60 min at +4°. As an example, stable HEK293 cell lines overexpressing Trx-R1 could be used to investigate the rate of reduction of ubiquinone compared with the amount of active TrxR-1 expressed in the cell lines and to study the selenium dependency of this reaction.[16] Homogenates and cytosol from the cell lines are prepared as described earlier, and the TrxR-1 activity is determined.[28] To the reaction mixture (100 μl) 0.3–1.7 mg of homogenate protein and 200 μg of cytosolic protein, respectively, are added to determine the rate of reduction of ubiquinone. The reaction mixtures are kept dark under

[28] A. Holmgren and M. Björnstedt, *Meth. Enzymol.* **252,** 199 (1995).

nitrogen at 37° during the incubation time continuing from 2.5–30 min. The reaction is terminated by the addition of methanol.

Extraction of Ubiquinone-6, Ubiquinone-10, and Ubiquinol-10 from the Reaction Mixture and Cell Extracts

Principle

To be able to calculate the concentrations of the reduced and oxidized forms of ubiquinone-10, 0.68 n*M* of ubiquinone-6 is added as internal standard to the mixture after the reaction has been terminated.

Reagents

Methanol, petroleum ether (40–60°), *n*-hexane, 2-propanol, and ubiqinone-6.

Procedure

The reaction is terminated by addition of 500 μl methanol to the mixture and, then, to achieve a higher degree of hydrophilicity during the extraction, 400 μl of distilled water is added. The ubiquinone-6 is vortexed with this mixture before the addition of 4.5 ml of methanol and 3 ml of petroleum ether. After shaking and centrifugation at 200*g* for 3 min at 25°, the upper phase is removed to a new tube and subsequently evaporated under nitrogen. The residue is dissolved in 25 μl methanol:*n*-hexane: 2-propanol, 2:1:1, and injected directly into the HPLC system. The entire procedure is completed within 10 min.

High-performance Liquid Chromatography Method

Principle

The extracts are analyzed with a reversed-phase column (Hypersil ODS 3-μm column) kept at 14° with a gradient high-performance liquid chromatography (HPLC) system equipped with an ultraviolet (UV) spectrophotometer.

Reagents

Distilled water, methanol, *n*-hexane, and 2-propanol.

Procedure

The A-solvent used contains methanol:water, 9:1, and the B-solvent contains methanol:*n*-hexane:2-propanol, 2:1:1. The gradient starts from 10% B-solvent for 2 min, then a linear gradient from 10–50% of B-solvent for 15 min, followed by 50–55% of B-solvent for 8 min, and, finally, 55–100% of B-solvent for an additional 3 min, all at a flow rate of 1.5 ml/min and at a column temperature of 14°. The absorbance of the eluates are monitored at 292 nm. Ubiquinone-6, ubiquinol-10, and ubiquinone-10 are eluated at 10–11, 19–20, and 25–26 min, respectively.

Concluding Remarks

We described the regeneration of the reduced form of ubiquinone, ubiquinol, by the three well-known enzymes lipoamide dehydrogenase, TrxR-1, and GR, as well as by homogenate and cytosol isolated from cells. Because these enzymes are located at overlapping but also at various different intracellular locations and *in vitro* studies show different enzymatic reaction characteristics in the reduction of ubiquinone, we believe that these reactions are important to protect the cell under various cellular conditions caused by, for example, oxidative stress. Obviously, these protecting mechanisms are correlated with other functions in the cell crucial for cell survival. The discovery that the multifunctional selenoenzyme TrxR-1 reduces ubiquinone by a selenium-dependent mechanism deserves a high degree of attention, because this enzyme connects the antioxidant ubiquinone to the essential trace element selenium. Further studies, including clinical trials, of this connection will be a highly interesting field of future research.

[9] Tissue Bioavailability and Detection of Coenzyme Q

By Igor Rebrin, Sergey Kamzalov, and Rajindar S. Sohal

Introduction

The coenzyme Q (CoQ) or ubiquinone molecule (2,3-dimethoxy-5-methyl-6-multiprenyl-1,4-benzoquinone) consists of a tyrosine-derived quinine ring attached to a polyisoprenoid side chain, which consists of 9 or 10 isoprene units in mammals. Coenzyme Q is synthesized endogenously in all cell types by the mevalonate pathway and has a half-life ranging from 49 h in the thyroid to 125 h in the kidney.[1] Coenzyme Q is localized in the

0076-6879/04 $35.00

middle region of the phospholipid bilayer in cellular membranes. It is believed to be involved in various cellular functions, such as electron transfer and proton transport in the mitochondrial respiratory chain, antioxidative quenching of peroxyl and superoxide anion radicals (O_2^-), and generation of O_2^- by autooxidation. All the various functions of CoQ depend on the capacity of its benzoate ring to assume three alternate redox states: the fully oxidized ubiquinone, the univalently reduced ($1e^- + 1H^+$) ubisemiquinone, and the fully reduced ($2e^- + 2H^+$) ubiquinol. The quinone group of CoQ is responsible for its redox properties, whereas the polyisoprenyl chain allows the insertion and mobility of the molecule within the hydrophobic interior of the phospholipid bilayer of cellular membranes.[2]

Coenzyme Q Content of Different Tissues and Mitochondria in Mouse and Rat

There are large variations in the distribution of CoQ among different cellular organelles, tissues, and species, albeit the functional significance of these variations is presently not well understood. Intracellular CoQ concentration is highest in the lysosomes and Golgi vesicles, followed by microsomes and mitochondria.[1]

The total amounts of CoQ and the percentages constituted by CoQ_9 and CoQ_{10} homologues in various tissues of the mouse[3,4] and rat[5], and their respective mitochondria, are presented in Table I. Total CoQ concentration in homogenates of different murine tissues varied almost 100-fold, with the rank order: kidney > heart > brain > liver > skeletal muscle.[3,4] Among mitochondria from various tissues of mouse, the variation in CoQ content was only about two-fold, with the rank order: heart > kidney > skeletal muscle > brain > liver. Compared with the homogenates, CoQ content ($CoQ_9 + CoQ_{10}$) of mitochondria was 6-fold higher in liver, 3-fold in kidney, 4-fold in the heart, 23-fold in skeletal muscle, and 5-fold in the brain. The percent amounts of CoQ_9 and CoQ_{10} also exhibited variation among different tissues (e.g., in the brain homogenate the CoQ_9/CoQ_{10} ratio was 3:1, whereas it was 16:1 in the liver homogenate). Such ratios in mitochondria tended to be similar to those found in their respective tissue homogenate. There was considerable variation in total CoQ concentration in mitochondria from different tissues. For instance, mitochondria from the

[1] G. Dallner and P. J. Sindelar, *Free Radic. Biol. Med.* **29,** 285 (2000).
[2] L. Ernster and G. Dallner, *Biochim. Biophys. Acta* **1271,** 195 (1995).
[3] A. Lass, M. J. Forster, and R. S. Sohal, *Free Radic. Biol. Med.* **26,** 1375 (1999).
[4] A. Lass and R. S. Sohal, *FASEB J.* **14,** 87 (2000).
[5] L. K. Kwong *et al., Free Radic. Biol. Med.* **33,** 627 (2002).

TABLE I
DISTRIBUTION OF CoQ IN TISSUES AND MITOCHONDRIA OF MOUSE AND RAT

Tissue	Species					
	Mouse			Rat[a]		
	$CoQ_9 + CoQ_{10}$ (pmol/mg protein)	CoQ_9 (%)	CoQ_{10} (%)	$CoQ_9 + CoQ_{10}$ (pmol/mg protein)	CoQ_9 (%)	CoQ_{10} (%)
Plasma	5.2 ± 0.8	60 ± 13	40 ± 7	4.1 ± 0.1	84 ± 7	16 ± 1
Liver	162 ± 32	94 ± 10	6 ± 0.1	296 ± 22	91 ± 8	9 ± 1
Homogenate						
Mitochondria	1307 ± 16	97 ± 0.1	3 ± 0.1	1140 ± 40	95 ± 3	5 ± 1
Kidney	1560 ± 115	88 ± 4	12 ± 1	1790 ± 200	82 ± 10	8 ± 3
Homogenate						
Mitochondria	4062 ± 111	89 ± 3	11 ± 1	4167 ± 70	89 ± 3	11 ± 2
Heart	1190 ± 38	93 ± 3	7 ± 1	1160 ± 40	86 ± 5	14 ± 3
Homogenate						
Mitochondria	4360 ± 166	91 ± 3	9 ± 1	3710 ± 90	87 ± 2	13 ± 1
Skeletal muscle	120 ± 5	83 ± 3	17 ± 1	132 ± 15	92 ± 9	8 ± 2
Homogenate						
Mitochondria	2830 ± 137	93 ± 4	7 ± 1	2710 ± 98	91 ± 4	9 ± 1
Brain	270 ± 26	75 ± 11	25 ± 2	335 ± 4	76 ± 8	24 ± 5
Homogenate						
Mitochondria	1573 ± 156	77 ± 8	23 ± 3	1730 ± 80	77 ± 3	23 ± 2

[a] Data are from L. K. Kwong *et al.*, *Free Radic. Biol. Med.* **33,** 627 (2002).

mouse heart contained 3.6-, 3.3-, 2.7-, and 1.5-times higher CoQ content than those from the kidney, liver, brain, and skeletal muscle, respectively.

Uptake of Supplemental Coenzyme Q_{10}

Results of the various studies on the uptake of exogenous CoQ by different organs of laboratory animals have been inconsistent. Initial studies suggested that CoQ administration in rats and mice resulted in a restricted uptake of CoQ into plasma, liver, and spleen, and their mitochondria, but not in other tissues, such as heart, skeletal muscles, kidney, and brain.[6] However, others have reported an elevation of the CoQ level in heart and skeletal muscle[3–5] and in brain cortex homogenates[7] after long-term supplementation in rats and mice, respectively. In general, the total CoQ concentration in mitochondria is ~1.5- to 5-fold higher than in tissue homogenates. In our experience, a more sensitive and accurate assessment of the uptake of exogenous CoQ is its concentration in mitochondria. In fact, measurements of CoQ solely in the homogenates may be inadequate to validate or refute the uptake of exogenous CoQ into the tissues. For instance, following 13 weeks of CoQ administration, no increase in CoQ concentration was observed in tissue homogenates of mice, except in the liver; however, significant increases were noted in mitochondria in heart, liver, and kidney.[3] Uptake of dietary CoQ does not seem to affect the synthesis of endogenous CoQ, suggesting the absence of feedback inhibition.[8] Additional studies have suggested that dosage and duration of intake are major factors determining the uptake of exogenous CoQ.[5]

Variations in the Distribution of Coenzyme Q Homologues in Mitochondria of Different Species

Comparisons of concentrations of CoQ_9 and CoQ_{10}, extracted from cardiac mitochondria from nine different mammalian species, namely, mouse, rat, guinea pig, rabbit, pig, goat, sheep, cow, and horse, indicated that the total, as well as the relative concentrations of, CoQ_9 and CoQ_{10} tends to vary among different species (Table II).[9] The total concentration of mitochondrial CoQ ($CoQ_9 + CoQ_{10}$) varied about two-fold in different species, with the rank order: horse = mouse = cow = sheep = goat > rat > pig = rabbit > guinea pig. Although cardiac mitochondria from all nine species

[6] Y. Zhang *et al., J. Nutr.* **126,** 2089 (1996).
[7] R. T. Matthews *et al., Proc. Natl. Acad. Sci. USA* **95,** 8892 (1998).
[8] Y. Zhang *et al., J. Nutr.* **125,** 446 (1995).
[9] A. Lass, S. Agarwal, and R. S. Sohal, *J. Biol. Chem.* **272,** 19199 (1997).

TABLE II

COMPARISON OF CoQ_9, CoQ_{10} CONTENT OF HEART MITOCHONDRIA, AND SUPEROXIDE RADICAL GENERATION BY SUBMITOCHONDRIAL PARTICLES BETWEEN NINE DIFFERENT MAMMALIAN SPECIES (VALUES ARE MEAN ± SE OF THREE DETERMINATIONS)[a]

Species	$CoQ_9 + CoQ_{10}$ (nmol/mg protein)	CoQ_9 (%)	CoQ_{10} (%)	Rate of O_2^- generation (nmol/mg protein/min)
Mouse	6.7 ± 0.1	89.3	11.7	2.8 ± 0.2
Rat	5.8 ± 0.1	89.2	11.8	2.4 ± 0.1
Guinea pig	3.6 ± 0.1	52.8	47.2	1.5 ± 0.1
Rabbit	4.9 ± 0.1	2.3	97.7	1.2 ± 0.1
Pig	5.0 ± 0.1	3.8	96.2	1.1 ± 0.1
Goat	6.4 ± 0.1	3.9	96.1	0.29 ± 0.04
Sheep	6.4 ± 0.1	2.3	97.7	0.43 ± 0.05
Cow	6.6 ± 0.2	1.7	98.3	0.72 ± 0.05
Horse	6.8 ± 0.1	2.5	97.5	0.32 ± 0.03

[a] Data are taken from A. Lass, S. Agarwal, and R. S. Sohal, *J. Biol. Chem.* **272,** 19199 (1997).

contained both CoQ_9 and CoQ_{10}, the ratio of CoQ_{10}/CoQ_9 varied up to 60-fold. In species such as the mouse and the rat, almost 90% of total mitochondrial CoQ was CoQ_9, whereas in the guinea pig, CoQ_9 and CoQ_{10} were present in roughly equal amounts. In mitochondria from the rabbit, pig, goat, sheep, cow, and horse, CoQ_{10} was the predominant form, whereas CoQ_9 constituted ~1.3–4% of the total CoQ content.

Relationship Between Coenzyme Q and Superoxide Anion Radical Generation

If the amounts of CoQ_9 and CoQ_{10} are plotted against the rate of superoxide anion radical (O_2^-) generation by submitochondrial particles (SMPs), CoQ_9 is found to be directly related and CoQ_{10} inversely related with the rate of O_2^- generation in different species; however, there was no correlation between the total amount of CoQ and the rate of O_2^- generation.[9] To determine whether these correlations between CoQ homologues and the rate of O_2^- generation were coincidental or causal, ox heart SMPs were experimentally depleted of their native CoQ and reconstituted with either CoQ_9 or CoQ_{10}.[9] After seven serial extractions with pentane, when the CoQ levels were below the detection threshold of 0.2 μM (i.e., 15 p*m*/mg SMP protein), the SMPs still exhibited succinate oxidation activity and were able to generate O_2^-, albeit at rates lower than the control, unprocessed SMPs. Augmentation of SMPs with relatively low amounts of CoQ_9 or CoQ_{10} caused a sharp increase in both the rate of oxygen consumption and O_2^- generation, but at higher concentrations of CoQ_9 or CoQ_{10}, these rates leveled off. If the SMPs were reconstituted with an amount of CoQ that equaled the in vivo concentration, there were no differences in the rate of O_2^- generation between SMPs reconstituted with CoQ_9 or CoQ_{10}. However, above physiological levels, the rate of O_2^- generation was up to 35% greater in the SMPs reconstituted with CoQ_9. CoQ_9 is, therefore, more susceptible to autooxidation than CoQ_{10}.

Another factor that may also affect mitochondrial O_2^- generation is the amount of CoQ associated with mitochondrial membrane proteins. In micelles of cardiac mitochondria from different mammalian species prepared by use of the detergent deoxycholate, the amount of CoQ bound to proteins in micelles was found to be higher in species in which CoQ_{10} was the predominant CoQ homologue.[10] Furthermore, the amounts of bound CoQ were found to be inversely related with the rates of mitochondrial O_2^- generation among these species.

Detection of Coenzyme Q

Analytical procedures for the measurement of CoQ, involving column chromatography, paper chromatography, and thin-layer chromatography, which were used in earlier studies, have in general been found to have unsatisfactory sensitivity and/or inadequate specificity and/or low reproducibility.[11] More recently (since 1971), procedures that use high-performance liquid chromatography (HPLC), in combination with ultraviolet (UV) absorbance, fluorescence detection, and electrochemical detection, have been used for measurement of CoQ content. Ultraviolet detection does not seem to provide sufficient sensitivity, whereas fluorescent detection requires additional complicated derivatization procedures. Gas chromatography has been used successfully, but its usefulness is limited only to vitamin E detection. Coulometric electrochemical detection is currently the most sensitive approach available and also offers selective detection of tocopherols on the basis of their specific redox properties.

Reagents and Materials

Vitamin E, CoQ_9, and CoQ_{10} can be obtained from Sigma Chemical Co. (St Louis, MO). Ethanol, methanol, and hexane can be procured from EM Science (Gibbstown, NJ). Deionized water is prepared using a

[10] A. Lass and R. S. Sohal, *Free Radic. Biol. Med.* **27,** 220 (1999).

[11] A. J. Sheppard *et al., in* "Vitamin E in Health and Disease" (L. Parker and J. Fuchs, eds.). Dekker, New York, 1993.

Millipore Milli-Q System. All other chemicals are HPLC grade or of the highest purity available. All solvents for HPLC are filtered using 0.2-μm filter and finally degassed before use.

Preparation of Homogenates and Plasma Samples and Isolation of Mitochondria

The organs of the animals are quickly removed and placed in ice-cold 50 m*M* potassium phosphate buffer, pH 7.4, containing 2 m*M* ethylene diamine tetra acetic acid (EDTA) and 0.1 m*M* butylated hydroxytoluene. All subsequent procedures are performed at 4°. Tissues are usually homogenized in 10 volumes (w/v) of the indicated tissue-specific isolation buffer: 0.25 *M* sucrose, 3 m*M* EDTA, 10 m*M* TRIS buffer, pH 7.4, for liver tissue; 0.22 *M* mannitol, 70 m*M* sucrose, 10 m*M* ethyleneglycol-bis-(β-aminoethylene ether)-*N,N,N′,N′*-tetracetic acid (EGTA), 2 m*M* HEPES, pH 7.4, for kidney; 0.3 *M* sucrose, 0.03 *M* nicotinamide, 0.02 *M* EDTA, pH 7.4, for heart; 0.32 *M* sucrose, 1 m*M* EDTA, 10 m*M* TRIS, pH 7.4, for brain. Isolation of mitochondria can be conducted by standard procedures of differential centrifugation.[12]

Blood plasma is obtained from EDTA-treated blood (~50 μl of 100 m*M* EDTA per each milliliter of blood) after separation from cells by centrifugation for 2 min at 2000*g*. Aliquots of tissue homogenates, mitochondria, and plasma for determination of CoQ may be stored at −80° until analysis.

Extraction of Coenzyme Q from Tissue Homogenates, Mitochondria, and Plasma

The first step in sample preparation for CoQ and/or vitamin E analysis is the extraction in ethanol/hexane solution. Ten to one hundred microliters of the sample is added to 750 μl of hexane/ethanol (5:2 v/v) solution containing 10 μl of 10% (w/v) EDTA. The mixture is vortexed vigorously for 1 min and centrifuged for 3 min at 4000*g*. Remove 400 μl of the top hexane layer and dry it under a stream of nitrogen. Dissolve the remaining residue in 50–100 μl of ethanol and inject 5–20 μl onto the HPLC column.

[12] L. H. Lash and D. P. Jones, "Mitochondrial Dysfunction." Academic Press, San Diego, 1993.

Preparation of Standards

Standards of vitamin E, CoQ_9, and CoQ_{10} are prepared by dissolving in ethanol. Ubiquinol 9 and ubiquinol 10 are prepared by reduction of the corresponding ubiquinones with sodium borohydride as follows: dissolve 0.25 mg of ubiquinone in 1 ml of ethanol; add 3 ml of water and 100 mg sodium borohydryde; vortex the mixture for 3 min and incubate at room temperature in the dark for 30 min. Extract ubiquinol with 4 ml of hexane. Centrifuge for 3 min at 4000*g* to separate the hexane phase. Carefully remove the top hexane layer and dry it under a stream of nitrogen. Dissolve the remaining residue in 1–2 ml of ethanol to yield the quinol stock solution. The concentration of each standard in stock solution is measured spectrophotometrically in a 1-cm quartz cuvette and calculated using the molar extinction coefficients $E_{290} = 4108$ ($CoQH_9$), $E_{290} = 4003$ ($CoQH_{10}$), $E_{275} = 14700$ (CoQ_9), and $E_{275} = 14600$ (CoQ_{10}).

Procedure of High-Performance Liquid Chromatography Separation and Coulometric Detection of Coenzyme Q

The following method allows the simultaneous determination of vitamin E, CoQ_9, and CoQ_{10} by use of isocratic HPLC and electrochemical detection. An HPLC unit equipped with solvent delivery system (Waters) and a reverse-phase C18 column (5 μm; 4.6 × 250 mm) obtained from Supelco Inc. (Bellefonte, PA) constitute the basic system. The mobile phase for isocratic elution consists of 0.7% $NaClO_4$ in ethanol methanol 70% $HClO_4$ (900:100:1, v/v/v). The flow rate is set at 1.0 ml/min. Under these conditions, the separation of vitamin E, CoQ_9, and CoQ_{10} is completed in 15 min; CoQ_{10} is the last eluting peak, with a retention time of 11 min. Samples are injected directly onto the column using an autosampler (Waters). Calibration standards of CoQ are prepared by dilution of stock solutions of CoQ in ethanol and injected at intervals to ensure uniform standardization. Each sample is injected twice, and the average of the peak areas is used for calculations of the CoQ concentrations. Comparison of the peak areas is made with the standard solutions of known concentrations. Following HPLC separation, CoQ is detected with a model Coulochem II electrochemical detector (ESA, Inc., Chelmsford, MA). The electrode potential settings of the detector are guard cell (upstream of the injector), +200 mV; conditioning cell (downstream of the column), −550 mV; analytical cell, +175 mV.

Protein content is determined by bicinchronic acid (BCA) protein assay according to the manufacturer's instructions (Pierce, Rockford, IL).

[10] Coenzyme Q and Vitamin E Interactions

By RAJINDAR S. SOHAL

Introduction

Coenzyme Q or ubiquinone (CoQ) and vitamin E or α-tocopheral are lipoidal substances, present within the hydrophobic domain of the phospholipid bilayer of cellular membranes. The CoQ molecule consists of a tyrosine-derived benzoate ring attached to a polyisoprenyl chain of 9 or 10 units in mammalian species. CoQ exists in three alternate redox states: fully oxidized or ubiquinone (Q); partially reduced, ubisemiquinone (•QH), which is a free radical; and fully reduced or ubiquinol (QH_2). Coenzyme Q plays several functional roles in cellular processes, among which the most well recognized is the translocation of protons across the inner mitochondrial membrane and the transfer of electrons from oxidoreductases, such as NADH-Q-, succinate-Q-, α-glycerophosphate-Q-, and dihydroorotate-Q-oxidoreductase, to Q-cytochrome *c* oxidoreductase in the mitochondrial electron transport chain.[1]

The α-tocopherol molecule consists of a hydroxyl-containing chromanol ring and a hydrophobic phytol chain. α-Tocopherol has been unambiguously demonstrated to have an antioxidative capability particularly for scavenging the lipid peroxyl radicals, thereby preventing the propagation of chain reactions during lipid peroxidation[2,3]:

$$k_1 = 3.3 \times 10^6\ M^{-1}\ \mathrm{s}^{-1} \qquad (1)$$

Coenzyme Q has also been shown to be an antioxidant in soluble systems and to inhibit lipid peroxidation in mitochondrial membranes from tissues such as liver, heart, and brain,[4,5] as well as in submitochondrial particles (SMPs) that have been depleted of native α-tocopherol.[6] However, in solutions,[7] the reactivity of ubiquinone toward peroxyl radicals [Eq. (2)] is much lower than that between ubiquinol and peroxyl [Eq. (3)].

[1] D. E. Nicholls and S. J. Ferguson, "Bioenergetics 3." Academic Press, London, 2002.
[2] E. Niki, *Chem. Phys. Lipids* **44,** 227 (1987).
[3] P. B. McCay, *Annu. Rev. Nutr.* **5,** 323 (1985).
[4] A. Mellors and A. L. Tappell, *J. Biol. Chem.* **214,** 4353 (1980).
[5] R. Takayanagi, K. Takeshige, and P. Minakami, *Biochem. J.* **192,** 853 (1980).
[6] P. Forsmark, F. Aberg, B. Norling, K. Nordenbrand, G. Dallner, and L. Ernster, *FEBS Lett.* **285,** 39 (1991).

METHODS IN ENZYMOLOGY, VOL. 378
0076-6879/04 $35.00

$$k_1 = 0.33 \times 10^3 \; M^{-1} \; s^{-1} \tag{2}$$

$$k_1 = 3.4 \times 10^5 \; M^{-1} \; s^{-1} \tag{3}$$

Electron spin resonance studies have shown that in homogeneous solutions,[8] as well as cellular membranes,[9] the presence of CoQ can prevent the oxidation of α-tocopherol. Furthermore, addition of NADH or succinate to SMPs can bring about the reduction of α-tocopheroxyl radicals, which suggests that ubiquinols may be involved in the recycling of α-tocopherol.[10] However, what remains ambiguous on the basis of such studies is whether α-tocopherol or CoQ is the primary radical scavenger in the inner mitochondrial membrane.

Interaction Between Coenzyme Q and α-Tocopherol in the Inner Mitochondrial Membrane

The nature of the interaction between CoQ and α-tocopherol in the inner mitochondrial membrane was further investigated[11] by comparing the autoxidizability of cardiac mitochondria from the rat and the ox. These two species contain similar levels of CoQ (CoQ_9 + CoQ_{10}), but the rat has a 15-fold higher concentration of α-tocopherol than the ox.[12] It was thought that specific antioxidative roles of α-tocopherol and CoQ in the attenuation of mitochondrial oxidative damage should be evident because of such a wide difference in their α-tocopherol concentrations. Mitochondria from the two species were incubated in the presence or absence of succinate for 6 h at 37°, and the kinetics of autoxidation were monitored as the rate of oxygen consumption and formation of thiobarbituric acid–reactive substances (TBARS) and protein carbonyls. The antioxidative roles of CoQ and α-tocopherol were followed by determining the amounts of reduced (ubiquinol) and oxidized (ubiquinone) forms of CoQ and the concentration of α-tocopherol.

In the ox heart mitochondria, which have a relatively low concentration of α-tocopherol, there was no loss in the total amount of CoQ (reduced plus oxidized forms) during *in vitro* autoxidation irrespective of the

[7] V. V. Naumov and N. G. Khrapova, *Biophysics* **28,** 774 (1983).

[8] D. A. Stoyanovsky, A. N. Osipov, P. J. Quinn, and V. E. Kagan, *Arch. Biochem. Biophys.* **323,** 343 (1995).

[9] V. Kagan, E. Serbinova, and L. Packer, *Biochem. Biophy. Res. Commun.* **169,** 851 (1990).

[10] J. J. Maguire, V. Kagan, B. A. Ackrell, E. Serbovina, and L. Packer, *Arch. Biochem. Biophys.* **292,** 47 (1992).

[11] A. Lass and R. S. Sohal, *Arch. Biochem. Biophys.* **352,** 229 (1998).

[12] R. S. Sohal, A. Lass, L.-J. Yan, and L. K. Kwong, *in* "Understanding the Process of Aging" (E. Cadenas and L. Packer, eds.), p. 119. Marcel Dekker, Inc., New York, 1999.

presence or absence of succinate. (α-Tocopherol was not monitored because of its very low concentration in the ox.) In contrast, in the rat heart mitochondria, there was a depletion of both ubiquinol and α-tocopherol during mitochondrial autoxidation in the absence of succinate. However, in the presence of succinate, where CoQ is mainly present as ubiquinol, there was no loss of α-tocopherol, but ubiquinol showed a time-associated depletion. The reason why the difference in rate of formation of autoxidation products, such as TBARS and protein carbonyls, between the rat and ox heart mitochondria was relatively minor, even though the latter contained 15-fold less α-tocopherol, seems to be that the rates of oxygen consumption and $O_2^{\bullet}$ and H_2O_2 generation are approximately 5-fold higher in the rat than in the ox,[13] which is suggestive of a relatively lower level of oxidative stress in the ox. Indeed, a comparison among various mammalian species indicates that the rate of mitochondrial $O_2^{\bullet}$ generation is inversely related with the mitochondrial α-tocopherol content.[12] Apparently, relatively low rates of $O_2^{\bullet}$ generation, such as those in the ox, lead to the formation of correspondingly low levels of peroxyl radicals, thereby decreasing the amount of mitochondrial α-tocopherol required to prevent lipid peroxidation.

Because under similar conditions of incubation, ubiquinol is depleted in the rat heart mitochondria, which have a relatively high concentration of α-tocopherol, but not in ox heart mitochondria, which have a relatively low amount of α-tocopherol, it was suggested that the loss of ubiquinol in the rat heart was due to its interaction with α-tocopherol. This interpretation was supported by studies on bovine heart SMPs augmented with varying amounts of exogenous α-tocopherol.[11] In the absence of succinate, the formation of TBARS was highest in the control ox heart SMPs that were not supplemented with exogenous α-tocopherol. Supplementation of SMPs with different amounts of α-tocopherol was found to retard the formation of TBARS in a dose-dependent manner. However, α-tocopherol was lost during this process in direct relationship to the supplemented amount and in inverse relationship to the inhibition of TBARS formation. Irrespective of the α-tocopherol content, CoQ occurred only in the oxidized form, which remained constant during the period of incubation.

In the presence of succinate, TBARS formation in the incubated ox heart SMPs was three times lower than that observed in the absence of succinate. Supplementation of SMPs with even a relatively small amount of α-tocopherol was found to be sufficient to prevent TBARS formation during the 6-h incubation period without incurring any loss of α-tocopherol. At the end of the incubation period, loss of ubiquinol was highest in the SMPs containing the lowest amounts of α-tocopherol and vice

[13] H.-H. Ku, U. T. Brunk, and R. S. Sohal, *Free Radic. Biol. Med.* **15,** 621 (1993).

versa. Collectively, these data supported the concept that the antioxidative role of ubiquinol in mitochondria is due to its interaction with α-tocopherol rather than its ability to directly scavenge the peroxyl radicals.

The recycling of α-tocopherol from tocopheroxyl radical is apparently dependent on the activity of mitochondrial succinate dehydrogenase, which provides the reducing equivalents for the conversion of ubiquinone to ubiquinols. Up to 80% of ubiquinone reduction is ascribable to succinate and 10% to NADH. The dependence of CoQ reduction on succinate-supported mitochondrial respiratory activity was demonstrated on the basis of the effect of 4,4,4-trifluoro-1-(2-thienyl)-1,3-butadione, a specific inhibitor of succinate dehydrogenase, which abolished the activity of succinate-CoQ oxidoreductase.[11]

An additional factor that suggests that α-tocopherol rather than ubiquinol acts as the direct scavenger of peroxyl radicals is that ubiquinol is a highly hydrophobic molecule, located in the middle of the phospholipid bilayer, and has considerably less intramembrane mobility than α-tocopherol,[14] which would tend to decrease its radical scavenging potential of the former. Considered together with the fact that the reactivity of α-tocopherol with peroxyl radical also far exceeds that of ubiquinol with peroxyl,[15] it can be proposed that CoQ is unlikely to be a direct scavenger of peroxyl radical in the mitochondrial inner membrane; instead, it seems to be involved in the regeneration of α-tocopherol from its phenoxyl radical.

To determine whether the "sparing/regeneration" effect of CoQ on α-tocopherol is also operative *in vivo,* mice were administered CoQ_{10} (123 mg/kg/d) or α-tocopherol (200 mg/kg/d) for 13 weeks. Experimental administration of CoQ_{10} and α-tocopherol to mice resulted in a corresponding rise in their amounts in the serum and mitochondria of various tissues.[15] CoQ_{10} intake was found to enhance the level of α-tocopherol in mitochondria, whereas α-tocopherol intake had no impact on CoQ content. These results thus supported the concept that CoQ intake can elevate α-tocopherol content in mitochondria, probably because of a "sparing/regeneration" effect that was originally demonstrated *in vitro.*

Relationship Between Amounts of Coenzyme Q and α-Tocopherol and the Rate of Superoxide Anion Radical Generation

Autoxidation of mitochondrial ubiquinone has been demonstrated to be the predominant source of superoxide anion radical ($O_2^{\bullet}$) generation in cells. The relationship between $O_2^{\bullet}$ production and CoQ content

[14] R. Fato, M. Battino, G. P. Castelli, and G. Lenaz, *FEBS Lett.* **179,** 238 (1985).

[15] A. Lass, M. J. Forster, and R. S. Sohal, *Free Radic. Biol. Med.* **26,** 1375 (1999).

(CoQ_9 plus CoQ_{10}) was examined in SMPs of nonprimate mammalian species such as mouse, rat, guinea pig, rabbit, goat, sheep, pig, ox, and horse, which have different rates of $O_2^{\bullet}$ generation, as well as maximum life spans (MLS). Mitochondrial CoQ_9 content was found to be positively correlated, whereas the CoQ_{10} amount was negatively correlated with the rate of mitochondrial $O_2^{\bullet}$ generation.[16] In turn, the rate of mitochondrial $O_2^{\bullet}$ generation exhibited an inverse relationship with the MLS of the different species. However, the total mitochondrial CoQ ($CoQ_9 + CoQ_{10}$) content was not correlated with the rate of $O_2^{\bullet}$ generation or MLS of the species.

To determine whether the rate of mitochondrial $O_2^{\bullet}$ production depended on the nature of the CoQ homologue, bovine SMPs were depleted of their native CoQ and reconstituted with different amounts of CoQ_9 or CoQ_{10}.[17] Coenzyme Q depletion was found to cause a decrease in $O_2^{\bullet}$ generation. Whereas augmentation of SMPs with CoQ_{10}, above the physiological level, had no effect on the rate of $O_2^{\bullet}$ generation, augmentation of SMPs with CoQ_9 above the normal level caused a small rise in the rate of $O_2^{\bullet}$ production. These results suggested that factors in addition to CoQ concentrations played a role in controlling the rate of mitochondrial $O_2^{\bullet}$ generation.

It was hypothesized that mitochondrial α-tocopherol may be one such factor, because α-tocopherol can react with $O_2^{\bullet}$ to form tocopheroxyl radical, which can in turn react with ubiquinol to regenerate α-tocopherol. Furthermore, as also discussed previously, sparing of α-tocopherol in the mitochondrial membrane is directly dependent on the molar ratio of CoQ and α-tocopherol.

The effects of CoQ_{10} and α-tocopherol on the rate of mitochondrial $O_2^{\bullet}$ generation were examined in skeletal muscle, liver, and kidney of mice whose diet was supplemented with α-tocopherol or CoQ.[17] Intake of α-tocopherol and CoQ caused elevations in mitochondrial α-tocopherol content, albeit to varying degrees in different tissues. In contrast, mitochondrial CoQ content (CoQ_9 and/or CoQ_{10}) could be increased only by the administration of CoQ_{10} and not of α-tocopherol. In control mice, the rates of $O_2^{\bullet}$ generation by SMPs from different tissues exhibited an inverse relationship with the amount of α-tocopherol. In the α-tocopherol–administered mice, the rate of $O_2^{\bullet}$ generation by SMPs was 15–35% lower in different tissues than in the comparable tissues of the controls. The rate of oxygen consumption was unaffected.

To establish directly whether elevation in the amounts of α-tocopherol could specifically cause a decrease in the rate of $O_2^{\bullet}$ generation, bovine

[16] A. Lass, S. Agarwal, and R. S. Sohal, *J. Biol. Chem.* **272,** 19199 (1997).

[17] A. Lass and R. S. Sohal, *FASEB J.* **14,** 87 (2000).

heart SMPs were reconstituted with different amounts of α-tocopherol.[4] Bovine heart SMPs were selected for this purpose, because they have a relatively low concentration of α-tocopherol (15-fold less than in the mouse or rat). Elevation in the α-tocopherol content of SMPs caused an up to 50% decrease in the rate of $O_2^{\bullet}$ generation, whereas the rate of oxygen consumption remained unaltered. Altogether, the results indicated that the elevation of α-tocopherol content of mitochondria *in vitro* or *in vivo* can cause a decrease in the rate of $O_2^{\bullet}$ generation by SMPs. In contrast, CoQ_{10} augmentation *in vitro* or *in vivo* had no effect on the rate of $O_2^{\bullet}$ generation.

Several lines of evidence suggest that α-tocopherol may act as an auxiliary to superoxide dismutase (SOD) in the elimination of $O_2^{\bullet}$ in mitochondria. For instance, it has been demonstrated that α-tocopherol can react directly with $O_2^{\bullet}$ in the liposomal membranes to form tocopheroxyl radical [Eq. (4)], with a rate constant of $4.5 \times 10^3\ M^{-1}\ s^{-1}$. Cadenas *et al.*[18] have shown that chromanoxyl radical of Trolox, a water-soluble analogue of α-tocopherol, can be reduced by $O_2^{\bullet}$ with a rate constant of $4.5 \times 10^8\ M^{-1}\ s^{-1}$. Because this reaction is thermodynamically more favorable than the reaction between $O_2^{\bullet}$ and α-tocopherol, these authors suggested that $O_2^{\bullet}$ might play a role in the reduction of α-tocopheroxyl radical.

$$\alpha - \mathrm{Toc} - \mathrm{O}^{\bullet} + \mathrm{O}_2^{\bullet} \rightarrow \alpha - \mathrm{Toc} - \mathrm{O}^{-} + \mathrm{O}_2 \qquad (4)$$

To summarize, an *in vivo* interaction between CoQ and α-tocopherol is indicated by the increase in α-tocopherol content of mitochondria in response to CoQ intake; however, α-tocopherol intake has no effect on CoQ content. Mitochondrial augmentation with α-tocopherol *in vivo* and *in vitro* results in a decrease in the rate of $O_2^{\bullet}$ generation by SMPs. The view that this relationship may be due to a "sparing/regeneration" effect of CoQ on α-tocopherol is supported by (1) *in vitro* studies showing that CoQ can react with tocopheroxyl radicals to regenerate α-tocopherol, (2) the fact that loss of α-tocopherol in the mitochondrial membranes is progressively stemmed by elevation of the molar ratio of CoQ to α-tocopherol, (3) the fact that the reactivity of ubiquinol with peroxyl radicals is much lower than that of α-tocopherol, and (4) the fact that elevation in CoQ level in mitochondria by dietary supplementation is linked to a corresponding rise in α-tocopherol content.

[18] E. Cadenas, G. Merenyi, and J. Lind, *FEBS Lett.* **253,** 235 (1989).

[11] Preparation of Tritium-Labeled 3-Methyl-3-buten-1[^{3}H]-yl Diphosphate (^{3}H-Isopentenyl Diphosphate)

By TADEUSZ CHOJNACKI

Introduction

Isopentenyl diphosphate is an important substrate in the isoprenoid pathway. The availability of its labeled form is crucial for a biochemical laboratory performing studies on the biosynthesis of a wide range of substances. The most commonly used form is the ^{14}C-labeled compound available (e.g., from The Radiochemical Centre, Amersham, Bucks). The 1-^{14}C-isopentenol serving as the substrate for chemical phosphorylation is prepared in a sequence of steps starting with carbonation with $^{14}CO_2$ methylallylmagnesium chloride and reduction of the resulting 3-methyl[1-^{14}C]butenoic acid to the required alcohol. The [4-^{14}C]isopentenyl diphosphate can be made enzymatically from [2-^{14}C]mevalonic acid available from the preceding source. The details of preparations are stated by Holloway and Popjak.[1] The preparation of tritium-labeled isopentenyl diphosphate is expected to provide the biochemical studies on isoprenoids with a more easily accessible and less-expensive labeled precursor.

The method described here gives a 10–20% yield (of the sodium borotritide used) of highly labeled isopentenyl diphosphate. The product obtained is an alternative to the commercially available ^{14}C-labeled substance. The presented procedure is a result of several trials undertaken to provide the researchers studying biosynthesis of isoprenoids with a simple way to make their own labeled intermediate. It can be performed without special equipment in every biochemical laboratory after a short training period with small amounts (10–100 mg) of unlabeled reagents. Special care will be required to be able to deal with the low amounts of reagents on each step of the procedure. The procedure consists of two parts: preparation of tritium-labeled isopentenol in which advantage is taken of the much higher boiling point of isopentenol (130–132°) and its vapor pressure compared with those of the other solvents used in the procedure (dichloromethane bp, 40° and ethanol bp, 78°). By standardizing the conditions (size of glass tubes, period of time required to remove most of the ethanol and dichloromethane from the sample containing isopentenol) milligram amounts of practically nonvolatile isopentenol can be prepared without a distillation process or an elaborate chromatographic procedure. The isotopically

[1] P. W. Holloway and G. Popjak, *Biochem. J.* **104,** 57 (1967).

0076-6879/04 $35.00

labeled isopentenol is available as a product of a custom synthesis (e.g., Moravek Biochemicals, Brea, CA), but the simplicity of preparing it from generally available sodium borotritide in every biochemical laboratory as described in this chapter can be a convincing argument to make it as described later. Usually, it takes no more than 4–5 days to complete the whole procedure. Following of the course of chemical reactions and of the chromatographic isolation of the labeled substances can well be done with the aid of staining the reaction products with iodine vapors. In a routine preparation with as little as 1–2 mg of labeled isopentenol, the iodine-positive spots on silica gel foil can be made well visible for quantities of approximately 0.05–0.1 μg of stainable substance (corresponding to approximately 2–5 μCi of tritium), provided the diameter of the spots does not exceed 1.0–1.5 mm. In the author's experience, the thin-layer chromatography (TLC) radioactive scanner (Berthold II scanner with G-M tube with methane gas flow (Bundoora, Australia)) was found to be the most convenient for following the course of the synthesis.

The overall reactions of the process are:

1. Oxidation of isopentenol to isopentenal with the aid of pyridinium chlorochromate and isolation of isopentenal by column chromatography on silica gel (or Florisil)[2]
2. Reduction of isopentenal to [1-T] isopentenol with $NaBT_4$[2]
3. Isolation of [1-T] isopentenol
4. Chemical phosphorylation of [1-T] isopentenol with the aid of ditriethylammonium phosphate[3] and trichloroacetonitrile
5. Isolation of isopentenyl diphosphate by column chromatography on a column of silica gel[4]

Reagents

1. Dichloromethane
2. Ditriethylammonium phosphate[3]
3. Ethanol, 99.5%
4. Isopentenol (3-methyl-3-buten-1-ol) (Aldrich Chemical Co., Milwaukee, WI)
5. Pyridinium chlorochromate (Aldrich Chemical Co.)
6. Silica gel 60 (0.040–0.063 mm) for column chromatography (E. Merck, Darmstadt, Germany)
7. Sodium boro[^{3}H]hydride ($NaBT_4$, TRK45, 5–10 Ci/mmole, Amersham, Little Chalfont, UK)

[2] W. L. Adair and S. Robertson, *Biochem. J.* **189,** 441 (1980).
[3] F. Cramer and W. Bohm, *Angew. Chem.* **71,** 775 (1959).
[4] R. H. Cornforth and G. Popjak, *Methods Enzymol.* **15,** 386 (1969).

8. Trichloroacetonitrile
9. TLC plastic sheets, silica gel 60 F_{254} (E. Merck)

1. Isopentenol (0.2 ml) in 2 ml of dichloromethane is mixed with an excess of pyridinium chlorochromate (300 mg) in a glass tube with a magnetic stirrer for 15–30 min. The course of oxidation is followed by TLC in a silica gel plate. Samples (0.5 μl) are spotted on a silica gel sheet and developed for approximately 5 min in dichloromethane up to the length of 4–5 cm. The spots of isopentenol ($R_F = 0.1$) and isopentenol ($R_F = 0.5$) are visible on staining in iodine vapors. Usually in 15–30 min, all isopentenol is converted to isopentenal. The reaction mixture is applied to a column of silica gel (0.8 × 8.0 cm), and 2- to 3-ml fractions are eluted with dichloromethane. The elution is followed by TLC as earlier, and polyprenal-positive fractions are pooled. Usually more than 80% of the expected polyprenal is recovered and stored until the next step in dichloromethane.

2. The glass vial with $NaBT_4$ (usually 100 or 250 mCi) is opened, and the contents on dissolving in approximately 0.1 ml of alkaline ethanol (1 drop of 1 *N* NaOH per 1 ml of 99.5% ethanol) are mixed with 10 molar equivalents of isopentenal (1 molecule of sodium borohydride may react with up to 4 molecules of the aldehyde) in dichloromethane (0.1–0.3 ml). The resulting mixture is left in a capped vessel for 4–5 h. The course of reduction of isopentenal is completed by that time as shown on checking by TLC a 0.5-μl sample of the reaction mixture (TLC on silica gel foil in dichloromethane as previously). The iodine-positive spot of $R_F = 0.1$ will appear, demonstrating the presence of radioactive isopentenol. The presence of radioactivity in this spot can be easily demonstrated with the TLC scanner of radioactivity or by autoradiography. A distinct spot of unreacted aldehyde should also be present.

3. The reaction mixture (solution in dichloromethane/ethanol mixture) containing tritium-labeled isopentenol is applied to a 0.6 × 5.0 column of silica gel in dichloromethane made in a Pasteur pipette. The column is eluted with dichloromethane, and the 0.5- to 1.0-ml fractions are collected and checked by TLC for the presence of labeled isopentenol (0.5-μl fractions are spotted on silica gel foil, developed in dichloromethane, and stained with iodine). In the course of chromatography, the unreacted, nonradioactive isopentenal is eluted first, and the radioactive isopentenol comes next from the column. The fractions containing radioactive isopentenol are pooled and left uncapped, so as the dichloromethane and traces of ethanol are evaporated, the labeled isopentenol is left at the bottom of the tube. The chromatography is repeated two to three times on the successive columns of silica gel as earlier (0.6 × 5.0 cm.) using

dichloromethane for elution. Owing to repeated column chromatography and leaving uncapped the collected fractions containing labeled isopentenol, the dichloromethane solvent together with traces of ethanol can evaporate with the radioactive isopentenol left on the bottom of the tube. When glass tubes of approximately 1 cm in diameter and a height of 10–12 cm are used, usually in 6–10 h at room temperature in the hood, almost all dichloromethane and ethanol can be removed because of the great differences of vapor pressure and boiling temperature among the dichlorometane, ethanol, and isopentenol (bp 40°, 78°, and 132°, respectively).

4. Chemical phosphorylation of radioactive isopentenol is performed by adding to the dichloromethane solution of isopentenol (approximately 2–10 mg in 0.5 ml of dichloromethane) an excess of ditriethyl ammonium phosphate (at least 100 mg) and 200 μl of trichloroacetonitrile. The mixture is stirred with a magnetic stirrer for at least 4 h, during that time it usually acquires a green color. It is then diluted with 2–3 ml of a mixture of *n*-propanol:25% aq.ammonia:water, 6:3:1 by volume, and applied directly with any solids to a silica gel column (0.8 $\times$ 30 cm) equilibrated with a mixture of *n*-propanol 25% aq. ammonia water, 6:3:1 by volume. Elution of the column was performed with the same solvent at the rate of approximately 1–2 ml/h. The elution of the radioactivity is checked by testing 0.5-μl portions (spotted on the bottom line on the silica gel foil) with the aid of a TLC scanner. First, the unreacted isopentenol is eluted (A: 20–25 ml of the elution volume), followed by isopentenyl monophosphate (B: 40–50 ml of the elution volume) and isopentenyl diphosphate (C: 70–80 ml of the elution volume). Equal amounts of radioactivity are found usually in the three preceding fractions, thus indicating that the yield of formation of the isopentenyl phosphate and isopentenyl diphosphate are similar (approximately 30%). The 0.5-μl spots on the silica gel foil are developed with the solvent consisting of *n*-propanol:25% aq.ammonia: water, 6:3:1 by volume, up to the length 12–15 cm, and the radioactivity of the chromatographic spots is localized with a TLC scanner or by fluorography. A standard unlabeled isopentenyl diphosphate[5] is used to identify the labeled substance present in fraction C and migrating on TLC with the $R_F = 0.3$. Fraction A contains unreacted isopentenol ($R_F = 1.0$), and fraction B contains isopentenyl monophosphate ($R_F = 0.6$). On spraying the chromatogram with phosphate spray reagent,[6] the presence of a single radioactive spot corresponding to isopentenyl diphosphate in

[5] V. J. Davisson, A. B. Woodside, and C. D. Poulter, *Methods Enzymol.* **110,** 130 (1985).
[6] V. E. Vaskovskii, E. Y. Kostetskii, and I. M. Vasendin, *Anal. Chim. Acta* **16,** 473 (1975).

fraction C can be detected. The radioactive isopentenyl diphosphate thus obtained has the calculated specific activity four times lower than that of the starting sodium borotritide. In our more than 15 years experience with the described procedure, the isopentenyl diphosphate of specific activity 1.25–2.5 Ci/m*M* (prepared from $NaBT_4$ of specific radioactivity 5–10 Ci/m*M*) can be stored for more than 5 y without radiochemical decomposition.

Storing of the product can be done either in the *n*-propanol:25% aq. ammonia:water solvent used for final chromatography or in 0.25 *M* NH_4OH in 50% aqueous ethanol at −20°.

Acknowledgment

A large part of the described procedure was worked out by the author in the laboratory of Professor Gustav Dallner in the Department of Biochemistry and Biophysics of the Stockholm University in Sweden.

This work was partly supported by the Polish State Committee for Scientific Research (KBN), grant No 6P04A 077 21.

[12] High-Performance Liquid Chromatography–EC Assay of Mitochondrial Coenzyme Q_9, Coenzyme Q_9H_2, Coenzyme Q_{10}, Coenzyme $Q_{10}H_2$, and Vitamin E with a Simplified On-Line Solid-Phase Extraction

By Maurizio Battino, Luciana Leone, and Stefano Bompadre

Introduction

Because lipid peroxidation may be a marker of tissue sufference and even pathology, there has been an increased interest in the determination of endogenous prooxidant and antioxidant compounds in tissue lipids. Coenzyme Q (CoQ), the essential hydrophobic molecule that links flavoproteins and cytochromes in the mitochondrial respiratory chain, as well as in other extramitochondrial redox chains, also possesses antioxidant properties.[1–11] The redox state of CoQ homologues in mitochondrial and

[1] R. E. Beyer, *Free Radic. Biol. Med.* **8,** 545 (1990).

[2] M. Battino, E. Ferri, S. Girotti, and G. Lenaz, *Anal. Chim. Acta* **255,** 367 (1991).

[3] R. Stocker, V. W. Bowry, and B. Frei, *Proc. Natl. Acad. Sci. USA* **88,** 1646 (1991).

[4] M. Battino, G. Lenaz, and G. P. Littarru, *in* "Free Radicals and Antioxidants in Nutrition" (F. Corongiu, S. Banni, M. A. Dessì, and C. Rice-Evans, eds.), p. 37. Richelieu Press, Cambridge, 1993.

0076-6879/04 $35.00

other membranes is important for the evaluation of cell welfare; unfortunately, because of their chemical features, these homologues are very sensitive and unstable compounds, especially in their reduced form, which displays the highest antioxidant potentiality. Thus, artifactual data may increase, depending on the different extraction methods and on the different analytical techniques used. All the valid analytical methods already described in peer-reviewed literature[12–19] require a large extraction procedure and, sometimes, complicated and time-consuming high-performance liquid chromatography (HPLC) analysis. The extraction procedure, before analytical determination, represents one of the possible limiting steps in analyzing actual membrane antioxidant contents and their redox status. The way to improve HPLC analysis resides in limiting, or possibly eliminating, the processing steps that take place before the chromatographic event. For these reasons, we set up a novel system requiring no preanalytical extraction procedure, very limited handling of the samples, and their direct injection into the HPLC apparatus. This method makes it possible to carry out, for the first time, a systematic and simultaneous determination of coenzyme Q_9, coenzyme Q_9H_2, coenzyme Q_{10}, coenzyme $Q_{10}H_2$, and vitamin E contents from rat liver mitochondria. The method

[5] L. Ernster and G. Dallner, *Biochim. Biophys. Acta* **1271,** 195 (1995).

[6] V. E. Kagan, H. Nohl, and P. J. Quinn, *in* "Handbook of Antioxidants" (E. Cadenas and L. Packer, eds.), p. 157. Marcel Dekker, New York, 1996.

[7] G. P. Littarru, M. Battino, and K. Folkers, *in* "Handbook of Antioxidants" (E. Cadenas and L. Packer, eds.), p. 203. Marcel Dekker, New York, 1996.

[8] E. Niki, *in* "Coenzyme Q: Molecular Mechanisms in Health and Disease" (V. E. Kagan and P. J. Quinn, eds.), p. 109. CRC Press, Boca Raton, 2001.

[9] V. E. Kagan, J. P. Fabisiak, and Yulia Tyurina, *in* "Coenzyme Q: Molecular Mechanisms in Health and Disease" (V. E. Kagan and P. J. Quinn, eds.), p. 119. CRC Press, Boca Raton, 2001.

[10] S. R. Thomas and R. Stocker, *in* "Coenzyme Q: Molecular Mechanisms in Health and Disease" (V. E. Kagan and P. J. Quinn, eds.), p. 131. CRC Press, Boca Raton, 2001.

[11] G. P. Littarru and M. Battino, *in* "Coenzyme Q: Molecular Mechanisms in Health and Disease" (V. E. Kagan and P. J. Quinn, eds.), p. 219. CRC Press, Boca Raton, 2001.

[12] J. K. Lang and L. Packer, *J. Chromatogr.* **385,** 109 (1987).

[13] T. Okamoto, Y. Fukunaga, Y. Ida, and T. Kishi, *J. Chromatogr.* (*Biomed. Appl.*) **430,** 11 (1988).

[14] G. Grossi, A. M. Bargossi, P. L. Fiorella, S. Piazzi, M. Battino, and G. P. Bianchi, *J. Chromatogr.* **593,** 217 (1992).

[15] F. Aberg, E.-L. Appelkvist, G. Dallner, and L. Ernster, *Arch. Biochem. Biophys.* **295,** 230 (1992).

[16] Y. Zhang, F. Aberg, E.-L. Appelkvist, G. Dallner, and L. Ernster, *J. Nutr.* **125,** 446 (1995).

[17] M. Podda, C. Weber, M. G. Traber, and L. Packer, *J. Lipid Res.* **37,** 893 (1996).

[18] Y. Zhang, M. Eriksson, G. Dallner, and E.-L. Appelkvist, *Lipids* **33,** 811 (1998).

[19] M. Podda, C. Weber, M. G. Traber, R. Milbradt, and L. Packer, *Meth. Enzymol.* **229,** 330 (1999).

used is an accurate, improved modification of methods successfully developed in the past few years.[20–22]

Reagents and Solvents

Only methanol, ethanol, and isopropanol of HPLC grade were used (Fluka, Neu-Ulm, Germany). Sodium perchlorate and sodium dithionite were from Sigma Chemical Co. (Deisenhofen, Germany). α-Tocopherol was kindly provided by Henkel (La Grange, IL), and CoQ_9 and CoQ_{10} were a generous gift from Eisai, Tokyo, Japan.

Preparation of Standard Solutions

Stock standard solutions of CoQ_9, CoQ_{10}, and vitamin E were obtained diluting pure substance with ethanol, reaching concentrations between 50 and 200 μM. These solutions were stored in sealed vials that were kept at $-80°$. Their actual concentrations were determined spectrophotometrically using the molar extinction coefficient according to Podda *et al.*[19] CoQ_9H_2 and $CoQ_{10}H_2$ standard solutions were prepared according to Lang *et al.*[23] A stock solution of oxidized CoQ_n in ethanol was reduced with 100 mg of sodium dithionite in water at room temperature and in the dark. CoQ_nH_2 was extracted with hexane, dried under nitrogen, and resuspended in ethanol. The solution concentrations were determined as indicated earlier.

Preparation and Treatment of the Mitochondrial Samples

The animals (6-month-old male Wistar rats) were killed by cervical dislocation after ether anesthesia and exsanguinated. The purification of mitochondria from rat liver was carried out as follows: aliquots of liver tissue (4–5 g) were homogenized 1:10 (w/v) in ice-cold buffer, pH 7.5, containing 75 m*M* sucrose, 225 m*M* mannitol, 1 m*M* EDTA, 5 m*M* HEPES, and 0.5 mg/ml fatty-acid–free bovine serum albumin. The homogenate was centrifuged for 10 min at 600*g* at 4°. The sediment was discarded, and the supernatant was centrifuged for 20 min at 1200*g* at 4°, obtaining a mitochondrial pellet and a supernatant. The mitochondrial pellet was washed

[20] S. Bompadre, L. Leone, L. Ferrante, F. P. Alò, and G. Ioannidis, *J. Liq. Chrom. Rel. Technol.* **21,** 417 (1998).

[21] S. Bompadre, L. Ferrante, M. De Martinis, and L. Leone, *J. Chromatogr. A* **812,** 249 (1998).

[22] M. Battino, S. Bompadre, L. Leone, R. F. Villa, and A. Gorini, *Biol. Chem.* **382,** 925 (2001).

[23] J. K. Lang, K. Gohil, and L. Packer, *Anal. Biochem.* **157,** 106 (1986).

twice (by careful resuspensions in ice-cold homogenization buffer and centrifugation, 10 min at 2800*g* at 4°), and purified mitochondria were carefully resuspended in 1 ml of ice-cold homogenization buffer.[24] Mitochondrial protein concentration was determined according to Lowry *et al.*[25] Rat liver mitochondria ranging a protein concentration of 2.0–15 mg/ml were used. Fifty microliters of the mitochondrial suspension was placed in an Eppendorf vial, and the protein precipitation was obtained by the addition of 150 μl of isopropanol. The Eppendorf vials were vortexed for 60 s and then were centrifuged at 10,000*g* for 10 min in a Sorvall MC 12 V bench top centrifuge. Finally, 50 μl of supernatant was injected into the HPLC device.

High-Performance Liquid Chromatography Assembly and Analysis

The HPLC system consisted of a Varian (Walnut Creek, CA) model Vista 5500 HPLC pump (Pump A), a Beckman 126 Programmable HPLC pump (Pump B), and a Coulochem 5100 A electrochemical detector from ESA (Bedford, MA) equipped with a model 5021 conditioning cell and a model 5011 analytical cell. The conditioning cell was set at +0.50 V. Potentials applied to the analytical cell were −0.50 V for the first electrode and +0.35 V for the second electrode. The chromatograms were integrated with Star 5.5 software (Varian, Valnut Creek, CA), using the signal from the positive electrode of the analytical cell (ESA, Bedford, MA). The injector was a Rheodyne Model 7125 manual injection valve equipped with a 50-μl sample loop. The coupled-column system was operated by a six-port, automated, switching valve (Valco, Schencon, Switzerland) controlled by the Vista 5500 HPLC pump.

Extraction was performed on a 20- × 4-mm column (column A), dry packed with 10 μm C:8 silica using a mobile phase of methanol/water 50:50 (mobile phase 1). Separation was performed on a 150- × 4.6-mm 3-μm C:18 column (column B) using a mobile phase constituted of 50 m*M* sodium perchlorate in methanol (mobile phase 2), at flow rate of 1 ml/min.

The extraction and analysis steps proceeded as follows (Fig. 1): after injection, the sample was loaded on the extraction column (column A) where the analytes were retained while the matrix, passing through the column, was directed to waste at a flow rate of 0.5 ml/min. At the same time, column B was conditioned with the mobile phase 2. After a period of 0.7 min, the valve was switched, and mobile phase 2 from pump B eluted

[24] T. Armeni, J. L. Quiles, C. Pieri, S. Bompadre, G. Principato, and M. Battino, *J. Bioenerg. Biomembr.* **35,** 181 (2003).

[25] O. H. Lowry, N. J. Rosebrough, A. L. Farr, and R. J. Randall, *J. Biol. Chem.* **193,** 265 (1951).

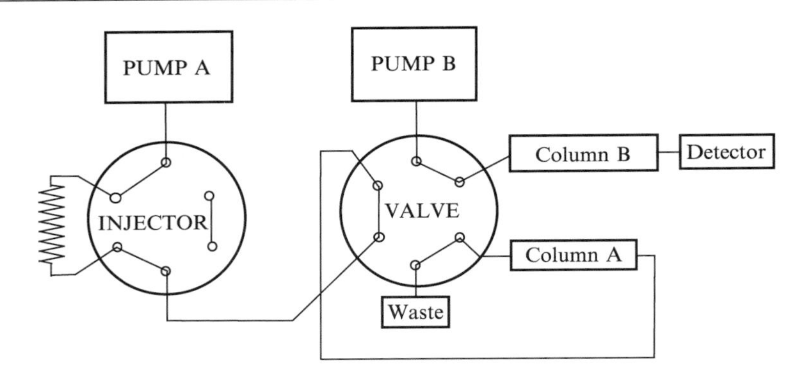

FIG. 1. Schematic diagram of the column switching system used. The sample was loaded on the extraction column (column A) where the analytes were retained; the matrix flowed through the column to waste. At the same time, column B was conditioned with the mobile phase 2; the valve was then switched, and mobile phase 2 from pump B eluted the analytes trapped on the column A to the analytical column B.

the analytes trapped on the column A to the analytical column B. Thereafter, the valve was switched to the initial position, and column A was conditioned again with mobile phase 1 to prepare it for the next sample. Simultaneously, pump B maintained the flow of mobile phase 2 through the analytical column, where the analytes were separated and detected by the electrochemical detector. Recoveries were determined comparing the areas of standard solutions extracted with the procedure described previously, with the peak areas obtained from the same solutions injected directly into the analytical column. Recovery values were 92% for vitamin E, 95% for CoQ_9H_2, 93% for CoQ_9, 96% for $CoQ_{10}H_2$, and 93% for CoQ_{10}.

The precision of the method was determined by repeating injection of the same sample six times and then calculating the corresponding mean and standard deviation. The variation coefficient ranged between 2.85% and 4.56%.

The results, chromatograms of standards and of rat liver mitochondria, are shown in Fig. 2A, B, respectively. Time of analysis was 18 min, and the respective retention times were 2.67 min for vitamin E, 6.63 min for CoQ_9H_2, 10.46 min for $CoQ_{10}H_2$, 11.12 min for CoQ_9, and 16.68 min for CoQ_{10}. The following collects the data of antioxidant concentrations found in rat liver mitochondria:

	Vitamin E (μmol/mg mit prot)	CoQ_9H_2 (nmol/mg mit prot)	CoQ_9 (nmol/mg mit prot)	$CoQ_{10}H_2$ (nmol/mg mit prot)	CoQ_{10} (nmol/mg mit prot)
Avg±SEM (N = 10)	0.45±0.06	2.11±0.28	0.77±0.08	0.17±0.03	0.04±0.01

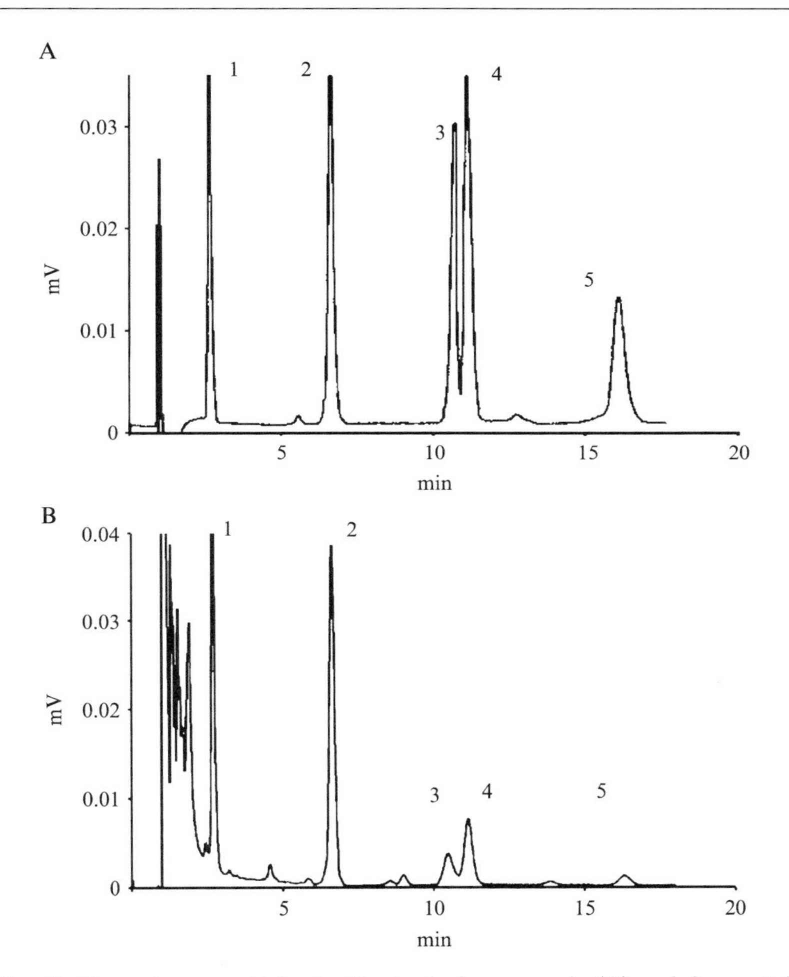

FIG. 2. Chromatograms obtained with standard compounds (A) and from rat liver mitochondria (B). HPLC-EC detection of (1) vitamin E, (2) CoQ_9H_2, (3) CoQ_9, (4) $CoQ_{10}H_2$, and (5) CoQ_{10}.

Conclusions

The described method allows the simultaneous detection of vitamin E, CoQ_9H_2, CoQ_9, $CoQ_{10}H_2$, and CoQ_{10} with good precision in rat liver mitochondria. The length of the assay is reasonably short (i.e., three assays per hour) and is performed with a simple isocratic HPLC separation. Our simple coulometric technique enables the analyst to use only different electrodes for the simultaneous determination of reduced and oxidized

CoQ_n instead of the concomitant use of different kinds of detectors (e.g., diode array and amperometric detectors).

Moreover, the limited handling of the sample has important consequences on the redox status of CoQ_n. In fact, although the ubiquinol amounts found with methods characterized by large extraction procedures are negligible, with the method we used at least 65% of CoQ resulted in the reduced form. This finding gives more reliable information on the actual concentrations of the CoQ_n redox forms in the mitochondrial membrane and can be a valuable tool for evaluating the efficiency of the respiratory chain. Such considerations may also have interesting applications in some diagnostic protocols, in which respiratory chain component deficiency syndromes are involved.

Acknowledgment

This work was supported in part by Ancona University. The helpful assistance of Ms M. Glebocki in editing the manuscript is gratefully acknowledged.

[13] Simultaneous Determination of Coenzyme Q_{10}, Cholesterol, and Major Cholesterylesters in Human Blood Plasma

By CRAIG A. GAY and ROLAND STOCKER

Introduction

Cardiovascular disease is the major cause of morbidity and mortality in the Western world, and a high plasma concentration of cholesterol associated with low-density lipoprotein (LDL) is a primary risk factor for the disease. Numerous primary and secondary prevention trials confirm clinical benefits with statins (see e.g., ref. 1) (i.e., inhibitors of hydroxymethylglutaryl coenzyme A [CoA] reductase that catalyze the formation of mevalonate, an intermediate in cholesterol biosynthesis). The beneficial effect of statins is attributed largely to their ability to inhibit cholesterol synthesis, leading to up-regulation of hepatic LDL receptors and corresponding reductions in plasma LDL. However, statins have pleiotropic effects, including anti-inflammatory and antioxidant actions that may

[1] The long-term intervention with pravastatin in ischaemic disease (LIPID) study group, *N. Engl. J. Med.* **339,** 1349 (1998).

0076-6879/04 $35.00

contribute to the clinical benefits of statin therapy. For example, statins inhibit the formation of myeloperoxidase-derived and nitric oxide-derived oxidants,[2] species that are implicated in atherogenesis.

Atherosclerosis, the major underlying cause of cardiovascular disease, is associated with heightened oxidative stress and increased oxidative damage to the vessel wall.[3] Much of the research related to the "oxidation hypothesis" of atherogenesis[4] has focused on LDL oxidation taking place in the vessel wall.[5] In addition, it is increasingly clear that oxidative events in atherosclerotic vascular disease are not limited to LDL oxidation but also include cellular oxidative events.[6] As a result, there has been much interest in the roles of antioxidants in cardiovascular disease.[7]

An antioxidant that has received relatively little attention is ubiquinol-10, the reduced form of coenzyme Q_{10} (CoQ_{10}), although evidence suggests that CoQ_{10} may protect against atherosclerosis (reviewed in ref. 8). Ubiquinol-10 is a primary antioxidant protecting LDL lipid against oxidation *in vitro*[9] and *in vivo*,[10] and pharmacological doses of CoQ_{10} supplements decrease atherosclerosis in apolipoprotein E-deficient mice.[10,11] CoQ_{10} supplementation may improve blood pressure and long-term glycemic control in subjects with type 2 diabetes.[12] In addition, CoQ_{10} is a cellular antioxidant[13] and a key component of membrane-associated electron transport systems, including that responsible for energy production by means of mitochondrial respiration.

Coenzyme Q_{10} and cholesterol are synthesized by pathways that share mevalonate as a common and important intermediate.[14] Of potential significance, statin therapy seems to lower plasma[15] and possibly also tissue

[2] M. H. Shishehbor, M. L. Brennan, R. J. Aviles, X. Fu, M. S. Penn, D. L. Sprecher, and S. L. Hazen, *Circulation* **108,** 426 (2003).

[3] G. M. Chisolm and D. Steinberg, *Free Radic. Biol. Med.* **28,** 1815 (2000).

[4] D. Steinberg, S. Parthasarathy, T. E. Carew, J. C. Khoo, and J. L. Witztum, *N. Engl. J. Med.* **320,** 915 (1989).

[5] J. M. Upston, X. Niu, A. J. Brown, R. Mashima, H. Wang, R. Senthilmohan, A. J. Kettle, R. T. Dean, and R. Stocker, *Am. J. Pathol.* **160,** 701 (2002).

[6] K. Chen, S. R. Thomas, and J. F. Keaney, Jr., *Free Radic. Biol. Med.* **35,** 117 (2003).

[7] L. Kritharides and R. Stocker, *Atherosclerosis* **164,** 211 (2002).

[8] S. R. Thomas, P. K. Witting, and R. Stocker, *BioFactors* **9,** 207 (1999).

[9] R. Stocker, V. W. Bowry, and B. Frei, *Proc. Natl. Acad. Sci. USA* **88,** 1646 (1991).

[10] P. K. Witting, K. Pettersson, J. Letters, and R. Stocker, *Free Radic. Biol. Med.* **29,** 295 (2000).

[11] S. R. Thomas, S. B. Leichtweis, K. Pettersson, K. D. Croft, T. A. Mori, A. J. Brown, and R. Stocker, *Arterioscler. Thromb. Vasc. Biol.* **21,** 585 (2001).

[12] J. M. Hodgson, G. F. Watts, D. A. Playford, V. Burke, and K. D. Croft, *Eur. J. Clin. Nutr.* **56,** 1137 (2002).

[13] L. Ernster and G. Dallner, *Biochim. Biophys. Acta* **1271,** 195 (1995).

[14] G. Dallner and P. J. Sindelar, *Free Radic. Biol. Med.* **29,** 285 (2000).

concentrations of CoQ_{10}.[16] This raises the possibility that the overall beneficial effect of statins may be reduced by inadvertently decreasing both the lipophilic antioxidant defense in LDL and vascular cells, as well as the cellular energy production that may relate to the occurrence of muscle-related side effects associated with statin therapy.[17]

The purpose of this chapter is to present a simple and robust method for the simultaneous analysis of CoQ_{10}, cholesterol, and major cholesteryl-esters in human plasma that can be applied to large numbers of samples. It is hoped that by analysis of existing, well-maintained samples from a suitable cohort of a high-profile clinical intervention study, this method will help resolve the controversial issue of whether a lowering of CoQ_{10} relates to well-documented, yet poorly understood, side effects of statin treatment.

Reagents and Materials

Unless stated otherwise, all solvents were stored and used in light-protected bottles. Extraction tubes and glass pipettes (5 ml) were used as received from the manufacturer and not recycled. Ethanol, methanol, hexane, and isopropylalcohol were of high-performance liquid chromatography (HPLC) grade, and these solvents, as well as HiperSolv water for HPLC and acetic acid, were purchased from BDH (Sydney, Australia); 1,4-benzoquinone, CoQ_{10}, cholesterol, cholesterylarachidonate (C20:4), cholesteryllinoleate (C18:2), cholesteryloleate (C18:1), and cholesteryl palmitate (C16:0) were obtained from Sigma (St. Louis, MO). A solution of 1,4-benzoquinone (2 mg/ml) was prepared fresh daily in HiperSolv water. Quality control plasma was obtained by combining the plasma isolated[18] from eight healthy volunteers (four women, four men, aged 26–46 years) and then stored at -80° until needed.

Methods and Applications

Extraction of Human Plasma

Total CoQ_{10} was determined by first oxidizing endogenous ubiquinol-10 to CoQ_{10} by treatment of plasma with 1,4-benzoquinone.[19] For this, 25 μl of 1,4-benzoquinone solution (2 mg/ml) was added to 100 μl plasma,

[15] P. G. Elmberger, A. Kalén, E. Lund, E. Reihnér, M. Eriksson, L. Berglund, B. Angelin, and G. Dallner, *J. Lipid Res.* **32,** 935 (1991).

[16] K. Folkers, P. Langsjoen, R. Willis, P. Richardson, L. J. Xia, C. Q. Ye, and H. Tamagawa, *Proc. Natl. Acad. Sci. USA* **87,** 8931 (1990).

[17] P. D. Thompson, P. Clarkson, and R. H. Karas, *JAMA* **289,** 1681 (2003).

[18] W. Sattler, D. Mohr, and R. Stocker, *Meth. Enzymol.* **233,** 469 (1994).

and the mixture was then incubated at room temperature for 30 min. Following incubation, 1 ml of methanol containing 0.02% acetic acid and 5 ml of water-washed hexane[18] was added, the tubes capped, and mixed vigorously for 30 s. The resulting biphasic plasma extract was then centrifuged at 4° and 1,430*g* for 5 min, the (top) hexane layer (4 ml) was removed and dried using an AES1010 Speedvac System (ThermoQuest, Holbrook, NY). The resulting lipid residue was redissolved completely in 200 μl ice-cold isopropylalcohol and transferred into appropriate HPLC vials that were used for HPLC analysis.

Preparation of Coenzyme Q_{10} and Lipid Standard

Standard of CoQ_{10} (10 μM) was prepared by weighing out approximately 1–3 mg CoQ_{10} directly into a scintillation vial and by dissolving the lipid in 2 ml ethanol. An aliquot of the resulting solution was diluted serially in ethanol, and the CoQ_{10} concentration was determined by measuring the absorbance at 275 nm, blanked against ethanol, and using $\varepsilon = 14{,}200\ M^{-1}\text{cm}^{-1}$.[20] The stock solution was then diluted to a 10-μM working solution of CoQ_{10}, aliquots of which were distributed into HPLC vials and stored at −80° until used. Working solutions of cholesterol and cholesterylesters (1 mM each, determined by weight) were prepared similarly in ethanol.

High-Performance Liquid Chromatography Assay

The HPLC system used was an Agilent 1100 series, composed of quaternary pump with degasser, a temperature-controlled autosampler, and a variable wavelength detector and using Agilent HPLC 2D ChemStation software. Analytes were separated on a 250 × 4.6 mm LC-18 column (5-μm particle size, Supelco, Bellefonte, PA) with 5-cm guard column (Supelco) eluted with ethanolmethanol (65:35, v/v).[19] The eluant was monitored at 210 nm from 0–9.5 and 19.5–40 min and at 275 nm from 9.5–19.5 min.

Linearity of and Detection Limit for Coenzyme Q_{10} Determination

Application of the method to standard solutions of CoQ_{10} indicated excellent linearity over the range tested, i.e., 0–100 pM ($R^2 = 0.9985$) (Fig. 1), with a detection limit (defined as signal-to-noise ratio ≥ 3) of 1 pM.

[19] F. Mosca, D. Fattorini, S. Bompadre, and G. P. Littarru, *Anal. Biochem.* **305,** 49 (2002).
[20] Y. Hatefi, *Adv. Enzymol.* **25,** 275 (1963).

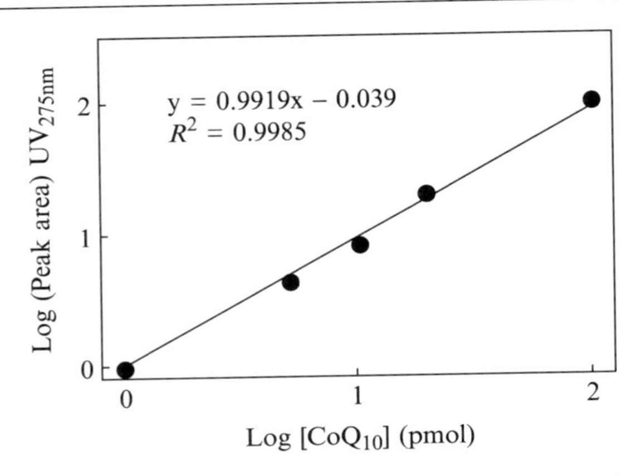

FIG. 1. Concentration response curve for CoQ_{10} standard.

Method Validation

Pretreatment of human plasma with 1,4-benzoquinone resulted in complete oxidation of endogenous ubiquinol-10 (not shown) as reported previously.[19] A representative HPLC chromatogram obtained from an extract of quality-control human plasma is shown in Fig. 2. Its characteristic features are the relatively large peaks representing cholesterol (RT = 6.4 min) and cholesterylesters (RT = 24.4, 27.0, 32.6, and 32.6 min for C20:4, C18:2, C18:1, and C16:0, respectively) compared with that of CoQ_{10} (RT = 14.7 min). In fact, CoQ_{10} was barely visible at the absorbance scale as that used for cholesterol and cholesterylesters. However, enlarging the chromatogram for the period during which the eluant was monitored at 275 nm revealed a clearly defined and separated CoQ_{10} peak (Fig. 2, inset) that could be integrated readily. We assume that the compounds eluting immediately before CoQ_{10}, which are not seen when using the method of Mosca *et al.*,[19] are derived from carotenoids that extract into hexane (used in our method) but not *n*-propanol (used by Mosca *et al.*[19]). Cholesterol, C20:4, and C18:2 were baseline separated, whereas under the conditions used, C18:1 and C16:0 were not separated completely. For quantification, peak area comparison with authentic standards was used for CoQ_{10}, cholesterol, C20:4, and C18:2. For quantification of C18:1 and C16:0, we assumed a 60/40% relative area distribution of the respective lipids.

Spiking experiments verified the identity of the peak with RT = 14.7 min as CoQ_{10}. For this, 100 μl of extract of quality control plasma without (Fig. 3, broken line) and with 20 μl of CoQ_{10} working standard (10 μM) was subjected to HPLC analysis (Fig. 3, solid line).

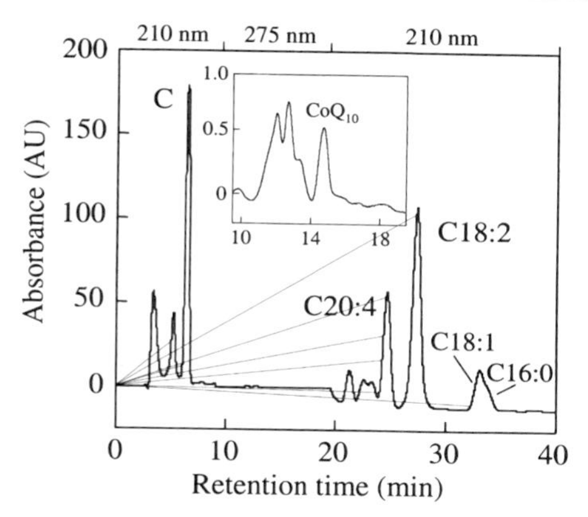

FIG. 2. Representative HPLC chromatogram of hexane-extracted lipids of quality control human plasma pretreated with 1,4-benzoquinone. Extracted lipids were separated by means of an LC-18 column eluted with ethanol/methanol (65/35, v/v) and monitored at 210 nm (0–9.5 min), 275 nm (9.5–19.5 min), and 210 nm (19.5–40 min). The main chromatogram shows cholesterol (C) eluting at 6.4 min and cholesterylarachidonate (C20:4), cholersteryllinoleate (C18:2), cholesteryloleate (C18:1), and cholesterylpalmitate (C16:0) at 24.4, 27.0, 32.6, and 32.6 min, respectively. CoQ_{10} is barely visible at this absorbance scale. Enlarging the chromatogram for the period during which the eluant was monitored at 275 nm reveals a clearly defined and separated CoQ_{10} peak at 14.7 min (inset).

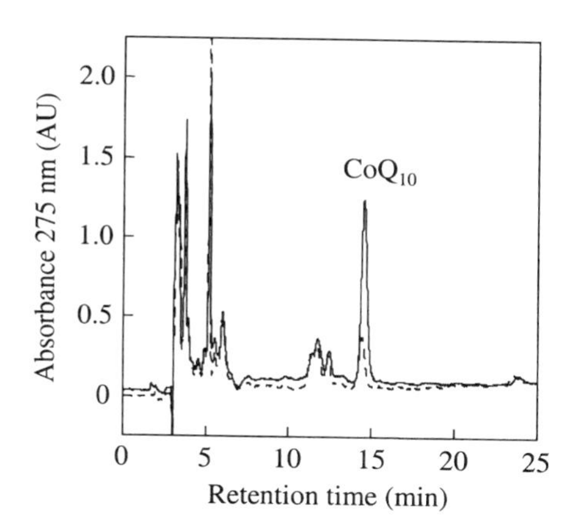

FIG. 3. Chromatographic identity of peak eluting at 14.7 min as CoQ_{10}. Hexane-extracted lipids of human plasma pretreated with 1,4-benzoquinone without (broken line) and with spiked CoQ_{10} standard (100 pmol) (solid line) were subjected to HPLC as described in Fig. 2.

TABLE I
INTRAASSAY VARIABILITY

	C	CoQ_{10}	C20:4	C18:2	C18:1	C16:0
Mean concentration (μM)[a]	1341	0.80	250	1166	613	975
SD (μM)	14	0.03	3.6	59.0	4.9	7.8
CV (%)	1.1	3.6	1.4	5.1	0.8	0.8

C, cholesterol; C20:4, cholesterylarachidonate; C18:2, cholesteryllinoleate; C18:1, cholesteryloleate; C16:0, cholesterylpalmitate; CoQ_{10}, coenzyme Q_{10}.
[a] $n = 4$.

TABLE II
INTERASSAY VARIABILITY

	C	CoQ_{10}	C20:4	C18:2	C18:1	C16:0
Mean concentration (μM)[a]	1328	0.75	235	1109	560	891
SD (μM)	61.8	0.08	20.9	100.5	59.4	94.6
CV (%)	4.7	10.9	8.9	9.1	10.6	10.6

For abbreviations see Table I.
[a] $n = 30$.

Recovery and Reproducibility of Method. Recovery of CoQ_{10} was determined by adding 100 or 200 pM CoQ_{10} (in ethanol) to quality-control plasma, followed by extraction and comparison of the peak area obtained for the spiked plasma samples with that of the corresponding standard. Recovery was documented in duplicates, and the experiment was repeated twice. The recovery obtained on both occasions was 75% and comparable for both amounts of CoQ_{10} added.

The intraassay precision was determined for CoQ_{10}, cholesterol, and cholesterylesters with four separate aliquots of quality-control plasma extracted and analyzed on the same day. The results obtained (Table I) show CV values of 1–5%. The day-to-day accuracy (interassay) was determined for CoQ_{10}, cholesterol, and cholesterylesters over a period of 45 days using a total of 30 quality-control plasma samples analyzed in duplicate on 15 separate occasions. The results obtained (Table II) show CV values of 4–11%.

Interferences

Possible interferences for the determination of CoQ_{10} are the putative carotenoids eluting immediately before CoQ_{10}. This could be particularly relevant for plasma samples rich in or supplemented with carotenoids.

However, we have applied our method to 600 different human plasma samples and have not yet come across a single sample in which CoQ_{10} could not be determined because of interference with other compounds. The method described is based on that published by Mosca *et al.*,[19] and hence optimized for CoQ_{10} determination. However, as shown here, essentially the same method can be used for the simultaneous determination of CoQ_{10}, cholesterol, and cholesterylesters if both a different, previously validated extraction method[18] and a simple change in detection wavelengths are applied. By using this method with two different HPLC systems, we noted that the precise timing of the wavelength change(s) may require some slight adjustment. It is perhaps not surprising that the separation of C18:1 and C16:0 is limited, although we believe reasonable quantification of these compounds is nevertheless possible. Obviously, if more accurate determination of all cholesterylesters is required, more appropriate HPLC methods may be used.

Conclusions

We have presented a procedure for the simultaneous determination of CoQ_{10}, cholesterol and major cholesterylesters in human blood plasma. The method outlined is simple, accurate, and robust. Although its sensitivity for CoQ_{10} determination does not match that of, e.g., the more complicated electrochemical detection, it nevertheless is sufficient to accurately determine the CoQ_{10} concentration in 100 μl plasma. Perhaps the strongest aspect of this method is its robustness, as we have verified by analyzing 600 human plasma samples without any problems. Also, if combined with post-column chemiluminescence detection,[18,21] this method has the potential to be extended to the analysis of cholesterylester hydroperoxides. We note, however, that other methods may be preferred for the determination of CoQ alone. For example, the method described by Mosca *et al.*,[19] uses a simpler sample workup, although it does not allow for the simultaneous determination of cholesterol and cholesterylesters.

Acknowledgments

We thank Sangeeta Ray for invaluable help in the initial stages of this project. Roland Stocker is supported by the Australian National Health and Medical Research Council. This work received support from the International CoQ_{10} Association.

[21] P. K. Witting, D. Mohr, and R. Stocker, *Meth. Enzymol.* **299,** 362 (1999).

[14] Assay of Coenzyme Q_{10} in Plasma by a Single Dilution Step

By GIAN PAOLO LITTARRU, FABRIZIO MOSCA, DANIELE FATTORINI, STEFANO BOMPADRE, and MAURIZIO BATTINO

Introduction

Coenzyme Q_{10} (CoQ_{10}) is the predominant homologue of ubiquinone in humans and is largely diffused in every organ.[1,2] It is also present in blood, both in plasma and in the cellular compartment,[3] where its actual significance is still debated; in fact, significant modifications of CoQ_{10} content in plasma are not always consistent with similar modifications at tissue level. This is proven by the fact that CoQ_{10} levels in human plasma vary from 0.35–1.49 μg/ml, depending on the healthy populations investigated and, presumably, on the analytical method used.[4] Moreover, it has been recognized that several factors such as aging, physical exercise and training, dietary habits, cardiovascular diseases, and several diseases affect CoQ_{10} contents. Plasma concentrations of CoQ_{10} probably reflect an overall metabolic demand[5]; furthermore, its presence in plasma lipoproteins, especially in LDL, places an important role in protecting them against oxidative damage.[6] Therefore, the concentration of CoQ_{10} in lipoproteins and plasma could assume a clinical relevance in the field of oxidative stress and antioxidant defense. Increased levels of CoQ_{10} in LDL lead to enhanced antioxidant protection[6]; regarding this point, one should keep in mind that antioxidant capacity *in vivo* depends on the redox status of CoQ_{10}, besides its overall concentration. Fresh plasma samples reflect the *in vivo* situation (i.e., contain almost exclusively reduced CoQ_{10}, $CoQ_{10}H_2$, ubiquinol-10). The content of CoQ_{10} in single classes of lipoproteins was found to strictly correlate with CoQ_{10} plasma concentration.[3] Besides

[1] A. Kalen, E. L. Appelkvist, and G. Dallner, *Lipids* **24,** 579 (1989).

[2] M. Soderberg, C. Edlund, K. Kristensson, and G. Dallner, *J. Neurochem.* **54,** 415 (1990).

[3] M. Tomasetti, R. Alleva, M. D. Solenghi, and G. P. Littarru, *BioFactors* **9,** 231 (1999).

[4] M. Battino, *in* "Mitochondrial Ubiquinone (Coenzyme Q_{10}): Biochemical, Functional, Medical and Therapeutic Aspects in Human Health and Disease" (M. Ebadi, J. Marwah, and R. Chopra eds.), p. 151. Prominent Press, Scottsdale AZ, 2001.

[5] G. P. Littarru, S. Lippa, A. Oradei, R. M. Fiorini, and L. Mazzanti, *in* "Biomedical and Clinical Aspects of Coenzyme Q" (K. Folkers, G. P. Littarru, and T. Yamagami eds.), Vol. VI, pp. 167. Elsevier North Holland, 1991.

[6] D. Mohr, V. W. Bowry, and R. Stocker, *Biochim Biophys. Acta* **1126,** 247 (1992).

0076-6879/04 $35.00

expressing CoQ_{10} concentration as micrograms per milliliter or as μM, it is useful to calculate the CoQ_{10}/cholesterol ratio, which takes into account the amount of particles that carry CoQ_{10}. CoQ_{10} is commonly used as a food supplement, or as a co-adjuvant to therapy in several diseases, and plasma levels achieved on oral administration have been found to sometimes correlate with clinical efficacy. Some effective, widely diffused hypocholesterolemic drugs, which usually act by inhibiting 3-hydroxy-3-methylglutaril coenzyme A (HMGCoA)-reductase, also lower plasma CoQ_{10},[7] on the basis of the common biosynthetic pathway of cholesterol and the isoprenoid side chain of CoQ. At present, it is debated whether these decreased plasma levels of CoQ_{10} also reflect impaired tissue levels.

Finally, plasma levels of CoQ_{10} are useful to evaluate the bioavailability of orally administered CoQ_{10}.

Different methods have been proposed for quantifying either total CoQ_{10} or the reduced and oxidized forms in plasma.[3,8–16]

This procedure is basically the one for which we gave a comprehensive account[17]; its novelty, which is also the key feature of the assay, is direct injection of a plasma extract into the high-performance liquid chromatography (HPLC) apparatus without bringing into dryness. Because plasma, *in vivo,* contains almost exclusively reduced CoQ_{10}, when a fresh sample is extracted, common procedures do not always lead to complete oxidation of ubiquinol. Because ultraviolet (UV) methods for assaying CoQ_{10} usually quantify the oxidized coenzyme at 275 nm, incomplete oxidation during the procedure would lead to underestimation of total CoQ_{10}. Therefore, the first phase of our method is chemical oxidation of CoQ_{10} present in the sample before propanol extraction.

[7] S. A. Mortensen, A. Leth, E. Agner, and M. Rohde, *Mol. Aspects Med.* **18,** s137 (1997).

[8] J. K. Lang and L. Packer, *J. Chromatogr.* **385,** 109 (1987).

[9] B. Finckh, A. Kontush, J. Commentz, C. Hubner, M. Burdeleski, and A. Kohlschutter, *in* "Methods in Enzymology" (L. Packer, ed.), Vol. 299, pp. 341. Academic Press, San Diego, 1999.

[10] M. Podda, C. Weber, M. G. Traber, R. Milbradt, and L. Packer, *in* "Methods in Enzymology" (L. Packer, ed.), Vol. 299, pp. 330. Academic Press, San Diego, (1999).

[11] B. Finckh, A. Kontush, J. Commentz, C. Hubner, M. Burdeleski, and A. Kohlschutter, *Anal. Biochem.* **232,** 210 (1995).

[12] T. Okamoto, Y. Fukunaga, Y. Ida, and T. Kishi, *J. Chromatogr.* **430,** 11 (1988).

[13] G. Grossi, A. M. Bargossi, P. L. Fiorella, M. Battino, and S. Piazzi, *J. Chromatogr.* **593,** 217 (1992).

[14] P. O. Edlund, *J. Chromatogr.* **425,** 87 (1988).

[15] J. Lagendijk, J. B. Ubbink, and W. J. Vermaak, *J. Lipid Res.* **37,** 67 (1996).

[16] S. Yamashita and Y. Yamamoto, *Anal. Biochem.* **250,** 66 (1997).

[17] F. Mosca, D. Fattorini, S. Bompadre, and G. P. Littarru, *Anal. Biochem.* **305,** 49 (2002).

Materials

Reagents

Methanol and *n*-propanol were R. S. type (Carlo Erba, Rodano, Milan, Italy). Ethanol was R. S. plus grade (Carlo Erba, Rodano, Milan, Italy). Benzoquinone was from Sigma (St. Louis, MO). Lithium perchlorate was from Aldrich (Steinheim, Germany).

Solutions for the electrochemical detection (ECD) chosen as a reference method were filtered through a nylon 66 membrane, 0.2 μm $\times$ 47 mm (Supelco, Bellafonte, PA) and degassed.

Preparation of Standards

Pure CoQ_{10} standard was kindly donated by Kaneka (Osaka, Japan). Coenzyme Q_{10} stock standard solutions were prepared by dissolving the pure compound in reagent ethanol to yield final concentrations of approximately 50 μM. The accurate concentration was determined spectrophotometrically at 275 nm using a molar extinction coefficient of 14,200.[18] Stock solution was divided into 0.5-ml aliquots and kept at -80°. Each working day, 33 μl of the stock solution was diluted to 2 ml with propanol/water 5:1. The different concentrations of the standards to be used for the calibration curve were obtained by diluting this solution, respectively, by 1/2, 1/3, 1/4, 1/6, and 1/12 with propanol/water 5:1.

High-Performance Liquid Chromatography Assembly

The HPLC apparatus consisted of a Beckman System pump model 126, a detector model 166 (Beckman Instruments, San Ramon, CA), and an injector equipped with a 200-μl loop (Rheodyne 7725i obtained from Supelco, Milano, Italy). The column was a Supelcosil LC 18 (Supelco, Milano, Italy), 25 cm $\times$ 0.46 cm i.d. 5 μm, precolumn LC 18S, 2 cm (Supelco, Milano, Italy). An in-line filter A-701 (Upchurch Scientific, Inc., Oak Harbor, WA) was placed between the injector and the precolumn. The photodiode array detector for the UV spectrum analysis of the CoQ_{10} peak was a SPD-M (Shimadzu, Tokyo).

Samples

Blood was drawn from the cubital vein of healthy blood donors from the local blood bank, after informed consent, and anticoagulated with lithium heparin. Plasma obtained after centrifugation at 4000*g* for

[18] Y. Hatefi, *Adv. Enzymol.* **25,** 275 (1963).

15 min at 4° was used fresh or after storage at −80° for the day-to-day precision assay.

To check the stability of total CoQ_{10} in plasma, aliquots of three different samples were kept for 3 days at 4°, room temperature (22°), and at −20°.

Description of the Method

Two hundred microliters of plasma was pipetted into an Eppendorf-type microtube, supplemented with 50 μl of a 1,4 benzoquinone solution (2 mg/ml), and vortexed for 10 s. After 10 min, 1 ml of *n*-propanol was added. The test tube was then vortexed for 10 s and centrifuged at 10,000 rpm for 2 min to spin down the protein precipitate.

Two hundred microliters of the supernatant was then injected into the HPLC (this supernatant, placed in a capped test tube, was found stable for up to 3 days when kept even at 22°).

Mobile phase was constituted by ethanol-methanol (65–35%), and the flux was 1 ml/min. UV detection was performed at 275 nm.

Two hundred microliters of different concentrations of pure oxidized CoQ_{10} was injected as standard. Peak area analysis was performed by a Beckman Gold Data System (DOS version).

Coulometric Analysis

Coulometric analysis of CoQ_{10} is not necessary to accomplish the whole procedure but can be used to quantify the extent of CoQ_{10} reduction. In developing our method, we used this procedure as a reference and to check the performance of our method with samples containing different $CoQ_{10}H_2/CoQ_{10}$ ratios. Ubiquinol/ubiquinone separation was performed on an ODS reversed-phase column (Supelcosil LC18, 15 × 4.6 mm i.d. 3 μm, Supelco, Milano, Italy) using a mobile phase constituted by 50 m*M* sodium perchlorate in methanolethanol (80:20), at flow rate of 1 ml/min. A Coulochem II, model 5200 electrochemical detector (ESA, Bedford, MA), with the analytical cell set at −0.5 V and +0.35 V was used to detect the oxidized and reduced forms of CoQ_{10}.

Recovery, Accuracy, and Precision. Recovery of CoQ_{10} was based on a comparison between the peaks obtained by spiking samples with increasing concentrations of oxidized CoQ_{10} and the corresponding peaks of the standard. Recovery was documented at three concentrations (1.16, 2.32, and 3.48 μM), with triplicate determinations for each concentration. Intra-assay accuracy and precision were determined using four samples, the value of which had been certified by a reference ECD method; each level was assayed five times for the intra-assay accuracy and precision test. Inter-assay accuracy and precision were determined over a 2-month period using a quality control sample ($n = 21$).

Chromatography and Recovery

Typical, representative chromatograms of both a standard and a plasma sample are shown in Fig. 1A. Diode array analysis of the peak with the same retention time as the CoQ_{10} gave the known CoQ_{10} spectrum. Spiking of a sample containing an initial concentration of 0.29 μM with 1.16, 2.32, or 3.48 μM CoQ_{10} yielded a recovery of 96.3%, 98.1%, and 98.5%.

Calibration Curves. Calibration curves constructed using propanol/water solutions of pure CoQ_{10} as described in "Materials and Methods" showed linearity over a concentration range of 7.9–579 nM, corresponding to a concentration of 47.4–3474 nM CoQ_{10} in plasma. Correlation coefficients (r^2) for 20 calibration curves obtained over a 2-month period ranged from 0.98–0.999. The limit of quantitation was 0.037 μM (1.23 nM in column) with a precision of 10.52%.

Accuracy and Precision. Four samples with different, decreasing concentrations of total CoQ_{10}, having a nominal value previously determined by electrochemical detection method chosen as reference, were analyzed five times each. Within-run (intra-assay) precision showed a CV% of 1.6 for a 911 nM concentration, 5.3 for for 432 nM, and 5.7 for 170 nM. For

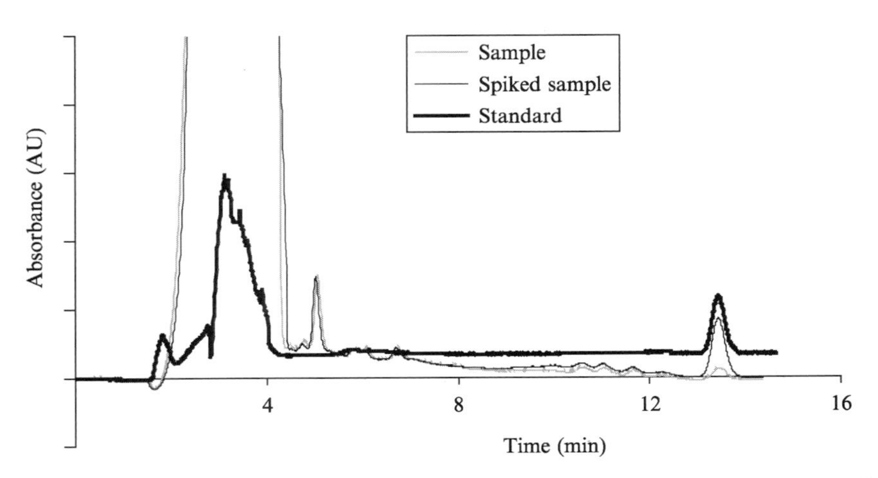

Fig. 1. A typical chromatogram of a plasma sample before and after spiking with a known amount of standard. The standard alone is also shown. From F. Mosca, D. Fattorini, S. Bompadre, and G. P. Littarru, *Anal. Biochem.* **305,** 49 (2002), with permission.

the lowest level, 37 nM, CV% was 10.5. Percent deviation from nominal value was between −4.7 and 2.6. CV% becomes consistently higher than five only for samples less than 0.06 μM. We should consider that normal plasma values of CoQ_{10} range between 0.75 and 0.98 μM. Deviation from nominal values was never higher than 4.7%.

Day-to-day precision was conducted over a 2-month period on a sample showing a normal plasma concentration (0.81 μM). CV% was 2.16, and the percent deviation from nominal value was 0.7.

Comparison with the Electrochemical Detection Method. Ten samples were analyzed, in parallel, both by the ECD and by the proposed method (UV), and the results showed very good correlation. These samples were chosen on the basis of their different levels of total CoQ_{10} and different extent of reduction. These data show that fully reduced samples are efficiently oxidized by the benzoquinone treatment.

Stability. Three different samples ranging from 1.06–2.68 μM were stored for 1, 2, or 3 days at room temperature, 0–4° and −20°. The levels of total CoQ_{10}, in the preceding conditions, were practically unchanged compared with the values obtained immediately after blood withdrawal. Therefore, CoQ_{10} was stable for at least 3 days, even when the sample was kept at room temperature.

Comments

Usual methods for evaluating CoQ_{10} in plasma and biological tissues are based on alcohol-hexane extraction; the extract is brought to dryness and injected into the HPLC apparatus, where CoQ_{10} is usually revealed and quantified at 275 nm or by ECD. An internal standard, such as CoQ_8 or CoQ_9, is often added to the sample before extraction to quantify a recovery.

In our method, the sample is only diluted with propanol and then injected into the HPLC. Direct injection of the propanol extract also makes the procedure particularly simple and fast. Prior oxidation of reduced CoQ_{10} with para-benzoquinone eliminates the possibility of underestimating total CoQ_{10} in fresh samples. In fact, fresh samples contain almost exclusively reduced CoQ_{10}, and the usual extraction procedures are often not sufficient to completely oxidize ubiquinol. On the other hand, UV detection at 275 only reveals ubiquinone. Our chromatographic features and the diode array analysis of the peak show a selective separation of CoQ_{10}, with no superimposed peaks. Reproducibility of the method is shown by a CV less than 2% for samples having normal values of plasma CoQ_{10} and about 5% for samples having a quarter of the normal amount.

The rather low retention time for CoQ_{10}, under our chromatographic conditions, makes the analysis fast enough, and our peak does not overlap with any other components, as shown by similar values obtained using the ECD method and by the diode array analysis of the peak.

The efficiency of CoQ_{10} extraction by our procedure seems to be very satisfactory, because the addition of 1, 2, or 3 μg of CoQ_{10} to a sample leads to a 96.3–98.5% quantitative recovery (see "Results"), which makes the use of an internal standard unnecessary. This is in agreement with data from Edlund's work,[13] where it seems that a dilution of the sample with *n*-propanol (1:4) leads to a 100% recovery.

Section II

Plasma Membrane Quinone Reductases

[15] Quinone Oxidoreductases of the Plasma Membrane

By D. JAMES MORRÉ

Transplasma Membrane Electron Transport

Quinone oxidoreductases of the plasma membrane relate functionally to the operation of a cell surface redox chain,[1] where cytosolic NAD(P)H is oxidized. Plasma membrane quinones serve as lipid-soluble transmembrane shuttles to transfer the $2H^+ + 2\ e^-$ from NAD(P)H to 1/2 O_2 to form water.[2] The reduction of 1/2 O_2 is at the expense of hydroquinone catalyzed by cell surface hydroquinone oxidases.[3,4] In this capacity, they function as terminal oxidases of plasma membrane electron transport.[5]

Plasma membrane electron transport is not a trivial cell function. Estimates are that up to 10% of the oxygen consumption of cultured mammalian cells is KCN insensitive and the result of plasma membrane electron transport.[6] In mitochondria-deficient r^o cells, survival depends on plasma membrane electron transport.[6,7] In these cells, plasma membrane electron transport is largely responsible for the life-sustaining regeneration of NAD^+ from NADH required for glycolytic ATP production. A number of authors have pointed out the potential importance of glycolytic ATP production even in cells containing mitochondria.[7,8] Mitochondria are widely separated in most cells. As a result, ATP-requiring reactions distant from mitochondria may be more reliant on glycolytic ATP than on mitochondrial ATP. Mitochondrial ATP production is particularly critical for transient high-energy needs, as for example, to support strenuous muscle activity.

The first demonstration of a redox-related plasma membrane enzyme was that of an NADH-ferricyanide reductase observed with purified fractions of plasma membranes isolated from rat liver.[9] A plasma membrane

[1] F. L. Crane, D. J. Morré, and H. Löw, "Oxidoreduction at the Plasma Membrane." Volumes I and II, CRC Press, Boca Raton, FL, 1990.
[2] Y. Hatefi, *Adv. Enzymol.* **25,** 275 (1963).
[3] T. Kishi, D. M. Morré, and D. J. Morré, *Biochim. Biophys. Acta* **1412,** 66 (1999).
[4] A. Bridge, R. Barr, and D. J. Morré, *Biochim. Biophys. Acta* **1463,** 448 (2000).
[5] D. J. Morré, R. Pogue, and D. M. Morré, *BioFactors* **9,** 179 (1999).
[6] J. A. Larm, F. Valliant, A. W. Linnane, and A. Lawen, *J. Biol. Chem.* **269,** 20097 (1994).
[7] A. Lawen, R. D. Martinus, G. L. McMullen, P. Nagley, F. Valliant, E. J. Wolvetang, and A. W. Linnane, *Mol. Aspects Med.* **15,** s13 (1994).
[8] K. Brand, *J. Bioenerg. Biomemb.* **29,** 355 (1997).
[9] E. L. Vigil, D. J. Morré, C. Frantz, and C. M. Huang, *J. Cell Biol.* **59,** 353a (1973).

0076-6879/04 $35.00

location was subsequently confirmed by electron microscope cytochemistry.[10] Involvement in plasma membrane electron transport was inferred from observations in which ferricyanide and other impermeant oxidants were reduced by intact cells.[11–13] A location at the cell surface was implicit from the inhibition of ferricyanide reduction by p-chloromercuri benzene sulfonate (PCMBS). PCMBS is impermeant and reacts only with thiols exported to or exposed at the cell surface.[13] However, the role of quinones and hydroquinone oxidases emerged somewhat later.[3–5] Initially, plasma membrane electron transport was considered to result more or less exclusively from the operation of an NADH ferricyanide reductase spanning the membrane bilayer (transplasma membrane ferricyanide reductase).[13]

Reduction of external ferricyanide is accompanied by oxidation of cellular NADH.[14,15] Oxidation of internal NADH is observed immediately on addition of ferricyanide or diferric transferrin to human cervical carcinoma (HeLa) cells, and the decrease reaches a maximum after about 15 to 20 min.[16] NAD^+ increases in parallel with the decrease in NADH, so the total pyridine nucleotide level in the cells does not change during short time periods. Insulin stimulates both the reduction of ferricyanide by HeLa cells and the oxidation of internal NADH.[14] On the other hand, doxorubicin (Adriamycin) inhibits both the transmembrane electron transport and the decline in NADH and increase in NAD^+ levels. In primary liver cells, ferricyanide induces a decrease in NADH of 33% in 5 min. With liver growth factor present, the ferricyanide-induced decline is 55%.[15]

The most likely donors for transmembrane electron transport were deduced to be substrates in the cytoplasm such as NADH, NADPH, and glutathione. Overall, NADH generated from glycolysis was regarded as the principal electron donor. Kruger and Böttger[17] point out that electron

[10] D. J. Morré, E. L. Vigil, C. Frantz, H. Goldenberg, and F. L. Crane, *Cytobiologie* **18,** 213 (1978).

[11] E. S. G. Barron and L. A. Hoffman, *J. Gen. Physiol.* **13,** 483 (1929).

[12] M. G. Clark, E. J. Patrick, G. S. Patten, F. L. Crane, H. Löw, and C. Grebing, *Biochem. J.* **200,** 565 (1981).

[13] F. L. Crane, I. L. Sun, M. G. Clark, C. Grebing, and H. Löw, *Biochim. Biophys. Acta* **811,** 233 (1985).

[14] P. Navas, I. L. Sun, D. J. Morré, and F. L. Crane, *Biochem. Biophys. Res. Commun.* **135,** 110 (1986).

[15] R. Garcia-Canero, *in* "Plasma Membrane Oxidoreductases in Control of Animal and Plant Growth" (F. L. Crane, D. J. Morré, and H. Löw, eds.), p. 27. Plenum, New York, 1988.

[16] H. Löw, F. L. Crane, D. J. Morré, and I. Sun, *in* "Oxidoreduction at the Plasma Membrane" (F. L. Crane, D. J. Morré, and H. Löw, eds.), p. 29. CRC Press, Boca Raton, FL, 1990.

[17] S. Kruger and M. Böttger, *in* "Plasma Membrane Oxidoreductases in Control of Animal and Plant Growth" (F. L. Crane, D. J. Morré, and H. Löw, eds.), p. 105. Plenum, New York, 1988.

transport across the plasma membrane will oxidize all the NADH in the cytoplasm in a matter of seconds, so that any sustained rate of transmembrane electron transport ultimately depends on pyridine nucleotide formation by glycolysis or by the hexose monophosphate shunt. This supposition is supported by studies in which electron transport by HeLa cells was blocked by inhibitors of glycolysis.[18] However, with primary liver cells, it was not possible to stimulate transmembrane ferricyanide reduction by adding glucose or other glycolytic substrates except for dihydroxyacetone.[19] Isolated Ehrlich ascites cells, on the other hand, did respond to glucose, lactate, and glutamine. These substances stimulated transmembrane electron transport within 10–20 min after addition.[20]

Plasma Membrane–Associated Ferricyanide Reductase

Initially, the nature of the electron acceptor for plasma membrane electron transport was a matter of speculation. Early studies used artificial electron acceptors such as ferricyanide, indophenol, or transferrin-bound iron as examples. The first ferricyanide reductases isolated and characterized from mammalian plasma membranes exhibited sequence homology or identity with NADH cytochrome b_5 oxidoreductases.[21,21a,22] Other natural electron acceptors other than molecular oxygen for plasma membranes include cytochrome *c*,[23,24] ascorbate free radical,[25–27] and protein disulfides.[28] Preparations consisting of proteins having prominent ferricyanide reductase activity with some capacity to reduce ferri-chelates have been reported for plant microsomes and plasma membranes.[29–32] These

[18] I. L. Sun, F. L. Crane, C. Grebing, and H. Löw, *Proc. Ind. Acad. Sci.* **95,** 137 (1986).

[19] K. Thorstensen, K. Thorstensen, I. Romslo, T. Nilsen, and O. Gederaas, *in* "Plasma Membrane Oxidoreductases in Control of Animal and Plant Growth" (F. L. Crane, D. J. Morré, and H. Löw, eds.), p. 127. Plenum, New York, 1988.

[20] M. A. Medina, F. Sanchez-Jimenez, J. A. Segura, and I. Nunerz de Castro, *Biochim. Biophys. Acta* **946,** 1 (1987).

[21] C. Kim, F. L. Crane, G. W. Becker, and D. J. Morré, *Protoplasma* **184,** 111 (1995).

[21a] J. M. Villaba, F. Navarro, F. Cordoba, A. Serrano, A. Arroyo, F. L. Crane, and P. Navas, *Proc. Natl. Acad. Sci. USA* **92,** 4887 (1995).

[22] T. M. Villalba, F. Navarro, C. Gomez-Diaz, A. Arroyo, R. I. Bello, and P. Navas, *Mol. Aspects. Med.* **18,** 57 (1997).

[23] T. Buckhout and T. C. Hrubec, *Protoplasma* **135,** 144 (1986).

[24] A. S. Sandelius, R. Barr, F. L. Crane, and D. J. Morré, *Plant Sci.* **48,** 1 (1987).

[25] D. J. Morré, P. Navas, C. Penel, and F. J. Castillo, *Protoplasma* **133,** 195 (1986).

[26] F. J. Alcain, M. I. Buron, J. M. Villalba, and P. Navas, *Biochim. Biophys. Acta* **1073,** 380 (1991).

[27] D. G. Luster and T. J. Buckhout, *Physiol. Plant.* **73,** 339 (1988).

[28] P.-J. Chueh, D. M. Morré, C. Penel, T. DeHahn, and D. J. Morré, *J. Biol. Chem.* **272,** 11221 (1997).

preparations were inactive with dQ, an artificial quinone used commonly in redox enzyme assays, and were distinct from both the intrinsic NAD(P)H:quinone oxidoreductase and the soluble NAD(P)H-quinone oxidoreductases of plasma membranes.

A Doxorubicin-Inhibited NADH-Coenzyme Q_0 Reductase (NADH-External Acceptor [Quinone] Reductase, EC 1.6.5) from Rat Liver Plasma Membranes

When cell growth was inhibited by the antitumor agent such as doxorubicin[33] or other tumor growth inhibitory anthracyclines, redox activities of the plasma membrane also were inhibited. Subsequently, a doxorubicin-inhibited NADH-quinone reductase was characterized and purified from plasma membranes of rat liver.[34] A doxorubicin-insensitive NADH-cytochrome b_5 reductase first was removed from the plasma membranes by the lysosomal protease, cathepsin D. After removal of the NADH-cytochrome b_5 reductase, the plasma membranes retained a doxorubicin-inhibited NADH-quinone reductase activity. The enzyme, potentially transmembrane, with an apparent molecular mass of 57 kDa, was purified 200-fold over the cathepsin D–treated plasma membranes. The purified enzyme also had an NADH-coenzyme Q_0 reductase (NADH:external acceptor [quinone] reductase; EC 1.6.5) activity. Partial amino acid sequence of the enzyme showed it to be unique, with no sequence homology to other known proteins. Antibody against the enzyme (peptide sequence) was produced and affinity purified. The purified antibody immunoprecipitated both the NADH-ferricyanide reducatase activity and NADH-coenzyme Q_0 reductase activity of plasma membranes and cross-reacted with cell surface determinants of human chronic myelogenous leukemia K562 cells and doxorubicin-resistant human chronic myelogenous leukemia K562R cells. Localization by fluorescence microscopy showed that the reaction was with the external surface of the plasma membranes. The doxorubicin-inhibited NADH-quinone reductase may provide a target for the anthracycline antitumor agents and remains a candidate ferricyanide reductase for plasma membrane electron transport.

---

[29] A. Serrano, F. Cordoba, J. A. Gonzales-Reyes, P. Navas, and J. M. Villalba, *Plant Physiol.* **106,** 87 (1994).

[30] P. Bagnaresi and P. Pupillo, *J. Exp. Bot.* **46,** 1497 (1995).

[31] A. Bérczi, B. Fredlund, and I. M. Møller, *Arch. Biochem. Biophys.* **320,** 65 (1995).

[32] P. Bagnaresi, B. Basso, and P. Pupillo, *Planta* **202,** 427 (1997).

[33] I. Sun, E. E. Sun, F. L. Crane, D. J. Morré, and P. Faulk, *Biochim. Biophys. Acta* **1105,** 84 (1992).

[34] C. Kim, F. L. Crane, W. P. Faulk, and D. J. Morré, *J. Biol. Chem.* **277,** 16441 (2002).

The Transplasma Membrane Electron Transport Chain

Although the possibility is not excluded of a physiological role for a transplasmic membrane NADH-ferricyanide or NADH-Q_0 reductase in plasma membrane electron transport from cytosolic NAD(P)H, artificial acceptors are required, and there is no possibility, except for ascorbate free radical, to transfer proton and electrons to natural acceptors such as protein thiols or molecular oxygen. This limitation was circumvented with the discovery of the ECTO-NOX proteins.[35] ECTO-NOX proteins are located on the cell surface, rather than being transmembrane. They are hydroquinone oxidases with protein disulfide–thiol interchange activity that also are capable of oxidizing external NAD(P)H, hence *N*AD(P)H *O*xidases, or NOX. The designation ECTO-NOX is given to distinguish the cell surface oxidases from the phox-NOX proteins of host defense.[36]

ECTO-NOX proteins were described initially as KCN insensitive NADH oxidases present in both animal and plant cells with protons and electrons being transferred to molecular oxygen to form water.[25,37,38] However, because there is little or no NADH at the cell surface, NADH must be regarded as a convenient but unphysiological substrate. Subsequently, terminal oxidase activities with the presumed natural substrate $CoQH_2$ (ubiquinol) or KH_2 (phylloquinol) were shown.[3–5]

The discovery of the ECTO-NOX family of proteins led ultimately to the formulation of a complete minimal electron transport chain for the plasma membrane depicted in Fig. 1.[5] The chain consists of three essential components: (1) One or more cytosolic or cytosolic surface located NAD(P)H:quinone reductases, (2) plasma membrane–associated quinone, and (3) an ECTO-NOX terminal oxidase. In addition, the capability of the plasma membrane redox system to reduce lipophilic quinones by means of the intrinsic NAD(P)H:quinone oxidoreductase complex might be extended to interactions with other redox components, for instance, *b*-type cytochromes, also present in plasma membranes[39–42] to comprise more

[35] D. J. Morré, *in* "Plasma Membrane Redox Systems and Their Role in Biological Stress and Disease" (H. Asard, A. Bérczi, and R. Caubergs, eds.), p. 121. Kluwer, Dordrecht, 1998.

[36] J. D. Lambeth, G. Cheng, R. S. Arnold, and W. A. Edens, *TIBS* **25,** 459 (2000).

[37] D. J. Morré, A. Brightman, J. Wang, R. Barr, and F. L. Crane, *in* "Plasma Membrane Oxidoreductases in Control of Animal and Plant Growth" (F. L. Crane, D. J. Morré, and H. Löw, eds.), p. 45. Plenum, New York, 1988.

[38] D. J. Morré, *J. Bioenerg. Biomemb.* **26,** 421 (1994).

[39] H. Asard, R. Caubergs, D. Renders, and J. A. De Greef, *Plant Sci.* **53,** 109 (1987).

[40] H. Goldenberg, *Biochim. Biophys. Acta* **694,** 203 (1982).

[41] H. Goldenberg, *Protoplasma* **205,** 3 (1998).

[42] S. Lüthje, O. Döring, S. Heuer, H. Lüthen, and M. Böttger, *Biochim. Biophys. Acta* **1331,** 81 (1997).

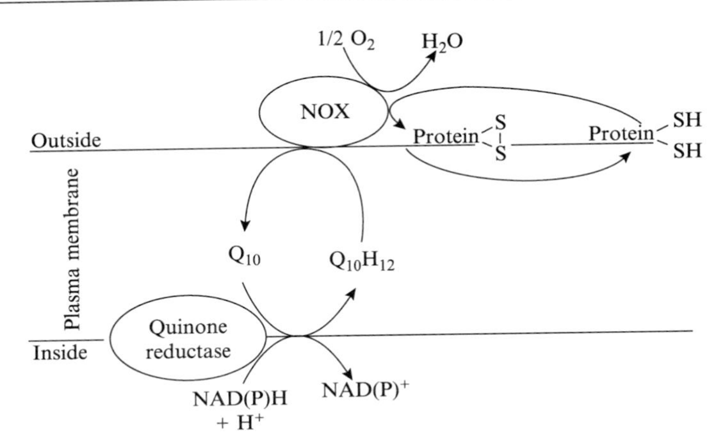

FIG. 1. Diagram illustrating the spatial relationships between quinones and the external NADH oxidase (ECTO-NOX) protein to donate electrons from cytosolic NAD(P)H to molecular oxygen.[3-5] The ECTO-NOX protein that functions as a terminal oxidase of plasma membrane electron transport alternates with an exogenous donor-independent and cell enlargement–related protein disulfide-thiol interchange activity depicted on the right.

extensive transmembrane electron transport, still with quinones shuttling reducing equivalents through the membrane.

Thus far, nicotinamide adenine dinucleotides (NADH or NADPH) are the prime cytosolic electron donors studied by most investigators.[43] However, the natural electron acceptors for the membrane redox systems have been a matter of controversy. Identification of electron acceptors has been validated from various laboratories, particularly for molecular oxygen. Besides both electron donors and acceptors, the intermediate carriers have also drawn extensive attention. Among those, coenzyme Q has been well investigated. Coenzyme Q is a lipid-soluble redox constituent that functions as an intermediate electron carrier.[44,45] Other cofactors, for instance, cytochrome *b*,[13] α-tocopherylquionone,[46] or flavins,[47,48] also have been suggested.

[43] P. C. Misra, *J. Bioenerg. Biomemb.* **23,** 425 (1991).

[44] J. M. Villalba, F. Navarro, F. Cordoba, A. Serrano, A. Arroyo, F. L. Crane, and P. Navas, *Proc. Natl. Acad. Sci. USA* **92,** 4887 (1995).

[45] I. L. Sun, E. E. Sun, F. L. Crane, D. J. Morré, A. Lindgren, and H. Löw, *Proc. Natl. Acad. Sci. USA* **89,** 11126 (1992).

[46] I. L. Sun, E. E. Sun, and F. L. Crane, *Biochem. Biophys. Res. Commun.* **189,** 8 (1992).

[47] D. G. Luster and T. J. Buckhout, *Plant Physiol.* **91,** 1014 (1989).

[48] I. M. Møller, K. M. Fredlund, and A. Bérczi, *Protoplasma* **184,** 124 (1995).

Lipophilic Quinones of the Plasma Membrane

Lipophilic quinones are essential lipid-soluble electron transport components distributed widely among cellular components.[49,50] Plasma membranes contain concentrations of quinones that equal or exceed those present in mitochondria.[49]

For most mammalian cells, the predominant quinone is coenzyme Q_{10} or coenzyme Q_9 for the rat.[2] Plant plasma membranes lack significant amounts of ubiquinones,[42,51] such that ubiquinones would not be expected to serve as significant sources of reducing equivalents for plasma membrane redox in plants. Rather than ubiquinones, plant plasma membranes contain phylloquinones (vitamin K_1).[42,51] Reduced phylloquinones serve as an electron donor for the plasma membrane ECTO-NOX proteins of plants,[4] with a specific activity comparable to that for ubiquionol in animals.[3]

At the plasma membrane, reduced ubiquionone (Q_{10}) was originally considered to function primarily as an antioxidant to inhibit lipid peroxidation or protein carboxylation.[50,52] Subsequently, hydroquinones of the plasma membranes were shown to participate in the plasma membrane electron transport chain and growth control.[45,46]

Cytosolic Quinone Reductases

Plasma membrane quinones most likely become reduced at the inner surface of the plasma membrane through the activity of an NAD(P)H coenzyme Q reductase or diaphorase.[52] A similar activity is located in the cytosol.[53] The QH_2 would be free to migrate through the lipid interior of the membrane, where it would become available for oxidation by ECTO-NOX proteins. The combination of NAD(P)H coenzyme Q reductase with plasma membrane quinones and ECTO-NOX proteins represents the first formulation of an electron transport chain for the plasma membrane based on known proteins and physiological electron donors, carriers, and acceptors.[5]

Diaphorase. The enzyme DT diaphorase [NAD(P)H:quinone-acceptor) oxidoreductase 1, (NQ01) [EC16.99.2] discovered shortly after coenzyme Q is largely cytosolic and inducible.[54] A concentration of 5 μM dicoumarol,

[49] A. Kalen, B. Norling, E. L. Appelkvist, and G. Dallner, *Biochim. Biophys. Acta* **926,** 70 (1987).

[50] F. Aberg, E. L. Appelkvist, G. Dallner, and L. Ernster, *Arch. Biochem. Biophys.* **295,** 230 (1992).

[51] O. Döring and S. Lüthje, *Mol. Member. Biol.* **19,** 127 (1996).

[52] P. Forsmark-Andree, G. Dallner, and L. Ernster, *Free Radic. Biol. Med.* **19,** 749 (1995).

[53] T. Takahashi, T. Okamoto, and T. Kishi, *J. Biochem.* **119,** 256 (1996).

[54] T. Iyanagi and I. Yamazaki, *Biochim. Biophys. Acta* **216,** 282 (1970).

an inhibitor of DT-diaphorase that does not inhibit cytosolic NAD(P)H–dependent UQ reductase *in vitro* did not inhibit reduction of endogenous CoQ_9 or CoQ_{10} added to cultured rat hepatocytes, suggesting that NQO1 is not involved in maintaining CoQ in the regeneration of cellular $CoQH_2$ and its antioxidant actions.[55] Furthermore, CoQ homologues with long isoprenoid chains were not good substrates for NQO1[56] compared with low molecular weight quinones such as menadione and 2,3-dimethyl-6-methyl-1,4 benzoquionone. Considering these points, NQO1 is unlikely to be the enzyme responsible for the reduction of plasma membrane conezyme Q.

In plants, an NAD(P)H-Q quinone reductase has been purified and characterized[57–59] that represents a functional equivalent of the animal DT-disphorase. It reduces short-chain quinones to quinols by two-electron donation without semiquinone intermediates. The result is enhanced quinone conjugation and a low probability of the formation of active oxygen species.[59] An FMN-binding oxidoreductase of the NAD(P)H-quinone reductase type involved in the detoxification of quionones produced naturally by the fungal degradation of lignin has been reported for the basidiomycete *Phanaerochete chrysosporium*.[60,61]

A number of properties of the plant oxidoreductase are different from animal-type dT-diaphorase. The NAD(P)H-quinone reductase from plants contains flavin mononucleotide (FMN) and has a mass of about 90 kDa with subunits of 21.4 kDa (by mass spectrometry). The hydride transfer from NAD(P)H to the flavin is β-stereospecific.[58] The animal DT-diaphorase is flavin adenine dinucleotide (FAD)-containing, astereospecific and dimeric with 26- to 31-kDa subunits.[62,63]

NADH-Quinone Reductases. Reportedly, enzymes do exist that serve as more likely candidates for cytosolic or plasma membrane–associated coenzyme Q reductases. In animal cells, these include both NADH cytochrome b_5 reductases[22,64] and NADH-NAD(P)H dehydrogenases.[22,44,64] Quinone reductases are flavoenzymes that catalyze the transfer of two

[55] T. Kishi, T. Takahashi, S. Mizobachi, K. Mori, and T. Okamoto, *Free Radical Res.* **36,** 413 (2002).
[56] L. Ernster, *Meth. Enzymol.* **10,** 309 (1967).
[57] A. Rescigno, F. Sollai, S. Masala, M. C. Porcu, E. Sanjust, A. C. Rinaldi, N. Curreli, D. Grifi, and A. Rinaldi, *Prep. Biochem.* **25,** 57 (1995).
[58] F. Sparla, G. Tedeschi, and P. Trost, *Plant Physiol.* **112,** 249 (1996).
[59] P. Trost, P. Bonora, S. Scagliarini, and P. Pupillo, *Eur. J. Biochem.* **234,** 452 (1995).
[60] B. J. Brock and M. H. Gold, *Arch. Biochem. Biophys.* **331,** 31 (1996).
[61] B. J. Brock, S. Rieble, and M. H. Gold, *Appl. Environ. Microbiol.* **61,** 3076 (1995).
[62] L. Ernster, *Chem. Scripta* **27A,** 1 (1987).
[63] A. K. Jaiswal, P. Burnett, M. Alesnik, and O. W. McBride, *Biochemistry* **29,** 1859 (1990).
[64] F. Navarro, J. M. Villalba, F. L. Crane, W. C. MacKellar, and P. Navas, *Biochem. Biophys. Res. Commun.* **212,** 138 (1995).

electrons from cellular donors (mainly NADH or NADPH) to a variety of quinone acceptors. These enzymes have been classified into two categories according to mechanisms by which quinones become reduced. Two electrons are sequentially donated with the concomitant generation of semiquinones in reactions catalyzed by those designated as *one-electron quinone reductases*. Included in this category are the microsomal enzymes cytochrome b_5 and cytochrome P450 reductases.[65] The resultant semiquinones react readily with molecular oxygen to generate superoxide and other highly reactive oxygen species (ROS) potentially involved in oxidative stress and damage to cellular structures.[66] Quinone reductases that use an obligatory two-electron reaction mechanism result in a direct hydride transfer to a variety of quinone substrates to generate the corresponding hydroquinones.[67,68,68a] These *two-electron quinone reductases* are mainly enzymes of the cytosol typified by NAD(P)H:(quinone acceptor) oxidoreductase 1 (NQO1, DT-diaphorase).[69,70] Because many hydroquinones are relatively stable, two-electron quionone reductases are generally regarded as antioxidant enzymes.[69] Cytosolic NQO1 activity maintains the reduced states of ubiquinones[70–73] as well as α-tocopherolquinones,[74] which might be expected to promote antioxidant functions in membranes.

The plant NAD(P)H-quinone reductases bear structural and kinetic resemblance to other enzymes purified from plant plasma membranes,[29,47] as well as rotenone-insensitive NAD(P)H dehydrogenases[75,76] from mitochondria. These enzymes are relatively unspecific toward pyridine

[65] M. Nakamura and T. Hayashi, *J. Biochem. (Tokoyo)* **115,** 1141 (1994).
[66] P. Joseph and A. K. Jaiswal, *Proc. Natl. Acad. Sci. USA* **91,** 8413 (1994).
[67] M. Faig, M. A. Bianchet, P. Talalay, S. Chen, S. Winski, D. Ross, and L. M. Amzel, *Proc. Natl. Acad. Sci. USA* **97,** 3177 (2000).
[68] R. Li, M. A. Bianchet, P. Talalay, and L. M. Amzel, *Proc. Natl. Acad. Sci. USA* **92,** 8846 (1995).
[68a] A. Strassburg, P. C. Strassburg, M. P. Manns, and R. H. Tukey, *Mol. Pharmacol.* **61,** 320 (2002).
[69] E. Cadenas, *Biochem. Pharmacol.* **49,** 127 (1995).
[70] A. K. Jaiswal, *Free Radic. Biol. Med.* **29,** 254 (2000).
[71] R. E. Beyer, J. Segura-Aguilar, S. Di Bernardo, M. Cavazzoni, R. Fato, D. Fiorentini, M. C. Galli, M. Setti, L. Landi, and G. Lenaz, *Proc. Natl. Acad. Sci. USA* **93,** 2528 (1996).
[72] L. Landi, D. Fiorentini, M. C. Galli, J. Segura-Aguilar, and R. E. Beyer, *Free Radic. Biol. Med.* **22,** 329 (1997).
[73] F. Navarro, P. Navas, J. R. Burgess, R. I. Bello, R. De Cabo, A. Arroyo, and J. M. Villalba, *FASEB J.* **12,** 1665 (1998).
[74] D. Siegel, E. M. Bolton, J. A. Burr, D. C. Liebler, and D. Ross, *Mol. Pharmacol.* **52,** 300 (1997).
[75] M. H. Luethy, M. K. Hayes, and T. E. Elthon, *Plant Physiol.* **97,** 1317 (1991).
[76] A. G. Rasmusson, K. M. Fredlund, and I. M. Møller, *Biochim. Biophys. Acta* **1141,** 107 (1993).

nucleotides and use hydrophilic quionones efficiently, whereas cytochrome *c*, oxygen, and ferri-chelates are not reduced to a significant extent. With the exception of the 43-kD, NAD(P)H dehydrogenase purified from sugar beet mitochondria by Luethy *et al.*,[75] these dehydrogenases are composed of subunits of 25–27 kDa,[29,47,76] similar to the apparent molecular mass of 22–24 kDa of the NAD(P)H-quinone reductase, as calculated from sodium dodecylsulfate–polyacrylamide gel electrophoresis (SDS–PAGE).[57–59,61] The mitochondrial dehydrogenases are extracted from the membrane without the aid of detergents[75,76] and, like NAD(P)H-quinone reductases, tend to be hydrophilic.

Another type of particulate pyridine nucleotide–dependent quinone oxidoreductase designated NADH-DQ reductase[77] is widespread in plant microsomal membranes. Unlike soluble NAD(P)H-quinone reductase, the NADH-DQ reductase is strongly bound to the membrane,[77a] and the molecular mass is larger (340 kDa).[78] NADH-DQ prefers NADH as donor in sugar beet microsomes,[78] whereas NADH and NADPH are more nearly equally preferred as electron donors in zucchini (*Cucurbita pepo* L.) plasma membranes. The NADH-DQ has been solubilized and purified from the plasma membrane of etiolated zucchini hypocotyls and retains properties of the intrinsic complex in the course of purification rather than acquiring properties of a soluble NAD(P)H-QR–like flavoprotein as might have been expected.[78a]

ECTO-NOX as Terminal Oxidases of Plasma Membrane Electron Transport

A major function of the family of cell surface ECTO-NOX proteins is as a terminal oxidase for plasma membrane electron transport (Fig. 1).[35] Their properties are summarized in Table I. Electrons are transferred with a stoichiometry of 1 mole of external NADH or quinol oxidized per 1/2 mole of O_2 reduced.[79] Superoxide is not a major product. With one exception, an age-related ECTO-NOX (arNOX), the ECTO-NOX proteins do not catalyze the reduction of ferricytochrome *c* or the tetrazolium dyes such as XTT used for detection of superoxide. Ferricyanide and other artificial electron acceptors are not cosubstrates with NAD(P)H.[38]

[77] P. Pupillo, V. Valenti, I. DeLuca, and R. Hertel, *Plant Physiol.* **80,** 384 (1986).
[77a] V. Valenti, F. Guerrini, and P. Pupillo, *J. Exp. Bot.* **41,** 183 (1990).
[78] F. Guerrini, A. Lombini, M. Bizarri, and P. Pupillo, *J. Exp. Bot.* **45,** 1227 (1994).
[78a] P. Trost, S. Foscarini, V. Proger, P. Bonora, L. Vitale, and P. Pupillo, *Plant Physiol.* **114,** 737 (1997).

TABLE I
PROPERTIES OF ECTO-NOX PROTEINS

Oxidative activity
Donor: Hydroquinone
NAD(P)H
Acceptor: Molecular oxygen → H_2O
Protein disulfides → protein thiols
Protein disulfide–thiol interchange
Restores activity to scrambled RNase (No exogenous donor required)
Dithiodipyridine substrates
Two activities alternate to generate a period length of 24 min (22 min for tNOX, 26 min for arNOX)
Low turnover number: 200–500
Specific activity: 10–20 μmoles/min/mg protein
Two moles of zinc and potentially up to 1 mole copper bound/mole of protein tNOX
No flavin, heme or non-heme iron
Refractory to *N*-terminal sequencing
Protease resistant
Located at external cell surface
No GPI anchor or membrane-spanning domains
No ancillary proteins required for activity

The ECTO-NOX proteins are widely distributed among animals and plant cells and tissues and bacteria. A diversity of ECTO-NOX forms is described on page 194. A constitutive ECTO-NOX, designated CNOX, seems to be highly conserved among species. A polyclonal antibody raised to human CNOX cross-reacts with CNOX from *Escherichia coli* and from plants.[80] Molecular weight estimates range from 57 kDa for tissue culture cells to approximately 24 kDa for a processed CNOX fragment shed into human sera.[80]

Although both NADH and NAD(P)H serve as donors for CNOX, specific activities with NADH are usually approximately twice those with NAD(P)H. Despite the absence of flavin, NAD(P)H oxidation by CNOX is inhibited by diphenyliodonium (a putative flavine site inhibitor), whereas as NADH oxidation is resistant to diphenyliodonium inhibition.[81]

Hydroquinone oxidation was demonstrated and characterized using an ECTO-NOX preparation from the surface of HeLa cells.[3] The

[79] D. J. Morré, P.-J. Chueh, J. Lawler, and D. M. Morré, *J. Bioenerg. Biomemb.* **30,** 477 (1998).

[80] D. Sedlak, D. M. Morré, and D. J. Morré, *Arch. Biochem. Biophys.* **386,** 106 (2001).

[81] D. J. Morré, *Antioxid. Redox Signal* **4,** 207 (2002).

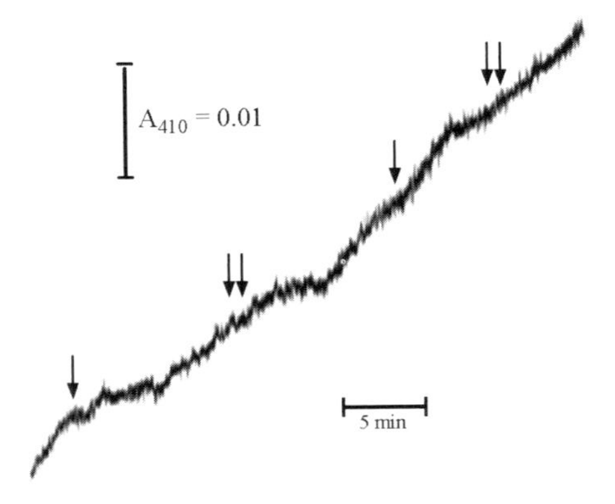

FIG. 2. Oxidation of $Q_{10}H_2$ as determined from the increase in absorbance at 410 nm monitored with an Aminco SLM 2000 in the dual wavelength mode of operation.[3] The enzyme source was a solubilized and partially purified ECTO-NOX preparation from the surface of human cervical carcinoma (Hela) cells that contained both CNOX (24-min period) and tNOX (22-min period). Periods of rapid oxidation (arrows) of $Q_{10}H_2$ (Tishcon, Westburg, NY) alternate with periods of much slower $Q_{10}H_2$ oxidation. The single arrows are separated by 24 min, whereas the double arrows are separated by 22 min. Results where $Q_{10}H_2$ oxidation was monitored from the decrease in A_{290} were similar.

preparations lacked NADH:ubiquinone reductase activity. Reduced coenzyme Q_{10} ($Q_{10}H_2$ = ubiquinol) was oxidized (Fig. 2) at an average rate of 15 nmol min^{-1} mg $protein^{-1}$, with an apparent K_m for $Q_{10}H_2$ oxidation of 33 μM.

With plants, vitamin K_1 hydroquinol (phylloquinol) oxidation was demonstrated and characterized by use of ECTO-NOX activities of plasma membrane vesicles and CNOX solubilized from plasma membranes of dark-grown soybeans *(Glycine max)*.[4] The reduced vitamin K_1 was oxidized at a rate of about 10 nmol min^{-1} mg $protein^{-1}$. Consumption of O_2 determined in parallel was 5 nmol min^{-1} mg protein.$^{-1}$ Both activities were stimulated at 0.1 μM by the synthetic auxin growth factor 2,4-dichlorophenoxyacetic acid (2,4-D).

ECTO-NOX Proteins Carry Out Protein Disulfide Thiol Interchange

ECTO-NOX proteins do transfer reducing equivalents from NADH to protein disulfides[28,79] as an alternative to oxygen and to sustain NADH oxidation by plasma membrane vesicles in an argon or nitrogen atmosphere. However, with plasma membrane preparations, at least NADH oxidation does take place under argon or nitrogen, suggesting that either the low levels of remaining dissolved oxygen are sufficient to sustain NADH oxidation or that other electron acceptors such as protein thiols are used under anaerobic conditions. Evidence for the latter has been provided by Chueh *et al.*[28] and Morré *et al.*[79]

In 1986, our laboratory first described from plants an NADH oxidase activity that was growth factor stimulated.[25] This activity was purified from a plant source[82] and was shown to be present in mammalian plasma membranes, where it also was hormone and growth factor stimulated,[83] and located at the external plasma membrane surface.[84] Both molecular oxygen[38] and protein disulfides[28,79] apparently serve as electron acceptors for electron transfer from NADH. In addition to an oxidative function, the enzyme also carried out protein disulfide–thiol interchange in the absence of NAD(P)H much in the manner of classical protein disulfide isomerases.[28,79,85,86] The interchange activity was estimated either from restoration of activity to reduced, denatured, and oxidized (scrambled) ribonuclease A[85–87] or from the cleavage of dithiodipyridine substrates.[88] Neither reduced nor oxidized glutathione served as ECTO-NOX substrates.

Although the protein disulfide–thiol interchange activity of ECTO-NOX proteins was similar to that of protein disulfide isomerases (PDI), ECTO-NOX proteins and PDI shared no other obvious features. Moreover, the cancer-specific ECTO-NOX form, tumor-associated NOX (tNOX), which has been cloned[89] and carries out protein disulfide thiol interchange typical of ECTO-NOX proteins, lacks the C-X-X-C motif characteristic of PDI.[90] In addition, protein disulfide–thiol interchange activities of CNOX were not inhibited either by antibodies to authentic PDI or to peptide antibodies to the C-X-X-C PDI motif.[86]

[82] A. O. Brightman, R. Barr, F. L. Crane, and D. J. Morré, *Plant Physiol.* **86,** 1264 (1988).

[83] A. O. Brightman, J. Wang, R. K. Miu, I. L. Sun, R. Barr, F. L. Crane, and D. J. Morré, *Biochim. Biophys. Acta* **1105,** 109 (1992).

[84] D. J. Morré, *Biochim. Biophys. Acta* **1240,** 201 (1995).

[85] D. J. Morré, D. de Cabo, E. Jacobs, and D. M. Morré, *Plant Physiol.* **109,** 573 (1995).

[86] D. J. Morré, E. Jacobs, M. Sweeting, R. de Cabo, and D. M. Morré, *Biochim. Biophys. Acta* **1325,** 117 (1997).

[87] M. M. Lyles and H. F. Gilbert, *Biochemistry* **30,** 613 (1991).

[88] D. J. Morré, M. L. Gomez-Rey, K. C. Schramke, O. Em, J. Lawler, J. Hobeck, and D. M. Morré, *Mol. Cell. Biochem.* **207,** 7 (1999).

[89] P.-J. Chueh, C. Kim, N. M. Cho, D. M. Morré, and D. J. Morré, *Biochemistry* **41,** 3732 (2002).

[90] K. Ohnishi, Y. Niimura, M. Hidaka, H. Masaln, H. Suzuki, T. Uozumi, and T. Nishino, *J. Biol. Chem.* **270,** 5812 (1995).

The Two ECTO-NOX Activities of Quinone Oxidation and Protein Disulfide–Thiol Interchange Alternate to Create Sustainable, Copper-Dependent Oscillatory Patterns

ECTO-NOX proteins exhibit both an oxidative function (quinol or NAD[P]H oxidation) and protein disulfide–thiol interchange as detailed in the preceding sections. What is unusual about the ECTO-NOX proteins and unprecedented in the biochemical literature is that maxima in rates of the two activities alternate to generate a regular and sustained pattern of oscillations with a period length of 24 min for the constitutive CNOX. The oscillations were first reported for oxidation of NADH[91] but have been subsequently observed for protein disulfide–thiol interchange,[81,92] the oxidation of ubiquinol by tNOX of HeLa cells (Fig. 3),[3] and the oxidation of K_1H_2 by soybean plasma membranes.[4]

The NOX oscillations are not simple sine functions.[89,92–94] In the oxidizing portion of the cycle, hydroquinone or NADH oxidation is normally represented by one or two maxima of variable ratios in amplitude followed

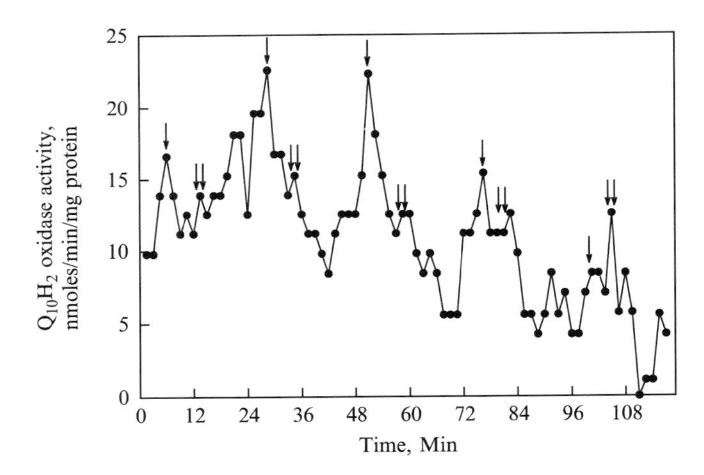

FIG. 3. Rate measurements over 1 min at 1.5-min intervals to illustrate the periodic variation in the rate of oxidation of $Q_{10}H_2$ as a function of time over 114 min. The enzyme source was the solubilized and partially purified ECTO-NOX preparation of Fig. 2 from the surface of human cervical carcinoma (Hela) cells that contained both a dominant CNOX component (24-min period, single arrows) and tNOX (22-min period, double arrows).[3]

[91] D. J. Morré and D. M. Morré, *Plant J.* **16,** 279 (1998).

[92] P. Sun, D. J. Morré, and D. M. Morré, *Biochim. Biophys. Acta* **1498,** 52 (2000).

[93] R. Pogue, D. M. Morré, and D. J. Morré, *Biochim. Biophys. Acta* **1498,** 44 (2000).

[94] S. Wang, R. Pogue, D. M. Morré, and D. J. Morré, *Biochim. Biophys. Acta* **1539,** 192 (2001).

by three minor oscillations. When disulfide–thiol interchange is measured, the two maxima of NADH oxidation coincide exactly with two minor oscillations, whereas disulfide–thiol interchange is now represented by three major oscillations seen as minor oscillations with NADH oxidation. Each of the components of the 2 + 3 pattern of oscillations repeats with a precise and temperature-compensated (independent of temperature) period length of approximately 24 min for CNOX, 22 min for tNOX, or 26 min for the age-related ECTO-NOX (arNOX).

Because blanks without NADH or without an enzyme source did not exhibit oscillatory absorbance changes,[91,94] the oscillations observed with the complete system are inherent in the cell surface NADH oxidase protein itself and are not a function of machine variation or more complex environments. The oscillatory behavior is similar for partially purified NOX proteins released from the HeLa cell surface as for the NOX activity of isolated plasma membrane vesicles and whole cells. As discussed later, the 2 + 3 pattern of oscillations is seen as well for cell enlargement (page 196), with enlargement being restricted to the protein disulfide–thiol interchange portion of the ECTO-NOX cycle of activity oscillations.

The 24-min periodic behavior is exhibited by pure recombinant tNOX protein and is accompanied by a recurring pattern of amide I/amide II Fourier transform infrared (FTIR) and Circular dichroism (CD) spectral changes suggestive of α-helix-β-sheet transformations.[95] Moreover, it now seems that a central role in the generation of the oscillations is played by ECTO-NOX–bound copper.

The ECTO-NOX periodic oscillations are copper dependent. H546A and H562A replacements in the copper site of tNOX conserved with superoxide dismutase[96] resulted in loss of enzymatic activity of the protein. The H546A replacement resulted as well in the loss of a characteristic pattern in the amide I to amide II ratio determined by FTIR consisting of five more or less equally spaced maxima within each 22-min tNOX period. The H546A replacement did not eliminate the pattern of amide I–amide II change but resulted in their random occurrence. The findings suggest that the bound copper of tNOX plays an essential role not only in the enzymatic activity of tNOX but in maintaining the structural changes that underlie the periodic alternations in activity that define the 24-min timekeeping cycle of the protein.

More recently, it has been possible to demonstrate the copper dependence of the periodic ECTO-NOX oscillations through direct complexation

[95] D. J. Morré and D. M. Morré, *Free Radic. Res.* **37,** 795 (2003).

[96] M. E. Schinina, P. Carlinin, F. Polticelli, F. Zappacosta, F. Bossa, and L. Calabrese, *Eur. J. Biochem.* **237,** 433 (1996).

of copper from recombinant tNOX by bathocuproene. Recombinant tNOX normally contains up to 0.8 moles of copper per mole of protein. The bound copper is resistant to removal by treatment with bathocuproene, a copper chelating agent, and cannot be increased by addition of copper. To make the tNOX copper accessible to chelation, a protocol was developed in which the protein was unfolded in the presence of 1% trifluoroacetic acid for 2 h also in the presence of a 100-fold excess of bathocuproene. The protein was then assayed directly and compared with protein unfolded in the absence of bathocuproene. The acid-treated tNOX still retained an oscillatory activity commensurate with its copper content. The period length remained at 22 min, as is characteristic of tNOX. However, when unfolded in the presence of the copper chelator, bathocuproene, the protein no longer exhibited discernible oscillatory activity.

Other ECTO-NOX Forms

At least two forms of ECTO-NOX activities have been distinguished on the basis of response to hormones,[83,96a] growth factors,[96a] capsaicin (8-methyl-*N*-vanillyl-6-noneamide),[97,98] and certain other quinone-site inhibitors or potential quinone-site inhibitors with anticancer activity.[35,99] The constitutive ECTO-NOX, designated CNOX, is hormone responsive and refractory to the quinone-site inhibitors.[35] A tNOX is unregulated, refractory to hormones and growth factors,[96a] and responds to inhibitors.[35,95,97] CNOX proteins are widely distributed and exhibit activity oscillations with a period length of 24 min.[35,80] Tumer-associated NOX proteins are cancer specific and exhibit oscillations with a period length of about 22 min, approximately 2 min shorter than that of CNOX.[89,100]

Tumor-Associated NOX

The cancer-associated and constitutively activated ECTO-NOX form has been cloned and expressed.[89] Designated tNOX, it is present at the cell surface of invasive human cancers.[84,101] Tumor-associated NOX is shed into the circulation[102] and, together with the cell surface form, provides a

[96a] M. Bruno, A. O. Brightman, J. Lawrence, D. Werderitsh, D. M. Morré, and D. J. Morré, *Biochem. J.* **284,** 625 (1992).

[97] D. J. Morré, P.-J. Chueh, and D. M. Morré, *Proc. Natl. Acad. Sci. USA* **92,** 1831 (1995).

[98] D. J. Morré, E. Sun, C. Geilin, L.-Y. Wu, R. de Cabo, K. Krasasakis, C. E. Orfanos, and D. M. Morré, *Eur. J. Cancer* **32A,** 1995 (1996).

[99] D. J. Morré, L.-Y. Wu, and D. M. Morré, *Biochim. Biophys. Acta* **1240,** 11 (1995).

[100] S. Wang, R. Pogue, D. M. Morré, and D. J. Morré, *Biochim. Biophys. Acta* **1539,** 192 (2001).

[101] N. M. Cho, P.-J. Chueh, C. Kim, S. Caldwell, D. M. Morré, and D. J. Morré, *Can. Immunol. Immunotheraphy* **51,** 121 (2002).

potential drug, vaccine, and diagnostic target for cancer. The cDNA sequence of 1830 bp is located within gene Xq25–26 and yields an open reading frame encoding 610 amino acids. The activities of the bacterially expressed tNOX oscillate with a period length of 22 min as is characteristic of tNOX activities *in situ*.[89] The activities are inhibited completely by capsaicin, which provides a defining characteristic of tNOX activity.[97,103,104] Functional motifs identified by site-directed mutagenesis within the C-terminal portion of the tNOX protein corresponding to the processed plasma membrane–associated form include quinone (capsaicin), copper and adenine nucleotide binding domains, and two cysteines essential for catalytic activity.[89,105] Four of the six cysteine to alanine replacements retained enzymatic activity, but the period lengths of the oscillations were increased (page 198).

Age-Related NOX

A distinctive age-related ECTO-NOX (arNOX) in which activity is blocked by coenzyme Q_{10} also has been described.[106,107] Age-related NOX occurs predominantly in aged cells and tissues. The period length of the oscillations is 26 min. Rather than reducing 1/2 O_2 to H_2O, electrons are transferred to O_2 to form superoxide. Both ubiquinol and NADH are oxidized by the activity with similar characteristics. Superoxide formation was demonstrated by superoxide dismutase–sensitive reduction of ferricytochrome *c*[108] and by reduction of a superoxide-specific tetrazolium salt such as XTT.[109] Both quinol and NADH reduction were inhibited by quinone forms.[107] Quinone inhibition was given by coenzymes Q_8, Q_9, and Q_{10} but not by Q_0, Q_2, Q_4, Q_6, or Q_7. The arNOX provides a mechanism to propagate ROS generated at the cell surface to surrounding cells and circulating lipoproteins of importance to atherogenesis.[110,111] Inhibition of arNOX by dietary coenzyme Q_{10} provides a rational basis for dietary coenzyme Q_{10} use to retard aging-related arterial lesions.[106]

[102] P.-J. Chueh, D. J. Morré, F. E. Wilkinson, J. Gibson, and D. M. Morré, *Arch. Biochem. Biophys.* **342,** 38 (1997).
[103] E. Yantiri and D. J. Morré, *Arch. Biochem. Biophys.* **391,** 149 (2001).
[104] A. del Castillo-Olivares, P.-J. Chueh, S. Wang, M. Sweeting, F. Yantiri, D. Sedlak, D. M. Morré, and D. J. Morré, *Arch. Biochem. Biophys.* **358,** 125 (1998).
[105] P.-J. Chueh, D. M. Morré, and D. J. Morré, *Biochim. Biophys. Acta* **1594,** 74 (2002).
[106] D. M. Morré, F. Guo, and D. J. Morré, *Mol. Cell. Biochem.,* In Press (2003).
[107] D. M. Morré and D. J. Morré, *BioFactors,* In Press (2003).
[108] J. Butler, W. H. Koppenol, and E. Margoliash, *J. Biol. Chem.* **257,** 10747 (1982).
[109] M. W. Sutherland and B. A. Learmonth, *Free Radic. Res.* **27,** 283 (1997).
[110] A. D. N. J. deGrey, "The Mitochondrial Free Radical Theory of Aging," Landes, Austin, 1999.
[111] D. M. Morré, G. Lenaz, and D. J. Morré, *J. Exp. Biol.* **203,** 1513 (2000).

Gorman *et al.*[112] reported earlier that ultraviolet (UV) irradiation of HL-60 cells resulted in an ability to generate superoxide. On the basis of inhibition by diphenyliodinium, a putative specific inhibitor of NADPH oxidases, Gorman *et al.*[112] further suggested that the UV target was a cell surface NADPH oxidase. These findings were confirmed in our studies[5] using superoxide dismutase–sensitive reduction of external ferricytochrome *c*[108] as an assay for superoxide generation.

Using the ferricytochrome *c* assay, an activity capable of generating superoxide was shown to be present with sera and buffy coats from aged patients but absent from sera and buffy coats of younger individuals.[106,104] A number of features identified the aging-related superoxide-generating oxidase (arNOX) as an ECTO-NOX. The activity was resistant to protease digestion (proteinase K) and resistant to heating to temperatures between 70° and 80°.[80] The superoxide-generating activity was not steady state but exhibited a pattern of oscillations with a characteristic period length as a defining characteristic of ECTO-NOX proteins (see previous section). However, the period length of the arNOX was about 26 min[5,106,107] rather than 24 min as is characteristic of CNOX.[80] The superoxide-generating activity has subsequently been demonstrated in aged cell cultures and associated with plasma membranes prepared from aged plant tissues (senescing spinach, skunk cabbage spathes). The arNOX activity was not inhibited by capsaicin, (−)-epigallocatechin gallate (EGCg),[113] nor other inhibitors of the cancer-associated tNOX.[35,95]

Role in Plasma Membrane Redox and ECTO-NOX Proteins in Growth

Plasma membrane electron transport has long been related to cell proliferation and cell differentiation[113] as first proposed by Kay and Ellem.[114] The constitutive ECTO-NOX proteins are activated by several ligands, including epidermal growth factor (EGF), platelet-derived growth factor (PDGF), and insulin.[83,115] Similar results were observed in the membrane redox system of a mouse neuroblastoma cell line NB41A3.[115] In addition, one of the mechanisms proposed for ascorbate free radical oxidoreductase modulation of cell growth was the control of the redox reactions involved in cell elongation mechanisms.[116,117] Conversely, inhibition of ascorbate oxidation caused a delay of ascorbate-mediated cell elongation.

[112] A. Gorman, A. McGowan, and T. G. Cotter, *FEBS Lett.* **404,** 27 (1997).

[113] D. J. Morré, A. Bridge, L.-Y. Wu, and D. M. Morré, *Biochem. Pharmacol. Pharmacol.* **60,** 937 (2000).

[114] G. F. Kay and K. A. Ellem, *J. Cell Physiol.* **126,** 275 (1986).

[115] R. Zurbriggen and J. L. Dryer, *Biochim. Biophys. Acta* **1183,** 513 (1994).

[116] A. Hidalgo, G. Garcia-Hendugo, J. A. Gonzáles-Reyes, D. J. Morré, and P. Navas, *Bot. Gaz.* **152,** 282 (1991).

Redox activities are altered in transformed cells,[118] including human retinoblastoma,[119] neuroblastoma,[119] HeLa,[120] HL-60 cells,[121] and in SV40-transformed 3T3 cells.[122]

NADH oxidation by ECTO-NOX proteins may be coupled to an amelioide-insensitive proton transport[123] that leads to alkalinization of the cytosol[124] and may contribute to an increased membrane potential. The released energy from the NADH oxidation might be coupled to growth, for example, or be used to drive proton transport.[110] Cell enlargement is an energy-requiring process, and conservation of energy of plasma membrane electron transport to drive the energy-requiring steps of cell enlargement would represent a new paradigm in biochemistry if correct. As described previously, the plasma membrane electron transport would serve to oxidize NADH accumulated from the glycolytic production of ATP, which is necessary to maintain NAD^+/NADH homeostasis essential for survival at points in the cytosol distant from mitochondria.

A relationship of the ECTO-NOX proteins to growth was first indicated from inhibitor studies.[35] Thus far, without exception, compounds and treatments that stimulate the NADH oxidase stimulate cell enlargement, and compounds that inhibit the NADH oxidase inhibit cell enlargement.

Vertebrate cells once having divided must reach some minimum size before they are able to divide again. By inhibiting tNOX in cancer cells, unregulated cell enlargement also is blocked, which is sufficient to restrict unregulated cell cycle entry characteristic of uncontrolled growth of cancer cells.

During the disulfide–thiol interchange portion of the 24-min ECTO-NOX cycle, the enlargement phase of cell growth occurs.[95] The rate of enlargement of plant, animal, and bacterial cells also oscillates with a period length of 24 min in parallel to the rate of hydroquinone (or NADH) oxidation. Maximum rates of cell enlargement correlate with the portion of the ECTO-NOX cycle involved in protein disulfide–thiol interchange.

[117] F. Cordoba and J. A. Gonzáles-Reyes, *J. Bioenerg. Biomemb.* **26,** 399 (1994).
[118] P.-J. Chueh, *Antiox. Redox Signal* **2,** 177 (2000).
[119] M. A. Medina and L. Schweigener, *Biochem. Mol. Biol. Int.* **31,** 997 (1992).
[120] I. L. Sun, E. E. Sun, F. L. Crane, D. J. Morré, and W. P. Faulk, *Biochim. Biophys. Acta* **1105,** 84 (1992).
[121] F. J. Alcain, M. I. Buron, J. M. Villalba, and P. Navas, *Biochim. Biophys. Acta* **1073,** 380 (1991).
[122] H. Löw, F. L. Crane, C. Grebing, M. Isaksson, A. Lindgren, and I. L. Sun, *J. Bioenerg. Biomemb.* **23,** 903 (1991).
[123] E. Sun, J. Lawrence, D. M. Morré, I. Sun, F. L. Crane, W. C. MacKellar, and D. J. Morré, *Biochem. Pharmacol.* **50,** 1461 (1995).
[124] D. M. Morré and D. J. Morré, *Photoplasma* **184,** 188 (1995).

During the electron transport phase of the cycle involving hydroquinone oxidation, cell enlargement rests, and the cells actually shrink in volume.[125,126,127,127a]

In actively enlarging cells, it is the net difference between the active and relaxation phases of cell enlargement that results in an incremental net increase in size with each 24-min cycle.[93,125,126] In nongrowing cells (cells that had reached a size optimal for division and were no longer enlarging), the cell volume continued to oscillate, but the net increase between the actual enlargement and relaxation phases was zero.[127a] As a result, enlargement rates were determined primarily by the extent to which enlargement during the active phase exceeded the shrinkage that occurred during the relaxation phase.[93]

Timekeeping (Clock) Function of ECTO-NOX Proteins

A unique feature of the ECTO-NOX proteins is that the two enzymatic activities they catalyze, hydroquinone (NADH) oxidation and disulfide–thiol interchange, alternate within a 24-min period.[3–5,89,91–95] The ECTO-NOX proteins carry out hydroquinone (NADH) oxidation for 12 min and then that activity rests. While the hydroquinone (NADH) oxidative activity rests, the protein engages in disulfide–thiol interchange activity for 12 min. That activity then rests as the cycle repeats. The alternation of activities imparts a timekeeping function to the protein.[117] The length of the period is temperature independent (temperature compensated) and entrainable, both features of the biological clock.[127] Site-directed mutagensis (cysteine to alanine replacements) of the tNOX cDNA has generated tNOX forms with period lengths of less than and greater than 24 min.[105] When COS cells were transfected with a tNOX protein with a period length greater than 24 min, (i.e., 36 or 40 min), the transfected cells exhibited a circadian period of 36 h or 40 h in the activity of glyceraldehyde-3-phosphate dehydrogenase, a common clock-regulated protein in addition to the normal circadian period of 24 h.[127] The 22-min tNOX period generated a 22-h circadian day,[127] so that in all four examples (CNOX, tNOX, and two cysteine to alanine replacements) the circadian period length was 60 × the ECTO-NOX period.[127] The fact that the expression of a single oscillatory ECTO-NOX protein determined the period length of a circadian biochemical marker (60 × the ECTO-NOX period

[125] D. J. Morré, R. Pogue, and D. M. Morré, *In Vitro Cell. Dev. Biol. Plant* **37,** 19 (2001).

[126] D. J. Morré, P. Ternes, and D. M. Morré, *In Vitro Cell. Dev. Biol. Plant* **38,** 18 (2002).

[127] D. J. Morré, P.-J. Chueh, J. Pletcher, X. Tang, L.-Y. Wu, and D. M. Morré, *Biochemistry* **30,** 613 (2002).

[127a] R. N. Pogue, D. M. Morré, and D. J. Morré, *Mol. Biol. Cell* **12,** 67a (2001).

length) provides compelling evidence that ECTO-NOX proteins are the biochemical ultradian core oscillators of the cellular biological clock.[127]

The period lengths of the activity oscillations of the ECTO-NOX proteins are independent of temperature (temperature compensated),[91,93,94,128] and their phases are entrainable.[128] These two characteristics, temperature compensation and entrainment (coupling the intrinsic clock to environmental cues), are two defining hallmarks of the biological clock.[129,130] Amplitude, on the other hand, doubles for each 10° rise in temperature, yielding a Q_{10} of 2 as is characteristic of most chemical reactions. ECTO-NOX synchrony through entrainment is achieved through autosynchrony in solution[128] by coupling to red (plants)[131] and blue (plants and animals)[132,133] light photoreceptors and in direct response to melatonin.[133]

Concluding Remarks

Quinone oxidoreductases of the plasma membrane have physiological significance as components in support of plasma membrane electron transport. Among these are the ECTO-NOX (because of their cell surface location) proteins that comprise a family of NAD(P)H oxidases of plants and animals that exhibit both oxidative and protein disulfide isomerase–like activities that alternate and serve as terminal oxidases for plasma membrane electron transport. Having two biochemical activities, hydroquinone [NAD(P)H] oxidation and protein disulfide–thiol interchange that alternate, is a property unprecedented in the biochemical literature. A tumor-associated ECTO-NOX is cancer specific and drug responsive. The constitutive ECTO-NOX (CNOX) is ubiquitous and refractory to drugs. The physiological substrate for the oxidative activity seems to be hydroquinones of the plasma membrane such as reduced coenzyme Q_{10}. ECTO-NOX proteins are growth related and drive cell enlargement. The regular pattern of oscillations seems to be related to α-helix-β-structure transitions[95] and may serve as the biochemical core oscillator of the cellular biological clock, where the period length is independent of temperature (temperature compensated) and where synchrony is achieved through entrainment.

[128] D. J. Morré, J. Lawler, S. Wang, T. W. Keenan, and D. M. Morré, *Biochim. Biophys. Acta* **1559,** 10 (2002).

[129] L. N. Edmunds, Jr., "Cellular and Molecular Basis of Biological Clocks." Springer-Verlag, New York-Berlin-Heidelberg, 1988.

[130] J. C. Dunlap, *Annu. Rev. Genet.* **30,** 579 (1996).

[131] D. J. Morré, D. M. Morré, C. Penel, and H. Greppin, *Int. J. Plant Sci.* **160,** 855 (1999).

[132] D. J. Morré, C. Penel, H. Greppin, and D. M. Morré, *Int. J. Plant Sci.* **163,** 543 (2002).

[133] D. J. Morré and D. J. Morré, *J. Photochem. Photobiol.* **70,** 7 (2003).

[16] Regulation of Ceramide Signaling by Plasma Membrane Coenzyme Q Reductases

By Plácido Navas and José Manuel Villalba

Introduction

Lipid signaling in mammalian cells involves sphingolipids such as sphingomyelin, sphingosine-1-phosphate, and GD3 ganglioside. Ceramide has received much attention, because an increase in ceramide levels initiates cellular responses for cell growth, differentiation, stress, and apoptosis.[1] Ceramide accumulation in the cell can occur either from de novo synthesis or from the hydrolysis of sphingomyelin by different sphingomyelinases (SMase, sphingomyelin phosphodiesterase; E.C. 3.1.4.12). A lysosomal acidic pH-optimum SMase (aSMase), and a Mg^{2+}-dependent neutral SMase (nSMase), which is an integral plasma membrane protein, have been related to apoptosis triggered by stress-inducing agents. Although a role for the aSMase in cell signaling has been questioned, the nSMase is apparently the most important in ceramide signaling according to the number of agents stimulating this enzyme and its favorable position at the plasma membrane, where sphingomyelin is highly concentrated.[2,3]

Because of its important role in signal transduction and the continuous availability of its substrate sphingomyelin, the activity of signaling SMase must be strictly controlled in cells.[4] Much is known about factors that activate nSMase such as TNF-α, arachinodate, and phosphatidylserine, but little is known about inhibitors of this enzyme. These inhibitory factors would be valuable tools to better understand the role of the enzyme and its product ceramide in signal transduction.[5]

Plasma membrane contains an electron transport system that reduces coenzyme Q (CoQ) in this membrane, which guarantees its antioxidant properties but also mediates the regeneration of other primary antioxidants such as ascorbate and α-tocopherol. Coenzyme Q seems to be the

[1] Y. A. Hannun, *Science* **274,** 1855 (1996).

[2] B. Ségui, C. Bezombes, E. Uro-Coste, J. A. Medin, N. Andrieu-Abadie, N. Augé, A. Brouchet, G. Laurent, R. Salvayre, J. P. Jaffrézou, and T. Levade, *FASEB J.* **14,** 36 (2000).

[3] C. Bezombes, B. Ségui, O. Cuvillier, A. P. Bruno, E. Uro-Coste, V. Gouazé, N. Andrieu-Abadie, S. Carpentier, G. Laurent, R. Salvayre, J. P. Jaffrézou, and T. Levade, *FASEB J.* **15,** 297 (2001).

[4] K. Hofmann and V. M. Dixit, *Trends Biochem. Sci.* **23,** 374 (1998).

[5] C. C. Lindsey, C. Gómez-Díaz, J. M. Villalba, and T. R. R. Pettus, *Tetrahedron* **58,** 4559 (2000).

0076-6879/04 $35.00

central component of this plasma membrane redox system[6] but also the key factor in preventing apoptosis triggered by externally induced oxidative stress.[7] Antagonists of CoQ such as short-chain ubiquinone analogues and capsaicin trigger the apoptotic program.[8,9] Furthermore, the activation of trans-plasma membrane redox system by the depletion of mitochondrial DNA is due to the increase of CoQ content in the plasma membrane,[10] which prevents both ceramide accumulation and apoptosis induced by serum removal.[7]

Ceramide acts as a mediator of stress responses modulating enzymes of the eicosanoid pathway, protein kinases, nuclear factors, and gene expression.[11] Ceramide is also able to induce cell death after its intracellular accumulation[1,12] by activating proteases of the caspase family such as caspase 3.[13] We have recently documented that CoQ is an efficient noncompetitive inhibitor of the Mg^{++}-dependent nSMase.[14] CoQ prevents both ceramide accumulation and activation of caspase-3, protecting cells from apoptosis induced by serum withdrawal.[15] Mitochondria have been related to a ceramide effect on caspase 3 activation[16] and could be responsible for the activation observed in ceramide-dependent apoptosis induced by serum withdrawal.[15] However, caspase 3 can be activated in mitochondria-free cytosol after its incubation with lipid extracts obtained from cells grown in serum-free media. When cells used to make the lipid extracts were grown either in the presence of 10% fetal calf serum or in a serum-free medium supplemented with CoQ, the activation *in vitro* of caspase 3 was not observed.[17]

[6] F. Navarro, P. Navas, J. R. Burgess, R. I. Bello, R. de Cabo, A. Arroyo, and J. M. Villalba, *FASEB J.* **12,** 1665 (1998).

[7] M. P. Barroso, C. Gómez-Díaz, J. M. Villalba, M. I. Burón, G. López-Lluch, and P. Navas, *J. Bioenerg. Biomembr.* **29,** 259 (1997).

[8] E. J. Wolvetang, J. A. Larm, P. Moutsoulas, and A. Lawen, *Cell Growth Differ.* **7,** 1315 (1996).

[9] A. Macho, M. A. Calzado, J. Muñoz-Blanco, C. Gómez-Díaz, C. Gajate, F. Mollinedo, P. Navas, and E. Muñoz, *Cell Death Differ.* **6,** 155 (1999).

[10] C. Gómez-Díaz, J. M. Villalba, R. Pérez-Vicente, F. L. Crane, and P. Navas, *Biochem. Biophys. Res. Commun.* **234,** 79 (1997).

[11] Y. A. Hannun, *J. Biol. Chem.* **269,** 3125 (1994).

[12] W. D. Jarvis, R. N. Kolesnick, F. A. Fornari, R. S. Traylor, D. A. Gewirtz, and S. Grant, *Proc. Natl. Acad. Sci. USA* **91,** 73 (1994).

[13] N. Mizushima, R. Koike, H. Kohsaka, Y. Kushi, S. Handa, H. Yagita, and N. Miyasaka, *FEBS Lett.* **395,** 267 (1996).

[14] S. F. Martín, F. Navarro, N. Forthoffer, P. Navas, and J. M. Villalba, *J. Bioenerg. Biomembr.* **33,** 143 (2001).

[15] P. Navas, D. M. Fernandez-Ayala, S. F. Martín, G. López-Lluch, R. de Cabo, J. C. Rodríguez-Aguilera, and J. M. Villalba, *Free Radic. Res.* **36,** 369 (2002).

[16] S. A. Susín, N. Zamzami, M. Castedo, E. Daugas, H.-G. Wang, S. Gelei, F. Fassy, J. C. Reed, and G. Kroemer, *J. Exp. Med.* **186,** 25 (1997).

These results suggest that CoQ may play a role in the regulation of the nSMase *in vivo*. The inhibition of nSMase in the plasma membrane is carried out more efficiently by the reduced form of CoQ ($CoQH_2$, ubiquinol) and also depends on the length of the isoprenoid side chain.[18] If inhibition of plasma membrane nSMase by ubiquinol has physiological significance, then endogenous levels of ubiquinol should also exert this regulatory action. Moreover, because endogenous CoQ can be reduced in plasma membranes by the intrinsic trans-membrane redox system, the activity of plasma membrane NAD(P)H-dependent dehydrogenases should also modulate the activity of nSMase. This function of CoQ occurs at the initiation phase of apoptosis by preventing the activation of the nSMase in the plasma membrane through the direct inhibition of this enzyme.

The analysis of the CoQ-dependent regulation of nSMase in the plasma membrane requires a highly purified membrane fraction free of endomembrane contamination and particularly of mitochondrial membranes. The procedure of two-phase partition has been shown to be the best to obtain plasma membranes without mitochondrial contamination from epithelia and cultured cells, and these membranes can be then used to purify nSMase and to analyze the relationship of the activity of this enzyme and the intrinsic content of CoQ. We review here the methods that we have found more useful to study the role of CoQ and its reductases in the regulation of liver plasma membrane nSMase.

Methods

Plasma Membrane Purification by Two-Phase Partition

Liver tissue is homogenized in a blender or similar equipment in a homogenization medium (2 ml/g fresh weight) containing 37 m*M* Tris-maleate, pH 6.4; 0.5 *M* sucrose; 5 m*M* $MgCl_2$; 5 m*M* β-mercaptoethanol; 1 m*M* polymethylsulfonyl fluoride (PMSF); and 20 $\mu g/\mu l$ each of chymostatin, leupeptin, antipain, and pepstatin A (CLAP). The homogenate is then centrifuged for 15 min at 5000*g*. The supernatant is discarded, and the light brown top portion of the pellet is resuspended in 5 ml of 1 m*M* sodium bicarbonate, a basic solution that has been found to be very effective to protect the functions of plasma membrane.[19] After homogenization in a

---

[17] D. J. M. Fernández-Ayala, S. A. Martín, M. P. Barroso, C. Gómez-Díaz, J. M. Villalba, J. C. Rodríguez-Aguilera, G. López-Lluch, and P. Navas, *Antiox. Redox Signal* **2,** 263 (2000).

[18] S. F. Martín, C. Gómez-Díaz, P. Navas, and J. M. Villalba, *Biochem. Biophys. Res. Commun.* **297,** 581 (2002).

[19] P. Navas, D. D. Nowack, and D. J. Morré, *Cancer Res.* **49,** 2147 (1989).

Teflon-glass potter, the mixture is centrifuged again for 15 min at 5000*g*, and the top half of the resulting pellet is recovered to obtain a crude membrane fraction. Plasma membranes were isolated from crude fractions by the two-phase partition method,[19] using a phase system composed of 6.0% (w/w) dextran T-500 (20% in water) (Pharmacia, Barcelona, Spain), 6.0% polyethylene glycol 3350 (40% water) (Fisher, Fair Lawn, NJ), in 0.1 *M* sucrose and 5 m*M* potassium phosphate, pH 7.2. The mixture must be inverted vigorously 40 times at 4° and centrifuged at 350*g* for 5 min to separate the phases. The upper phase, containing the plasma membrane, is then diluted in 25 ml of 1 m*M* sodium bicarbonate and recovered by centrifugation at 20,000*g* for 30 min. After this last centrifugation, plasma membranes are resuspended in 50 m*M* TRIS-HCl, pH 7.5, containing 10% glycerol, 1 m*M* EDTA, 0.1 m*M* dithiothreitol (DTT), 1 m*M* PMSF, and 20 μg/μl CLAP, and stored at −80°.

Purification of SMase from Plasma Membrane

The entire purification process must be performed at 4°. Plasma membranes (about 23 mg in 11.5 ml) are extracted with 0.5 *M* KCl and 5 m*M* ethylenediaminetetraacetic acid (EDTA) for 30 min to remove peripheral proteins, and then the membrane residue is separated by ultracentrifugation at 100,000*g* for 90 min. The resulting pellet is then resuspended in 12 ml of 50 m*M* TRIS-HCl, pH 7.4, containing 2 m*M* EDTA, 0.25% 3[3-chloraminopropyl diethylammonio]-1-propane sulfonate (CHAPS), 5 m*M* dithiothreitol (DTT), 1 m*M* PMSF, and 20 μg/ml CLAP. The detergent/plasma membrane mixture is rocked overnight and then centrifuged for 60 min at 100,000*g* to separate solubilized integral proteins. The proteins contained in the supernatant are used in a series of four-column chromatographies following the activity of nSMase in the elution profile. These chromatographies are as follows.

Gel Filtration Chromatography. Solubilized proteins are fractionated in a column (2.6 × 60 cm) of Sephacryl S-300 HR. The gel is pre-equilibrated with 900 ml buffer A (5 m*M* TRIS-HCl, pH 7.4, 2 m*M* EDTA, 0.25% CHAPS, 5 m*M* DTT, 1 m*M* PMSF, and 1 μg/ml CLAP). A volume of 13 ml of the sample is loaded, and the column is eluted with 1 volume buffer A at 20 ml/h. Fractions of 6 ml are collected. The most active fractions containing nSMase activity that eluted into the included volume are pooled and used directly for the following chromatography step.

Heparin-Sepharose Chromatography. Fractions from Sephacryl S-300 HR are loaded onto a 5-ml column of heparin sepharose-high performance, which has been pre-equilibrated with 50 ml buffer A. The column is washed with 50 ml of buffer A after sample loading, and then bound proteins are eluted with a 20-ml linear gradient of 0–100% buffer B (50 m*M* TRIS-HCl, pH 7.4, 2 *M* NaCl, 2 m*M* EDTA, 0.25% CHAPS, 5 m*M* DTT, 1 m*M* PMSF, and 1 μg/ml CLAP) at a flow rate of 30 ml/h. Fractions of 4 ml (2 ml during gradient elution) are now collected during sample loading and column washing.

Anion-Exchange Chromatography. Pooled fractions from the heparin-sepharose chromatography are previously desalted when needed by gel filtration in Sephadex G-25 (PD-10 columns) or directly added at 4 ml/h to a 1-ml HiTrap Q column pre-equilibrated with 20 ml buffer A. After washing the column with at least 10 column volumes of buffer A, bound proteins are eluted by a 20-ml linear gradient of 0–100% buffer C (50 m*M* TRIS-HCl, pH 7.4, 1 *M* NaCl, 2 m*M* EDTA, 0.25% CHAPS, 5 m*M* DTT, 1 m*M* PMSF, and 1 μg/ml CLAP). Fractions of 0.7 ml are collected during gradient elution.

Hydrophobic Interaction Chromatography. Active fractions collected from anion exchange chromatography are finally applied to a 1-ml phenyl-sepharose high-performance column, which has been pre-equilibrated with 30 ml buffer C. After sample loading, the column is washed with buffer C, and then bound proteins are eluted with a 20-ml linear gradient of 0–100% buffer A. After this step, the column is washed again with 4 ml of buffer A, and bound proteins are then eluted with buffer D (50 m*M* TRIS-HCl pH 7.4, 2 *M* NaCl, 2 m*M* EDTA, 0.5% CHAPS, 5 m*M* DTT, 1 m*M* PMSF, and 1 μg/ml CLAP) at a flow rate of 30 ml/h. Fractions of 3 ml are collected during sample loading and column washing, of 0.6 ml during gradient elution, and of 1 ml during the last elution step.

Purification must be monitored by both the determination of the nSMase activity in each chromatographical separation, following the protocol described in the next section, and by discontinuous sodium dodecylsulfate–polyacrylamide gel electrophoresis (SDS–PAGE) (10% polyacrylamide) as described elsewhere.[20] The analysis of the protein fraction with nSMase activity purified by this procedure has shown that it is mostly composed by the magnesium-dependent isoform,[14] which is the only inhibited one by CoQ.[18]

Assay for nSMase Activity and Coenzyme Q Inhibition

Mg^{2+}-dependent neutral SMase activity is assayed in a 50 m*M* TRIS-HCl buffer, pH 7.4, containing 0.05% Triton X-100 and 10 m*M* $MgCl_2$. To exclude putative inhibitions caused by a shift from the optimal pH of the enzyme, the pH of assay buffer must be carefully controlled in each

[20] U. K. Laemmli, *Nature* **277,** 680 (1970).

experiment in which the compounds used can modify the pH. Samples (50–100 μg plasma membrane fraction in 5–10 μl, or 15–30 μg purified nSMase, in 25–50 μl of column eluates) are mixed with assay buffer plus 10 nmol of a mixture of cold sphingomyelin and [methyl-^{14}C]-sphingomyelin (specific radioactivity 10,000 cpm/nmol). After incubation for 30–60 min at 37°, the reaction is stopped by adding 900 μl chloroform/methanol (2:1) and 200 μl distilled water. Tubes are vortexed and then centrifuged at 1500*g* for 5 min to achieve separation of phases. [^{14}C]-phosphorylcholine present in the aqueous phase is quantified using a liquid scintillation counter. nSMase activity was expressed as cpm μg protein^{-1} min^{-1}.

Studies of nSMase inhibition by CoQ/$CoQH_2$ in plasma membrane were carried out by modifying the concentrations of these compounds in membrane samples.[7,14,15,18] To obtain plasma membranes free of quinones, membrane samples (about 3 mg protein) are extracted after lyophilization with 1 ml of heptane for 6 h at 20° in the dark. Heptane is then decanted and evaporated, and membranes are resuspended in the nSMase assay buffer. The analysis with HPLC will be used to demonstrate that quinones have been removed. Supplementation with quinones is carried out by the incubation of membranes with ethanolic solutions of the corresponding quinone. The oxidized form is dissolved directly from a commercial source in pure ethanol, and a yellow solution is obtained. This is also used to produce the reduced form ubiquinol. To prepare $CoQ_{10}H_2$, oxidized CoQ_{10} is dissolved in ethanol at a 1 m*M* concentration, and then microliter amounts of a solution of sodium borohydride (10 mg/ml in water) are added until the yellow color of the CoQ_{10} solution is lost and a colorless mixture is obtained. A volume of 1 *M* NaCl and hexane are added, and the mixture is then vortexed for some seconds. Phases are separated by a brief centrifugation in a microfuge, and the upper hexane phase (containing the hydroquinone) is withdrawn and dried under a nitrogen stream. The resulting dried samples are resuspended in ethanol and have to be used immediately to prevent reoxidation.

To test the effect of CoQ(H_2) addition on nSMase activity, the required amount of oxidized or reduced CoQ (in ethanol) is added to a 1.5-ml tube, and the solvent is evaporated under a N_2 stream. The dried antioxidant is dissolved in 50 μl of assay buffer (which contains Triton X-100, see preceding), and the sample is then added. A stronger inhibition is achieved if membranes or purified nSMase are preincubated (up to 60 min at 37°) in the presence of the antioxidant. Afterwards, another 50 μl of assay buffer containing the substrate sphingomyelin is incorporated into the reaction mixture, and the assay is carried out as described previously.

Conclusion

Ceramide is an important lipid-signaling molecule, because its intracellular accumulation triggers cell growth arrest and cell death.[1,12] Ceramide can be accumulated by synthesis by ceramide synthase, but most is produced from the hydrolysis of sphingomyelin by a number of SMases.[21] This sphingomyelin is an abundant component of the outer leaflet of the plasma membrane bilayer, and it is the most important candidate to be the source of ceramide as a consequence of an externally induced oxidative stress.[15,22] It has been important to develop a procedure to purify the Mg^{++}-dependent nSMase from plasma membrane with the guarantee that enzyme isoforms from other cellular membranes are not present.[14,18]

The use of fractions highly enriched in plasma membrane to study the properties of this enzyme has facilitated the understanding of its properties and also the tentative redox mechanism of its regulation, mainly its inhibition by $CoQH_2$,[18] which is recycled by NAD(P)H-dependent reductases at this membrane.[6] Also, the purification of this enzyme has further clarified its inhibition by ubiquinol and opened new perspectives for the study of enzyme regulation mechanisms.

Recently, inhibitors of the plasma membrane nSMase have attracted increasing interest because of their potential clinical relevance as therapeutic agents in the treatment of several diseases, such as immune and cardiovascular diseases. These inhibitors include scyphostatin,[23] manumycin A,[24] and F11334s,[25] which interestingly display structural similarities with CoQ, increasing the importance of this quinone as a natural and specific regulator of the plasma membrane nSMase.

Acknowledgment

This work has been partially supported by the Spanish MCyT grants numbers BMC2002–01602 and BMC2002–01078.

[21] G. S. Dbaibo, W. El-Assad, A. Krikorian, B. Liu, K. Diab, N. Z. Idriss, M. El-Sabban, T. A. Driscoll, D. K. Perry, and Y. A. Hannun, *FEBS Lett.* **503,** 7 (2001).

[22] N. Andrieu-Abadie, V. Gouazé, R. Salvayre, and T. Levade, *Free Radic. Biol. Med.* **31,** 717 (2001).

[23] F. Nara, M. Tanaka, S. Masuda-Inoue, Y. Yamasato, H. Doi-Yoshioka, K. Sukuki-Konagai, S. Kumakura, and T. Ogita, *J. Antibiot. (Tokyo)* **52,** 531 (1999).

[24] C. Arenz, M. Thutewohl, O. Block, H. Waldmann, H. J. Altenbach, and A. Giannis, *Chembiochem.* **2,** 141 (2001).

[25] M. Tanaka, F. Nara, Y. Yamasato, Y. Ono, and T. Ogita, *J. Antibiot. (Tokyo)* **52,** 827 (1999).

[17] Stabilization of Extracellular Ascorbate Mediated by Coenzyme Q Transmembrane Electron Transport

By ANTONIO ARROYO, JUAN C. RODRÍGUEZ-AGUILERA, CARLOS SANTOS-OCAÑA, JOSÉ MANUEL VILLALBA, and PLÁCIDO NAVAS

Introduction

Ascorbate is the major water-soluble antioxidant found in body fluids of mammals, where it acts as a scavenger of noxious soluble free radicals or lipid peroxidation-initiating radicals in the aqueous phase, as well as a regenerator of other antioxidants (e.g., vitamin E) in the lipid phase of biological membranes and lipoproteins. Thus, the maintenance of extracellular ascorbate in its reduced, antioxidative, form results in great importance for organisms like humans, who do not synthesize it de novo, and it should be provided as a vitamin, vitamin C, through the diet. Ascorbate oxidation is a two-step process involving the one-electron intermediate ascorbate free radical (AFR) or semidehydroascorbate, and the fully, two-electron, oxidized form dehydroascorbate (DHA). Intracellular reduction of ascorbate has been documented to occur from both AFR and DHA through different enzymatic reactions with the participation of reduced nicotinamide adenine dinucleotide (NADH) and reduced glutathione (GSH) as electron donors, as well as from ascorbate itself acting at different cell locations.[1] Among these mechanisms are the AFR reductase found in the chromaffin-vesicle membrane[2] and in the outer mitochondrial membrane,[3] the thioredoxin reductase,[4] and glutaredoxin.[5]

Reductive mechanisms for maintaining extracellular ascorbate in its reduced form necessarily imply an electron transport across the plasma membrane. Alternative mechanisms such as an export of ascorbate from the cell would not explain the stabilization of extracellular ascorbate by cells in short-term experiments.[6] Our studies have revealed that extracellular ascorbate stabilization is accomplished by a trans-plasma membrane redox system. At least three components of this redox system are involved,

[1] L. M. Wakefield, A. E. Cass, and G. K. Radda, *J. Biol. Chem.* **261,** 9746 (1986).
[2] D. Njus, J. Knoth, C. Cook, and P. M. Kelley, *J. Biol. Chem.* **258,** 27 (1983).
[3] E. J. Diliberto, Jr., G. Dean, C. Carter, and P. L. Allen, *J. Neurochem.* **39,** 563 (1982).
[4] J. M. May, S. Mendiratta, K. E. Hill, and R. F. Burk, *J. Biol. Chem.* **272,** 22607 (1997).
[5] W. W. Wells, D. P. Xu, Y. F. Yang, and P. A. Rocque, *J. Biol. Chem.* **265,** 15361 (1990).
[6] J. C. Rodríguez-Aguilera and P. Navas, *J. Bioenerg. Biomembr.* **26,** 379 (1994).

0076-6879/04 $35.00

namely (1) a NAD(P)H-coenzyme Q reductase that transfers the electrons from intracellular NAD(P)H to (2) the pool of coenzyme Q (Q) present in the plasma membrane, acting as an electron carrier, and (3) a glucidic-rich environment (or a glycoprotein still to be identified) at the cell surface that favors the electron transfer from reduced Q (ubiquinol) to AFR in the vicinity of plasma membrane to recycle ascorbate. Therefore, from the point of view of the overall reaction, stabilization of extracellular ascorbate by cells can be defined as a Q-dependent NAD(P)H:AFR reductase activity. Three different strategies have been used for determining such activity: (1) to monitor the disappearance of the substrate NADH, (2) to measure the ascorbate oxidation in the absence and in the presence of cells to determine ascorbate stabilization, and (3) to determine the amount of the substrate AFR in the absence and in the presence of cells to assess their capability to scavenge these radicals.

The plasma membrane NAD(P)H:Q reductase has been characterized as a cytochrome b_5 reductase,[7] which has been shown to regenerate ubiquinol at this cell location.[8] According to the one-electron reaction mechanism described for the cytochrome b_5 reductase, the production of semiquinone radicals has also been detected during the enzymatic reduction of Q by the plasma membrane NAD(P)H:Q reductase.[9] Other Q reductases can also be involved as, for example, the cytosolic DT-diaphorase (NAD(P)H:quinone oxidoreductase 1, NQO1) that is bound to the plasma membrane under situations of high oxidative stress.[10] The involvement of a glucidic-rich environment at the cell surface has been evidenced by the inhibition of the cell-dependent extracellular ascorbate stabilization by the lectin wheat germ agglutinin (WGA)[11] and by tunicamycin, an inhibitor of protein *N*-glycosylation.[12] The role of Q in extracellular ascorbate stabilization by the trans-plasma membrane redox system is supported by a series of experiments; the method is described in this chapter.

[7] F. Navarro, J. M. Villalba, F. L. Crane, W. C. Mackellar, and P. Navas, *Biochem. Biophys. Res. Commun.* **212,** 138 (1995).

[8] A. Arroyo, F. Navarro, P. Navas, and J. M. Villalba, *Protoplasma* **205,** 107 (1998).

[9] V. E. Kagan, A. Arroyo, V. A. Tyurin, Y. Y. Tyurina, J. M. Villalba, and P. Navas, *FEBS Lett.* **428,** 43 (1998).

[10] F. Navarro, P. Navas, J. R. Burgess, R. I. Bello, R. de Cabo, A. Arroyo, and J. M. Villalba, *FASEB J.* **12,** 1665 (1998).

[11] P. Navas, A. Estévez, M. I. Burón, J. M. Villalba, and F. L. Crane, *Biochem. Biophys. Res. Commun.* **154,** 1029 (1988).

[12] J. C. Rodríguez-Aguilera, F. Navarro, A. Arroyo, J. M. Villalba, and P. Navas, *J. Biol. Chem.* **268,** 26346 (1993).

Methods

Generation and Measurement of the Substrate Ascorbate Free Radicals

Ascorbate free radicals can be generated by either one of the following systems: (1) an equimolecular mixture of ascorbate plus DHA, (2) oxidation of ascorbate by metal ions, mainly Cu^{2+}, and (3) oxidation of ascorbate by ascorbate oxidase (AO) as depicted in reaction 1. We have chosen the latter one to control the production of AFR by adding different amounts of AO and to avoid other sources of AFR by keeping a metal ion–free buffer.

$$\text{Ascorbate} + O_2 \xrightarrow{\text{AO}} \text{AFR} + O_2^{\bullet -} + H^+ \qquad \text{(reaction 1)}$$

The production of steady-state levels of AFR can be measured by the increase in absorbance at 360 nm, a method previously reported for the detection of these radicals.[13] We have taken advantage of the strong pH dependency of the AFR lifetime (see reaction 2) to further validate this method by studying steady-state levels of AFR at a different pH, measured as the absorbance increase at 360 nm on addition of AO to a reaction mixture containing ascorbate.

$$2\text{AFR} + H^+ \longleftrightarrow \text{Ascorbate} + \text{DHA} \qquad \text{(reaction 2)}$$

However, changing the pH in the reaction mixture of ascorbate plus AO to influence the stability of AFR (and thus, steady-state levels of the radicals) would affect the activity of AO itself to oxidize ascorbate. To avoid this side effect, different amounts of AO are added to give the same ascorbate oxidation (measured at 265 nm) at any given pH. The buffers used for these experiments are 20 m*M* citrate-phosphate buffer (pH range, 3.7–7.5), 20 m*M* phosphate buffer (pH range, 6.0–8.2), 20 m*M* TRIS-HCl buffer (pH range, 7.9–9.4), and 20 m*M* NaOH-glycine buffer (pH range, 8.7–9.8). Ascorbate at a final concentration of 0.1 m*M* is added to 1 ml buffer at 37°, and the signal at 360 nm is recorded in a Beckman (Los Angeles, CA) DU-650 UV/vis spectrophotometer for 0.5–1 min for monitoring the baseline. The reaction is started by addition of AO (the amount added should be adjusted to yield an effective activity of 22 mU at any given pH), and the signal is recorded for up to 5 min (total time for each assay). The increase in absorbance on addition of AO is taken as the concentration of AFR generated (ε: 5 mM^{-1} cm^{-1}). The change in absorbance caused by AO itself was tested in the absence of ascorbate to correct the estimation of the AFR concentration. As predicted by the mechanism of

[13] T. Skotland and T. Ljones, *Biochim. Biophys. Acta* **630,** 30 (1980).

reaction 2, the AFR signal markedly raises when the pH exceeds values of 7.4–7.8.[14] The profile of AFR signal versus pH range obtained is almost identical to that obtained by Buettner and Jurkiewicz who used electron paramagnetic resonance (EPR) techniques for AFR detection.[14,15] Whenever possible, the use of EPR techniques is desirable. However, when EPR instrumentation is not available, the measurement of absorbance changes at 360 nm could be considered as a valid and convenient alternative for detecting AFR.

Extracellular Ascorbate Stabilization by Cells Is Correlated with a Quenching in the Ascorbate Free Radical Signal

Cells (the human promyelocytic HL-60 cell line or the human erythroleukemic K-562 cell line) are cultured under standard conditions, harvested by centrifugation, washed in serum-free RPMI medium, resuspended in 100 m*M* TRIS-HCl, pH 7.4, and kept in ice for no longer than 2–3 h. Short-term ascorbate oxidation is followed for 10 min at 37° by a direct reading at 265 nm in a Beckman DU-650 UV/vis spectrophotometer under constant gentle stirring in a final volume of 1 ml adjusted with 100 m*M* TRIS-HCl, pH 7.4, either in the presence or the absence of cells. Ascorbate (0.1 m*M*) oxidation is initiated by adding AO (22 mU), and specific activity of the ascorbate stabilization is calculated from the difference between the initial rates of ascorbate oxidation with and without cells. An extinction coefficient of 11.2 mM^{-1} cm^{-1} is used in calculations of specific activities. The addition of cells to the reaction mixture prevents ascorbate oxidation (i.e., increases ascorbate stabilization) in a cell number–dependent manner as previously reported.[16] Under our assay conditions, ascorbate stabilization activity of K-562 cells (at 10^6 cells/ml) is about 1.4 nmoles/min.[17]

For AFR determinations, cells are collected by centrifugation, washed, and resuspended in phosphate-buffered saline (PBS). The assay mixture (taken up to 1 ml with PBS, pH 7.4) contains cells (up to 10^6 cells/ml), 1 m*M* ascorbate in PBS, pH 7.4, and 150 m*M* NaCl to counteract the cell-surface charge, thus favoring the access of anionic AFR to the cell

[14] A. Arroyo, F. Navarro, C. Gómez-Díaz, F. L. Crane, F. J. Alcaín, P. Navas, and J. M. Villalba, *J. Bioenerg. Biomembr.* **32,** 199 (2000).

[15] G. R. Buettner and B. A. Jurkiewicz, *in* "Handbook of Antioxidants" (E. Cadenas and L. Packer, eds.), p. 91. Marcel Dekker Inc., New York, 1996.

[16] F. J. Alcaín, M. I. Burón, J. M. Villalba, and P. Navas, *Biochim. Biophys. Acta* **1073,** 380 (1991).

[17] C. Gómez-Díaz, J. C. Rodríguez-Aguilera, M. P. Barroso, J. M. Villalba, F. Navarro, F. L. Crane, and P. Navas, *J. Bioenerg. Biomembr.* **29,** 251 (1997).

surface.[18] Ascorbate free radical is generated by the addition of 100 mU AO. The signal detected at 360 nm is quenched by cells as a function of the cell number added to the reaction mixture.[14] All the assays are carried out at 37° with constant gentle stirring. To calculate the AFR scavenging activity, the rise in absorbance at 360 nm in the presence of cells (plus/minus any additive) is subtracted from the maximal increase in absorbance, obtained by adding ascorbate plus AO in the absence of cells. Similar results have been obtained by May and coworkers,[19] who used erythrocytes as a cell model system and EPR techniques to detect AFR, thus further supporting the validity of measuring AFR by the increase in absorbance at 360 nm. Under our assay conditions, approximately 1 μM AFR is scavenged per 10^6 K-562 cells (Fig. 1).

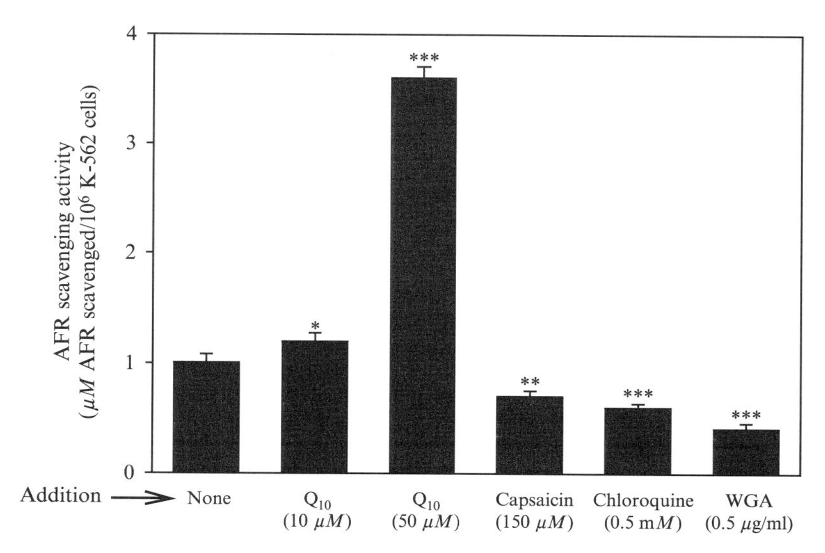

FIG. 1. Effect of different compounds on AFR scavenging by K-562 cells. Listed compounds were added to K-562 cells (250,000–500,000 cells/ml) in assay buffer and preincubated for 3 min before assay for AFR reduction. Absorbance changes at 360 nm on addition of 100 mU ascorbate oxidase were then recorded; AFR concentrations were calculated using an extinction coefficient of 5 mM^{-1} cm^{-1}. Scavenging activity was calculated from the difference in steady-state levels of AFR in the absence and in the presence of cells and referred to 10^6 cells. Data represent mean ± SD ($n = 3$). $^{*}p < 0.05$, $^{**}p < 0.01$, and $^{***}p < 0.001$ vs. none addition, respectively.

[18] R. Pethig, P. R. Gascoyne, J. A. McLaughlin, and A. Szent-Györgyi, *Proc. Natl. Acad. Sci. USA* **81,** 2088 (1984).

[19] J. M. May, Z.-C. Qu, and C. E. Cobb, *Biochem. Biophys. Res. Commun.* **267,** 118 (2000).

Together, these results show that both stabilization of extracellular ascorbate and quenching of AFR signal by cells represent the same phenomenon: the recycling of extracellular ascorbate by cells from the AFR generated during ascorbate oxidation.

Coenzyme Q Participates in the Extracellular Ascorbate Stabilization by Cells

To evidence the role of Q in extracellular ascorbate stabilization by cells, exogenous Q_{10} in an ethanolic solution is added to cells, incubated for 3 min at 37°, and the excess of Q_{10} is washed out by spinning down the cells before starting the readings for both ascorbate and AFR signal. The addition of exogenous Q_{10} to K-562 cells increases their ability to stabilize ascorbate in a dose-dependent manner, the difference being significant at Q_{10} concentrations greater than 20 μM.[17] In the same manner, addition of exogenous Q_{10} to K-562 cells increases their ability to quench the AFR signal (i.e., it increases their AFR scavenging activity in a dose-dependent manner, being significant even at lower Q_{10} concentrations [Fig. 1]). In accordance, two Q antagonists, such as capsaicin and chloroquine inhibit the AFR scavenging activity of K-562 cells by 30% and 40%, respectively (Fig. 1). In addition, the pretreatment of cells with the lectin wheat germ agglutinin (WGA) inhibits AFR scavenging activity by 60% (Fig. 1), supporting the involvement of a rich-glucidic environment at the cell surface in the extracellular ascorbate stabilization. The activity of AO used for AFR generation was not affected by the compounds tested, because no significant effects on steady-state levels of AFR in the absence of cells were observed.

Coenzyme Q–Dependent Extracellular Ascorbate Stabilization by Cells Takes Place at the Plasma Membrane

Extracellular ascorbate stabilization by whole cells could be interpreted not solely by the effect of the electron transport across the plasma membrane but also by alternative mechanisms. Some of these alternative explanations, as a release of ascorbate from the cell, are unlikely owing to the short time of our experiments. In addition, even if ascorbate is released from the cell, this would not explain the results obtained with the quenching of the AFR signal. Nonetheless, to further demonstrate that Q-dependent extracellular ascorbate stabilization is due to the trans-plasma membrane redox system, we used purified plasma membrane fractions. Plasma membrane is purified by the two-phase partition method in dextran T-500/polyethylene glycol from crude membrane fractions obtained from pig liver homogenates by differential centrifugation.[20] A critical point for obtaining a good preparation of plasma membrane is the choice of polymer concentration. We have found 6.0% (w/w) dextran T-500 and 6.0% (w/w) polyethylene glycol to be the best for purifying plasma membrane from pig liver. Other organs, tissues, or even animal sources could require different concentrations. Dextran T-500 is a heterogeneous polymer of α-1,6-glucose, with variable water content from batch to batch of product. The exact concentration should be accurately determined by polarimetry.[21] Plasma membrane is resuspended in 50 m*M* TRIS-HCl, pH 7.6, containing 10% glycerol, 1 m*M* PMSF, and 1 m*M* EDTA and stored at $-70°$ until needed. Purity is checked by marker enzyme analysis.[22]

Plasma membrane itself does not have the capability of either stabilizing ascorbate or quenching the AFR signal. An electron donor is needed, in our case NADH, to accomplish these functions. The reaction mixture is the same as used previously for AFR determination in the presence of whole cells, except that plasma membrane (0.05–0.25 mg/ml) is used instead, and 100 μM NADH is added to the assays. AFR scavenging activity is also calculated by subtracting the increase in absorbance at 360 nm in the presence of plasma membrane from that obtained in its absence. As previously observed with whole cells, AFR scavenging activity is a function of the amount of plasma membrane added to the assay.[14] In the same way, AFR scavenging activity is inhibited by Q antagonists, chloroquine, and capsaicin, as well as WGA (Fig. 2). All these compounds are incubated with plasma membranes for 3 min before the start of the experiment. Interestingly, inhibition by capsaicin is partially reversed by an excess of Q_{10} added to the reaction mixture, whereas inhibition by WGA is not affected by the addition of exogenous Q_{10} (Fig. 2). In addition, corroborating the importance of a rich glucidic environment at the extracellular leaflet of plasma membrane, these results support that reduction of AFR at the plasma membrane is a two-step process involving the NADH-dependent transfer of reducing equivalents to the plasma membrane by Q reductase and the membrane-mediated reduction of AFR by ubiquinol, resulting in ascorbate regeneration.

Ascorbate stabilization by isolated plasma membranes plus NADH may not be so accurately measured under the conditions used for whole cells, because NADH strongly interferes with absorbance readings at 265

[20] F. J. Alcaín, J. M. Villalba, H. Löw, F. L. Crane, and P. Navas, *Biochem. Biophys. Res. Commun.* **186,** 951 (1992).

[21] S. Bamberger, D. E. Brooks, K. A. Sharp, J. M. Van Alsine, and T. J. Weber, *in* "Partitioning in Aqueous Two-Phase Systems. Theory, Methods, Uses, and Applications to Biotechnology" (H. Walters, D. E. Brooks, and E. Fisher, eds.), p. 85. Academic Press, Orlando, FL, 1985.

[22] P. Navas, D. D. Nowack, and D. J. Morré, *Cancer Res.* **49,** 2147 (1994).

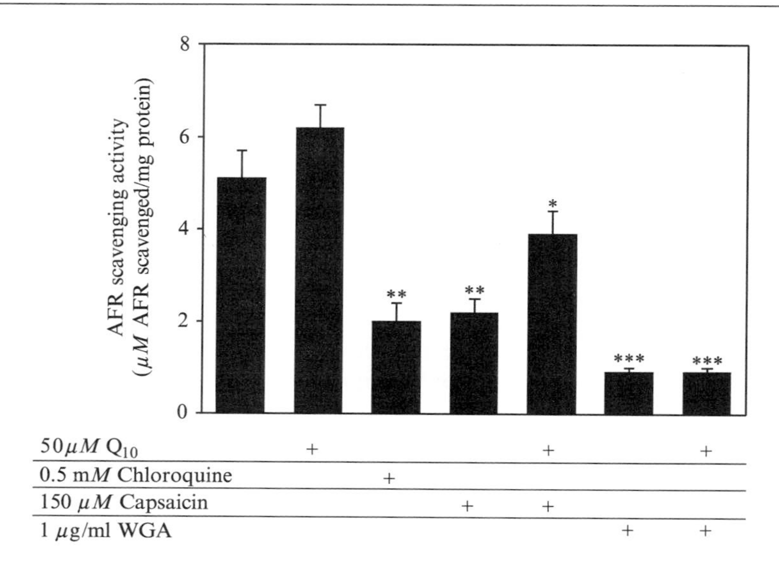

FIG. 2. Effect of various compounds on the AFR scavenging activity of plasma membranes in the presence of NADH. All assays contained plasma membranes (0.2 mg/ml) and NADH (100 μM) in assay buffer at a final volume of 1 ml. Membranes were preincubated with the listed compounds for 3 min and then, 100 mU ascorbate oxidase was added and the absorbance change at 360 nm recorded. Data were calculated from steady-state concentration of AFR measured in the absence and in the presence of plasma membranes plus NADH and referred to milligrams protein. Data represent mean ± SD ($n = 3$). The sign (+) indicates addition of the corresponding compound. $^{**}p < 0.01$ and $^{***}p < 0.001$ versus no addition, respectively. $^{*}p < 0.05$ versus plasma membrane plus capsaicin treatment.

nm for measuring ascorbate oxidation. The Q-dependent NAD(P)H:AFR reductase activity, however, can be measured as the consumption of NADH at 340 nm. Results obtained by this means are in agreement with the measurement of the activity by the other two methods (i.e., ascorbate stabilization at 265 nm and the quenching of AFR signal at 360 nm by whole cells),[11,16,17] as well as with the measurement of the activity by the quenching of AFR signal by isolated plasma membranes.[10,17] All spectrophotometric readings should be corrected for background caused by cells, membranes, and/or NADH. In our case, background-corrected readings were 5-fold to 10-fold higher than the noise signal of the instrument.

The addition of an extra amount of Q_{10} to plasma membrane slightly increases its AFR scavenging activity, although not in a statistically significant manner (Fig. 2). This phenomenon might represent a saturation of Q in the plasma membrane and/or in the activity of the Q reductase under our assay conditions. To further clarify this aspect, different sets of plasma membranes were subjected to different treatments leading to contain different amounts of Q. To remove Q from plasma membranes, these are lyophilized (control membrane, 20 mg) and then incubated with 15 ml of heptane for 6 h at 20° in the dark. The solvent is then decanted and evaporated to obtain Q-extracted membranes. Exogenous Q_{10} in heptane is added to either control or Q-extracted membranes to obtain Q-supplemented and Q-reconstituted membranes, respectively. The content of Q in lipid extracts of each plasma membrane type can be determined by reversed-phase high-performance liquid chromatography (HPLC).[14] As a requisite for AFR scavenging activity, none of the four plasma membrane types displayed activity in the absence of NADH. The dependency on Q of the AFR scavenging activity is revealed when NADH is added to the different plasma membranes (Fig. 3). The ability of plasma membrane to quench the AFR signal is almost completely abolished in Q-extracted membranes and restored after its reconstitution with exogenous Q_{10} (Fig. 3). The extra addition of Q_{10} over control membranes (Q-supplemented membranes) results in a slight increase in its AFR-scavenging activity as previously observed.

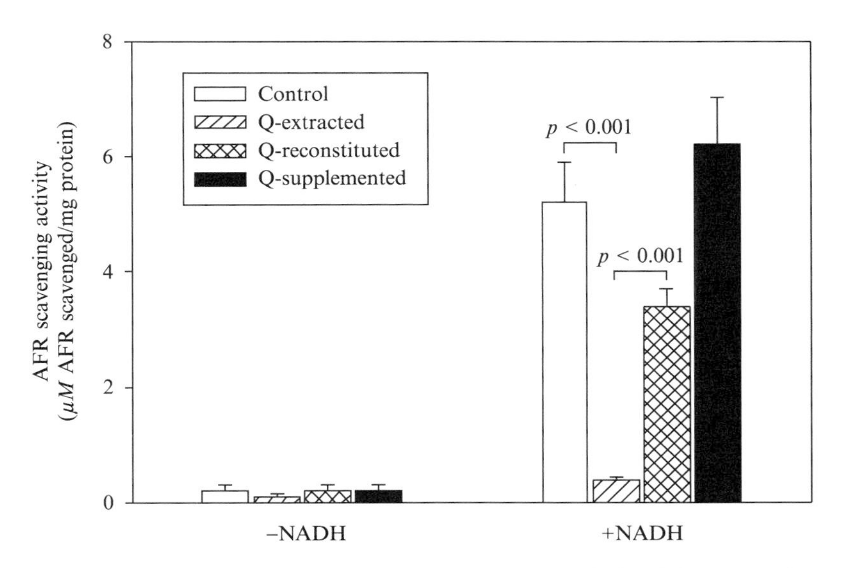

FIG. 3. Coenzyme Q_{10} requirement for the AFR reduction by isolated plasma membranes. Plasma membranes were lyophilized (control), and Q_{10} was extracted with heptane for 6 h at 20° in the dark (Q-extracted). Coenzyme Q_{10} (50 nmol) was then added to extracted membrane samples (Q-reconstituted) and to control membranes (Q-supplemented). AFR-reduction activity was calculated from the amount of AFR scavenged (μM) and was referred to milligrams protein. Assays were carried out in the absence and in the presence of 100 μM NADH. Data represent mean ± SD ($n = 3$).

Genetic Evidence for Q Requirement in Extracellular Ascorbate Stabilization Mediated by the Trans-Plasma Membrane Redox System

Genetic analysis requires a more flexible model system other than human-derived cell lines. To this end, the yeast *Saccharomyces cerevisiae* can be used on the basis of its ability to stabilize extracellular ascorbate.[23] Several mutant strains (*coq2*Δ, *coq3*Δ, and *coq7*Δ), which are defective in Q biosynthesis and completely lack Q_6, the natural Q isoform in these yeasts, as well as the wild-type strain, are grown in YPD media (1% yeast extract, 2% peptone, 2% dextrose) at 30° with shaking until the end of the log phase ($OD_{660nm} = 2–3$). Yeasts are washed twice in 5 m*M* EDTA, once in cold water, and resuspended at 10^7 cells/ml in 100 m*M* TRIS-HCl buffer, pH 7.4, containing 0.15 m*M* ascorbate and 0.06 μM $CuSO_4$ as ascorbate-oxidizing agent. After 1 h incubation at 30° and continuous gentle stirring, yeast cells are discarded by centrifugation, and the supernatants are used to measure the ascorbate concentration at 265 nm, and hence to calculate the rates of ascorbate oxidation/stabilization.[23]

Plasma membrane isolation from yeast cells is carried out by applying a total membranous fraction to a discontinuous sucrose gradient made of 4 ml 43% (w/w) sucrose and 2 ml 53% (w/w) sucrose. After centrifugation at 100,000*g* for 4 h, plasma membranes are recovered at the 43/53 interface. Plasma membranes are then diluted with sucrose buffer containing 20% (w/w) sucrose, 10 m*M* TRIS-HCl, pH 7.6, and 1 m*M* EDTA, washed by centrifugation, and resuspended in 1 ml of sucrose buffer.[24]

Coenzyme Q–deficient mutant strains, as well as plasma membranes derived from them, contained greatly diminished levels of NADH:AFR reductase activity measured as NADH consumption (e.g., about 10% of the wild-type for plasma membrane of the *coq3*Δ mutant) and, as a consequence, a reduced ability to stabilize the extracellular ascorbate added to the medium.[24,25] Restoration of the wild-type genotype by transformation with a single-copy plasmid bearing the corresponding deleted fragment fully restored both NADH:AFR reductase activity and ascorbate stabilization.[24,25] This genetic analysis also showed that, at least in *S. cerevisiae,* part of the ascorbate stabilization takes place by Q-independent mechanisms, being dependent on the iron-regulated ferric reductase complex.[25]

[23] C. Santos-Ocaña, P. Navas, F. L. Crane, and F. Córdoba, *J. Bioenerg. Biomembr.* **27,** 597 (1995).

[24] C. Santos-Ocaña, J. M. Villalba, F. Córdoba, S. Padilla, F. L. Crane, C. F. Clarke, and P. Navas, *J. Bioenerg. Biomembr.* **30,** 465 (1998).

[25] C. Santos-Ocaña, F. Córdoba, F. L. Crane, C. F. Clarke, and P. Navas, *J. Biol. Chem.* **273,** 8099 (1998).

Concluding Remarks

Our method has clearly demonstrated the role of Q on trans-plasma membrane stabilization of extracellular ascorbate by cell by modifying the Q content on plasma membrane, as well as by the addition of different Q antagonists. Moreover, the use of *S. cerevisiae* mutants for the biosynthesis of Q establishes a model that intrinsically possesses different Q contents. The two spectrophotometric assays described here (i.e., ascorbate measurement at 265 nm and AFR signal at 360 nm), together with the measurement of NADH consumption at 340 nm, are valuable and valid tools to successfully address this issue.

Acknowledgments

Supported by Grants No. BMC2002-01602 and BMC2002-01078 of the Spanish Ministerio de Ciencia y Tecnología.

Section III

Quinones, Cellular Signaling, and Modulation of Gene Expression

[18] Regulation of Antioxidant Response Element–Dependent Induction of Detoxifying Enzyme Synthesis

By ANIL K. JAISWAL

Introduction

An increase in reactive oxygen species (ROS) generated from the various sources is known to cause oxidative stress and have a profound impact on the survival of all living organisms.[1,2] A variety of oxidases, cellular respiration, physiological responses including neutrophil activation, produce ROS as by-products.[3–5] The oxidative metabolism of xenobiotics, drugs, and chemicals catalyzed by cytochrome P450 generate ROS.[6–8] Long-wavelength ultraviolet (UV) light (UVA and UVB) and radiolysis of water, caused by exposure to environmental radiation, leads to the formation of ROS.[9,10] Therefore, it is clear that all cells must keep the levels of ROS in check for survival and growth. ROS attack cellular macromolecules including DNA, which leads to oxidative stress, premature aging, neurodegenerative diseases, arthritis, arterosclerosis, inflammatory responses, and tumor induction and promotion.[3,10–12] Interestingly, it is now clearly established that ROS and hydrogen peroxide activate a group of cellular enzymes that either prevent the generation of ROS or detoxify ROS, leading to protection of cells against damage caused by oxidative stress. It may be noteworthy that the concentration at which ROS provide protection to a cell might be only slightly lower than the concentration at which damage occurs. Therefore, the cellular defense mechanisms must be tightly regulated by cytosolic and nuclear factors.

[1] L. H. Breimer, *Mol. Carcinogenesis* **3,** 188 (1990).
[2] R. Meneghini, *Free Radic. Biol. Med.* **23,** 783 (1997).
[3] M. B. Grisham and J. M. McCord, *in* "Physiology of Oxygen Radicals," p. 1. Waverly Press, Baltimore, 1986.
[4] M. Thelen, B. Dewald, and M. Baggiolini, *Physiol. Rev.* **73,** 797 (1993).
[5] K. S. Kasprzak, *Cancer Invest.* **13,** 411 (1995).
[6] H. V. Gelboin, *Physiol. Rev.* **60,** 1107 (1980).
[7] F. P. Guengerich, *Pharmacol. Ther.* **54,** 17 (1980).
[8] A. Balmain and K. Brown, *Adv. Can. Res.* **51,** 147 (1988).
[9] J. A. Last, W. M. Sun, and H. Witschi, *Environ. Health Perspect.* **102** (Suppl 10), 179 (1994).
[10] J. F. Ward, *Int. J. Rad. Biol.* **66,** 427 (1994).
[11] A. P. Breen and J. A. Murphy, *Free Radic. Biol. Med.* **18,** 1033 (1995).
[12] G. M. Rosen, S. Pou, C. L. Ramos, M. S. Cohen, and B. E. Britigan, *FASEB J.* **9,** 200 (1995).

0076-6879/04 $35.00

All the cell types, either prokaryotic or eukaryotic, face oxidative stress and must contain defensive mechanisms for their protection. Most of the studies on the mechanisms of protection against oxidative stress have come from the study of the bacteria *Escherichia coli*.[13,14] Two transcription factors OxyR and SoxRS are used by prokaryotes to sense the redox state of the cell, and in times of oxidative stress, these factors induce the expression of a battery of defensive genes.[14] OxyR contains a thio-disulfide switch that is sensitive to hydrogen peroxide. SoxRS contains a 2Fe-2S cluster that is sensitive to superoxide. Both of these sensors can be turned on and off very quickly, which allows the cell to respond promptly to subtle changes in the concentrations of ROS. The existence of similar mechanisms has been found in eukaryotic cells.[15,16] More than one dozen genes have been found to be involved in the response to oxidative and/or electrophilic stress caused by exposure of cells to chemicals. The products of these genes regulate a wide variety of cellular activities, including signal transduction, proliferation, and immunological defense reactions. The signal transduction pathways responsible for sensing oxidative and/or electrophilic stress and activating these responses are still not understood in eukaryotes and are an area of intense study. Only a few known transcription factors, which include NF-kB and NF-E2–related factors (Nrfs), are activated by ROS and/or electrophiles generated because of exposure of cells.[15,16] In this chapter, we have focused our methods, results, and discussion on Nrfs activation of detoxifying/antioxidant enzyme genes expression.

Exposure of cells to xenobiotics and antioxidants leads to the induction of a battery of genes that include antioxidant enzymes and antioxidants. The induced enzymes/proteins provide critical protection against oxidative and electrophilic stress.[15,16] The antioxidant enzymes that are induced include glutathione *S*-transferase Ya (GST Ya), γ-glutamylcysteine synthetase (γ-GCS), heme oxygenase 1 (HO-1), NAD(P)H:quinone oxidoreductase 1 (NQO1), and NRH:quinone oxidoreductase 2 (NQO2). Glutathione *S*-transferase Ya conjugates hydrophobic electrophiles and ROS with glutathione, aiding in their excretion.[17,18] γ-Glutamylcysteine synthetase plays

[13] C. E. Bauer, S. Elsen, and T. H. Bird, *Annu. Rev. Microbiol.* **53,** 495 (1999).

[14] M. Zheng and G. Storz, *Biochem. Pharm.* **59,** 1 (2000).

[15] R. Venugopal, P. Joseph, and A. K. Jaiswal, *in* "Oxidative stress and signal transduction" (H. J. Forman and E. Cadenas, eds.), p. 441. Chapman & Hall, New York, 1997.

[16] S. Dhakshinamoorthy, D. J. Long II, and A. K. Jaiswal, *Curr. Top. Cell. Regul.* **36,** 201 (2000).

[17] C. B. Pickett and A. Y. H. Lu, *Ann. Rev. Biochem.* **58,** 743 (1989).

[18] S. Tsuchida and K. Sato, *Crit. Rev. Biochem. Mol. Biol.* **27,** 337 (1992).

a role in the metabolism of glutathione.[19] Heme oxygenase 1 catalyzes the first and rate-limiting step in heme catabolism.[20] NAD(P)H:quinone oxidoreductase 1 is a flavoprotein that competes with cytochrome P450 reductase and catalyzes two-electron reduction and detoxification of quinones and other redox cycling compounds.[21] NRH:quinone oxidoreductase 2 also metabolically reduces quinones.[22,23] Over the years, our laboratory has cloned and sequenced human and mouse NQO1 and NQO2 genes and studied the mechanism of coordinated induction of these genes along with other detoxifying enzyme genes.[6,24–29]

NQO1 and NQO2 activity is ubiquitously present in all tissue types.[27–29] Significant variations in the levels of NQO1 have been observed among human tissues.[27] Higher levels of NQO1 gene expression were observed in liver, lung, colon, and breast tumors compared with normal tissues of the same origin.[27] NQO1-null and NQO2-null mice have been generated.[28,30] NQO1-null mice demonstrated altered intracellular redox status; altered carbohydrate, lipid, and pyridine metabolism; and myelogenous hyperplasia of bone marrow.[31,32] The NQO1-null mice lacking NQO1 also demonstrated increased sensitivity to menadione-induced oxidative stress and hepatic damage compared with wild-type mice expressing NQO1.[28] In addition, NQO1-null mice showed increased sensitivity to benzo(a)pyrene and 7,12-dimethylbenzanthracene–induced skin carcinogenicity.[33,34]

[19] M. Kretzschmar, D. Reinhardt, J. Schlechtweg, G. Machnik, W. Klinger, and W. Schirrmeister, *Exp. Toxicol. Pathol.* **44,** 344 (1992).

[20] A. M. Choi and J. Alam, *Am. J. Respir. Cell. Mol. Biol.* **15,** 9 (1996).

[21] P. Joseph and A. K. Jaiswal, *Proc. Natl. Acad. Sci. USA* **91,** 8413 (1994).

[22] K. B. Wu, R. Knox, X. Z. Sun, P. Joseph, A. K. Jaiswal, D. Zhang, P. K. S. Deng, and S. Chen, *Arch. Biochem. Biophys.* **347,** 221 (1997).

[23] Q. Zhao, X. L. Yang, W. D. Holtzclaw, and P. Talalay, *Proc. Natl. Acad. Sci. USA* **94,** 1669 (1997).

[24] A. K. Jaiswal, O. W. McBride, M. Adensik, and D. W. Nebert, *J. Biol. Chem.* **263,** 13572 (1988).

[25] A. K. Jaiswal, P. Burnett, M. Adesnik, and O. W. McBride, *Biochemistry* **29,** 1899 (1990).

[26] A. K. Jaiswal, *Biochemistry* **30,** 10647 (1991).

[27] A. K. Jaiswal, *J. Biol. Chem.* **269,** 14502 (1994).

[28] V. Radjendirane, P. Joseph, Y. H. Lee, S. Kimura, A. J. P. Klein-Sazanto, F. J. Gonzalez, and A. K. Jaiswal, *J. Biol. Chem.* **273,** 7382 (1998).

[29] D. J. Long II and A. K. Jaiswal, *Gene* **252,** 107 (2000).

[30] D. J. Long II, K. Iskander, A. Gaikwad, M. Arin, D. R. Roop, R. Knox, R. Barrios, and A. K. Jaiswal, *J. Biol. Chem.* **277,** 46131 (2002).

[31] A. Gaikwad, D. J. Long II, J. L. Stringer, and Anil K. Jaiswal, *J. Biol. Chem.* **276,** 22559 (2001).

[32] D. J. Long II, A. Gaikwad, A. Multani, S. Pathak, C. A. Montgomery, F. J. Gonzalez, and A. K. Jaiswal, *Cancer Res.* **62,** 3030 (2002).

NQO2-null mice also showed myelogenous hyperplasia of bone marrow.[30] However, NQO2-null mice showed protection against menadione-induced hepatic damage, indicating that NQO2 might activate menadione.[30]

Methods and Results

Expression and Coordinated Induction of Detoxifying Enzyme Genes

Northern analysis has shown that NQO1 mRNA was significantly more abundant than NQO2 mRNA in liver-derived carcinoma cells.[35] Transcription of the NQO1 and NQO2 genes is activated in response to xenobiotics (e.g., β-naphthoflavone [β-NF]); antioxidants (e.g., tert-butylhydroquinone [t-BHQ]); oxidants (e.g., hydrogen peroxide [H_2O_2]); 2,3,7,8-tetrachlorodibenzo-*p*-dioxin (TCDD); heavy metals (e.g., arsenic); ultraviolet (UV) light; and ionizing radiation.[16] These agents also induce the expression of other detoxifying enzyme genes, including GST Ya, γ-GCS, HO-1, and L-ferritin.

We used polymerase chain reaction (PCR) and standard cloning procedures and cloned 1.55 kb of human NQO1 gene promoter and a similar length of human NQO2 gene promoter along with first exon and a small portion of the first intron in pBLCAT3 vector to generate reporter plasmids pNQO1-CAT and pNQO2-CAT.[36] These reporter plasmids on transfection were expected to express chloramphenicol acetyl transferase (CAT) activity under the control of 1.55 kb of NQO1 gene promoter or 1.5 kb of NQO2 gene promoter. The reporter plasmids pNQO1-CAT and pNQO2-CAT, along with control plasmid RSV-β-gal, were transfected in human hepatoblastoma (Hep-G2 cells) by the calcium phosphate method and analyzed for CAT activity by procedures as described later. The CAT activities for NQO1 and NQO2 were compared to determine the relative strengths of the two promoters in transfected cells.

Human Hepatoblastama Cell Culture

The Hep-G2 cells were grown in monolayers up to 80% confluence in 90% air and 10% carbon dioxide in minimal essential medium supplemented with 10% fetal calf serum, penicillin (40 units/ml), streptomycin (40 μ/ml), and mycostatin (25 μ/ml) before transfection.

---

[33] D. J. Long II, R. L. Waikel, X. Wang, L. Perlaky, D. R. Roop, and A. K. Jaiswal, *Cancer Res.* **60,** 5913 (2000).

[34] D. J. Long II, R. L. Waikel, X. Wang, D. R. Roop, and A. K. Jaiswal, *J. Natl. Cancer Inst.* **93,** 1166 (2001).

[35] V. Radjendirane and A. K. Jaiswal, *Biochem. Pharm.* **58,** 597 (1999).

[36] Y. Li and A. K. Jaiswal, *J. Biol. Chem.* **267,** 15097 (1992).

Transfection by Calcium Phosphate Procedure and Treatment with Xenobiotic and Antioxidant

Human hepatoblastoma cells were plated at approximately 3×10^5 cells per 60-mm culture dish 16–20 h before transfection. The cells were grown to 70–80% confluence. At 2–4 h before transfection, the culture medium was replaced with 5 ml of fresh medium and continued incubation at 37° in a CO_2 incubator. Ten micrograms of DNA per dish was used for transfection. DNA + water (438 μl) was mixed with 62 μl of 2 *M* $CaCl_2$. Five hundred microliters of 2× DNA precipitation buffer (50 m*M* HEPES, pH 7.05, 1.5 m*M* Na_2HPO4, 10 m*M* KCl, 280 m*M* NaCl, and 12 m*M* glucose and filtered through 0.2-μm filter) was added slowly with mixing very gently by either bubbling or mild swirling. The sample was kept without shaking for 15–20 min at room temperature, during which time a fine precipitation was visible as cloudiness appeared. The calcium phosphate/DNA precipitate was mixed gently by inverting the tube and slowly added to the cell culture medium. The medium was gently swirled to mix calcium phosphate/DNA precipitate with medium and incubated at 37° for 4 h without disturbance. It may be noteworthy that a fine precipitate can be seen on top of the cells by light microscopy within 15 min of adding DNA to culture medium. The sample and medium were removed by aspiration and each washed in a 60-mm culture dish once with 5 ml of serum-free medium. The wash was removed by aspiration. Each dish was covered with 3 ml of 15% glycerol shocking buffer (glycerol in DNA precipitation buffer) and incubated for 2 min at room temperature. The shocking buffer was removed, and cells were washed with 5 ml of serum-free media. The wash was removed by aspiration. Five milliliters of complete culture medium with serum was added to each dish and incubated at 37° in a CO_2 incubator. Thirty-six hours after the transfection, the cells were treated with 50 μM tert-butyl hydroquinone (t-BHQ) or 50 μM β-naphthoflavone (β-NF) for 12 h. The cells were scraped and analyzed for CAT activity by the procedures as described in the following.

Measurement of Chloramphenicol Acetyl Transferase Activity

The transfected cells were washed with 1× PBS, scraped, and transferred to a 10-ml plastic tube. The cells were spun down by centrifugation at 2000 rpm for 5 min in cold. The supernatant was discarded, and cells were suspended in 1.0 ml of 0.25 *M* TRIS, pH 7.5, and transferred to an Eppendorf tube. Cells were spun down for 5 min in an Eppendorf centrifuge in cold. The pellet was collected and suspended in 500 μl of 0.2 *M* TRIS, pH 7.5, and sonicated for 10 s three times (each time at the interval of 1/5 min on ice). The tubes were centrifuged for 5 min in an Eppendorf centrifuge in cold. The supernatant was transferred to a new tube. A small quantity of the supernatant was analyzed for protein content and β-gacatosidase enzyme by procedures as described.[37,38] One hundred microliters of supernatant or suitable amount after normalizing for transfection efficiency was pipetted in a new tube. The volume was completed to 100 μl and mixed with 20 μl of acetyl CoA (6 mg/ml in cold water) and 3 μl of ^{14}C-chloramphenicol (NEN) and incubated at 37° for 1 h. One milliliter ethyl acetate was added, vortexed for 10 s, and centrifuged for 5 min. The upper ethyl acetate layer was collected, dried in vacuum dessicator, and dissolved in 30 μl of ethyl acetate. All the 30 μl was spotted on a thin layer chromatography (TLC) plate (Sigma 20 $\times$ 20-cm silica gel TLC plates [Sigma catalog # T-6520]) about 1 inch from the bottom of the plate. The TLC plate was run in 95 ml chloroform:5 ml methanol in a TLC chamber. Let it migrate up to 1/2 inch to the top. The plate was taken out, air dried, marked with radioactive ink, and wrapped in sceran wrap and autoradiographed overnight. The faster moving product (acetylated chloramphenicol) and slower moving leftover substrate (chloramphenicol) spots were cut out and counted into 5 ml of liquid scintillation cocktail.

The results are shown in Fig. 1. pNQO1-CAT plasmid expressed CAT activity in transfected Hep-G2 cells that was induced fourfold to fivefold in response to xenobiotic (β-NF) and antioxidant (t-BHQ). Similar results were also observed with pNQO2-CAT plasmid. However, the expression of CAT activity from pNQO2-CAT was significantly lower than pNQO1-CAT. These data led to two conclusions. First, 1.5 kb of NQO1 and a similar length of NQO2 promoter contain the *cis*-elements required for basal expression and induction in response to xenobiotics and antioxidants. Second, NQO2 gene is expressed at a significantly lower level in hepatoblastoma cells compared with NQO1 gene. Similar techniques were used to perform deletion mapping in the human NQO1 gene promoter.

Antioxidant Response Element

Deletion, internal deletion, and nucleotide mutagenesis of the human NQO1 gene promoter identified several *cis*-elements that are essential for the expression and induction of the NQO1 gene (Fig. 1). One of these elements is an ARE between nucleotides −470 and −447.[36] This region is required for basal expression, as well as induction, of NQO1 in response to

[37] M. M. Bradford, *Anal. Biochem.* **72,** 248 (1976).
[38] L. Guarente, *Meth. Enzymol.* **101,** 181 (1983).

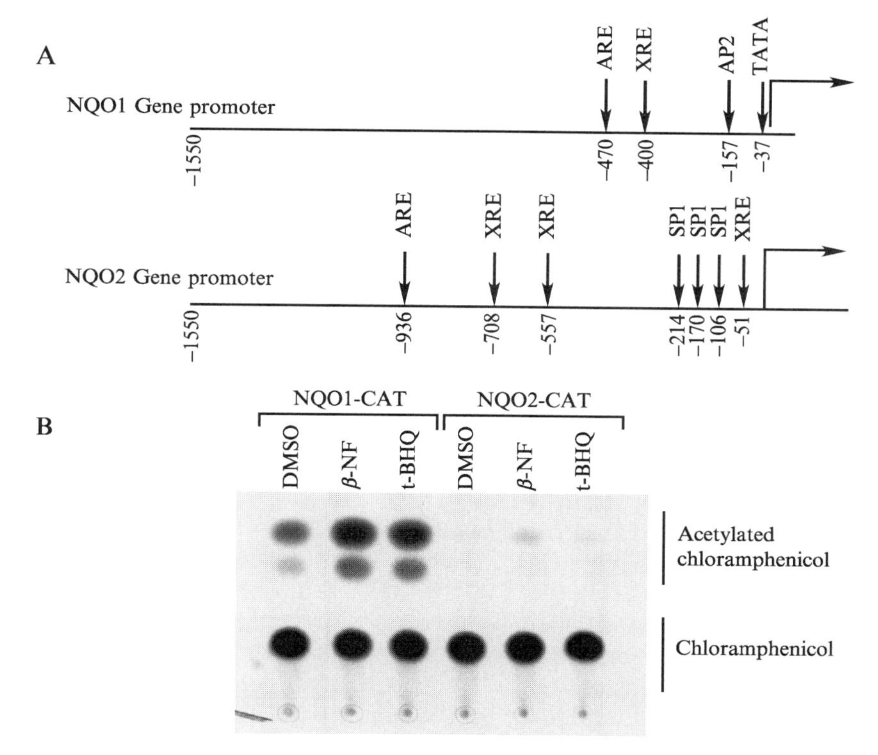

FIG. 1. Human NQO1 and NQO2 gene promoter regulated chloramphenicol acetyl transferase activity. (A) Human NQO1 and NQO2 gene promoter. The nucleotide positions of the various *cis*-elements are shown. Deletion mutagenesis was used to identify the *cis*-elements in NQO1 gene. The NQO2 *cis*-elements are putative binding sites and have not been experimentally confirmed. ARE, Antioxidant response element, XRE, xenobiotic response element, AP2, binding site for AP2, SP1, binding site for SP1. (B) Comparison of the relative strengths of human NQO1 and NQO2 gene promoters in transfected Hep-G2 cells; 1.55 kb of NQO1 and a similar length of the NQO2 gene promoters were attached to the CAT gene in separate construction plasmids and transfected in Hep-G2 cells by procedures as described in "Methods." Thirty-six hours after the transfection, the cells were treated with either DMSO, β-NF, or t-BHQ and analyzed for CAT activity.

β-NF, tBHQ, and hydrogen peroxide. Other elements include a xenobiotic response element (XRE) at nucleotide position −400,[39] which is required for 2,3,7,8-tetrachlorodibenzo-p-dioxin (TCDD) induction of NQO1 gene expression, the basal elements (between −837 and −560 and 130 and

[39] V. Radjendirane and A. K. Jaiswal, *Biochem. Pharm.* **58,** 1649 (1999).

		AP1-Like		AP1/ AP1-Like	GC Box	
Human NQO1	AAATC	GCAGTCA	CAG	TGACTCA	GCA	GAATC
Human NQO2	AGG	TGACTGC	AAA	TGAGGTG	GCA	GAAGC
Rat GST P	CAAAAGTAG	TCAGTCA	CTA	TGATTCA	GCA	ACAAA
Rat GST Ya	GAGCTTGGAAA	TGGCATT	GCTAATGG	TGACAAA	GCA	ACTTT
Human γ-GCS			CTCCCCG	TGACTCA	GCT	TTG
ARE CORE SEQUENCE			**G**	**TGACNNN**	**GCN**	

FIG. 2. Alignment of antioxidant response element (ARE) from the various detoxifying enzyme genes. The conserved sequences are boxed. NQO1, NAD(P)H:quinone oxidoreductase 1 gene; NQO2, NRH:quinone oxidoreductase 2 gene; GST P, glutathione S-transferase P gene; GST Ya, glutathione S-transferase Ya subunit gene; γ-GCS, γ-glutamyl cysteinyl synthetase gene.

−47), and an AP2 element (at nucleotide position −157) essential for cylic adenosine monophosphate (cAMP)–induced expression of the NQO1 gene.[40] The nucleotide sequence analysis of human NQO2 gene promoter indicated the presence of several putative SP1 and XRE elements and a single copy of ARE (Fig. 1). The role of these elements in regulation of NQO2 gene expression remains unknown.

Deletion mutagenesis and transfection studies also identified ARE in the promoter regions of several other detoxifying genes, including GST P, GST Ya, γ-GCS, and HO-1, that regulated expression and coordinated induction of these genes in response to xenobiotics and antioxidants.[15,16] The NQO1 and NQO2 genes ARE are aligned with the AREs from these other detoxifying enzyme genes (Fig. 2). Nucleotide sequence analysis of the human NQO1 gene ARE revealed that it contains one perfect and one imperfect activator protein 1 (AP1) (TPA response element) element. These elements are arranged as inverse repeats that are separated by three base pairs and are followed by a "GC" box.[15,16] The collagenase and metallothionein genes contain the seven base pair AP1 element in their promoter regions. This AP1 element is responsible for their induction in response to TPA.[41] It has been shown that the ARE is a unique element, responding independently from the AP1 element. It is the ARE, and not AP1, that is used for the induction of human NQO1 and other detoxifying genes in response to antioxidants and xenobiotics.[42] Mutational analysis revealed GTGACA***GC to be the core sequence of the ARE.[42–44]

[40] T. Xie and A. K. Jaiswal, *Biochem. Pharm.* **51,** 771 (1996).

[41] P. Angel and M. Karin, *Biochem. Biophys. Acta* **1072,** 129 (1991).

[42] T. Xie, M. Belinsky, Y. Xu, and A. K. Jaiswal, *J. Biol. Chem.* **270,** 6894 (1995).

However, other neighboring sequences and elements also affect the ARE-mediated expression and induction of detoxifying genes.[36,45,46]

NF-E2 and NF-E2–Related Factors

NF-E2–related factors are known to bind to ARE and regulate ARE-mediated antioxidant enzyme gene expression and induction in response to a variety of stimuli, including xenobiotics and antioxidant metals and UV irradiation.[15,16,47–50] NF-E2 was first cloned in 1993.[51,52] The mouse NF-E2 protein contains 373 amino acids and has an apparent molecular mass of 45 kDa; it is therefore also referred to as p45. NF-E2 is expressed only in erythroid cells, megakaryocytes, and mast cells. It binds to an AP1-like, NF-E2 recognition site (GCTGAGTCA) and regulates tissue-specific expression of the globin genes.[53–55] NF-E2 functions as a heterodimer with the ubiquitously expressed small Maf proteins.[56] NF-E2 −/− mice have no circulating platelets, and most die from hemorrhage.[57] The loss of NF-E2 has only a mild effect on erythroid cells, so mice that survive to adulthood are normal, except for exhibiting conditions consistent with a minor decrease in hemoglobin.

NF-E2–related factors Nrf1 and Nrf2, both 66–68 kDa proteins, were cloned using a yeast complementation assay (Fig. 3).[58,59] Both display a significant amount of homology to NF-E2, but unlike NF-E2 both are ubiquitously expressed. More recently, a third family member of the Nrfs,

[43] T. H. Rushmore, M. R. Morton, and C. B. Pickett, *J. Biol. Chem.* **266,** 11632 (1991).

[44] T. H. Rushmore and C. B. Pickett, *J. Biol. Chem.* **268,** 11475 (1993).

[45] T. Prestera, W. D. Holtzclaw, Y. Zhang, and P. Talalay, *Proc. Natl. Acad. Sci. USA* **90,** 2965 (1993).

[46] W. Wasserman and W. E. Fahl, *Proc. Natl. Acad. Sci. USA* **94,** 5361 (1997).

[47] R. Venugopal and A. K. Jaiswal, *Proc. Natl. Acad. Sci. USA* **93,** 14960 (1996).

[48] J. Alam, D. Stewart, C. Touchard, S. Boinapally, M. K. Choi, and J. L. Cook, *J. Biol. Chem.* **274,** 26071 (1999).

[49] A. C. Wild, H. R. Moinova, and R. T. Mulcahy, *J. Biol. Chem.* **274,** 33627 (1999).

[50] T. Nguyen, H. C. Huang, and C. B. Pickett, *J. Biol. Chem.* **275,** 15466 (2000).

[51] N. C. Andrews, H. Erdjument-Bromage, M. B. Davidson, P. Tempst, and S. H. Orikin, *Nature* **339,** 722 (1993).

[52] P. A. Ney, B. P. Sorrentino, K. T. McDonaugh, and A. W. Nienhuis, *Genes Dev.* **4,** 993 (1990).

[53] P. Moi and Y. W. Kan, *Proc. Natl. Acad. Sci. USA* **87,** 9000 (1990).

[54] D. Liu, J. C. Chang, P. Moi, W. Liu, Y. W. Kan, and P. T. Curtin, *Proc. Natl. Acad. Sci. USA* **89,** 3899 (1992).

[55] V. Mignotte, J. F. Eleouet, N. Raich, and P. H. Romeo, *Proc. Natl. Acad. Sci. USA* **86,** 6548 (1989).

[56] K. Igarashi, K. Kataoka, M. Nishizawa, and M. Yamamoto, *Nature* **367,** 568 (1994).

[57] R. Shivdasani and S. H. Orkin, *Proc. Natl. Acad. Sci. USA* **92,** 8690 (1995).

[58] J. Y. Chan, X. Han, and Y. W. Kan, *Proc. Natl. Acad. Sci. USA* **90,** 11371 (1993).

[59] P. Moi, K. Chan, I. Asunis, A. Cao, and Y. W. Kan, *Proc. Natl. Acad. Sci. USA* **91,** 9926 (1994).

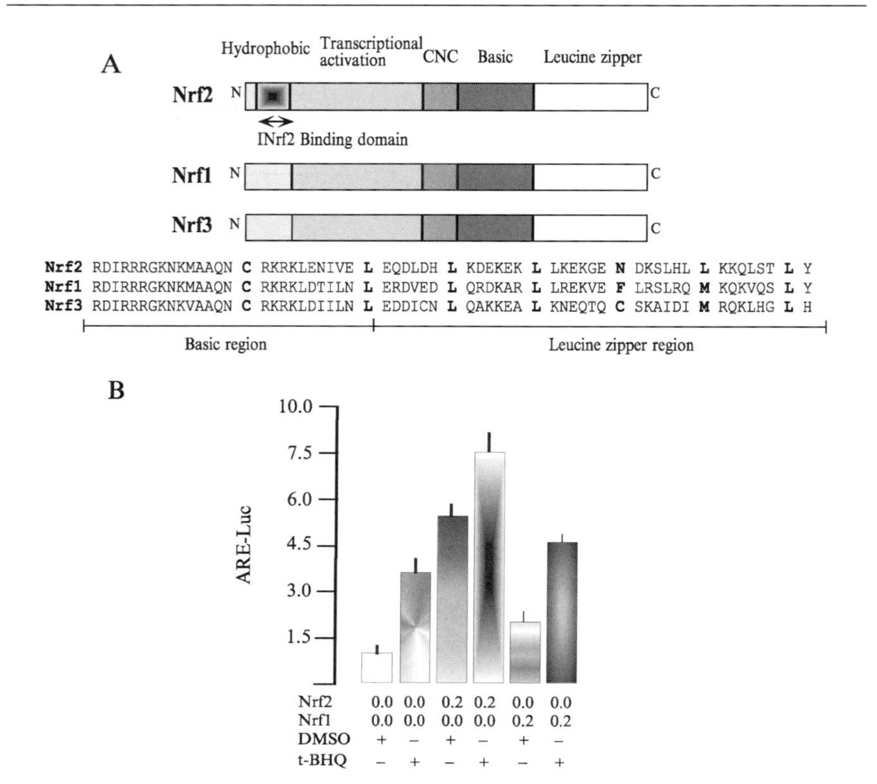

FIG. 3. NF-E2–related factors and their role in upregulation of ARE-mediated NQO1 gene expression. (A) NF-E2–related factors. The various protein domains of NF-E2–related factors (Nrf2, Nrf1, and Nrf3) are shown. INrf2 binding site in Nrf2 is identified. The amino acid sequence of basic and leucine zipper regions of Nrf2, Nrf1, and Nrf3 are aligned to demonstrate the conserved sequences and leucines. Nrf2 contains one, and Nrf1 and Nrf3 each contain two mutated leucines. (B) Comparative strengths of Nrf2 and Nrf1 in upregulation of NQO1 gene ARE-mediated expression and t-BHQ induction of luciferase gene expression in transfected Hep-G2 cells. The Hep-G2 cells were cotransfected with reporter plasmid NQO1 gene ARE-Luc and expression plasmids pcDNA-Nrf2 or pcDNA-Nrf1 in microgram concentration as shown. The cells were incubated for 36 h and then treated with either DMSO (vehicle control) or t-BHQ for 12 h. The cells were analyzed for luciferase activity. The luciferase activity is shown in numbers × million.

Nrf3, was cloned and sequenced.[60] The Nrfs belong to the family of basic leucine zipper proteins (bZIP). The basic region, just upstream of the leucine zipper region, is responsible for DNA binding. The acidic region is

[60] A. Kobayashi, E. Ito, T. Toki, K. Kogame, S. Takahashi, K. Igarashi, N. Hayashi, and M. Yamamoto, *J. Biol. Chem.* **274,** 6443 (1999).

required for transcriptional activation. The cap'n'collar region, so called because of its homology to the *Drosophila* cap'n'collar protein, is highly conserved among the Nrfs, but the function of this region remains unknown. Nrf1 and Nrf2 have been shown to regulate β-globin gene expression by binding to the AP1-like NF-E2 recognition site.[58,59] Nrf1 −/− mice die in utero because of a decreased number of enucleated red blood cells and severe anemia.[61] Nrf2−/− mice are viable and live to adulthood.[62] Nrf2 is therefore not required for erythropoeisis, development, or growth.[62]

The first evidence demonstrating the role of Nrf1 and Nrf2 in protection against oxidative and/or electrophilic stress came from our studies on the role of Nrf2 and Nrf1 in ARE-mediated regulation of NQO1 gene expression (Fig. 3).[47] The details of the experimental protocols and results that led to this conclusion are described in the following.

Construction of Reporter (NQO1 Gene ARE-Luciferase) and Expression (pcDNA-Nrf1 and pcDNA-Nrf2) Plasmids

The sense and antisense oligonucleotides corresponding to the NQO1 gene ARE (5′ CAGTCACAGTGACTCAGCAGAATCT3′) were synthesized with either NheI/XhoI ends. The oligonucleotides were annealed, phosphorylated using T4 polynucleotide kinase, and cloned at the respective sites in the pGL2 promoter vector to generate NQO1 gene ARE-Luc plasmid. The presence of ARE was checked by DNA sequencing. The mouse Nrf2 and Nrf1 cDNAs were kindly provided by Drs. Jefferson Y. Chan and Y. T. Kan (University of California, San Francisco, CA). The full-length Nrf2 and Nrf1 cDNAs were amplified by PCR and subcloned separately into the mammalian expression vector pcDNA3.1 to make the expression plasmids pcDNA-Nrf2 and pcDNA-Nrf1. Both these expression plasmids were confirmed by DNA sequencing.

Transfection and Measurement of Luciferase Activity

Human hepatoblastoma cells were grown in six-well monolayer cultures containing α-modified Eagle's medium (MEM) supplemented with Fetal Bovine Serum (FBS) by procedures as described previously. The Effectene Transfection Reagent Kit from Qiagen was used. We followed the procedures as described in the manufacturer's protocol. Briefly, 0.1 μg of reporter construct NQO1 gene ARE-Luc was mixed with 0.2 μg expression plasmid pcDNA-Nrf2 or pcDNA-Nrf1 and 0.2 μg of control plasmid

[61] J. Y. Chan, M. Kwong, R. Lu, J. Chang, B. Wang, T. S. Yen, and Y. W. Kan, *EMBO J.* **17,** 1779 (1998).

[62] K. Chan, R. Lu, J. C. Chang, and Y. W. Kan, *Proc. Natl. Acad. Sci. USA* **93,** 13943 (1996).

pRL-TK encoding *Renilla* luciferase and transfected into Hep-G2 cells. The combined plasmids were mixed with DNA Condensation Buffer and Enhancer Solution from the kit and incubated at room temperature for 5 min. This was followed by the addition of Effectene reagent to the mixture and incubated for 7 min at room temperature. The DNA-Enhancer-Effectene mixture was added drop by drop onto the Hep-G2 cells. The cells were incubated at 37° with 5% CO_2. Forty-eight hours after the transfection, the cells were washed with 1× PBS and lysed in 1× Passive Lysis buffer from the kit. The Dual-Luciferase Reporter Assay System from Promega was used to assay the samples for luciferase activity as described in the manufacturer's protocol. First, the cell lysate was assayed for the firefly luciferase activity using 100 μl of the substrate LARII. Then 100 μl of the STOP & GLO reagent was added to quench the firefly luciferase activity and activate the renilla luciferase activity, which was also measured. The assays were carried out in a Packard luminometer, and the relative luciferase activity was calculated as follows: 100,000/activity of renilla luciferase (in units) × activity of firefly luciferase (in units). Each set of transfections was repeated three times. For induction studies, the cells were treated with 50 μM t-BHQ, dissolved in DMSO for 12 h, and analyzed for luciferase activity by procedures as described previously.

The results of the transfection experiments are shown in Fig. 3. The transfection of Hep-G2 cells with expression plasmids pcDNA-Nrf2 and pcDNA-Nrf1 resulted in overexpression of respective proteins (data not shown). Overexpression of Nrf2 and Nrf1 led to upregulation of the expression and induction of the NQO1 gene in response to xenobiotics and antioxidants (Fig. 3).[47] This was supported by observations that mice lacking the Nrf2 gene exhibited a marked decrease in the expression and induction of NQO1, indicating that Nrf2 plays an essential role in the in vivo regulation of NQO1 in response to oxidative stress.[63] Further studies have shown that Nrf2 is also a prevailing factor in the regulation of ARE-mediated activation of other defensive genes, including GST Ya, γ-GCS, and HO-1.[48–50] Recent studies have shown that the presence of Nrf2 and Nrf1 is critical for cells to cope with oxidative stress. Nrf1-deficient fibroblasts have lower levels of glutathione and are more sensitive to oxidative stress–producing compounds.[64] These data demonstrate that Nrf2 and Nrf1 have a significant role in the regulation of antioxidant genes and their induction by oxidative stress–inducing agents (e.g., xenobiotics,

[63] K. Itoh, T. Chiba, S. Takahashi, T. Ishii, K. Igarashi, K. Katoh, T. Oyake, N. Hayashi, K. Satoh, I. Hatayama, M. Yamamoto, and Y. Nabeshima, *Biochem. Biophys. Res. Commun.* **236,** 313 (1997).

[64] M. Kwong, Y. W. Kan, and J. Y. Chan, *J. Biol. Chem.* **274,** 37491 (1999).

antioxidants, heavy metals, UV light, and ionizing radiations). The coordinated induction of this battery of antioxidant genes provides the cell with the necessary protection against oxidative stress. Studies have also demonstrated that Nrf2 is more efficient/potent in activation of ARE-mediated gene expression and induction compared with Nrf1.[47] Nrf3 is also expected to play a part in the response to oxidative stress, although no results have been reported yet.

Nrf1 and Nrf2 do not bind to ARE as homodimers or heterodimers with each other.[47] Therefore, Nrf1 and Nrf2 require another bzip protein to heterodimerize and bind with ARE. Jun (c-jun, jun-D, and jun-B) and small Maf (MafG, MafK, and MafF) proteins have all been shown to heterodimerize with both Nrf1 and Nrf2, and these complexes are capable of altering ARE-mediated expression of NQO1 and GST Ya in response to antioxidants and xenobiotics.[48–50,63,65] Interestingly, heterodimerization of Nrf2 and c-jun requires unknown cytosolic factors.[65] The small Maf proteins are a family of nuclear transcription factors that act as both activators and repressors of a number of eukaryotic genes.[66–69] The small Mafs are homologous to viral-Maf (v-Maf), especially in the DNA-binding domain and the leucine zipper domain. However, the small Mafs lack the transactivation domain that is present in v-Maf.[66–69] Small Mafs form homodimers, as well as heterodimers, with Nrf2. The homodimers repress whereas small Maf-Nrf2 heterodimers activate βglobin expression.[70] Recently, it has been shown that overexpression of small Maf proteins stimulate Maf homodimers that repress the ARE-mediated expression and antioxidant induction of NQO1, GST Ya, and γ-GCS genes.[49,50,71] The studies have also shown a role of Nrf2-MafK heterodimers in the activation of ARE-mediated gene expression.[63] However, others have reported a negative role of Nrf2-MafK in the regulation of ARE-mediated gene expression.[50] Therefore, although a role for Nrf2 in upregulation of ARE-mediated genes expression and induction is very well established,[47–50] the search is still on to find heterodimeric partners of Nrf2 for similar functions. In addition to small Maf proteins, overexpression of c-Fos or Fral also represses ARE-mediated

[65] R. Venugopal and A. K. Jaiswal, *Oncogene* **17,** 3145 (1998).

[66] K. T. Fujiwara, K. Ashida, H. Nishina, H. Iba, N. Miyajima, M. Nishizawa, and S. Kawai, *Oncogene* **8,** 2371 (1993).

[67] K. Kataoka, M. Nishizawa, and S. Kawai, *J. Virol.* **67,** 2133 (1993).

[68] K. Kataoka, M. Noda, and M. Nishizawa, *Mol. Cell. Biol.* **14,** 700 (1994).

[69] M. J. Kim and N. C. Andrews, *Blood* **89,** 3925 (1997).

[70] M. G. Marini, K. Chan, L. Casula, Y. W. Kan, A. Cao, and P. Moi, *J. Biol. Chem.* **272,** 16490 (1997).

[71] S. Dhakshinamoorthy and A. K. Jaiswal, *J. Biol. Chem.* **275,** 40134 (2000).

gene expression.[47] Mice lacking c-Fos have increased expression of NQO1 and GST Ya, substantiating a negative role for c-Fos in ARE-mediated gene expression *in vivo*.[72] As with many response elements, it seems that ARE-mediated gene expression is a balance between positive and negative regulatory factors. However, this remains to be further studied.

INrf2, a Cytosolic Inhibitor of Nrf2

Recently a cytosolic inhibitor of Nrf2, INrf2 (inhibitor of Nrf2) or KEAP1 (*K*elch-like *E*CH-*a*ssociated *p*rotein1) was discovered (Fig. 4).[73,74] Analysis of the INrf2 amino acid sequence revealed a BTB/POZ BTB (broad complex, tramtrack, bric-a-brac)/POZ (poxvirus, zinc finger) domain and a Kelch domain (Fig. 4).[73,74] The role of the BTB/POZ domain remains unknown in INrf2, but in other proteins it has been shown to be a protein–protein interaction domain. In the *Drosophila* Kelch protein, and in PIP, the Kelch domain binds to actin.[75,76] Therefore, it is expected that INrf2 binds to actin in the cytoskeleton. INrf2 retains Nrf2 in the cytoplasm. The treatment of cells with xenobiotics and antioxidants leads to the release of Nrf2 from INrf2. Nrf2 translocates into the nucleus and induces the expression of a battery of defensive genes. Therefore, overexpression of INrf2 in Hep-G2 cells led to repression of ARE-mediated luciferase gene expression (Fig. 4). Overexpression of INrf2 also significantly inhibited Nrf2 activation of ARE-mediated gene expression in transfected Hep-G2 cells (Fig. 4). Interestingly, the treatment of transfected cells with t-BHQ resulted in activation of ARE-mediated luciferase gene expression. This induction was irrespective of presence/absense of INrf2. In other words, t-BHQ relieved the effect of INrf2. This presumably was due to antioxidant-induced release of Nrf2 from INrf2. Therefore, the release of Nrf2 from INrf2 is the most significant process required for xenobiotic and antioxidant induction of ARE-mediated detoxifying enzyme genes, including NQO1 and NQO2. The mechanism of dissociation of INrf2 from Nrf2, and its fate in the cytosol, remains unknown.

[72] J. Wilkinson IV, V. Radjendirane, G. R. Pfeiffer, A. K. Jaiswal, and M. L. Clapper, *Biochem. Biophys. Res. Commun.* **253,** 855 (1998).

[73] K. Itoh, N. Wakabayashi, Y. Katoh, T. Ishii, K. Igarashi, J. D. Engel, and M. Yamamoto, *Genes Dev.* **13,** 76 (1999).

[74] S. Dhakshinamoorthy and A. K. Jaiswal, *Oncogene* **20,** 3906 (2001).

[75] Y. Li and A. K. Jaiswal, *Eur. J. Biochem.* **226,** 31 (1994).

[76] O. Albagli, P. Dhordain, C. Deweindt, G. Lecocq, and D. Leprince, *Cell Growth Differentiation* **6,** 1193 (1995).

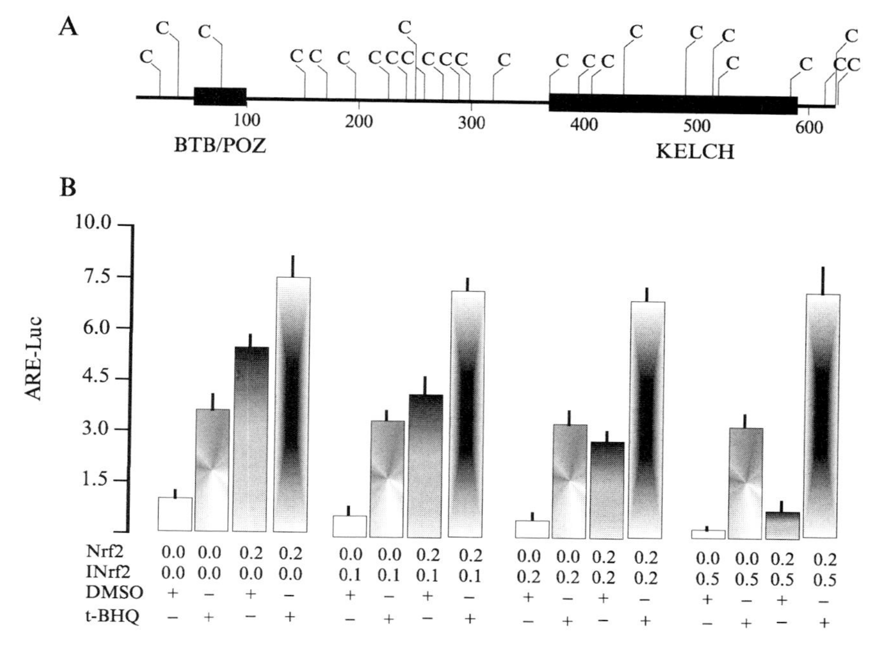

FIG. 4. Role of INrf2 in regulation of ARE-mediated detoxifying enzyme genes expression and induction in response to t-BHQ. (A) INrf2 structure. BTB/POZ, Kelch, and cysteines C are shown. (B) Effect of overexpression of INrf2 on Nrf2-mediated expression and induction of ARE-regulated detoxifying genes expression in response to t-BHQ. Hep-G2 cells were cotransfected with reporter plasmid NQO1 gene ARE-Luc, expression plasmids pcDNA-INrf2, and pcDNA-Nrf2 alone and in combination in microgram concentrations as shown. Thirty-six hours after transfection, the transfected cells were treated with either DMSO or t-BHQ for 12 h. The cells were harvested and analyzed for luciferase activity as described in "Methods." The luciferase activity is shown in numbers × million.

Discussion and Conclusions

A model illustrating the role of INrf2, Nrf2, and other ARE-binding factors in activation of detoxifying enzyme genes by xenobiotics and antioxidants is shown in Fig. 5. The signal transduction pathway that leads from xenobiotics and antioxidants to the proteins that regulate ARE-mediated expression of detoxifying enzyme genes remains largely unknown. The metabolites of both xenobiotics and antioxidants result in the generation of superoxides and electrophiles.[77] It has been suggested

[77] M. J. De Long, A. B. Santamaria, and P. Talalay, *Carcinogenesis* **8,** 1549 (1987).

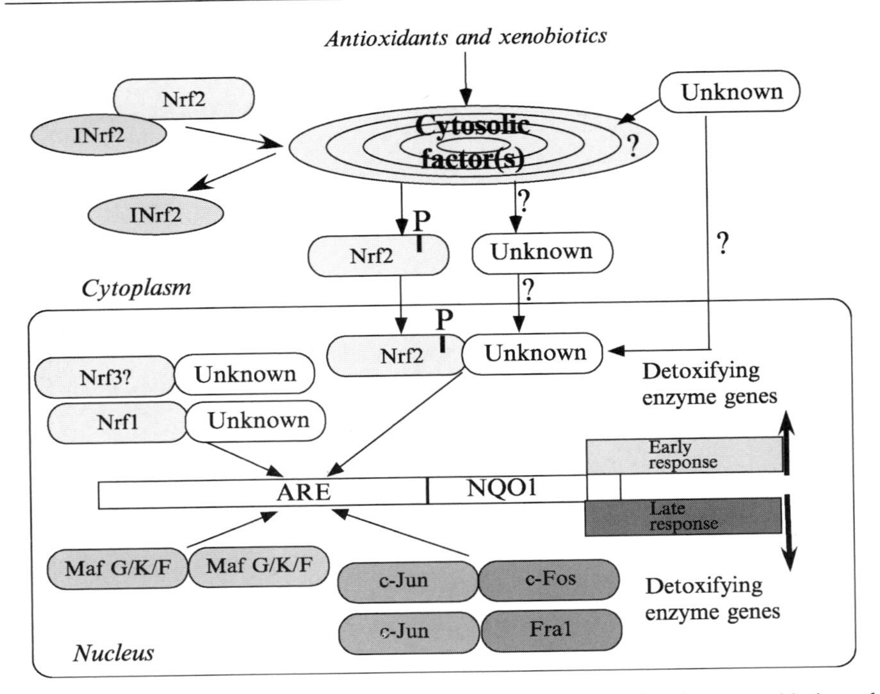

FIG. 5. A model showing the mechanism of signal transduction from xenobiotic and antioxidant to protein factors that regulate ARE-mediated genes expression and induction in response to xenobiotics and antioxidants.

that these molecules might act as second messengers, activating ARE-mediated expression of a host of detoxifying enzyme genes, including NQO1 and NQO2.[15,16] To support this, hydroxyl radicals and hydrogen peroxide are known to induce ARE-mediated expression of GST Ya and NQO1 genes.[78–80] Even though electrophiles are thought to be possible messengers in the oxidative stress pathway, their role, if any, has not been demonstrated. The superoxide and/or electrophilic signal presumably passes through unknown cytosolic factor(s). This factor(s) then catalyzes the modification and/or transcriptional activation of INrf2 and/or Nrf2. As a result, Nrf2 is released from INrf2. Nrf2 then translocates to the nucleus, where it heterodimerizes with c-Jun/Small Maf/Unknown heterodimeric partner(s) and induces the expression of NQO1 and other

[78] R. Pinkus, L. M. Weiner, and V. Daniel, *J. Biol. Chem.* **271,** 13422 (1996).
[79] L. V. Favreau and C. B. Pickett, *J. Biol. Chem.* **266,** 4556 (1991).
[80] I. F. Kim, E. Mohammadi, and R. C. C. Huang, *Gene* **228,** 73 (1999).

ARE-regulated genes. It may be noteworthy that we have shown Nrf2 heterodimeric partner as "unknown" in Fig. 5. This is because the roles of c-Jun and/or small Maf proteins in heterodimerization with Nrf2 leading to upregulation of ARE-mediated gene expression have not yet been clearly established and are under investigation. The role of Nrf2, however, in upregulation of ARE-mediated gene expression and induction is clearly established.[47–50] The cytosolic factor(s) that catalyzes xenobiotic and antioxidant-induced modifications of Nrf2 and/or INrf2 may be kinases and/or redox factors. Several cytosolic kinases that include MAPK, p38, PI3K, and PKC have been shown to modify Nrf2 and participate in the mechanism of signal transduction from xenobiotics and antioxidants to the ARE.[16] However, little information is available on the modifications of INrf2 in response to xenobiotics and antioxidants. Therefore, it is largely unknown whether xenobiotics and antioxidants induce phosphorylation/dephosphorylation and/or redox regulation of INrf2 in response to xenobiotics and antioxidants. Therefore, it is largely unknown if xenobiotics and antioxidants induce phosphorylation/dephosphorylation and/or redox regulation of INrf2 leading to dissociation of INrf2 and Nrf2. INrf2 contains potential serine/threonines and tyrosine phosphorylation sites and 25 cysteine residues that are conserved across species and might be involved in antioxidant signaling and dissociation of the INrf2–Nrf2 complex (unpublished). A recent report indicated that sulfhydryl groups of Keap1 (INrf2) act as sensors of inducers of the detoxifying enzyme genes.[81] This suggested that redox modification of INrf2 may play a role in INrf2–Nrf2 signaling. However, a second report showed that INrf2 existed as dimers inside the cells.[82] This report also demonstrated that mutation of serine 104 to alanine led to disruption of Keap1 (INrf2) dimerization and the loss of INrf2 ability to sequester Nrf2 in the cytoplasm.[82] However, it is not known whether serine 104 in INrf2 is phosphorylated in wild-type INrf2 or is phosphorylated in response to antioxidants.

Interestingly, several ARE-binding negative transcription factors, in addition to positive factors, have also been identified, which rapidly bring down the induced enzymes to their normal level (Fig. 5).[16] This may be because it is always necessary to have a small amount of ROS present to keep the cellular defenses active. Because activation of detoxifying enzymes and other defensive proteins leads to significant reduction in the levels of superoxide and other free radicals, the cell may require negative regulatory

[81] A. T. Dinkova-Kostova, W. D. Holtzclaw, R. N. Cole, K. Itoh, N. Wakabayashi, Y. Katoh, M. Yamamoto, and P. Talalay, *Proc. Natl. Acad. Sci. USA* **99,** 11908 (2002).
[82] L. M. Zipper and R. T. Mulcahy, *J. Biol. Chem.* **277,** 36544 (2002).

factors such as the small Mafs and c-Fos to keep the expression of defensive proteins "in check."

Acknowledgments

We thank our colleagues from Baylor College of Medicine, Houston, Texas, for valuable suggestions. This work was supported by NIH grants RO1 GM47644, RO1 ES07943, and RO1 CA81057.

[19] Antioxidant Responsive Element Activation by Quinones: Antioxidant Responsive Element Target Genes, Role of PI3 Kinase in Activation

By JIANG LI, JONG-MIN LEE, DELINDA A. JOHNSON, and JEFFREY A. JOHNSON

Antioxidant Responsive Element Activation by Quinones

Phase II detoxification enzymes include NAD(P)H:quinone oxidoreductase (NQO1), epoxide hydrolases, glutathione *S*-transferases (GSTs), *N*-acetyltransferases, sulfotransferases, UDP-glucuronosyltransferases (UGT), and other enzyme superfamilies. The ability of these enzymes to conjugate redox-cycling chemicals is an important protective mechanism against electrophiles and oxidative stress. Prototypical redox-cycling chemicals are phenolic antioxidants, Michael reaction acceptors, isothiocyanates, trivalent arsenicals, redox-cycling polycyclic aromatic hydrocarbons (PAH), and quinones.[1,2] The transcriptional activation of the phase II detoxification enzymes and/or antioxidant genes by redox-cycling chemicals has been traced to a *cis*-acting element called the antioxidant responsive element (ARE) or the electrophile response element (EpRE) that regulates either or both constitutive and inducible gene expression. ARE sequences have been detected in the promoter region of genes, including rat and mouse GST-Ya,[3–6] rat GST-P,[7] rat and human NQO1,[8–11] murine heme oxygenase-1 (HO1),[12,13] murine ferritin heavy chain,[14,15] as well as the

[1] A. T. Dinkova-Kostova, W. D. Holtzclaw, R. N. Cole, K. Itoh, N. Wakabayashi, Y. Katoh, M. Yamamoto, and P. Talalay, *Proc. Natl. Acad. Sci. USA* **99,** 11908 (2002).
[2] T. Prestera and P. Talalay, *Proc. Natl. Acad. Sci. USA* **92,** 8965 (1995).
[3] J. D. Hayes, D. J. Pulford, E. M. Ellis, R. McLeod, R. F. James, J. Seidegard, E. Mosialou, B. Jernstrom, and G. E. Neal, *Chem. Biol. Interact.* **1112,** 51 (1998).
[4] J. Y. Chan and M. Kwong, *Biochim. Biophys. Acta* **1517,** 19 (2000).

0076-6879/04 $35.00

γ-glutamylcysteine ligase catalytic (GCLC)[16,17] and regulatory (GCLR)[18–20] subunits. In the early studies of promoter regions in phase II detoxification enzymes, it was noted that all of the AREs share a common RTGACnnnGC motif ("core sequence") originally identified by mutagenesis of the rat GST-Ya ARE.[6] However, this sequence alone is not sufficient to mediate induction. Through systematic mutation analysis of the murine GST Ya ARE and GST Mu ARE, a second corelike sequence adjacent to the primary ARE core was identified as necessary to maintain a sufficient and functional ARE.[21] For example, the ARE sequence of hNQO1 contains one perfect (5′TGACTCA-3′) and one imperfect AP-1 binding sequence (5′GCAGTCA-3′) arranged as inverse orientation and shows strong similarity to the NF-E2 binding sequence (5′-TGAGTCA-3′). On the basis of mutational analysis of the hNQO1-ARE, we have demonstrated that the palindromic sequence (5′ to the ARE core sequence) and the GC box are required for maximal activation of the ARE.[22]

Quinones Activate Antioxidant Responsive Element Through Oxidative Stress-Dependent or Independent Mechanisms

Oxidative stress has been considered to be a main driving force for ARE activation. Most chemical inducing agents of phase II enzymes have the potential to induce oxidative stress and deplete GSH. Original work from Rushmore and coworkers[21] demonstrated that catechol and

[5] D. J. Pulford and J. D. Hayes, *Biochem. J.* **318** (Pt 1), 75 (1996).
[6] T. H. Rushmore and C. B. Pickett, *J. Biol. Chem.* **265,** 14648 (1990).
[7] J. D. Hayes and D. J. Pulford, *Crit. Rev. Biochem. Mol. Biol.* **30,** 445 (1995).
[8] R. Venugopal and A. K. Jaiswal, *Proc. Natl. Acad. Sci. USA* **93,** 14960 (1996).
[9] A. Wang and G. Williamson, *Biochim. Biophys. Acta* **1219,** 645 (1994).
[10] D. H. Barch, L. M. Rundhaugen, and N. S. Pillay, *Carcinogenesis* **16,** 665 (1995).
[11] D. Bloom, S. Dhakshinamoorthy, and A. K. Jaiswal, *Oncogene* **21,** 2191 (2002).
[12] T. Prestera, P. Talalay, J. Alam, Y. I. Ahn, P. J. Lee, and A. M. Choi, *Mol. Med.* **1,** 827 (1995).
[13] J. Alam, D. Stewart, C. Touchard, S. Boinapally, A. M. Choi, and J. L. Cook, *J. Biol. Chem.* **274,** 26071 (1999).
[14] Y. Tsuji, H. Ayaki, S. P. Whitman, C. S. Morrow, S. V. Torti, and F. M. Torti, *Mol. Cell. Biol.* **20,** 5818 (2000).
[15] E. C. Pietsch, J. Y. Chan, F. M. Torti, and S. V. Torti, *J. Biol. Chem.* **278,** 2361 (2003).
[16] R. T. Mulcahy and J. J. Gipp, *Biochem. Biophys. Res. Commun.* **209,** 227 (1995).
[17] D. C. Galloway, D. G. Blake, A. G. Shepherd, and L. I. McLellan, *Biochem. J.* **328** (Pt 1), 99 (1997).
[18] D. C. Galloway and L. I. McLellan, *Biochem. J.* **336** (Pt 3), 535 (1998).
[19] W. A. Solis, T. P. Dalton, M. Z. Dieter, S. Freshwater, J. M. Harrer, L. He, H. G. Shertzer, and D. W. Nebert, *Biochem. Pharmacol.* **63,** 1739 (2002).
[20] H. R. Moinova and R. T. Mulcahy, *Biochem. Biophys. Res. Commun.* **261,** 661 (1999).
[21] T. H. Rushmore, M. R. Morton, and C. B. Pickett, *J. Biol. Chem.* **266,** 11632 (1991).
[22] J. D. Moehlenkamp and J. A. Johnson, *Arch. Biochem. Biophys.* **363,** 98 (1999).

hydroquinone, but not phenol or resorcinol, activate the ARE. Prester and Talalay[2] concluded from their studies of the induction of phase II enzymes that oxidative liability was essential for an inducer's activity, because catechol (1,2-diphenols) and hydroquinone (1,4-diphenols) derivatives undergo facile oxidation to quinones, whereas 1,3-diphenols, inactive for induction of phase II enzymes, cannot participate in such an oxidation. Although these experiments did not establish whether the oxidation products or oxidation processes (potentially involving a radical scavenging reaction, multiple one- and two-electron oxidoreduction, and redox-dependent reactive oxygen species [ROS] generation) were an inductive signal, electrophilic quinone oxidation products were presumed to be the ultimate inducer, because electrophiles inducing Michael reaction acceptors and isothiocyanates potentially induce phase II enzyme expression.[23,24] Pinkus *et al.*[25] demonstrated the autooxidation of tert-butylhydroquinone (tBHQ) to the semiquinone radical or 1,4-benzoquinone and the generation of hydroxyl radical using the electron spin resonance spectroscopy technique. They also showed that the induction of an endogenous GST alpha class gene (rGSTA1) in hepatoma cells by tBHQ was inhibited by the antioxidants *N*-acetylcysteine, GSH, and exogenous catalase. It was expected that the intermediate formation of H_2O_2 during the metabolism of tBHQ might be a critical step for induction of phase II enzymes. However, the dose-dependency of the hydroxyl radical generation correlated poorly with GST gene expression. As the concentration of tBHQ increased, a lower amount of hydroxyl radical was generated. Kong *et al.*[26] reported that butylated hydroxyanisole (BHA) and tBHQ strongly activate ERK2, JNK1, and p38 MAP kinase in a time-dependent and dose-dependent fashion in multiple epithelial cell lines. Free radical scavengers *N*-acetyl cysteine (NAC) and glutathione (GSH) inhibited ERK2 activation and, to a much lesser extent, JNK1 activation by tBHQ/BHA, implicating the role of oxidative stress in tBHQ-induced ARE activation. We observed that compounds with catechol or hydroquinone structure activate the ARE leading to NQO1 gene expression in IMR-32 human neuroblastoma cells. Interestingly, dtBHQ (di-tert-butylhydroquinone), which is identical to tBHQ except for the substitution of another tert-butyl group in the 5′-position on the phenol

[23] T. Prestera, Y. Zhang, S. R. Spencer, C. A. Wilczak, and P. Talalay, *Adv. Enzyme Regul.* **33,** 281 (1993).

[24] Y. Nakamura, T. Kumagai, C. Yoshida, Y. Naito, M. Miyamoto, H. Ohigashi, T. Osawa, and K. Uchida, *Biochemistry* **42,** 4300 (2003).

[25] R. Pinkus, L. M. Weiner, and V. Daniel, *J. Biol. Chem.* **271,** 13422 (1996).

[26] A. N. Kong, E. Owuor, R. Yu, V. Hebbar, C. Chen, R. Hu, and S. Mandlekar, *Drug Metab. Rev.* **33,** 255 (2001).

ring, did not increase NQO1 gene expression (Lee Jong-Min *et al.* unpublished data, 2003). These observations imply a structure-activity relationship for quinone-mediated NQO1 gene expression. However, the exact molecular mechanism for this has not been clear until now. Preliminary data from our laboratory suggest that quinone-mediated ARE activation and NQO1 gene expression is reactive oxygen species–independent. We are currently investigating the hypothesis that a possible chemical–protein interaction initiates the signaling pathway leading to increased NQO1 gene expression.

To further clarify whether the involvement of oxidative stress in ARE activation is indispensable, we pretreated IMR-32 human neuroblastoma cells with antioxidants or antioxidant enzymes.[27] Pretreatment with antioxidants such as GSH, glutathionemonoethyl ester (GSHEE), or NAC did not inhibit ARE activation by tBHQ (10 μM). In contrast diethyl maleate, (DEM)–mediated ARE activation was completely inhibited by pretreatment with GSH, GSHEE, or NAC. In addition, the increase in NQO1 protein level by tBHQ was not inhibited, whereas the increase by DEM was significantly inhibited by these antioxidants. Pretreatment with either catalase or superoxide dismutase (SOD) did not inhibit ARE activation by tBHQ. However, ARE activation by DEM was decreased significantly by catalase in a dose-dependent manner. Western blot demonstrated that catalase decreased NQO1 protein induced by DEM but had no effect on NQO1 induction by tBHQ. SOD had no effect on ARE activation by either tBHQ or DEM.

Other strong evidence showing the oxidative stress–independent ARE activation comes from transgenic reporter mice that carry the core ARE coupled to the human placental alkaline phosphatase (hPAP) reporter gene in their genome.[28] Primary cortical cultures derived from these mice were treated with tBHQ, resulting in a dose-dependent increase in hPAP activity. Histochemical staining for hPAP activity was observed in both astrocytes and neurons from tBHQ (30 μM)-treated cultures. The tBHQ-mediated increase in hPAP was not affected by the GSHEE, whereas the increase in hPAP following DEM treatment was completely blocked by GSHEE. Activation of the ARE by tBHQ was independent of oxidative stress, whereas activation of the ARE by DEM was dependent on oxidative stress. This implies that tBHQ may activate the ARE through oxidative stress–independent mechanisms.

[27] J. M. Lee, J. D. Moehlenkamp, J. M. Hanson, and J. A. Johnson, *Biochem. Biophys. Res. Commun.* **280,** 286 (2001).

[28] D. A. Johnson, G. K. Andrews, W. Xu, and J. A. Johnson, *J. Neurochem.* **81,** 1233 (2002).

Nrf2 Is One of the Major Transcriptional Factors Involved in the Regulation of Antioxidant Responsive Element–Driven Gene Expression

The mechanisms that regulate phase II detoxification gene expression through ARE activation are under intense investigation. Several ARE-binding proteins have been proposed and/or identified, including AP1 family proteins (c-Jun, Jun-B, Jun-D, Fra1 and Fra2, and c-Fos), members of the "cap 'n' collar" (CNC) basic leucine zipper transcription factor (bZIP) family (Nrf1, Nrf2, Nrf3, Bach1, Bach2, and small Maf proteins [MafG, MafF, and MafK]).[29] Identification of the transcriptional activator mediating ARE activation has been controversial: early suggestions that AP-1 factors mediate the response have been disputed by contrary findings.[29,30] However, Nrf2 has been demonstrated to play a central role in the gene expression of phase II detoxification enzymes and some antioxidant genes.

Human Nrf2 (*N*F-E2–*r*elated *f*actor *2*), 605 amino acids (aa) and approximately 67 KDa, contains regions of strong homology with defined domains such as Neh2, "CNC," DNA binding, and leucine zipper.[31] Several research groups reported that Nrf2 is the actual transactivation factor responsible for upregulating ARE-driven gene expression. For instance, Venugopal and Jaiswal[8] reported that Nrf1 and Nrf2 positively regulate human ARE-mediated expression of NQO1 gene, whereas c-Fos and Fra1 act as negative regulators. Nguyen *et al.*[32] reported that transcriptional regulation of the ARE is activated by Nrf2 and repressed by MafK. Recent studies with Nrf2 knockout mice support this conclusion. Chan *et al.*[33] reported that Nrf2-deficient mice have lower basal levels of expression of phase II enzymes and lack the ability to induce them. They correlated this with an increased susceptibility to carcinogensis and oxidative stress–induced cell injuries that could not be blocked by pretreatment with chemicals known to protect in wild-type mice.

Nrf2 is localized mainly in the cytoplasm bound to a chaperone, KIAA0132 (the human homologue to Keap1). Keap1 (*K*elch-like *E*CH *a*ssociating *p*rotein *1*), 624 amino acids (~72 KDa), contains 25 cysteine residues, 9 of which are predicted to have highly reactive sulfhydryl groups (low pKa values), because they are flanked by one or more basic amino

[29] H. Motohashi, T. O'Connor, F. Katsuoka, J. Engel, and M. Yamamoto, *Gene* **294,** 1 (2002).
[30] W. W. Wasserman and W. E. Fahl, *Proc. Natl. Acad. Sci. USA* **94,** 5361 (1997).
[31] P. Moi, K. Chan, I. Asunis, A. Cao, and Y. W. Kan, *Proc. Natl. Acad. Sci. USA* **91,** 9926 (1994).
[32] T. Nguyen, H. C. Huang, and C. B. Pickett, *J. Biol. Chem.* **275,** 15466 (2000).
[33] K. Chan, X. D. Han, and Y. W. Kan, *Proc. Natl. Acad. Sci. USA* **98,** 4611 (2001).

acid residues.[34] Because all inducers react with sulfhydryl groups, researchers proposed that oxidative stress or electrophiles induce the release of Nrf2 from the Nrf2–Keap1 complex resulting in nuclear translocation of Nrf2, where it binds (in heterodimeric forms with other transcription factors) to the ARE of phase II genes and stimulates transcription.[1,34] The Keap1–Nrf2 complex is a plausible candidate for the cytoplasmic sensor system that recognizes and reacts with inducers.

Antioxidant Responsive Element Activation, a Potential Chemoprotective Response?

Although this antioxidant defense mechanism has been studied extensively as a hepatic detoxification mechanism, it also may contribute to antioxidant defenses in the lung and central nervous system. Others have shown that treating cells with tBHQ, a strong inducer of phase II detoxification enzymes by means of activation of ARE, can protect cells from oxidative stress.[35,36] Induction of NQO1 in N18-RE-105 neuronal cells by tBHQ before glutamate treatment was correlated with a significant decrease in glutamate toxicity. Glutamate toxicity in these cells is not due to *N*-methyl-D-aspartate receptor activation and calcium influx. Rather, it is a result of competitive inhibition of cystine uptake, depletion of GSH, increased oxidative stress, and apoptosis. Subsequent studies by Murphy and colleagues[37] using H_2O_2 and dopamine to induce oxidative stress, however, demonstrated that N18-RE-105 cells overexpressing NQO1 were not resistant to cytotoxicity. These data suggest that the protective effect conferred by tBHQ may not simply be due to an increase in one gene but the coordinate upregulation of many genes. We also used H_2O_2 to generate oxidative stress and subsequent apoptosis in cultured IMR-32 cells.[38] The apoptotic process and development of cell injury in H_2O_2-treated IMR32 cells was prevented through tBHQ-induced ARE activation. We recently reported the differential sensitivity for the primary cortical astrocytes from Nrf2+/+ and Nrf2−/− mice to H_2O_2-induced cytotoxicity.[39] Nrf2−/− astrocytes were more sensitive to H_2O_2-induced cytotoxicity compared with

[34] K. Itoh, N. Wakabayashi, Y. Katoh, T. Ishii, K. Igarashi, J. D. Engel, and M. Yamamoto, *Genes Dev.* **13,** 76 (1999).

[35] S. Duffy, A. So, and T. H. Murphy, *J. Neurochem.* **71,** 69 (1998).

[36] T. H. Murphy, M. Miyamoto, A. Sastre, R. L. Schnaar, and J. T. Coyle, *Neuron* **2,** 1547 (1989).

[37] T. H. Murphy, M. J. De Long, and J. T. Coyle, *J. Neurochem.* **56,** 990 (1991).

[38] J. Li, J. M. Lee, and J. A. Johnson, *J. Biol. Chem.* **277,** 388 (2002).

[39] J. M. Lee, M. J. Calkins, K. Chan, Y. W. Kan, and J. A. Johnson, *J. Biol. Chem.* **278,** 12029 (2003).

Nrf2+/+ astrocytes. tBHQ pretreatment significantly increased cell viability in Nrf2+/+ astrocytes but not in Nrf2 −/− astrocytes. These observations imply that the coordinate upregulation of ARE-driven genes by tBHQ in Nrf2 +/+ astrocytes is more efficient in protecting cells from oxidative damage. Chan and Kan[40] demonstrated the importance of the transcription factor Nrf2 when Nrf2 knockout mice fed butylated hydroxytoluene showed extreme pulmonary toxicity. Gene expression of NQO1, HO1, and GCLR were significantly reduced in the lungs of these animals, suggesting that these ARE-containing genes, regulated by Nrf2, play a critical role in protection from oxidative stress and that the lack of induction of phase II–detoxifying enzymes made Nrf2-deficient mice highly sensitive to cytotoxic electrophiles. It is also known that treating mice with nontoxic phase II enzyme inducers, such as BHA and oltipraz, protects them from carcinogen attack.[41,42] Taken together, these data suggest ARE activation elicits a potential chemoprotective response. The identification of the ARE-targeting genes was an initial step in the explanation of the molecular mechanism of this chemoprotective response.

Antioxidant Responsive Element Target Genes

High throughput microarray technology allows us to monitor the expression level of thousands of genes in parallel and has the potential to surpass traditional approaches in terms of sensitivity and speed. Compared with cDNA microarrays, oligonucleotide microarrays have been proven to provide highly reproducible and reliable data in terms of correlation with quantitative reverse transcriptase–polymerase chain reaction (RT–PCR).[43] Recently, several groups have actively participated in dissecting ARE-driven gene expression through oligonucleotide microarrays.

Time-Dependent Antioxidant Responsive Element–Driven Gene Expression Profile Induced by Tert-Butylhydroquinone

It is a fast and simple way to dissect ARE-driven genes through treatment with a strong activator of the ARE such as tBHQ. Time-dependent and dose-dependent gene expression profiles may quickly be revealed by

[40] K. Chan and Y. W. Kan, *Proc. Natl. Acad. Sci. USA* **96,** 12731 (1999).

[41] M. McMahon, K. Itoh, M. Yamamoto, S. A. Chanas, C. J. Henderson, L. I. McLellan, C. R. Wolf, C. Cavin, and J. D. Hayes, *Cancer Res.* **61,** 3299 (2001).

[42] M. K. Kwak, P. A. Egner, P. M. Dolan, M. Ramos-Gomez, J. D. Groopman, K. Itoh, M. Yamamoto, and T. W. Kensler, *Mutat. Res.* **480,** 305 (2001).

[43] J. Li, M. Pankratz, and J. A. Johnson, *Toxicol. Sci.* **69,** 383 (2002).

microarray analysis, and a variety of biostatistical methods also are available to cluster genes on the basis of their expression patterns.

We reported that time-dependent gene expression profiles of ARE-driven genes are induced by tBHQ (Table I).[38,44] Using an efficient noise-filtering process for microarray data generated from tBHQ versus vehicle-treated IMR32 cells, we quickly narrowed down the final gene list from a total of 9670 genes to 101, which demonstrated dynamic changes in response to the tBHQ (10 μM) treatment from 4 h to 48 h. In this study, one of the major clusters of transcriptionally activated genes in IMR-32 cells was the phase II detoxification enzymes. Within this category of genes, there were early-response genes (NQO1, HO1, glutathione reductase [GR], glutathione transferase M3 [GSTM3], GCLR, and thioredoxin reductase [TR]) and late-response genes (ferritin heavy and light chains). Interestingly, unlike other phase II detoxification enzymes, HO1 was transiently upregulated at 8 h treatment and not changed at 24 or 48 h. NAD(P)H regeneration enzymes like hepatic dihydrodiol dehydrogenase and its isoform KIAA0119, malate NADP oxidoreductase, and breast cancer cytosolic NADP(+)–dependent malic enzyme also were found to be upregulated by tBHQ. Of all the genes induced by tBHQ, these have the most significant changes, suggesting their importance in the tBHQ-mediated antioxidant effect. Nuclear transcription factors, including c-Jun, c-Fos, Jun-B, Jun-D, Fra-1, Fra-2, ATF-3, ATF-4, NF-κB, Nrf1, Nrf2, Nrf3, and small Maf proteins (MafK and MafF), which have been reported to influence the Nrf2-ARE interaction, did not change at the mRNA level after treatment with tBHQ. Interestingly, KIAA0132, the human homologue of mouse Keap1, was increased. Keap1, as mentioned earlier, is a cytosolic chaperone of Nrf2. Disruption of this Keap1–Nrf2 complex allows Nrf2 to translocate into the nucleus, where it binds to the ARE and stimulates transcription.

Dissection of Antioxidant Responsive Element–Driven Genes through Nrf2, KO vs. WT; Nrf2 vs. DN Overexpression

Microarray analysis can be applied to dissect direct Nrf2-dependent gene expression profiles in isolated tissues or primary cultures derived from Nrf2 −/− versus Nrf2 +/+ mice.

Liver. Comparative studies of gene expression changes between Nrf2 +/+ and Nrf2 −/− mice were performed on liver samples using murine U74AV2 oligonucleotide arrays (Table II). Numerous genes regulated by Nrf2 were identified, including the phase II detoxification enzymes,

[44] J. Li and J. A. Johnson, *Physiol. Genomics* **9,** 137 (2002).

TABLE I
OVERVIEW OF TRANSCRIPTIONAL UPREGULATION INDUCED BY tBHQ IN IMR-32 CELLS

Category of genes	Genbank accession number	Description of genes
Cell death/apoptosis	D83699	Brain 3UTR of mRNA for neuronal death protein (harakiri)
	N/A	Bax delta
Chaperones/heat shock proteins	N/A	Heat shock protein hsp40 homolog
	X87949	BiP protein (a member of Hsp70 family of chaperones)
	W28493	Heat shock 70 kD protein 8
CNS specific function	U79299	Neuronal olfactomedin-related ER localized protein
	AF009674	Axin (AXIN)
	U40572	Beta$_2$-syntrophin (SNT B2)
	M25756	Secretogranin II gene
	Z48054	Peroxisomal targeting signal 1
	U73304	CB1 cannabinoid receptor (CNR1)
	AB023209	KIAA0992 (palladin)
	N/A	Neurofibromatosis 2 tumor suppressor
	L27745	Voltage-operated calcium channel, alpha-1 subunit
	AB012851	Musashi (RNA-binding protein)
Cytoskeleton	AB002323	Dynein, cytoplasmic, heavy polypeptide 1 (KIAA0325)
	W26631	Microtuble-associated protein 1A
	X15306	NF-H gene, exon 1
	S67247	Smooth muscle myosin heavy chain isoform (Smemb)
	M22299	T-plastin polypeptide
Detoxification and antioxidative stress	M81600	NAD(P)H-quinone oxidoreductase
	X15722	Glutathione reductase
	N/A	Glutathione transferase M3 (GSTM3)
	Z82244	Heme oxygenase 1
	N/A	Breast cancer cytosolic NADP(+) –dependent malic enzyme
	AL049699	Malic enzyme 1, soluble (NADP-dependent malic enzyme, malate oxidoreductase)
	D17793	KIAA0119 (aldo-keto reductase family 1)
	U05861	Hepatic dihydrodiol dehydrogenase
	X91247	Thioredoxin reductase
	L35546	Gamma-glutamylcysteine ligase regulatory subunit
	AL031670	Ferritin, light polypeptide-like 1
	J04755	Ferritin H processed pseudogene

(continued)

TABLE I (*continued*)

Category of genes	Genbank accession number	Description of genes
	L20941	Ferritin heavy chain
DNA repair	N/A	Human growth arrest and DNA-damage–inducible protein (gadd45) mRNA
	N/A	ERCC5 excision repair protein
Extracellular matrix	AL050138	Elastin microfibril interface located protein
	M92642	Alpha-1 type XVI collagen
Glycoprocess	U84007	Glycogen debranching enzyme isoform 1 (AGL)
	U84011	Glycogen debranching enzyme isoform 6 (AGL)
	L12711	Transketolase (tk)
Immunosystem	D32129	HLA class-I (HLA-A26) heavy chain
RNA processing/ modification	L22009	hnRNP H
	AL03168	Splicing factor, arginine/serine-rich 6 (SRP55-2) (isoform 2)
Signaling	L20861	Wnt-5a
	N/A	Guanine nucleotide-binding protein Rap2, Ras-oncogene related
	AJ011679	Rab6 GTPase activating protein
	D79990	Ras association domain family 2 (KIAA0168)
	L36870	MAP kinase kinase 4 (MKK4)
	U35113	Metastasis-associated mta 1
	M88714	Bradykinin receptor (BK-2)
	U46751	Phosphotyrosine independent ligand p62 for the Lck SH2 domain
	Y13493	Protein kinase; Dyrk2
	Z85986	Clone 108K11 on chromosome 6p21 contains SRP20 (protein serine/threnione kinase)
	N/A	Interferon-inducible RNA–dependent protein kinase (Pkr)
	X59656	Crk-like gene CRKL (protein tyrosine kinase)
	N/A	Ptdins 4-kinase (P14kb)
	L35594	Autotaxin(ectonucleotide pyrophosphatase/ phosphodiesterase)
Transcription regulation	D50922	KIAA0132 (kelch-like ECH-associated protein 1)
	U66561	Kruppel-related zinc finger protein (ZNF 184)
	X78992	ERF-2
	AF078096	Forkhead/winged helix-like transcription factor 7 (FKHL7)
	AF040963	Mad4 homolog (Mad4)
	AF096870	Estrogen-responsive B box protein (EBBP)
	X87838	Beta-catenin
	U19969	Two-handed zinc finger protein ZEB
	N/A	DNA-binding protein (APRF)
	D88827	Zinc finger protein FPM3

(*continued*)

TABLE I *(continued)*

Category of genes	Genbank accession number	Description of genes
	U10324	NF-90
	X96381	Erm
Translation/ posttranslation regulation	U20180	Iron-regulatory protein 2 (IRP2)
	D26600	Proteasome subunit HsN3
Others	U29332	Heart protein (FHL-2)
	X64728	CHML
	AF055001	Homocysteine-inducible, endoplasmic reticulum stress inducible, ubiquitin-like domain member 1
	AB007865	Fibronectin leucine-rich transmembrane protein 2 (KIAA0405)
	X54232	Heparan sulfate proteaglycan (glypican)
	Z25535	Nuclear pore complex protein hnup 153
	D19878	Transmembrane protein
	U23070	Putative transmembrane protein (nma)
	AL031781	Human orthologue of zebrafish Quaking protein homolog ZKQ-1 (isoform 1)
	AF070598	Clone 24410 ABC transporter
	U60644	HU-K4 (phospholipase D)
	AJ131581	Latrophilin-2
	AF061573	Protocadherin (PCDH8)

From J. Li *et al. Physiol. Genomics* **9,** 137 (2002), with permission.

NAD(P)H-regenerating enzymes, and multiple antioxidant genes. The decreases in detoxification enzymes and the antioxidant defense network observed in data from Nrf2 −/− mice are presumed to make these mice more vulnerable to a variety of oxidative stressors, and subsequently the mice were prone to have chronic inflammation, cancer, and autoimmune diseases develop.[45]

Kwak *et al.*[46] reported modulation of gene expression by the cancer chemopreventive dithiolethiones through the Keap1–Nrf2 pathway. To identify those genes regulated by the Nrf2 pathway, hepatic gene expression profiles were examined by oligonucleotide microarray analysis in

[45] R. K. Thimmulappa, K. H. Mai, S. Srisuma, T. W. Kensler, M. Yamamoto, and S. Biswal, *Cancer Res.* **62,** 5196 (2002).

[46] M. K. Kwak, N. Wakabayashi, K. Itoh, H. Motohashi, M. Yamamoto, and T. W. Kensler, *J. Biol. Chem.* **278,** 8135 (2003).

TABLE II
DIFFERENTIALLY EXPRESSED DETOXIFICATION-RELATED GENES IN THE LIVERS OF MALE AND FEMALE Nrf2 KNOCKOUT MICE

Gene name	Male +/+ vs −/− (2 × 2)			Female +/+ vs −/− (2 × 2)		
	FC	CV	R	FC	CV	R
Aflatoxin aldehyde reductase	−1.71 ± 0.18	−0.21	−8	—	—	—
Aldehyde dehydrogenase 2, mitochondrial	—	—	—	−1.24 ± 0.08	−0.13	−4
Aldehyde dehydrogenase family 1, subfamily A1	−1.45 ± 0.08	−0.12	−8	−1.37 ± 0.15	−0.22	−4
Aldehyde dehydrogenase family 3	−2.25 ± 0.22	−0.20	−8	—	—	—
Aromatic amino acid decarboxylase	2.09 ± 0.27	0.26	8	—	—	—
ATP-binding cassette, subfamily C (CFTR/MDR), member 6	−1.77 ± 0.20	−0.23	−8	−1.49 ± 0.03	−0.03	−4
Carboxyl esterase	−8.61 ± 0.44	−0.10	−8	−2.32 ± 1.36	−1.17	−6
Dual-specificity protein tyrosine phosphatase	−2.93 ± 0.32	−0.22	−8	−1.91 ± 0.31	−0.32	−6
Epoxide hydrolase	−3.13 ± 0.33	−0.21	−8	−1.47 ± 0.09	−0.13	−6
Ferritin heavy chain	−1.45 ± 0.10	−0.14	0	−1.35 ± 0.11	−0.16	−4
Ferritin light chain 1	−1.99 ± 0.18	−0.18	−8	—	—	—
Flavin-containing monooxygenase	−1.82 ± 0.13	−0.14	−8	−1.52 ± 0.10	−0.13	−8
Flavin-containing monooxygenase 3	—	—	—	−1.43 ± 0.10	−0.14	−6
Glucose-6-phosphate dehydrogenase X-linked	—	—	—	−2.00 ± 0.23	−0.23	−6
γ-Glutamylcysteine synthetase, catalytic	−1.98 ± 0.14	−0.14	−8	—	—	—
Glucocorticoid-regulated kinase	—	—	—	−1.51 ± 0.21	−0.28	−4
Glutathione S-transferase alpha 1(Ya)	−2.01 ± 0.48	−0.48	−5	−1.44 ± 0.06	−0.09	−6
Glutathione S-transferase alpha 2(Yc2)	−2.59 ± 0.38	−0.29	−8	—	—	—
Glutathione S-transferase alpha 3	−1.54 ± 0.16	−0.20	−6	−1.59 ± 0.16	−0.20	−6
Glutathione S-transferase mu 1	−2.75 ± 0.20	−0.14	−8	−1.75 ± 0.10	−0.11	−8
Glutathione S-transferase mu 1	−2.74 ± 0.12	−0.09	−8	−1.72 ± 0.10	−0.12	−8
Glutathione S-transferase mu 2	−2.17 ± 0.46	−0.43	−6	—	—	—
Glutathione S-transferase mu 3	−3.57 ± 0.25	−0.14	−8	−2.40 ± 0.31	−0.26	−8
Glutathione S-transferase mu 6	−2.37 ± 0.33	−0.28	−8	—	—	—

(continued)

TABLE II (*continued*)

Gene name	Male +/+ vs −/− (2 × 2)			Female +/+ vs −/− (2 × 2)		
	FC	CV	R	FC	CV	R
Glutathione S-transferase mu 6	−2.50 ± 0.25	−0.20	−8	—	—	—
Haptaglobin	—	—	—	−1.65 ± 0.03	−0.04	−8
Malic enzyme	−2.06 ± 0.29	−0.28	−8	−2.09 ± 0.29	−0.28	−6
Membrane-associated progesterone receptor component	−1.30 ± 0.07	−0.10	−5	—	—	—
Multidrug resistance protein	−9.38 ± 0.42	−0.09	−8	−1.68 ± 0.03	−0.04	−7
NAD(P)H quinone oxidoreductase	−4.30 ± 0.98	−0.46	−4	−7.54 ± 2.42	−0.64	−6
Nucleoside diphosphatase (ER-UDPase gene)	−2.01 ± 0.24	−0.23	−6	—	—	—
Ornithine aminotransferase	−2.39 ± 0.11	−0.09	−8	−1.37 ± 0.06	−0.09	−4
Sequestosome 1 (A170)	−1.46 ± 0.03	−0.04	−7	—	—	—
Thioredoxin reductase 1	−1.35 ± 0.10	−0.14	−4	—	—	—
Transketolase	−1.33 ± 0.08	−0.13	−4	—	—	—
UDP-glucuronosyl-transferase 1 family, polypeptide A6	−1.40 ± 0.11	−0.15	−8	—	—	—
UDP-glucuronosyl-transferase 2 family	−1.92 ± 0.17	−0.18	−8	−1.24 ± 0.07	−0.11	−6
Nuclear, factor, erythroid derived 2, like 1	−1.46 ± 0.03	−0.04	−8	—	—	—
Nuclear, factor, erythroid derived 2, like 2	−40.62 ± 18.99	−0.93	−8	−18.84 ± 1.33	−0.14	−8

Total RNA was extracted from liver samples isolated from male and female Nrf2 wild-type and knockout mice at 5 months of age (two samples in each group). Affymetrix platform was used to generate microarray data. Output from the microarray analysis was merged with the Unigene or GenBank descriptor and stored as an Excel data spreadsheet. Significantly changed genes were determined using the Wilcoxon signed rank test for each comparison. Probe sets with p values < 0.0025 were called increased/decreased; probe sets with p values in the range $0.0025 < p$ value < 0.003 were called marginally increased/decreased; and the remaining probe sets were called no change. An additional level of ranking was used to incorporate multiple comparisons such that no change = 0, marginal increase/decrease = 1/−1, and increase/decrease = 2/−2. The final rank (R) equaled the sum of the ranks from the four comparisons, and the value varied from −8 to 8 for a 2 × 2 comparison. R values ranging from 4–8 indicate significantly increased gene expression, and values from −4 to −8 indicate significantly decreased genes. Dashes (–) indicate no significant change. FC, fold change; CV, coefficient of variance. FC is expressed as mean ± SEM.

vehicle-treated or D3T(3*H*-1,2-dithiole-3-thione)-treated wild-type mice, as well as in Nrf2 single and Keap1–Nrf2 double knockout mice. Transcript levels of 292 genes were elevated in wild-type mice 24 h after treatment with D3T; 79% of these genes were induced in wild-type, but not Nrf2-deficient, mice. These Nrf2-dependent, D3T-inducible genes included known detoxification and antioxidant enzymes. Unexpected clusters included genes for chaperones, protein trafficking, ubiquitin/26S proteasome subunits, and signaling molecules. Gene expression patterns in Keap1–Nrf2 double knockout mice were similar to those in Nrf2 single knockout mice. D3T also led to Nrf2-dependent repression of 31 genes at 24 h, principally genes related to cholesterol/lipid biosynthesis. Collectively, D3T increases the expression of genes, through the Keap1–Nrf2 signaling pathway, that directly detoxify toxins and generate essential cofactors such as glutathione and reducing equivalents. Induction of Nrf2-dependent genes involved in the recognition and repair/removal of damaged proteins expands the role of this pathway beyond primary control of electrophilic and oxidative stresses and into secondary protective actions that enhance cell survival.

Cortical Culture. Astrocytes have a higher antioxidant potential than neurons that, in part, seems to be due to the lower or lack of expression of Nrf2 in differentiated neurons. Thus, the Nrf2 pathway is primarily localized to cells of glial origin under nonpathological conditions. We identified Nrf2-dependent ARE-driven genes encoding detoxification enzymes, glutathione-related proteins, antioxidant proteins, and NADPH-producing enzymes in mouse and rat primary astrocyte cultures.[39,47] We also observed that these Nrf2-dependent genes protected primary astrocytes from H_2O_2 or platelet activating factor–induced apoptosis. Proteins within these functional categories are vital to the maintenance and responsiveness of a cellular defense system, suggesting that an orchestrated change in expression by means of Nrf2 and the ARE gives a synergistic protective effect.

Shih *et al.*[47] demonstrated that Nrf2 overexpression through an adenovirus-mediated infection can re-engineer neurons to express this glial pathway and enhance antioxidant gene expression. Nrf2-mediated protection from oxidative stress is conferred primarily by glia in mixed rat cultures. The antioxidant properties of Nrf2-overexpressing glia are more pronounced than those of neurons, and a relatively small number of these glia ($< 1\%$ of total cell number added) could protect fully co-cultured naive neurons from oxidative glutamate toxicity associated with glutathione depletion. Microarray and biochemical analyses indicate that Nrf2 overexpression coordinately upregulates enzymes involved in GSH biosynthesis

[47] A. Y. Shih, D. A. Johnson, G. Wong, A. D. Kraft, L. Jiang, H. Erb, J. A. Johnson, and T. H. Murphy, *J. Neurosci.* **23,** 3394 (2003).

(xCT cystine antiporter, gamma-glutamylcysteine synthetase, and GSH synthase), GSH use (GST and GR), and GSH export (multidrug resistance protein 1). This leads to an increase in both secreted and intracellular GSH. Selective inhibition of glial GSH synthesis and the supplementation of medium with GSH indicated that an Nrf2-dependent increase in glial GSH synthesis was both necessary and sufficient for the protection of neurons. Neuroprotection was not limited to overexpression of Nrf2, because activation of endogenous glial Nrf2 by the small molecule ARE inducer, tert-butylhydroquinone, also protected against oxidative glutamate toxicity.

From the multiple systems including human, mouse, and rat, both in vivo and in vitro as described here, it is clearly demonstrated that Nrf2-dependent ARE activation plays a central role in ARE-driven gene expression. This can confer to cells or organs a protective and/or survival effect from oxidative stress–induced cell injuries.

Role of Phosphatidylinositol 3-Kinase in Antioxidant Responsive Element Activation

Several protein kinase pathways have been implicated in transducing signals to gene expression mediated by ARE activation. Kong *et al.*[26] reported that BHA and tBHQ strongly activated c-Jun N terminal kinase 1 (JNK1), extracellular signal-regulated protein kinase 2 (ERK2), or p38 MAP kinase, in a time-dependent and dose-dependent fashion in several epithelial cell lines. Meanwhile, Huang *et al.*[48] reported regulation of the ARE by protein kinase C (PKC)–mediated phophorylation of Nrf2 at Ser-40 in HepG2 cells. Numazawa *et al.*[49] reported that the atypical PKC subfamily is responsible for the phosphorylation of Nrf2 Ser40. Phosphorylation of wild-type Nrf2 by PKC promoted its dissociation from Keap1, whereas the Nrf2(S40A) mutant remained associated, suggesting that the PKC-catalyzed phosphorylation of Nrf2 at Ser40 is a critical signaling event promoting Nrf2 nuclear translocation and leading to the ARE-mediated cellular antioxidant response.

We recently reported that the phosphatidylinositol 3-kinase (PI3-kinase) pathway, a classical cell survival and proliferation pathway, also may be involved in ARE activation.[50] PI3 kinase phosphorylates phosphatidylinositol at the D3-position of the inositol ring and has been shown to

[48] H. C. Huang, T. Nguyen, and C. B. Pickett, *J. Biol. Chem.* **277,** 42769 (2002).

[49] S. Numazawa, M. Ishikawa, A. Yoshida, S. Tanaka, and T. Yoshida, *Am. J. Physiol. Cell. Physiol.* **285,** C334 (2003).

[50] J. M. Lee, J. M. Hanson, W. A. Chu, and J. A. Johnson, *J. Biol. Chem.* **276,** 20011 (2001).

form a heterodimer consisting of an 85-KDa (adapter protein) and a 110-KDa (catalytic) subunit. The role of PI3-kinase in intracellular signaling has been underscored by its implication in a plethora of biological responses such as cell growth, differentiation, apoptosis, calcium signaling, and insulin signaling. Among the downstream targets of PI3-kinase are phospholipase C, serine/threonine kinase Akt, and ribosomal S6-kinase. Akt (protein kinase B), one of the best-known downstream targets of PI3 kinase, protects cells from apoptosis by the phosphorylation and inhibition of the proapoptotic protein, BAD. Our study shows that PI3-kinase contributes to the tBHQ-induced ARE response.

Phosphatidylinositol 3-Kinase Is Involved in Antioxidant Responsive Element Activation

Pretreatment with PI3-kinase inhibitor LY294002 demonstrated a dose-dependent decrease in tBHQ-induced NQO1-ARE–hPAP activity or NQO1-ARE–luciferase expression. Primary cortical cultures derived from ARE-hPAP reporter mice were pretreated (20 min) with vehicle or increasing doses of LY 294002 before tBHQ treatment (30 μM) for 24 and 72 h. At 24 h, 15 μM and 30 μM LY294002 completely blocked tBHQ activation of the ARE, and the relative hPAP activities were at or near the activity observed in vehicle-treated samples. At 72 h, LY294002 in a dose-dependent fashion inhibited the activation of ARE from 26-fold to 10-fold. In addition, the tBHQ-mediated expression of ARE-driven genes such as NQO1 and GCLR in primary cortical cultures was blocked by LY294002. This effect is also observed in IMR32 neuroblastoma cells. Pretreatment of IMR32 cells with LY294002 inhibited both tBHQ-mediated NQO1-ARE–luciferase expression and endogenous NQO1 increase in a dose-dependent manner, suggesting a positive role for PI3-kinase in ARE activation. In contrast, pretreatment with PD 98059 (50 μM), a selective inhibitor of MAP/Erk kinase, did not inhibit ARE activation by tBHQ. Similarly, endogenous NQO1 protein induction by tBHQ was not blocked by PD98059. These concentrations of LY294002 had no effect on cell viability as determined by the MTS cytotoxicity assay.

Direct evidence that PI3-kinase is involved in ARE activation was obtained through transfection of IMR32 cells with constitutively active PI3-kinase p110*(CA PI3 kinase) or kinase-deficient PI3 kinase p110* Δkin (KD PI3-kinase). Only the CA PI3-kinase increased hNQO1-ARE reporter gene expression, and this induction was inhibited completely by treatment with LY294002. ARE activation mediated by CA PI3-kinase was blocked completely by DN Nrf2, suggesting that Nrf2 is downstream of PI3 kinase in IMR32 cells. KD PI3-kinase did not show a dominant negative effect on endogenous PI3-kinase, and ARE activation by tBHQ was not inhibited by KD PI3-kinase.

Recent findings linking PI3-kinase to Akt and subsequent phosphorylation of GSK-3β led us to speculate that tBHQ may activate the same pathway. By the use of insulin as a positive control, the data show that tBHQ does not increase Akt activity or lead to increased phosphorylation of GSK-3β. As expected, the effect of insulin on the phosphorylation of GSK-3β was blocked completely by PI3-kinase inhibitors. Finally, insulin did not increase either hNQO1-ARE–luciferase expression or hNQO1 protein. These findings indicate ARE activation by tBHQ is not mediated by the characterized PI3-kinase-Akt–GSK3 pathway.

However, Kang *et al.*[51] reported that activation of PI3-kinase and Akt were increased from 10 min to 6 h after tBHQ (30 μM) treatment in H4IIE cells, a rat hepatoma cell line, whereas wortmannin or LY294002 completely abolished ARE binding activity and increases in rGSTA2 mRNA and protein. They also reported that ERK, p38 MAP kinase, and JNK were activated by tBHQ in H4IIE cells. This result demonstrated that tBHQ activated PI3 kinase and Akt, which was responsible for ARE-mediated rGSTA2 induction in H4IIE cells.

Numazawa *et al.*[49] recently reported ARE-mediated transactivation by overexpression of aPKCι and aPKCζ, two members of atypical PKC subfamily. Because it has been shown that the aPKC isoforms are activated by products of PI3-kinase, such as PI-3,4,5-triphosphate,[52] and its downstream kinase phosphoinositide-dependent kinase, PDK1,[53] the author proposed that aPKC transmit signals originating from PI3-kinase to cause nuclear translocation of Nrf2.

Phosphatidylinositol 3-Kinase Pathway Contributes to NF-E2–Related Factor Nuclear Translocation

Pretreatment of IMR-32 cells with the PI3-kinase inhibitors LY294002 or wortmannin significantly decreased Nrf2 nuclear translocation induced by tBHQ.[50] This conclusion was based on comparison of Nrf2 immunoblotting results from whole cell lysates and nuclear extracts with or without inhibitor pretreatment. PI3-kinase inhibitors had no influence on the whole Nrf2 protein level after tBHQ treatment, whereas they significantly decreased the Nrf2 protein level in nuclear extracts suggesting that

[51] K. W. Kang, M. K. Cho, C. H. Lee, and S. G. Kim, *Mol. Pharmacol.* **59,** 1147 (2001).

[52] H. Nakanishi, K. A. Brewer, and J. H. Exton, *J. Biol. Chem.* **268,** 13 (1993).

[53] A. Balendran, R. M. Biondi, P. C. Cheung, A. Casamayor, M. Deak, and D. R. Alessi, *J. Biol. Chem.* **275,** 20806 (2000).

Nrf2 nuclear translocation has been attenuated through PI3-kinase inhibition (Li Jiang *et al.*, unpublished data, 2003). In support of these findings, Kang *et al.*[54] recently reported the PI3-kinase mediated nuclear translocation of Nrf2 and the inhibition of Nrf2 nuclear translocation by pretreatment with PI3-kinase inhibitors (wortmannin/LY294002) in H4IIE cells. tBHQ relocalized Nrf2 in concert with changes in actin microfilament architecture, as visualized by superposition of immunochemically stained Nrf2 and fluorescent phalloidin-stained actin. Furthermore, tBHQ increased the level of nuclear actin, which returned to that of control by pretreatment of the cells with PI3-kinase inhibitors. Cytochalasin B, an actin disruptor, stimulated actin-mediated nuclear translocation of Nrf2 and induced rGSTA2. In contrast, phalloidin, an agent that prevents actin filaments from depolymerizing, inhibited Nrf2 translocation and rGSTA2 induction by tBHQ. This study demonstrated that PI3-kinase may regulate nuclear translocation of Nrf2 through actin rearrangement in response to oxidative stress.

Stabilization of NF-E2–Related Factor Is Independent on Phosphatidylinosital 3-Kinase Pathway

Sekhar *et al.*[55] reported Nrf2 as a substrate for the ubiquitin proteasome pathway. This conclusion was based on their following observations: (1) immunoprecipitation of a transfected ubiquitin polymer coprecipitated Nrf2; (2) incubation of a mutant ts20 cell line at the nonpermissive temperature of 39° that inactivates the ubiquitin-activating enzyme E1 resulted in Nrf2 accumulation; (3) inhibition of 26S proteasome function using lactacystin resulted in Nrf2 accumulation. Nguyen *et al.*[56] and Stewart *et al.*[57] reported a physically high rate of intracellular Nrf2 turnover ($T_{1/2} = 15$ min or 3 h) in hepatoma cells. Unpublished data from our laboratory show an increase in protein levels of Nrf2 in response to tBHQ in whole cell extracts, without an increase in Nrf2 mRNA level. This finding is consistent with other studies demonstrating that Nrf2 mRNA levels are unaffected by various inducers of ARE activity. All the preceding evidence leads us to propose that quinone compounds, like tBHQ, may stabilize Nrf2 by inhibition of the ubiquitin–proteasome pathway. However, how tBHQ targets the ubiquitin–proteasome pathway is still unknown. Through ubiquitin antibody immunoprecipitation or HA-tagged polyubiquitin pull-down, immunoblots

[54] K. W. Kang, S. J. Lee, J. W. Park, and S. G. Kim, *Mol. Pharmacol.* **62,** 1001 (2002).

[55] K. R. Sekhar, X. X. Yan, and M. L. Freeman, *Oncogene* **21,** 6829 (2002).

[56] T. Nguyen, P. J. Sherratt, H. C. Huang, C. S. Yang, and C. B. Pickett, *J. Biol. Chem.* **278,** 4536 (2003).

[57] D. Stewart, E. Killeen, R. Naquin, S. Alam, and J. Alam, *J. Biol. Chem.* **278,** 2396 (2003).

with Nrf2 antibody demonstrated that Nrf2 physically binds the polyubiquitin molecule *in vivo*. This observation was confirmed by overexpression of Nrf2 through advenovirus-mediated infection (Li Jiang *et al.*, unpublished data, 2003).

Ubiquitin is a small protein with 76 amino acids (MW ~8.6 KDa). The way in which ubiquitin is coupled to a substrate has a direct impact on how the cell interprets the ubiquitin signal. When a chain of four or more ubiquitin monomers is linked to a substrate through Lys48 of ubiquitin, the substrate is marked for proteolysis by the 26S proteasome.[58] This mechanism leads us to deduce that after tetraubiquitin covalently binds Nrf2, the theoretical MW is ~100 KDa. This is precisely the Nrf2 band we find in SDS–PAGE gels. This polyubiquitin-conjugated Nrf2 is presumably more sensitive to proteasome-mediated degradation.

We have found that tBHQ does not prevent Nrf2 ubiquitination, and in fact it seems to stabilize the ubiquitinated Nrf2 (Li Jiang *et al.*, unpublished data, 2003). The questions then arise as to how tBHQ prevents Nrf2 degradation and does tBHQ act as 26S proteasome inhibitor? *In vitro* proteasome activity assay clearly demonstrated that tBHQ was not a general 20S/26S proteasome inhibitor (Li Jiang *et al.*, unpublished data, 2003). Additional evidence showed that, unlike the reversible 26S proteasome inhibitor MG132, tBHQ could not stabilize other short half-life transcriptional factors like c-fos, suggesting its specificity in the stabilization of Nrf2. Recent reports indicated that Keap1/KIAA0132 may be involved in Nrf2 stabilization. McMahon *et al.*[59] reported that Keap1-dependent proteasomal degradation of Nrf2 contributes to the negative regulation of ARE-driven gene expression. All these findings suggest that tBHQ may regulate Nrf2 stabilization through manipulation of its binding partner(s) and that this occurs upstream of the ubiquitin-proteasome pathway.

Nguyen *et al.*[56] reported that the MAP kinase phosphorylation pathway has a positive effect on Nrf2 stabilization. We also were curious about whether the PI3-kinase pathway is involved in Nrf2 stabilization by tBHQ. On the basis of the whole cell level of Nrf2 protein and its level after ubiquitin antibody pull-down with or without LY294002 pretreatment, we demonstrated a lack of involvement by the PI3-kinase pathway in Nrf2 stabilization by tBHQ in IMR-32 cells (Li Jiang *et al.*, unpublished data, 2003). However, we maintain that PI3-kinase is involved in Nrf2 nuclear translocation, because pretreatment with LY294002 could significantly decrease Nrf2 levels in nuclear extract without altering levels in the whole cell. The PI3-kinase inhibitor, LY294002, had no influence on the whole

[58] J. S. Thrower, L. Hoffman, M. Rechsteiner, and C. M. Pickart, *EMBO J.* **19,** 94 (2000).
[59] M. McMahon, K. Itoh, M. Yamamoto, and J. D. Hayes, *J. Biol. Chem.* **278,** 21592 (2003).

cell Nrf2 protein level after tBHQ treatment. However, there was significantly decreased Nrf2 protein level in the nuclear extract, indicating that the Nrf2 nuclear translocation has been attenuated by PI3-kinase inhibitors (Li Jiang *et al.*, unpublished data, 2003).

Phosphatidylinositol 3-Kinase NF-E2–Related Factors–Antioxidant Responsive Element–Pathway Contributes to the Protective Response against Oxidative Stress Induced Apoptosis

On the basis of recent work in our laboratory characterizing a PI3-kinase–dependent mechanism of ARE activation for tBHQ in IMR-32 cells,[50] we hypothesize that tBHQ-mediated activation of the ARE is critical for generating the protective response. Using the MTS cell viability assay, we determined that addition of LY294004 (20 μM) 30 min before tBHQ treatment completely reversed the protective effect of tBHQ, suggesting that PI3-kinase–dependent, ARE-driven genes are responsible for the cytoprotective effect manifested by tBHQ. We were interested in identifying the ARE-driven gene set increased by tBHQ and correlating this information with the protective effect of tBHQ against H_2O_2-induced cytotoxicity. Oligonucleotide microarrays were used to analyze the gene expression profile associated with tBHQ treatment in the absence and presence of LY294002. Inhibition of PI3-kinase significantly blocked the expression of 49 of the 63 genes induced by tBHQ. These data are the first to show a set of potential ARE-driven genes involved in conferring protection against an oxidative stress–induced apoptosis and suggest that the PI3-kinase-Nrf2–ARE pathway may contribute to this chemprotective response.[38]

Conclusion

The transcriptional activation of phase II detoxification enzymes and/or antioxidant genes by quinone compounds has been traced to a *cis*-acting element called the ARE or the EpRE that regulates either or both constitutive and inducible gene expressions. Antioxidant responsive element activation signals from quinone compound through either or both oxidative stress–dependent or independent mechanisms. Antioxidant responsive element activation elicits a potential chemoprotective response, and the identification of the ARE-targeting genes is an initial step in the explanation of the molecular mechanism of this chemoprotective response. Multiple transcription factors have been reported to mediate chemical-induced ARE activation, and Nrf2 has been demonstrated to play a central role in the process. Microarray analysis reveals a cluster of phase 2 detoxification enzymes and some antioxidant genes that were coordinately upregulated through Nrf2-dependent ARE activation and were responsible for tBHQ protective effect against oxidative damage in the multiple culture systems. Several protein kinase pathways, including MAP kinases and PKC pathways, have been implicated in transducing signals to initiate gene expression mediated by ARE activation. Accumulated evidence indicates that PI3-kinase also is involved in ARE activation. The PI3-kinase pathway contributes to Nrf2 nuclear translocation, but there is a lack of evidence to show that the PI3-kinase pathway also is involved in tBHQ induced, Nrf2 protein stabilization. More detailed information is expected in the future to explain how quinone compounds like tBHQ encourage Nrf2 nuclear translocation and subsequent ARE activation through a PI3-kinase pathway.

[20] Signaling Effects of Menadione: From Tyrosine Phosphatase Inactivation to Connexin Phosphorylation

By KOTB ABDELMOHSEN, PAULINE PATAK, CLAUDIA VON MONTFORT, IRA MELCHHEIER, HELMUT SIES, and LARS-OLIVER KLOTZ

Introduction

Menadione is a naphthoquinone derivative (2-methyl-1,4-naphthoquinone) that has been used clinically because of its vitamin K–like properties (it is also termed vitamin K_3). It is enzymatically converted to menaquinone-4 (2-methyl-3-geranyl-geranyl-1,4-naphthoquinone, a form of vitamin K_2) by mammals.[1,2] Because it is easily synthesized chemically,[3] it had been regarded as a suitable source of vitamin K. Because of its unsubstituted position at C-3, however, menadione has alkylating properties, accounting for side effects such as thiol depletion that may ultimately mediate cytotoxicity. For this reason, it is no longer recommended as a dietary vitamin K supplement, although menadione is still commonly used in animal diets, partly in water-soluble form (e.g., as menadione sodium bisulfite).

It is the cytotoxicity of menadione, however, that has rendered it an interesting lead compound for the development of chemotherapeutics.[4]

[1] W. V. Taggart and J. T. Matschiner, *Biochemistry* **8,** 1141 (1969).
[2] G. H. Dialameh, K. G. Yekundi, and R. E. Olson, *Biochim. Biophys. Acta* **223,** 332 (1970).
[3] H. Mayer and O. Isler, *Methods Enzymol.* **18,** 491 (1971).

0076-6879/04 $35.00

Indeed, menadione was demonstrated to potently kill cultured cancer cells and was used in animal studies, as well as in preliminary clinical studies (see refs. 5 and 6 and references therein).

Cellular responses to stressful stimuli such as the exposure to menadione range from adaptation reactions, including stress signaling processes, to growth arrest and cell death. Menadione is a potent activator of mitogen-activated protein kinases (MAPK) such as the extracellular signal-regulated kinases (ERK) 1 and ERK 2.[7] ERK 1/2, on activation by dual phosphorylation, phosphorylate transcription factors as well as proteins regulating protein and nucleotide biosynthesis and thus are involved in the regulation of cellular proliferation and growth (see ref. 8 for review). Further substrates include the connexins, the building blocks of gap junctions. Gap junctions consist of two semi-channels (the connexons), each built of six connexin molecules, which connect the cytoplasms of adjacent cells, allowing for the diffusion of low molecular weight compounds of less than approximately 1 kDa, such as cyclic adenosine monophosphate (cAMP), nutrients, and others.[9] To date, 19 and 20 distinct connexin genes, in part putative, are known for mouse and humans, respectively.[10] Connexin phosphorylation is a means of regulating gap junctional channel conductance; it was shown for connexin43 (Cx43) that it is phosphorylated by protein kinase C, the nonreceptor tyrosine kinase c-Src, as well as by ERK 1 and ERK 2, usually resulting in closure of gap junctional channels.[11,12]

Menadione is capable of inducing phosphorylation of ERK 1/2 and Cx43, resulting in attenuated gap junctional communication (GJC), which is reversed in the presence of inhibitors of MAPK/ERK kinase (MEK) 1 and MEK 2 (the kinases directly upstream of ERK 1/2) and the epidermal growth factor receptor (EGFR) tyrosine kinase. Tyrosine phosphorylation

[4] B. I. Carr, Z. Wang, and S. Kar, *J. Cell. Physiol.* **193,** 263 (2002).

[5] L. M. Nutter, A. L. Cheng, H. L. Hung, R. K. Hsieh, E. O. Ngo, and T. W. Liu, *Biochem. Pharmacol.* **41,** 1283 (1991).

[6] M. Tetef, K. Margolin, C. Ahn, S. Akman, W. Chow, P. Coluzzi, L. Leong, R. J. Morgan, Jr., J. Raschko, S. Shibata *et al.*, *J. Cancer Res. Clin. Oncol.* **121,** 103 (1995).

[7] L. O. Klotz, P. Patak, N. Ale-Agha, D. P. Buchczyk, K. Abdelmohsen, P. A. Gerber, C. von Montfort, and H. Sies, *Cancer Res.* **62,** 4922 (2002).

[8] A. J. Whitmarsh and R. J. Davis, *Nature* **403,** 255 (2000).

[9] D. A. Goodenough and D. L. Paul, *Nat. Rev. Mol. Cell. Biol.* **4,** 285 (2003).

[10] K. Willecke, J. Eiberger, J. Degen, D. Eckardt, A. Romualdi, M. Guldenagel, U. Deutsch, and G. Söhl, *Biol. Chem.* **383,** 725 (2002).

[11] B. J. Warn-Cramer, P. D. Lampe, W. E. Kurata, M. Y. Kanemitsu, L. W. Loo, W. Eckhart, and A. F. Lau, *J. Biol. Chem.* **271,** 3779 (1996).

[12] P. D. Lampe and A. F. Lau, *Arch. Biochem. Biophys.* **384,** 205 (2000).

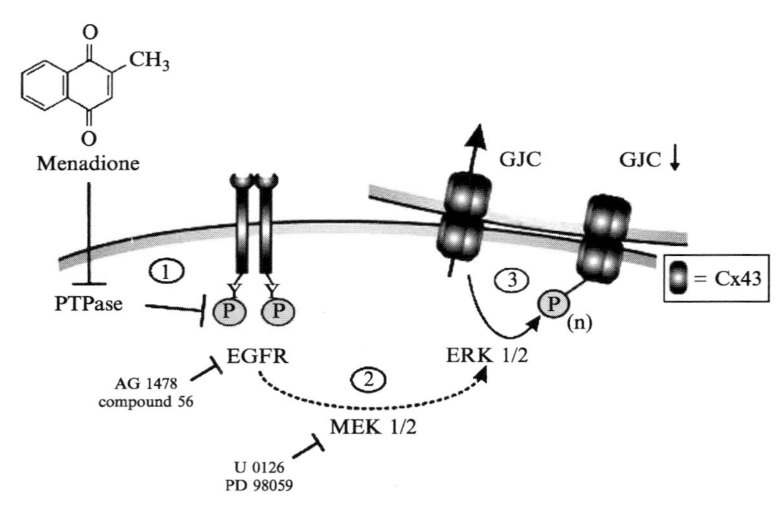

FIG. 1. Exposure of cells to menadione in rat liver epithelial cells leads to an activation of extracellular signal–regulated kinases (ERK) by means of activation of the epidermal growth factor receptor (EGFR) and MAPK/ERK kinases (MEK) and results in phosphorylation of connexin43 (Cx43). Phosphorylation entails attenuation of intercellular coupling and of gap junctional communication (GJC). Activation of the EGFR by menadione is thought to be brought about by inactivation of a regulating protein tyrosine phosphatase (PTPase). Numbers in circles represent the order in which the respective methods and results are discussed in the "Methods" section.

of the EGFR was postulated to be brought about by inactivation of a tyrosine phosphatase regulating the EGFR by menadione[7] (Fig. 1).

According to the numbers in Fig. 1, the methods used to delineate the preceding pathway leading from protein tyrosine phosphatase (PTPase) inhibition to Cx43 phosphorylation will be explained in detail in three sections dealing with (1) receptor tyrosine kinase phosphorylation and PTPase inhibition, (2) ERK activation and the use of pharmacological inhibitors, and (3) Cx43 phosphorylation and GJC.

Methods

Stock solutions of menadione (100 mM) should be prepared in dimethyl sulfoxide (DMSO) and kept at −20° in the dark. Working solutions are kept in the dark until use. Exposure of cells to menadione usually is in serum-free cell culture medium. Human primary skin fibroblasts (Clonetics/BioWhittaker Europe, Taufkirchen, Germany), HeLa cells (a human cervix carcinoma cell line, European Collection of Cell Cultures,

Salisbury, UK), or WB-F344 rat liver epithelial cells[13] (a cell line with stem cell–like properties[14]; a kind gift from Dr. James E. Trosko, East Lansing, MI) were held in Dulbecco's modified Eagle's medium (DMEM; Sigma-Aldrich, Deisenhofen, Germany) supplemented with (final concentrations) 10% (v/v) fetal calf serum (FCS; Greiner-BioWest, Frickenhausen, Germany), 2 m*M* L-glutamine, and penicillin/streptomycin in a humidified atmosphere with 5% (v/v) CO_2 at 37°.

Tyrosine Phosphorylation of Receptor Tyrosine Kinases and Protein Tyrosine Phosphatase Inhibition

Exposure of cultured cells to menadione results in an enhanced general tyrosine phosphorylation (Fig. 2A) readily detectable by Western blotting. In human skin fibroblasts relatively high menadione concentrations are required to induce this effect (Fig. 2). In contrast, exposure to menadione is usually done at 50–100 μM in HeLa or WB-F344 cells. Among the proteins tyrosine phosphorylated on exposure to menadione are the receptor tyrosine kinases EGFR and the platelet-derived growth factor receptor B (PDGFR), as demonstrated in Fig. 2B.

Detection of Tyrosine Phosphorylation: Western Blot, Immunoprecipitation. For detection of general tyrosine phosphorylation by Western blotting, cells grown and exposed to menadione in 6-well plates are lysed directly by collecting cells in 100 μl/well of 2 × sodium dodecyl sulfate–polyacrylamide gel electrophoresis (SDS–PAGE) buffer (125 m*M* TRIS/HCl, 4% [w/v] SDS, 20% [w/v] glycerol, 100 m*M* DTT, 0.2% [w/v] bromophenol blue, pH 6.8) with a cell lifter, followed by brief sonication (1–5 s) to lower viscosity if required. After boiling and brief centrifugation, samples are applied to SDS–polyacrylamide gels of 10% (w/v) acrylamide, followed by electrophoresis and blotting onto nitrocellulose or polyvinylidene difluoride (PVDF) membranes. Immunodetection of tyrosine phosphorylated proteins may be performed with a monoclonal anti-phosphotyrosine antibody (4G10) from Upstate Biotechnology (Lake Placid, NY). To this end, membranes are blocked with 5% (w/v) of BSA in TBST (TRIS-buffered saline [50 m*M* Tris/Cl, 150 m*M* NaCl, pH 7.4] containing 0.1% [v/v] Tween-20) for 1 h at room temperature and incubated with the primary antibody diluted 1:1000 in 1% BSA/TBST at 4° overnight. After washing in TBST (three times for at least 10 min at room temperature), the membrane is incubated with an anti-mouse secondary antibody

[13] M. S. Tsao, J. D. Smith, K. G. Nelson, and J. W. Grisham, *Exp. Cell Res.* **154,** 38 (1984).

[14] W. B. Coleman, K. D. McCullough, G. L. Esch, R. A. Faris, D. C. Hixson, G. J. Smith, and J. W. Grisham, *Am. J. Pathol.* **151,** 353 (1997).

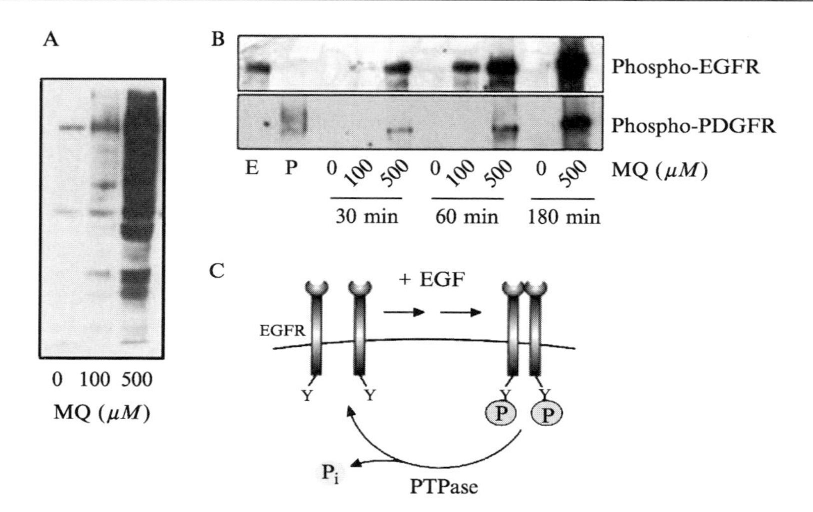

FIG. 2. (A) Tyrosine phosphorylation induced by exposure of human skin fibroblasts to menadione at the given concentrations for 1 h. DMSO was taken as control ("0 μM"). (B) Tyrosine phosphorylation of epidermal growth factor receptor (EGFR) and platelet-derived growth factor receptor-B (PDGFR) as induced by menadione. DMSO (0.5% [v/v]) was taken as control ("0 μM"). Cells treated for 10 min with 100 ng/ml of human recombinant EGF ("E") or PDGF-AB ("P"), respectively, were taken as positive controls. (C) Tyrosine phosphorylation of the EGFR is negatively regulated by action of (a) protein tyrosine phosphatase(s) (PTPase).

coupled to horseradish peroxidase diluted in TBST. Detection of immunolabeled proteins may be accomplished by enhanced chemiluminescence (e.g., "ECLplus" from Amersham, Buckinghamshire, UK, or "SuperSignal pico" substrate from Pierce/Perbio, Rockford, IL).

Detection of tyrosine phosphorylation of EGFR or PDGFR is done by Western blotting as described previously after immunoprecipitation (IP) of the respective protein (Fig. 2B). To immunoprecipitate EGFR or PDGFR from human skin fibroblasts, cells grown to confluence are exposed to menadione in 60-mm (diameter) dishes. After treatment, cells are washed once with PBS and lysed in 250 μl IP lysis buffer (20 mM TRIS [pH 7.5], 150 mM NaCl, 1 mM ethylenediamine tetraacetic acid [EDTA], 1 mM ethyleneglycol-bis-(β-aminoethyl ether)-N, N, N', N'-tetraacetic acid [EGTA], 1% Triton X-100, 1 mM Na_3VO_4, 2.5 mM sodium pyrophosphate, 1 mM β-glycerolphosphate, 1 mM phenylmethyl sulfonyl fluoride [PMSF], 1 μg/ml leupeptin) per dish by collecting them with a cell lifter, transferring them to Eppendorf cups, and by incubation on ice for up to 1 h. The crude lysates of at least three dishes are combined for one sample; less is sufficient if HeLa or WB-F344 cells are used. Lysates are frozen at −80°, thawed, and centrifuged (20,000*g* 20 min at 4°) to pellet nonsoluble fractions. The resulting supernatants are transferred to fresh Eppendorf cups, and protein is determined with a detergent-compatible protein assay, such as the BioRad Protein DC assay (Bio Rad, Hercules, CA).

EGFR may be immunoprecipitated from 50–200 μg of total protein in a total volume of 100–200 μl (adjusted with IP lysis buffer) by the addition of 0.7 to 1 μg of rabbit or sheep polyclonal anti-EGFR antibody (both Upstate Biotechnology) per 100 μg of total protein. Immunoprecipitation of PDGFR-B is performed similarly but using a rabbit polyclonal antibody against PDGFR-B (Upstate Biotechnology). After an optional incubation on ice for up to 2 h, 40 μl of a 50% slurry of protein A-agarose (if primary antibody is from rabbit; Upstate Biotechnology) or protein G-agarose (if primary antibody is from sheep; Upstate Biotechnology) is added, and the suspension rotated end-over-end at 4° overnight. Protein A/G-agarose should be pre-equilibrated beforehand by washing the supplied suspensions in IP lysis buffer at least twice (i.e., centrifugation of the beads at 20,000*g* for 10 s and resuspension of the pellet in fresh IP lysis buffer). After incubation, the samples are centrifuged at 20,000*g* at 4° for 20 s and washed twice with lysis buffer and once with PBS. After addition of 50 μl of 2 × SDS-PAGE sample buffer to the washed protein A/G-pellets, samples are boiled and applied to TRIS-glycine gels of 8% (w/v) acrylamide, followed by blotting and immunodetection of phosphorylated tyrosines as described previously.

A signal in the 170–200 kDa region is to be expected. As positive controls, lysates of cells exposed to EGF or PDGF, respectively, should be taken (see Fig. 2B). Furthermore, the membranes need to be stripped (e.g., by incubation in stripping buffer [100 m*M* 2-mercaptoethanol, 2% (w/v) SDS, 62.5 m*M* TRIS, pH 6.8] for 30 min at 55°, followed by intense washing with tap water and re-equilibration in TBST) and reprobed with the antibodies recognizing the respective immunoprecipitated receptor to (1) demonstrate that equal amounts of EGFR or PDGFR were precipitated and loaded onto the gels (thus excluding differences in tyrosine phosphorylation as being due to different amounts of total protein) and (2) to show that the phosphotyrosine signal is indeed at the correct position in the gel.

How can menadione, which is not a specific ligand for either EGFR or PDGFR, induce the tyrosine phosphorylation of these receptor tyrosine kinases indicative of their activation? A common hypothesis is based on the assumption that activity and autophosphorylation of the respective receptor tyrosine kinase are negatively regulated by a PTPase. In the presence of the respective ligand, kinase activation may transiently outcompete PTPase activity, resulting in a net increase in tyrosine phosphorylation of the receptor (Fig. 2C). The presence of a PTPase inhibitor, however, would lead to an enhanced response of the receptor to a ligand, because negative regulation is blocked.

Inhibition of Isolated CD45–Protein Tyrosine Phosphatase by Menadione. To demonstrate the ability of menadione to inactivate isolated tyrosine phosphatases, its reaction with the cytoplasmic domain of human recombinant CD45 (EC 3.1.3.4; Calbiochem, La Jolla, CA), a transmembrane PTPase that is expressed by all nucleated hematopoietic cells,[15] was investigated. Indeed, menadione at 50 μM strongly lowered PTPase activity of CD45.[7]

Tyrosine phosphatase activity is measured using *p*-nitrophenyl phosphate (pNPP; Sigma-Aldrich) as substrate. CD45 (0.1 μM in 50 mM HEPES buffer, pH 6.8) is preincubated with either DMSO or menadione at the desired concentrations for 15 min in a volume of 50 μl and then added to 750 μl of pNPP/HEPES (2 mM pNPP in 50 mM HEPES buffer, pH 6.8). The subsequent linear increase in absorbance at 405 nm (associated with the formation of *p*-nitrophenolate) is monitored, and the formation of *p*-nitrophenol(ate) per minute is calculated using the pK_a of 7.15 for *p*-nitrophenol and ε_{405} of *p*-nitrophenolate of 18,000 M^{-1} cm^{-1} [16] after correction for spontaneous hydrolysis of pNPP.

Protein Tyrosine Phosphatase Inhibition by Menadione: Net Activation of Epidermal Growth Factor Receptor. Epidermal growth factor receptor dephosphorylation and the impairment thereof may serve as an indicator of the presence of a PTPase inhibitor, as shown in Fig. 3. Exposing cells that under nonstimulated conditions do not exhibit significant EGFR tyrosine phosphorylation (lane 1 in Fig. 3A) to EGF will result in EGFR tyrosine phosphorylation (lane 2 in Fig. 3A, 3B). If an inhibitor of the EGFR tyrosine kinase is added, the extent of EGFR tyrosine phosphorylation will go back to control levels as a result of the action of a regulatory PTPase (lane 6 in Fig. 3A, 3B). In the presence of a PTPase inhibitor, however, the EGFR will remain tyrosine phosphorylated (lane 7 in Fig. 3A, 3B). In the case of menadione, such tyrosine phosphorylation is still visible after the addition of the EGFR tyrosine kinase inhibitor "compound 56" or the tyrphostin AG 1478 (Fig. 3A, lane 7), indicative of menadione inhibiting a PTPase regulating the EGFR. Moreover, even in the absence of kinase inhibitor, EGF-induced EGFR phosphorylation is enhanced by menadione

[15] T. Sasaki, J. Sasaki-Irie, and J. M. Penninger, *Int. J. Biochem. Cell Biol.* **33,** 1041 (2001).

[16] M. M. Fickling, A. Fischer, B. R. Mann, J. Packer, and J. Vaughan, *J. Am. Chem. Soc.* **81,** 4226 (1959).

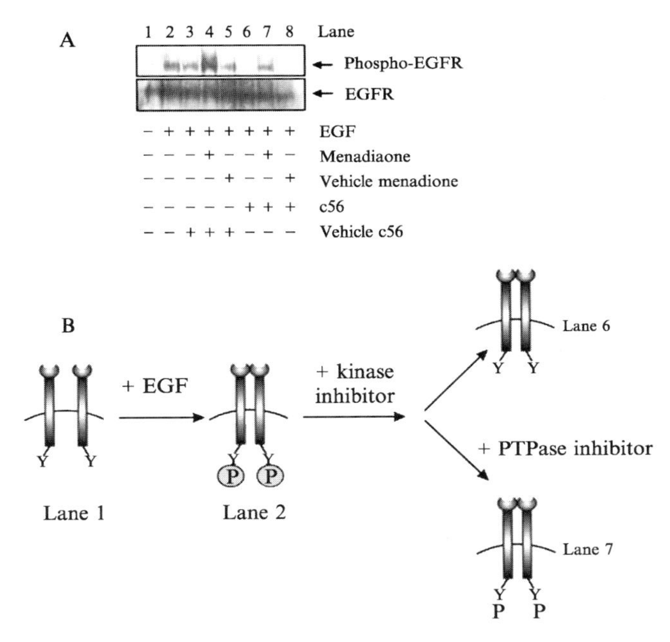

FIG. 3. Inactivation of a protein tyrosine phosphatase (PTPase) regulating the epidermal growth factor receptor (EGFR) by menadione. See text for detailed explanation. (A) Result of a representative experiment in HeLa cells; (B) interpretation. DMSO was used as vehicle for both the EGFR tyrosine kinase inhibitor compound 56 (c56) and menadione.

(Fig. 3A, lane 4), again pointing to an interruption of negative regulation of EGFR tyrosine phosphorylation.

The assay is essentially performed as described by Knebel *et al.*[17] HeLa cells are grown to 80–100% confluency on 3-cm (diameter) culture dishes and serum-starved (i.e., held in serum-free DMEM overnight). EGF receptor tyrosine phosphorylation is then stimulated by incubation in the presence of EGF (100 ng/ml) for 5 min. The cells are washed with PBS and exposed to menadione (100 μM in serum-free DMEM) for 15 min. The quinone solution is aspirated, and fresh serum-free medium containing the EGFR tyrosine kinase inhibitor compound 56 or, alternatively, AG1478 (both Calbiochem; 10–20 μM) is added to prevent any further autophosphorylation of the receptor. After 30 s, medium is quickly removed and cells lysed in 2 × SDS–PAGE sample buffer, followed by electrophoresis on a gel of 8% (w/v) acrylamide and Western blotting with detection of

[17] A. Knebel, F. D. Böhmer, and P. Herrlich, *Methods Enzymol.* **319,** 255 (2000).

phosphorylated tyrosine residues as described previously. The EGFR signal is to be expected at about 170 kDa: reprobing the membrane with an anti-EGFR antibody (see earlier) is required to ensure that the correct band is taken for analysis (Fig. 3A). Furthermore both the EGFR kinase inhibitor used and menadione have to be controlled for by use of the respective vehicle (here: DMSO) instead of the compound (Fig. 3A).

Menadione turned out to be a very effective PTPase inhibitor in this assay so that we now frequently use it as an inexpensive and easy-to-handle positive control.

Menadione-Induced Phosphorylation of Extracellular Signal-Regulated Kinases 1 and 2

A pathway emanating from the EGFR results in activation of ERK 1/2 by means of recruitment of the small G-protein Ras, activation of the serine-threonine kinase Raf that phosphorylates and activates MEK 1 and MEK 2, dual-specificity (i.e., Tyr- and Ser/Thr-specific) kinases that phosphorylate ERK 1 and ERK 2 at a Thr-Glu-Tyr motif, thereby activating the kinases (for review, see ref. 18).

The classical assay for ERK activity is based on immunoprecipitation of the kinase from cell lysate, followed by incubation of the collected immune complex in the presence of [γ-^{32}P]ATP and an ERK substrate, which may be either specifically recognized by ERK or a more general kinase substrate, such as myelin basic protein. Substrate phosphorylation is then analyzed by gel electrophoresis and autoradiography. Instead of determination of ^{32}P labeling of the substrate, antibodies specifically recognizing the phosphorylated forms of the respective substrate have been used to analyze kinase activity. Because full activation of ERKs is correlated with their being dually phosphorylated at the mentioned TEY motif, antibodies specifically recognizing the dual (Thr- and Tyr-) phosphorylation have been developed and are in use in Western blotting (Fig. 4A) and enzyme-linked immuno sorbent assays (ELISAs)(e.g., Fig. 4B). Before phospho-specific antibodies became generally available, one approach in addition to immunocomplex kinase assays was to analyze ERKs for changes in electrophoretic mobility that are due to phosphorylation. As seen in Fig. 4C, exposure of cells to menadione results in an electrophoretic mobility shift of ERK 1 and ERK 2. The methods for the immunocomplex kinase assays, as well as Western analysis of ERK phosphorylation and activity, have been described in ref. 19. For analysis of ERK phosphorylation by ELISA

[18] J. Pouyssegur, V. Volmat, and P. Lenormand, *Biochem. Pharmacol.* **64,** 755 (2002).
[19] L. O. Klotz, K. Briviba, and H. Sies, *Methods Enzymol.* **319,** 130 (2000).

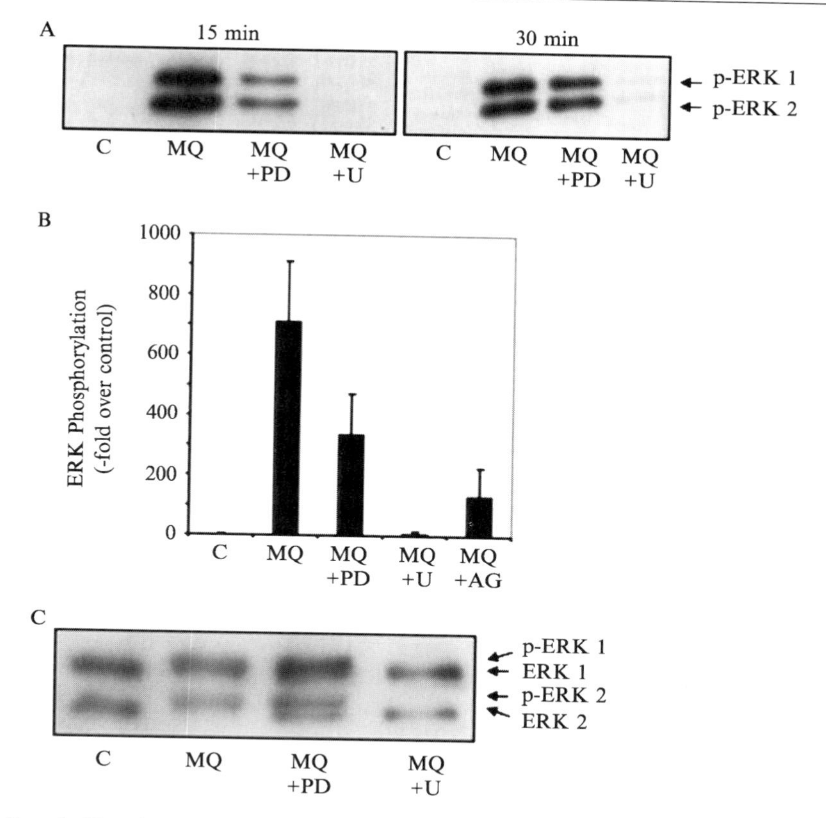

FIG. 4. Phosphorylation of ERK 1/2 is induced by menadione in WB-F344 rat liver epithelial cells, as detected with phospho-specific (anti-phospho-ERK 1/2) antibodies by Western blotting (A) by ELISA (B) or by analysis of changes in electrophoretic mobility (C). Cells were exposed to menadione (MQ, 50 μM) for 15 (A, B) and 30 min (A, C) in the absence or presence of inhibitors of MEK 1/2 activation, PD 98059 (PD, 50 μM) or U 0126 (U, 10 μM), as well as an inhibitor of the EGFR tyrosine kinase, AG 1478 (AG, 10 μM). DMSO was taken as vehicle control ("C"). Addition of inhibitors alone did not elicit any effects significantly different from "C". In (B), control was set equal to 1 and data are means ($n = 3$) $\pm$ SD.

(Fig. 4B), two commercially available ELISA kits (BioSource International, Camarillo, CA) were used in combination, one designed to specifically recognize dually phosphorylated ERK 1/2, the other one for detection of total ERK 1/2, which was used for normalization of ERK 1/2 levels. After exposure to menadione, cell lysates were prepared according to the supplier's instructions; relative increases in phosphorylation as in

Fig. 4B were calculated from the phospho-ERK/total-ERK ratios that were related to the phospho-ERK/total-ERK ratio of the control sample. Anti-phospho-ERK 1/2, as well as anti-total ERK 1/2 antibodies that were used for the blots in Figs. 4A and 4C, were from Cell Signaling Technology (Beverly, MA). All antibodies used for Western analysis were diluted in 5% (w/v) skim milk powder (ICN Biomedicals, Aurora, OH) in TBST and used at dilutions recommended by the supplier.

Exposure of human skin fibroblasts (not shown) or rat liver epithelial cells (Fig. 4) to menadione results in activation of ERK 1/2 as seen from the enhanced dual phosphorylation of ERK 1 and ERK 2 (Fig. 4A, B) and the shift in electrophoretic mobility (Fig. 4C). Both effects are reversed in the presence of inhibitors of the activation of MEK 1/2, the direct upstream kinases of ERK 1/2. Under the conditions chosen, PD 98059 ("PD"; Calbiochem, San Diego, CA; 50 μM) is less efficient in preventing ERK phosphorylation than U 0126 (Calbiochem; 10 μM); different from U 0126, dual phosphorylation is only partially blocked with PD. Also, four bands are still visible in Fig. 4C, indicating that both ERK 1/2 and phospho-ERK 1/2 are present. ERK activation by menadione is also blocked by inhibitors of the EGFR tyrosine kinase, AG 1478 (Fig. 4B) and compound 56 (not shown).[7]

All mentioned inhibitors are stored as stocks in DMSO at -20° (stock concentrations of PD 98059: 50 mM, of U 0126, AG 1478, compound 56: 10 mM). Cells are pretreated with the respective inhibitor at the given concentration in serum-free medium for 30 min, followed by aspiration and addition of fresh serum-free medium with menadione plus inhibitor. DMSO is taken instead as vehicle control (usually at 0.1% [v/v] for the inhibitor, plus DMSO at the appropriate concentration to control for menadione). If Western blotting is to be performed, media are discarded after exposure to menadione, cells washed once with PBS, and then lysed in $2 \times$ SDS–PAGE sample buffer (see earlier).

To control for an interaction of the inhibitor with menadione, the activation of another kinase, known to not usually be prevented by the inhibitor, may be tested for. For example, p38 activation by menadione is unaffected by AG 1478, implying that inhibitor and menadione do not interact (not shown).[7] Alternately, such an interaction may be tested for by UV/Vis spectrometry.

Connexin Phosphorylation

Connexin43 is a known substrate of ERK 1/2,[11] its phosphorylation usually resulting in decreased intercellular communication. Exposure of WB-F344 rat liver epithelial cells to menadione causes the phosphorylation of

Cx43 (Fig. 5A), accompanied by a loss of intercellular communication of 50% (at 50 μM menadione). In the presence of any of the aforementioned MEK or EGFR inhibitors (PD 98059, U 0126, AG 1478, compound 56), Cx43 phosphorylation is largely abolished (Fig. 5B) and GJC restored (not shown).[7] This section will focus on methods of determination of Cx43 phosphorylation in WB-F344 rat liver epithelial cells rather than the methods of evaluation of gap junctional intercellular communication, which have been described elsewhere.[7]

Western and Dot Blotting. As with ERK phosphorylation, Cx43 phosphorylation may be detected as a shift in electrophoretic mobility in an SDS–polyacrylamide gel. As seen in Fig. 5, three connexin bands, the nonphosphorylated (P0), singly (P1), and doubly (P2) phosphorylated forms, can be detected in unstimulated cells. On stimulation of the cells with menadione, the P0 band disappears, whereas P1 and P2 amounts increase. Furthermore, hyperphosphorylated forms of Cx43 (Pn) may be detected.

Samples for Western blotting are prepared as described previously by lysing cells grown to confluence on a 3-cm diameter cell culture dish in 100 μl of 2 × SDS–PAGE sample buffer. Samples are applied to TRIS-glycine SDS–polyacrylamide gels of 10% (w/v) acrylamide, followed by electrophoresis and blotting. To visualize more than one Cx43 band on the blot, it is crucial not to apply too much protein to the gel; only 3 μl of the lysates usually suffices to yield satisfactory results.

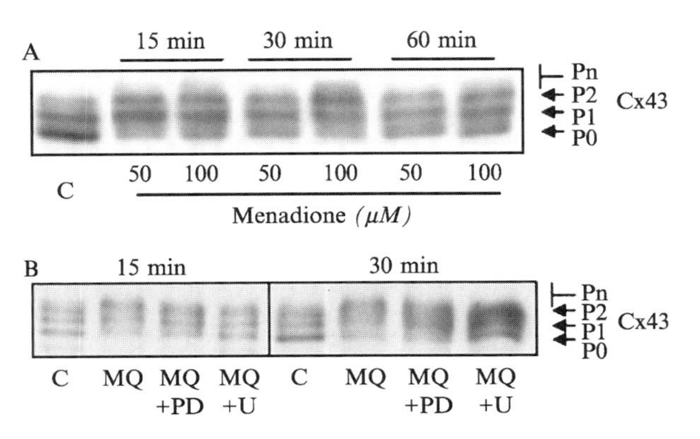

FIG. 5. Phosphorylation of connexin43 (Cx43) after exposure of WB-F344 rat liver epithelial cells to menadione (MQ, 50 μM) for the given times. Cx43 phosphorylation was detected as a shift in electrophoretic mobility of Cx43. P0, P1, P2, Pn: nonphosphorylated, singly, doubly, and hyperphosphorylated forms of Cx43. (A) Time course (B) inhibition of Cx43 phosphorylation by inhibitors of MEK 1/2 activation, PD 98059 (PD, 50 μM) or U 0126 (U, 10 μM). DMSO was taken as vehicle control (C). Addition of inhibitors alone did not elicit any effects different from C.

Immunodetection of connexin43 is performed with rabbit polyclonal anti-Cx43 antibodies from Zymed Laboratories (San Francisco, CA; #71–0700; diluted 1:2000 in 5% milk powder/TBST) or from Sigma (Deisenhofen, Germany; #C6219; diluted 1:1000 in 5% milk powder/TBST). A horseradish peroxidase–coupled anti-rabbit antibody is used as secondary antibody. Best results for detection of connexin shift/phosphorylation were obtained by blocking the PVDF membrane with 5% milk powder/TBST at 4° overnight, followed by an incubation with primary antibody at room temperature for 2 h; the rest of the procedure is the same as the preceding. As with all Western blotting procedures, the dilution of the secondary antibody has to be optimized. A dilution of 1:3000 to 1:10,000 usually is a good first guess for Cx43 blots.

Antibodies specifically recognizing the phosphorylated forms of Cx43 are available and can be used in Western or dot blotting. For dot blotting, 2 μl of the lysates in SDS-PAGE sample buffer are carefully applied to nitrocellulose membrane and air-dried. For immunodetection, these membranes are now treated like membranes after electroblotting; blocking in milk powder/TBST for 1 h at room temperature is followed by incubation with primary and secondary antibodies each for 1 h at room temperature. As primary antibody, we use a rabbit polyclonal anti-phospho-Cx43 antibody that specifically recognizes Cx43 phosphorylated at Ser279 and Ser282, the sites phosphorylated by ERK 1/2. The antibody (SA226P) was a kind gift from Dr. Kerstin Leykauf and Dr. Angel Alonso from the German Cancer Research Center, Heidelberg.[20] As can be seen in Fig. 6A, menadione strongly induces Cx43 phosphorylation. The presence of equal amounts of Cx43 in both dots is controlled for by stripping and reprobing the membrane with anti-Cx43 antibody (Sigma C6219).

Phosphorylation of Cx43 at Ser 279/282 can also be analyzed in immunohistochemical studies. Again, an antibody specifically recognizing phospho-Cx43(Ser279/282) is taken for these experiments (sc-12900-R from Santa Cruz Biotechnology, Santa Cruz, CA).

Immunohistochemistry. For immunohistochemistry, WB-F344 cells are grown to confluence on coverslips in 3-cm diameter plastic dishes, briefly washed in PBS, and kept in serum-free medium overnight before exposure to menadione or DMSO (vehicle control). After treatment, cells are washed twice with cold PBS and fixed with 5 ml per dish of methanol for 15 min at −20°. Fixed cells are then washed with ice-cold PBS another five times, followed by blocking of nonspecific binding sites with 5% (v/v)

[20] K. Leykauf, M. Durst, and A. Alonso, *Cell Tissue Res.* **311,** 23 (2003).

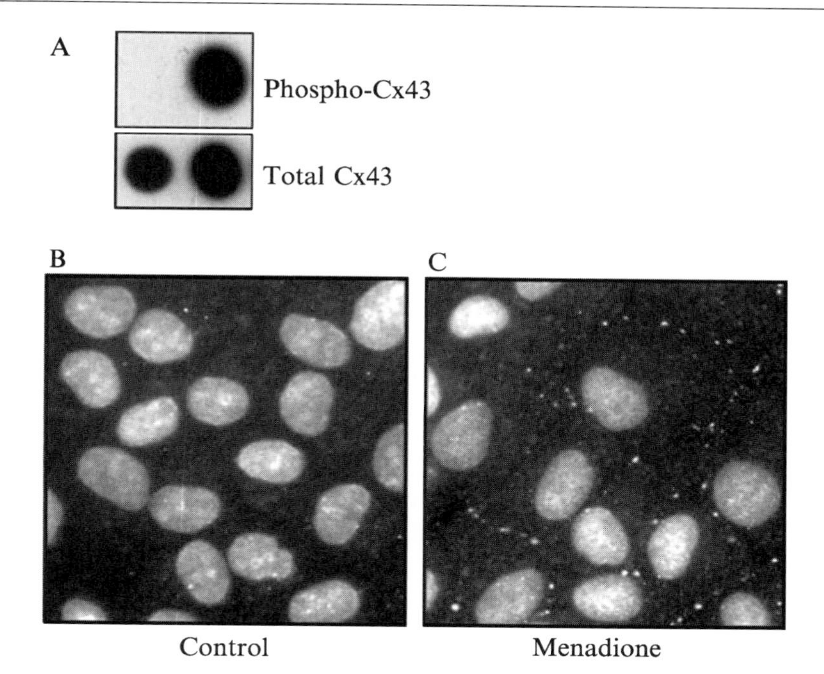

FIG. 6. Phosphorylation of connexin43 (Cx43) after exposure of WB-F344 rat liver epithelial cells to menadione (MQ, 50 μM) for 30 min. Antibodies directed against Cx43 phosphorylated at Ser279/282, sites phosphorylated by ERK 1/2, were used as described in the text. (A) Dot blot; control (left) and menadione (right) treatment, (B) immunohistochemistry: phospho-Cx43 is visible in the cell membranes of cells exposed to menadione, nuclei are stained with DAPI. DMSO was taken as vehicle control (C) left, control; right, MQ.

normal goat serum (Gibco BRL, Rockville, MD) in PBS containing 0.3% (v/v) Triton X-100 for 90 min at room temperature. For detection of connexin 43, cells are incubated at 4° overnight under slight agitation with rabbit polyclonal anti-phospho-connexin 43 antibody (sc-12900-R) diluted 1:1500 in PBS containing 1% (v/v) goat serum. Cells are then washed five times (about 30 min total time) with PBS and incubated with an Alexa 488-coupled goat anti-rabbit IgG (H + L) antibody (Molecular Probes, Eugene, OR) for 1 h at 37°. After intense washing with PBS (five times, about 30 min in total), nuclear staining is performed by addition of 4′,6-diamidino-2-phenylindole (DAPI, 0.2 μg/ml final concentration) in PBS for 15 min, followed again by intense washing and embedding with Fluoromount-G (Southern Biotechnology Associates, Birmingham, AL). The images in Fig. 6B were taken with a Zeiss Axiovert fluorescent microscope coupled to a CCD camera (ORCA II, Hamamatsu, Japan).

Detection of total Cx43 is performed accordingly, and the aforementioned polyclonal anti-Cx43 from Zymed may be taken as primary antibody at a dilution of 1:1500. Cx43 is detectable in both control and menadione-treated cells, with Cx43 molecules accumulated in distinct spots in the cell membrane after exposure to menadione.[7]

Conclusions

The methods described were helpful in analyzing menadione-induced signaling pathways in rat liver epithelial cells. As shown in Fig. 1, exposure of cells to menadione leads to activation of a signaling pathway that results in the activation of ERK 1/2, entailing the phosphorylation of Cx43 and a decrease in gap junctional intercellular communication in rat liver epithelial cells. These effects are brought about by the activation of the EGFR, probably because of the inactivation of a not yet identified PTPase regulating the receptor.

The described findings may be of interest for chemotherapeutic approaches based on the use of menadione or other quinones; chemotherapy of cancerous tissue should be most effective with diffusion of the chemotherapeutic quinone from cell to cell ("bystander effect"). Because this diffusion is hampered by the cellular reaction to the quinone itself (i.e., by the induced Cx43 phosphorylation and decreased GJC), a possible approach to enhance efficiency of quinone-based chemotherapy may be to pharmacologically block the EGFR-ERK-Cx43 pathway in cells exposed to the quinoid agent.

Acknowledgments

We thank Elisabeth Sauerbier for expert technical assistance. This work was supported by Deutsche Forschungsgemeinschaft, Bonn, Germany (SFB 503/B1 and SFB 575/B4).

[21] Unique Function of the Nrf2–Keap1 Pathway in the Inducible Expression of Antioxidant and Detoxifying Enzymes

By AKIRA KOBAYASHI, TSUTOMU OHTA, and MASAYUKI YAMAMOTO

Introduction

Cellular homeostasis against oxidative stresses and xenobiotic insults is accomplished by the coordinate expression of antioxidant and drug detoxifying genes. These cytoprotective genes include the phase 2 detoxification enzyme genes, such as NAD(P)H:quinone oxidoreductase (NQO1) and glutathione *S*-transferases (GSTs), and antioxidant genes, such as heme oxygenase 1 (HO-1) and γ-glutamylcysteine synthetase (γ-GCS). Extensive analyses of the regulatory mechanisms for the phase 2 enzyme genes revealed that the inducible expression of these enzymes is attained at the transcriptional level through a *cis*-acting element called antioxidant responsive element (ARE) or electrophile responsive elements (EpRE).[1–5] Although several candidate regulators, which bind to ARE/EpRE and regulate the expression of these cytoprotective genes, had been suggested through transfection studies, the transcription factor(s) that actually binds to and transduces the activation signals for the expression of these genes *in vivo* remained elusive.

We noticed, through a consensus binding sequence comparison with ARE, that transcription factor Nrf2 (NF-E2–related factor 2) might be an important regulatory factor, which activates transcription of these cytoprotective genes.[6,7] Nrf2 belongs to the cap 'n' collar (CNC) family of transcription factors, which includes p45/NF-E2, Nrf1, Nrf2, Nrf3, Bach1, and

[1] T. H. Rushmore, M. R. Morton, and C. B. Pickett, *J. Biol. Chem.* **266,** 11632 (1991).
[2] R. S. Friling, S. Bensimon, and V. Danile, *Proc. Natl. Acad. Sci. USA* **87,** 6258 (1970).
[3] T. Prestera, J. Talalay, J. Alam, Y, Ahn, P. J. Lee, and A. M. K. Choi, *Mol. Med.* **1,** 827 (1995).
[4] R. T. Mulcahy, M. A. Wartman, H. H. Bailey, and J. J. Gipp, *J. Biol. Chem.* **272,** 7445 (1997).
[5] T. Primiano, T. R. Sutter, and T. W. Kensler, *Adv. Pharmacol.* **38,** 293 (1997).
[6] K. Itoh, T. Chiba, S. Takahashi, T. Ishii, K. Igarashi, Y. Katoh, T. Oyake, N. Hayashi, K. Satoh, I. Hatayama, M. Yamamoto, and Y. Nabeshima, *Biochem. Biophys. Res. Commun.* **236,** 313 (1997).
[7] T. Ishii, K. Itoh, S. Takahashi, H. Sato, T. Yanagawa, Y. Katoh, S. Bannai, and M. Yamamoto, *J. Biol. Chem.* **275,** 16023 (2000).

0076-6879/04 $35.00

Bach2.[8–14] All these factors possess a conserved basic region–leucine zipper (b-Zip) structure and form a heterodimer with one of the three small Maf proteins, MafK, MafF, or MafG. These CNC–Maf complexes regulate gene expression through binding to *Ma*f recognition *e*lement (MARE, TGCTGA(G/C)TCAGC).[15–17] Because the MARE consensus sequence shows a high level similarity to that of ARE (TGA(G/C)NNNGC),[1,2] we hypothesized that one of the CNC family factors might regulate the cytoprotective gene expression through binding to the *cis*-acting ARE element. Considering the expression profile of each CNC factor, we surmised that Nrf2 is the most likely candidate for this regulation.

We, therefore, prepared a germ line mutant mouse line of Nrf2 by gene targeting and analyzed the cytoprotective enzyme expression. Nrf2-deficient mice lack substantially the inducible expression of the detoxification enzymes in response to xenobiotic inducers. The absence of the detoxification enzyme expression renders the mutant mouse highly sensitive to carcinogens and oxidative stresses, such that the administration of benzo[a]pyrene or the exposure to the diesel exhaust results in forestomach tumors and DNA adduct formation in the lung epithelial cells, respectively.[18,19] These results thus demonstrate that the enzymes under the influence of the Nrf2 regulatory system play indispensable roles in the defense against chemical carcinogenesis.

[8] N. C. Andrews, H. Erdjument-Bromage, M. B. Davidson, P. Tempst, and S. H. Orkin, *Nature* **362,** 722 (1993).

[9] J. Y. Chan, X. L. Han, and Y. W. Kan, *Proc. Natl. Acad. Sci. USA* **90,** 11371 (1993).

[10] P. Moi, K. Chan, I. Asunis, A. Cao, and Y. W. Kan, *Proc. Natl. Acad. Sci. USA* **91,** 9926 (1994).

[11] K. Igarashi, K. Kataoka, K. Itoh, N. Hayashi, M. Nishizawa, and M. Yamamoto, *Nature* **367,** 568 (1994).

[12] K. Itoh, K. Igarashi, N. Hayashi, M. Nishizawa, and M. Yamamoto, *Mol. Cell. Biol.* **15,** 4184 (1995).

[13] A. Kobayashi, E. Ito, T. Toki, K. Kogame, S. Takahashi, K. Igarashi, N. Hayashi, and M. Yamamoto, *J. Biol. Chem.* **274,** 6443 (1999).

[14] T. Oyake, K. Itoh, H. Motohashi, N. Hayashi, H. Hoshino, M. Nishizawa, M. Yamamoto, and K. Igarashi, *Mol. Cell. Biol.* **16,** 6083 (1996).

[15] K. Kataoka, M. Noda, and M. Nishizawa, *Mol. Cell. Biol.* **14,** 700 (1994).

[16] K. Kataoka, K. Igarashi, K. Itoh, T. Fujiwara, M. Noda, M. Yamamoto, and M. Nishizawa, *Mol. Cell. Biol.* **15,** 2180 (1995).

[17] H. Motohashi, T. O'Connor, F. Katsuoka, J. D. Engel, and M. Yamamoto, *Gene* **294,** 1 (2002).

[18] M. Ramos-Gomez, M. K. Kwak, P. M. Dolan, K. Itoh, M. Yamamoto, P. Talalay, and T. W. Kensler, *Proc. Natl. Acad. Sci. USA* **98,** 3410 (2001).

[19] Y. Aoki, H. Sato, N. Nishimura, S. Takahashi, K. Itoh, and M. Yamamoto, *Toxicol. Appl. Pharmacol.* **173,** 154 (2001).

The CNC family can be divided into two subfamilies on the basis of transcriptional activity, that is, the Nrf-related family (p45, Nrf1, Nrf2, and Nrf3) and Bach family (Bach1 and Bach2).[17] The former were shown to act as activators, whereas the latter were shown to act as repressors, suggesting the presence of elaborate positive and negative regulations for the cytoprotective gene expression by the Nrf-related and Bach family factors. Supporting this contention, the analysis of the Bach1-deficient mice revealed that Bach 1 normally represses the *ho-1* gene expression in the liver and heart under normal conditions, but disruption of the *bach1* gene provokes upregulation of the *ho-1* gene expression.[20] The important observation here is that a simultaneous mutation of *nrf2* gene diminished the elevation of *ho-1* gene expression provoked by the *bach1* gene mutation, suggesting that the balance between Nrf2 activation and Bach1 repression is crucial for the *ho-1* gene expression.

To explore regulatory mechanisms of the cytoprotective gene transcription by Nrf2 in response to oxidative stresses and xenobiotics, we carried out a structure–function analysis of Nrf2.[21] We found six domains conserved between human Nrf2 and chicken ECH (an Nrf2 orthologue) molecules.[22] Of the six domains, Neh2 (Nrf2-ECH homology 2) domain negatively regulates the transcriptional activity of Nrf2. To clarify the mechanisms of how Neh2 represses the Nrf2 activity, we undertook a yeast two-hybrid screen. Using the entire Neh2 domain as a bait, we isolated a new molecule Keap1 (*K*elch-like *E*CH *a*ssociated *p*rotein 1). Keap1 belongs to the Kelch family proteins. The founding member of this family is *Drosophila* Kelch protein, which was reported to be essential for the formation of actin-rich ring canals in the egg chambers.[23,24] The Kelch family proteins usually contain two characteristic motifs in common; one is the BTB/POZ (*b*road complex, *t*ramtrack, *b*ric-a-brac/*po*x virus and *z*inc finger) domain and the other is the DGR (*d*ouble *g*lycine *r*epeat or Kelch repeats) domain at the N-terminal and C-terminal regions, respectively. The BTB/POZ domain is known to promote a specific protein–protein

[20] J. Sun, H. Hoshino, K. Takaku, O. Nakajima, A. Muto, H. Suzuki, S. Tashiro, S. Takahashi, S. Shibahara, J. Alam, M. M. Taketo, M. Yamamoto, and K. Igarashi, *EMBO J.* **21,** 5216 (2002).

[21] K. Itoh, N. Wakabayashi, Y. Katoh, T. Ishii, K. Igarashi, J. D. Engel, and M. Yamamoto, *Genes Dev.* **13,** 76 (1999).

[22] K. Itoh, K. Igarashi, N. Hayashi, M. Nishizawa, and M. Yamamoto, *Mol. Cell. Biol.* **15,** 4184 (1995).

[23] F. Xue and L. Cooley, *Cell* **72,** 681 (1993).

[24] L. Cooley and W. E. Theurkauf, *Science* **266,** 590 (1994).

interaction in GAGA factor, Bach factor, and PLZF.[25–29] The DGR domain is composed of six Kelch motif repeats, and these repeats form a β-propeller structure.[30] Because DGR domain was reported to associate with actin filaments, the Kelch family proteins are speculated to play roles in the formation or maintenance of the cytoskeleton.[31]

Further analyses of the Nrf2–Keap1 regulatory pathway, which were executed extensively by many laboratories and by us, revealed that this pathway presents a unique mode of biological regulation.[32–35] In the absence of oxidative or xenobiotic inducers, Nrf2 is retained in the cytoplasm through sequestration by Keap1 and during which Nrf2 is degraded rapidly by the proteasome system (Fig. 1). On the other hand, inducers liberate Nrf2 from Keap1 and thereby Nrf2 enters into the nucleus, resulting in nuclear accumulation of Nrf2. Nrf2 then activates the transcription of the cytoprotective enzyme gene.[21,36] Investigation of the Nrf2–Keap1 system with gene-disrupted mouse lines further supports this contention.[37] In the Keap1-null mice, Nrf2 accumulates constitutively in the nucleus and activates the expression of cytoprotective genes. On the other hand, a cross of the Keap1-null mice to the Nrf2-deficient mice completely reversed the phenotype of Keap1-deficient mice, indicating that the aberrant expression of cytoprotective enzymes depends on the unregulated expression of Nrf2. Taken together, these results uncover the fact that the expression of a subset of cytoprotective enzyme genes is induced in response to oxidative and xenobiotic insults through the Nrf2–Keap1 system, which is a

[25] O. Albagli, P. Dhordain, C. Deweindt, G. Lecocq, and D. Leprince, *Cell Growth Differ.* **6,** 1193 (1995).

[26] D. Read, M. J. Buttle, A. F. Dernburg, M. Frasch, and T. G. Kornberg, *Nucleic Acids Res.* **28,** 3864 (2000).

[27] K. Igarashi, H. Hoshino, A. Muto, N. Suwabe, S. Nishikawa, H. Nakauchi, and M. Yamamoto, *J. Biol. Chem.* **273,** 11783 (1998).

[28] C. Yoshida, F. Tokumasa, K. I. Hohmura, J. Bungert, N. Hayashi, T. Nagasawa, J. D. Engel, M. Yamamoto, K. Takeyasu, and K. Igarashi, *Genes Cells* **4,** 643 (1999).

[29] K. Ahmad, C. K. Engel, and G. G. Prive, *Proc. Natl. Acad. Sci. USA* **95,** 12123 (1998).

[30] N. Ito, S. E. V. Phillips, K. D. S Yadav, and P. F. Knowles, *J. Mol. Biol.* **238,** 794 (1994).

[31] J. Adams, R. Kelso, and L. Cooley, *Trends Cell Biol.* **10,** 17 (2000).

[32] K. Itoh, N. Wakabayashi, Y. Katoh, T. Ishii, T. O'Connor, and M. Yamamoto, *Genes Cells* **8,** 379 (2003).

[33] T. Nguyen, P. J. Sherratt, H. C. Huang, C. S. Yang, and C. B. Pickett, *J. Biol. Chem.* **278,** 8135 (2003).

[34] D. Stewart, E. Killeen, R. Naquin, S. Alam, and J. Alam, *J. Biol. Chem.* **278,** 2396 (2003).

[35] K. R. Sekhar, R. R. Soltaninassab, M. J. Borrelli, Z. Q. Xu, M. J. Meredith, F. E. Domann, and M. L. Freeman, *Biochem. Biophys. Res. Commun.* **270,** 311 (2000).

[36] K. Itoh, T. Ishii, N. Wakabayashi, and M. Yamamoto, *Free Radic. Res.* **31,** 319 (1999).

[37] N. Wakabayashi, K. Itoh, J. Wakabayashi, H. Motohashi, S. Noda, S. Takahashi, S. Imakado, T. Kotsuji, F. Otsuka, D. R. Roop, T. Harada, J. D. Engel, and M. Yamamoto, *Nature Genetics* **35,** 238 (2003).

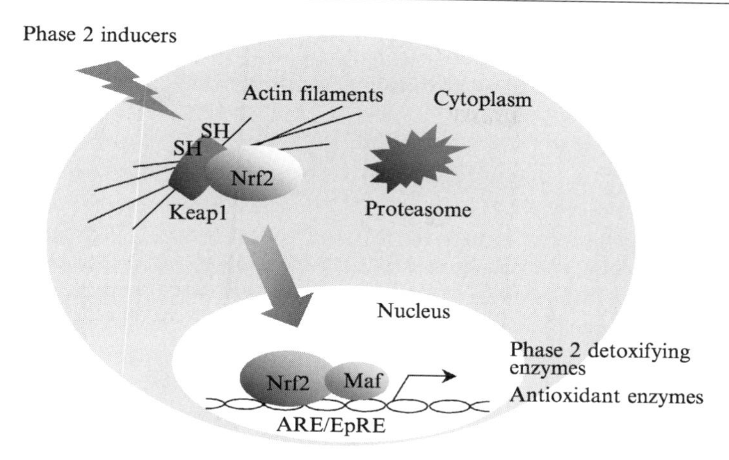

FIG. 1. The Keap1–Nrf2 system is composed of the sensor for the cellular response to oxidative stress. The actin-bound cytosolic Keap1 sequestrates Nrf2 in the cytoplasm through the interaction with the Neh2 domain of Nrf2. Exposure to phase 2 inducers liberates Nrf2 from Keap1, resulting in the translocation of Nrf2 into the nucleus. Nrf2 then activates the phase 2 enzyme gene expression.

unique regulatory system fundamentally different from the NF-κB-IκB system or the other stress response systems.

An important question still remains as to how signals of oxidants and xenobiotics are transmitted to Nrf2–Keap1 and induce the expression of cytoprotective genes. Although no structural similarity is known for the inducers, the phase 2 inducers share a common chemical property as electrophiles, implying that the xenobiotic inducers may interact with a sulfhydryl group of cysteine residues.[38] Consistent with this notion, Keap1 possesses 25 cysteine residues, which are evolutionally conserved between human and mouse Keap1. Five of the 25 cysteines could be reacted with dexamethasone mesylate, a phase 2 inducer, resulting in the disruption of the Keap1–Nrf2 complex.[39] Hence, Keap1 may function as one of the oxidative stress sensors, although the precise mechanism of the Nrf2–Keap1 activation needs to be explained.

[38] A. T. Dinkova-Kostova, M. A. Massiah, R. E. Bozak, R. J. Hicks, and P. Talalay, *Proc. Natl. Acad. Sci. USA* **98,** 3404 (2001).

[39] A. T. Dinkova-Kostova, W. D. Holtzclaw, R. N. Cole, K. Itoh, N. Wakabayashi, Y. Katoh, M. Yamamoto, and P. Talalay, *Proc. Natl. Acad. Sci. USA* **99,** 11908 (2002).

```
Nrf2    LPPPGLQSQQDMDLIDILWRQDIDLGVSREVFDFSQRQKDYELEKQKKLEK
                   .**************..*****.*.***. ...*. . ..
Nrf1    PPVSGDLTKEDIDLIDILWRQDIDLGAGREVFDYSHRQKEQDVDKELQDGR

        ERQEQLQKEQEKAFFAQFQLDE[ETGE]FLPIQPAQHIQ
        **..    * ..*.  .. .*.[****] .* * ....
        EREDTWSGEGAEALARDLLVDG[ETGE]SFPAQFPADVS
```

FIG. 2. Sequence similarity between the N-terminal region of Nrf1 and the Neh2 domain, which is crucial for Nrf2 to interact with Keap1. This observation suggests possible regulatory control by Keap1 on Nrf1. Identical or similar amino acids are shown with asterisks and dots, respectively. The ETGE motif, which is an essential motif in the Neh2 domain for association with Keap1, is boxed.

A microarray expression analysis of the cytoprotective enzymes was conducted using the Nrf2-null mutant mice administered with phase 2 inducers.[40–43] The results suggest that these enzymes can be divided into two categories: Nrf2 dependent and Nrf2 independent. Consistent with these data, it was reported that Nrf1 plays roles in the response to oxidative stimuli.[44] Mouse embryonic fibroblasts (MEF) derived from Nrf1-null mutant mice are sensitive to oxidants such as paraquat, because of the loss of γ-GCS induction. γ-GCS is known to catalyze the rate-limiting reaction in the glutathione biosynthesis pathway, and the expression of γ-GCS subunit genes has been shown to be ARE dependent. As shown in Fig. 2, Nrf1 contains a domain that shares partial similarity with the Neh2 domain of Nrf2, suggesting that Keap1 may also regulate Nrf1. We also identified Nrf3 as the fourth member of the p45-related CNC subfamily,[13] but Nrf3 as well as p45 lack the Neh2-like domain.

These observations give rise to a question as to how the CNC family factors contribute differentially to the cytoprotective gene expression. This article describes the functional examination of the Nrf2–Keap1 system with special attention to two other related factors Nrf1 and Nrf3; that is, whether these factors cross-talk with each other in response to oxidative and electrophilic stress.

[40] M. K. Kwak, K. Itoh, M. Yamamoto, T. R. Sutter, and T. W. Kensler, *Mol. Med.* **7,** 135 (2001).

[41] R. K. Thimmulappa, K. H. Mai, S. Srisuma, T. W. Kensler, M. Yamamoto, and S. Biswal, *Cancer Res.* **62,** 5196 (2002).

[42] M. K. Kwak, N. Wakabayashi, K. Itoh, H. Motohashi, M. Yamamoto, and T. W. Kensler, *J. Biol. Chem.* **278,** 8135 (2003).

[43] J. M. Lee, M. J. Calkins, K. Chan, Y. W. Kan, and J. A. Johnson, *J. Biol. Chem.* **278,** 12029 (2003).

[44] M. Kwong, Y. W. Kan, and J. Y. Chan, *J. Biol. Chem.* **274,** 37491 (1999).

Materials

Chemical Reagents

Butylated hydroxianisole (BHA) and *o*-nitrophenyl galactoside (ONPG) were purchased from Sigma (St. Louis, MO) and WAKO Pure Chemical (Osaka, Japan), respectively.

Plasmids

The yeast expression vectors pGAD424 and pGBT9 are from Clontech (Palo Alto, CA). The expression vector pEFBos is a generous gift from Dr. Sigekazu Nagata (Osaka University)[45] and pcDNA3 is from Invitrogen (Carlsbad, CA).

Transformed Yeast Strains

Yeast was cultured in YPAD medium (1% bacto-yeast extract, 2% bacto-pepetone, and 2% glucose) at 30° until OD600 became 1. This culture was then diluted with YPAD medium to OD600 = 0.2 and further incubated for 3 h. Cells were collected by centrifugation and resuspended with 1 × TE/LiOAc solution (10 m*M* TRIS-HCl (pH 7.5), 1 m*M* EDTA, and 100 m*M* lithium acetate). Cell suspension (100 μl) was mixed with plasmid DNAs as described in Table I, salmon sperm DNA (100 μg), and 1 × PEG/LiOAc solution (600 μl, 40% PEG4000 in LiOAc solution), and further incubated for 30 min at 30°. After the dimethyl sulfoxide (DMSO) treatment at 42° for 10 min, yeast cells were plated on an agar plate containing SD medium (yeast nitrogen base without amino acids [Beckton, San Jose, CA] and 2% glucose, trp/leu dropout supplement). Z buffer is 62 m*M* Na_2HPO_4-$12H_2O$, 37.8 m*M* NaH_2PO_4-$2H_2O$, 10 m*M* KCl, and 1 m*M*

TABLE I
Keap1 Can Associate with Nrf1 as Well as Nrf2 in Yeast Two Hybrid System

	GAD	GAD Nrf1	GAD Nrf2
GBD	−0.6 ± 0.1	−0.7 ± 0.1	−0.9 ± 0.1
GBD Keap1	5.8 ± 2.8	61.0 ± 4.6	220.0 ± 29.0

Yeast strain Mav203 containing *lacZ* gene, under the control of GAL1 promoter, was transformed with the plasmids as indicated. Degree of Nrf2–Keap1 association is indicated by β-galactosidase activity, which was measured as described in the text. GBD-Keap1 served as bait, whereas GAD-Nrf1 and GAD-Nrf2 were prey. GBD and GAD were negative controls.

[45] S. Mizushima and S. Nagata, *Nucleic Acid. Res.* **18,** 5332 (1990).

$MgSO_4$-$7H_2O$, whereas ONPG solution is 4 mg/ml ONPG in Z buffer supplemented with 0.27% β-mercaptoethanol.

Cell Culture and Luciferase Assay

Dulbecco's modified Eagle's medium (DMEM) was purchased from Sigma (St. Louis, MO) and used for NIH3T3 cell culture with the supplementation of 10% fetal bovine serum (FBS). Plasmid DNAs were transfected with Fugene 6 (Roche, Indianopolis, IN). After transfection, cells were lysed with the Passive Lysis Buffer (Promega, Madison, WI). Luciferase assay was performed with dual-luciferase reporter assay system kit (Promega, Madison, WI) and a Biolumat Luminometer (Berthold, Bad Wildbad, Germany).

RNA Analysis

Total RNA samples were prepared using Isogen (Nippon Gene, Tokyo, Japan). RNAs were transferred to Zeta-Probe membranes (Bio-Rad, Hercules, CA) and prehybridized with ExpressHyb Solution (Clontech, Palo Alto, CA). Biotinylated UTP and CTP (Enzo Diagnostics, New York, NY) and RNA probes were hybridized to oligonucleotide array (Murine Genome U74Av2, Affymetrix, Santa Clara, CA).

Nrf2 and Keap1-Deficient–Mutant Mice

Germ line mutant mice were generated by replacing the coding region of *nrf2* and *keap1* genes with β-galactosidase gene fused with the SV40 (simian virus 40) nuclear localization signal.[6,37] Homozygotic *nrf2* (–/–) mutant mice are viable and fertile, whereas those of *keap1* (–/–) mutant mice die within 3 weeks after birth.

Methods

Nrf1 Can Interact with Keap1 in the Yeast Two-Hybrid System

Keap1 is known to interact directly with the Neh2 domain of Nrf2 and, henceforth, negatively regulates Nrf2 transcriptional activity. As shown in Fig. 2, the N-terminal sequence of Nrf1 shares partial similarity to that of the Neh2 domain, suggesting that Keap1 may also regulate Nrf1. We therefore examined the association between Keap1 and Nrf1 by a yeast two-hybrid system. Expression plasmids, in which Nrf1 and Nrf2 were fused with the activation domain of yeast transcription factor GAL4 (GAD), were generated by inserting each cDNA fragment into pGAD424 vector

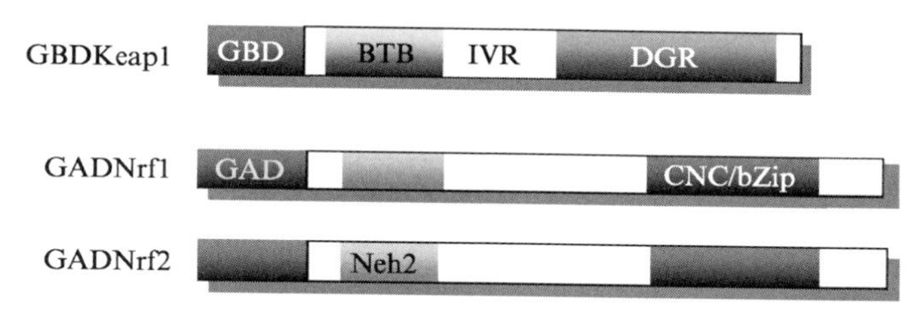

FIG. 3. A schematic presentation of the fusion proteins of Keap1, Nrf1, and Nrf2 for the yeast two-hybrid screening. GBD and GAD is the DNA binding domain and the transcriptional activation domain of GAL4, respectively.

(Fig. 3, GADNrf1 and GADNrf2, respectively). The resulting expression vectors served as prey. In addition, expression plasmid of Keap1 was also generated, as the bait, by inserting Keap1 cDNA fragment into pGBT9 of the DNA-binding domain of GAL4 (Fig. 3, GBDKeap1). Yeast strain Mav203 was transformed with several combinations of plasmids as indicated in Table I by the lithium acetate method.

To examine the interaction between the bait and the prey, β-galactosidase activity was measured. To this end, a single colony was inoculated in triplicate into SD medium supplemented with essential amino acids, except tryptophan and leucine, and cultured until absorbance at 600 nm (OD600) became 1. This culture (500 μl) was centrifuged, and the cell pellet was resuspended with 100 μl of Z buffer. After three freeze and thaw cycles, ONPG solution (700 μl) was added to the cell extracts and incubated at 30° until the solution color turned to yellow. Reaction was terminated by neutralization with 160 μl of 1 *M* Na_2CO_3. After centrifugation, OD420 of the supernatant was measured. β-Galactosidase activity was calculated by the equation as follows: 1000 $\times$ OD420/(incubation time [min] $\times$ culture volume [ml] $\times$ OD600).

When GBDKeap1 was expressed alone with GAD as the negative control, β-galactosidase expression was slightly activated. This may be due to an association of Keap1 with other endogenous transcription factors and/or coactivators (Table I). On the contrary, coexpression of GADNrf2 with GBDKeap1 robustly induced the β-galactosidase expression (220 units), indicating that Keap1 interacts strongly with Nrf2 in yeast cells. Under the same condition, coexpression of GADNrf1 along with GBDKeap1 increased the β-galactosidase activity, albeit the increase is moderate (61 units), indicating that Nrf1 can also interact with Keap1 in yeast cells. This fact shows that the affinity of Nrf1 to Keap1 is much lower than that of Nrf2. Taken together, these data have suggested the possibility that Keap1 may also regulate the Nrf1 activity in a similar manner as in Nrf2.

Keap1 Does Not Affect Transcriptional Activity of Nrf1 in a Reporter Transfection Analysis

To examine whether Keap1 represses the transcriptional activity of Nrf1 in mammalian cells, we performed a reporter transfection analysis using NIH3T3 mouse fibroblasts. The mammalian expression plasmids of Nrf1, Nrf2, and Keap1 were generated by inserting the full-length open reading frames (ORFs) into pEFBos[30] or pcDNA3. NIH3T3 cells were maintained in DMEM supplemented with 10% FBS and transfected with expression plasmids of Keap1 and Nrf1 or Nrf2. The day before transfection, NIH3T3 cells were seeded at a density of 1×10^5 cells per well of a 12-well plate. Twenty-four hours after transfection, cells were washed with ice-cold phosphate-buffered saline (PBS) and were collected by scraping with passive lysis buffer.

Nrf2 strongly activates the expression of luciferase reporter gene, but coexpression of Keap1 represses the transcriptional activity of Nrf2 in a dose-dependent manner. These results are quite consistent with our previous study.[21] In contrast, transactivation activity of Nrf1 is much weaker than that of Nrf2 (Fig. 4, compare lanes 3 and 7). Importantly, Keap1 does

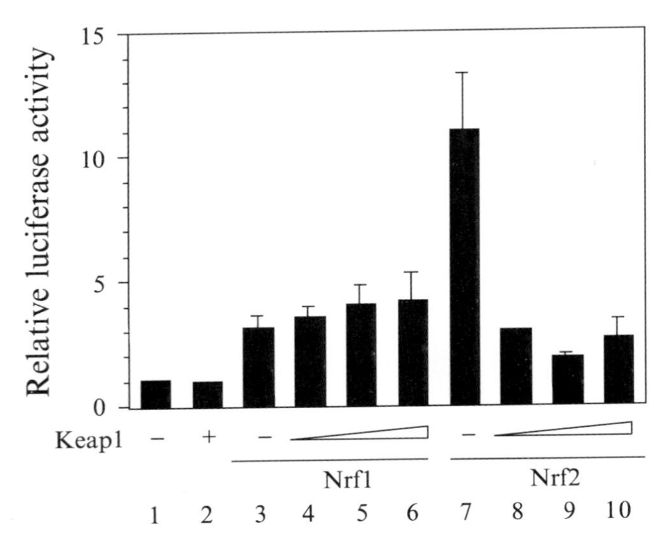

FIG. 4. Keap1 cannot repress the Nrf1 transcriptional activity in NIH3T3 cells. Expression plasmid of Nrf1 (30 ng, lanes 3–6), Nrf2 (30 ng, lanes 7–10), and Keap1 (0, 10, 30, and 100 ng in lanes 3–6 and in lanes 7–10, respectively) were transfected into NIH3T3 cells along with pRBGP2 reporter (100 ng), which contains three copies of MARE sequence upstream of the luciferase gene. Lanes 1 and 2 are controls without and with Keap1, respectively.

not repress the transcriptional activity of Nrf1 (Fig. 4, lanes 3 to 6). These results are reproducible in the transfection analysis using COS7 cells (data not shown). Thus, the results of transfection analysis suggest that, although Keap1 interacts with Nrf1 in yeast cells, the affinity between Nrf1 and Keap1 may be much weaker than that of Nrf2 and Keap1 (approximately one fourth) in mammalian cells.

Transcriptome Analysis in the Relationship Between Keap1 and Nrf1

To examine further the relationship between Keap1 and Nrf1 *in vivo,* we performed microarray analyses and compared transcriptome of *keap1::nrf2* compound knockout mice with that of *nrf2* gene single knockout mice. If Keap1 sequesters Nrf1 within the cytoplasm through physical association, ablation of Keap1 protein would liberate Nrf1 and result in the nuclear translocation of Nrf1. This may in turn activate the phase 2 and antioxidant enzyme gene expression. In this case, to see whether the Nrf1 activity will be specifically induced by Keap1 deficiency, we need to abolish the Nrf2 activity. We therefore examined transcriptome alteration by the loss of Keap1 under the Nrf2-deficient condition (i.e., analysis with *keap1::nrf2* compound knockout mouse versus *nrf2* single knockout mouse).

Total RNA samples were isolated from the liver and brain of mice, and the RNA samples were subjected to *in vitro* transcription in the presence of biotinylated UTP and CTP to produce biotin-labeled cRNA. The cRNA probes were hybridized to Affymetrix oligonucleotide array of Murine Genome U74Av2. Fluorescence intensity was captured by a laser confocal scanner (Hewlett-Packard, Palo Alto, CA) and was analyzed by Microarray Suite Version 4.0 software (Affymetrix, Santa Clara, CA). Intriguingly, comparison of transcriptome between *keap1::nrf2* compound-null mutant mice and *nrf2*-null mutant mice does not show a significant alteration in the phase 2 gene expression. These data thus demonstrate that the primary target of Keap1 is Nrf2, at least in the mouse liver and brain, and Keap1 or its defect does not affect the activity of Nrf1.

Expression Profile of Nrf3 and Generation of Nrf3-Deficient Mutant Mouse

We then examined whether Nrf3 plays any roles in the ARE-mediated induction of the cytoprotective enzyme genes. Nrf3 was isolated as the fourth member of the Nrf-related CNC subfamily. Nrf3, like the other CNC family members, binds to MARE consensus sequence through forming a heterodimer with one of the small Maf factors. This binding activity of Nrf3 to MARE implies the possibility that Nrf3 may also regulate the expression of cytoprotective enzyme genes, as in the case of Nrf2. In our preliminary analysis, Nrf3 was expressed abundantly in the placenta in human tissues.[13] The placenta is a metabolic organ, and the expression of cytoprotective enzymes is important in the tissue.

To explore physiological function of Nrf3, therefore, we examined the expression of Nrf3 in various mouse tissues. The RNA samples of 6-week-old ICR female mice were electrophoresed and transferred to Zeta-Probe membrane. Nrf3 cDNA was used as a probe for hybridization. The hybridization condition was as follows: prehybridization at 60° for 30 min, hybridization at 60° for 90 min, and washing twice with 2 × SSC and 0.1 × SSC at 60° for 30 min. We found that Nrf3 mRNA is abundantly expressed in thymus, skin, eye, stomach, and intestine in mice (data not shown).

We then generated Nrf3-deficient mice by the gene-targeting technique. To this end, Nrf3 genomic clones were isolated from 129/SV mouse genomic DNA library. A mouse Nrf3 targeting vector was designed to replace the coding region of Nrf3 with neomycin-resistant gene (detail will be published elsewhere). E14 ES cells were cultured on mouse embryonic feeder layers and selected with G418 for neomycin resistance. Drug-resistant and polymerase chain reaction (PCR)–positive colonies were amplified, and a Southern blot analysis was executed to verify homologous recombination. Positive clones were then injected into blastocysts from C57BL6/J mice and transferred into pseudopregnant ICR recipients. Male chimeric mice were bred with C57BL6/J female mice. Germline transmission was determined in F1 animals by Southern blot analysis using genomic DNA.

We then produced homozygous Nrf3 knockout mice by intercrossing. These mice grew normally and were fertile, indicating that Nrf3 deficiency was dispensable for the development and reproduction of mice. We are currently analyzing the Nrf2 contribution to aging (i.e., senile mice) and behavior.

Nrf3-Deficiency Does Not Affect the Expression of Phase 2 Enzyme Genes

To examine the contribution of Nrf3 to the regulation of phase 2 enzyme genes, we generated a compound *nrf2::nrf3* null mutant mouse line by cross-breeding *nrf2* and *nrf3* single knockout mice. We administered BHA to Nrf3-deficient mice. In addition to examining differences between Nrf3 and Nrf2 in the regulation of phase 2 enzyme genes, we also administered BHA to *nrf2* single mutant and *nrf2::nrf3* compound mutant mice. Three pairs of wild-type, homozygous *nrf2*, and homozygous *nrf2::nrf3*

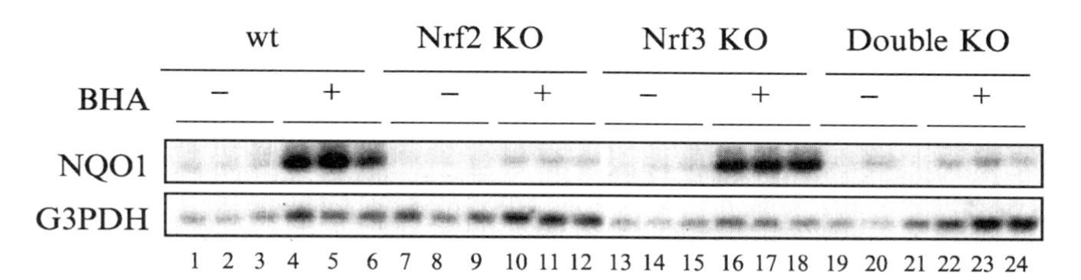

FIG. 5. Nrf3 does not contribute to the inducible expression of *nqo1* gene in the liver. BHA was administered into wild-type mice, *nrf2* knockout mice, *nrf3* knockout mice, and *nrf2::nrf3* compound knockout (double KO) mice. Expression of *nqo1* gene was examined by RNA blot analysis using liver RNA samples. *Gapdh* is a control for relative RNA concentration.

mutant mice, all of which were 6 ~ 10-week-old female mice, were administered BHA at LD_{10} (i.e., 40 mg/kg weight of mouse) by gavage. Twenty-four hours after administration, mice were killed, and RNA samples from the liver and intestine were isolated immediately. Expression levels of the phase 2 enzyme mRNAs were examined by RNA blot analysis, with NQO1 as the cDNA probe.

As shown in Fig. 5, BHA markedly induced the expression of NQO1 mRNA in the wild-type mice. Similarly, BHA also induced NQO1 expression in the Nrf3-deficient mice (lanes 16–18). In contrast, in the Nrf2-null mutant mice, a significant decrease of the *nqo1* gene induction was observed regardless of BHA administration. An interesting observation here is that some residual induction of the *nqo1* gene by BHA was observed in the Nrf2-deficient mice (lanes 10–12), suggesting that the low level induction of *nqo1* gene might be due to the other transcription factors, including Nrf3. Therefore, we investigated the Nrf3 contribution to the *nqo1* induction using *nrf2::nrf3* compound null mutant mice. Contrary to our expectation, however, comparable low-level induction was also observed in the compound mutant mice (lanes 22–24), indicating that Nrf3 does not contribute to the *nqo1* gene expression under this condition.

Conclusion

To date, the research field related to the regulation of antioxidant and detoxifying enzyme genes is expanding. The emergence of the Nrf2–Keap1 regulatory system attracts a wide range of interests from a variety of fields in medical and biological sciences. We have examined in this article whether, besides Nrf2, other CNC family factors, namely Nrf1 and Nrf3, also cross-talk with the Nrf2–Keap1 regulatory pathway. The results strongly suggest that, although Nrf1 acts to regulate a subset of antioxidant and detoxifying genes, Nrf1 works in a different manner from Nrf2. In other words, Nrf1 functions independently from the Keap1 regulatory system. Examination of Nrf3-deficient mice also supports the contention that Nrf3 does not contribute to the inducible expression of the *nqo1* gene in the liver and intestine. These results, taken together, argue that the Nrf2–Keap1 regulatory system is the mainstream pathway for the regulation of antioxidant and phase 2 detoxifying enzyme gene expression.

Acknowledgments

We are grateful to Drs. Nobunao Wakabayashi, Makoto Kobayashi, Hozumi Motohashi, and Ken Ito for stimulating discussion and help for this work. We also thank Ms. Kit Tong for her critical reading of the manuscript. This work was supported in part by grants-in-aid from the JST-ERATO environmental response project, Ministry of Education, Sciences, Sports and Culture of Japan (AK and MY), Ministry of Welfare and Labor of Japan (MY), and the Naito Foundation.

[22] Role of Protein Phosphorylation in the Regulation of NF-E2–Related Factor 2 Activity

By PHILIP J. SHERRATT, H.-C. HUANG, TRUYEN NGUYEN, and CECIL B. PICKETT

Introduction

The antioxidant response element (ARE) is a *cis*-acting enhancer that coordinates the induction of a battery of genes in response to oxidative stress and electrophilic compounds.[1] These genes contain the consensus ARE sequence within their promoter region and encode phase II detoxification enzymes or antioxidant proteins. The ARE consensus sequence, 5′-TGACnnnGC-3′,[2] is recognized by the cap 'n' collar basic region–leucine zipper transcription factor NF-E2–related factor 2 (Nrf2), which binds to the ARE as a heterodimer with a small Maf protein.[3–5] A critical role for Nrf2 in enhancing transcription through the ARE has been documented

[1] T. Nguyen, P. J. Sherratt, and C. B. Pickett, *Annu. Rev. Pharmacol.* **43,** 233 (2003).
[2] T. H. Rushmore, M. R. Morton, and C. B. Pickett, *J. Biol. Chem.* **266,** 11632 (1991).
[3] K. Itoh, T. Chiba, S. Takahashi, T. Ishii, K. Igarashi, Y. Katoh, T. Oyake, N. Hayashi, K. Satoh, I. Hatayama, M. Yamamoto, and Y.-I. Nabeshima, *Biochem. Biophys. Res. Commun.* **236,** 313 (1997).
[4] R. Venugopal and A. K. Jaiswal, *Oncogene* **17,** 3145 (1998).
[5] T. Nguyen, H. C. Huang, and C. B. Pickett, *J. Biol. Chem.* **275,** 15466 (2000).

0076-6879/04 $35.00

by studies using engineered mice and cell lines deficient in Nrf2, which showed that loss of the transcription factor results in a substantial decrease in the expression and inducibility of the ARE gene battery.[6–8] The decreased expression level of these genes has been associated with the increased sensitivity of these mice to toxic insults that promote carcinogenesis and other inflammatory diseases.[9,10] It is important therefore to determine the mechanisms that regulate cellular Nrf2 activity with the aim of improving future chemopreventive strategies.[11]

Current evidence suggests a model in which Nrf2 interacts with a cytoplasmic repressor protein, designated Kelch-like ECH-associated protein 1 (Keap1), which is associated with the actin cytoskeleton.[12] In response to oxidative stress or electrophiles, Nrf2 becomes dissociated from Keap1 and is predominantly localized in the nucleus, where it enhances gene transcription by binding to the ARE in the promoter of target genes.

A number of mechanisms have been proposed that may contribute to maintaining active nuclear Nrf2 in response to a toxic insult. One mechanism involves reactive oxygen species (ROS) and electrophiles directly modifying the sulfhydryl groups of cysteine residues on the Keap1 repressor protein.[13] This may cause a change in the repressor that results in the Keap1 protein having a lower affinity for Nrf2. In addition, there is mounting evidence pointing to the importance of protein phosphorylation in activating and repressing the activity of Nrf2.[1] Initial reports suggested that the mitogen-activated protein kinase (MAPK) and the stress-activated p38 cascades were important regulators of the ARE.[14–19] Subsequently,

[6] M. McMahon, K. Itoh, M. Yamamoto, S. A. Chanas, C. J. Henderson, L. I. McLellan, C. R. Wolf, C. Cavin, and J. D. Hayes, *Cancer Res.* **61,** 3299 (2001).

[7] S. A. Chanas, Q. Jiang, M. McMahon, G. K. McWalter, L. I. McLellan, C. R. Elcombe, C. J. Henderson, C. R. Wolf, G. J. Moffat, K. Itoh, M. Yamamoto, and J. D. Hayes, *Biochem. J.* **365,** 405 (2002).

[8] T. Ishii, K. Itoh, S. Takahashi, H. Sato, T. Yanagawa, Y. Katoh, S. Bannai, and M. Yamamoto, *J. Biol. Chem.* **275,** 16023 (2000).

[9] A. Enomoto, K. Itoh, E. Nagayoshi, J. Haruta, T. Kimura, T. O'Connor, T. Harada, and M. Yamamoto, *Toxicol. Sci.* **59,** 169 (2001).

[10] M. Ramos-Gomez, M-K, Kwak, P. M. Dolan, K. Itoh, M. Yamamoto, P. Talalay, and T. W. Kensler, *Proc. Natl. Acad. Sci. USA* **98,** 3410 (2001).

[11] J. D. Hayes and M. McMahon, *Cancer Lett.* **174,** 103 (2001).

[12] K. Itoh, N. Wakabayashi, Y. Katoh, T. Ishii, K. Igarashi, J. D. Engel, and M. Yamamoto, *Genes Dev.* **13,** 76 (1999).

[13] A. T. Dinkova-Kostova, W. D. Holtzclaw, R. N. Cole, K. Itoh, N. Wakabayashi, Y. Katoh, M. Yamamoto, and P. Talalay, *Proc. Natl. Acad. Sci. USA* **99,** 11908 (2002).

[14] R. Yu, W. Lei, S. Mandlekar, M. J. Weber, C. J. Der, J. Wu, and A.-N. T. Kong, *J. Biol. Chem.* **274,** 27545 (1999).

[15] R. Yu, S. Mandlekar, W. Lei, W. E. Fahl, T.-H. Tan, and A.-N. T. Kong, *J. Biol. Chem.* **275,** 2322 (2000).

other signal transduction pathways have been implicated as having effects on the ARE, including phosphatidylinositol 3-kinase (PI3K)[19–23] and protein kinase C (PKC) isoenzymes.[24] Of these kinases, only the PKC isoenzymes have been shown to be capable of acting directly on Nrf2.[24,25]

A role for PKC was first suggested following the observation that phorbol esters, as well as ROS and electrophiles, can stimulate ARE-driven transcription.[26] In reporter gene assays, ARE induction stimulated by phorbol esters and the redox cycling compound *tert*-butylhydroquinone (*t*BHQ) could be impaired by pharmacological inhibitors that block PKC phosphorylating activity.[24] In addition, immunocytochemical and subcellular fractionation experiments demonstrated that *t*BHQ and phorbol ester treatment could also increase the amount of Nrf2 in the nucleus of a cell. This accumulation of Nrf2 in the nucleus could also be impaired by protein kinase C inhibitors.

To address further whether PKC was acting on the Nrf2 protein itself or in an indirect manner, experiments were performed to show that the Nrf2 protein is indeed a phosphoprotein.[25] Cells were metabolically labeled with radioactive inorganic phosphate, and Nrf2 was immunoprecipitated from their lysates. Radiolabeled Nrf2 was detected from the cells, and the amount of radiolabeling was increased when the cells were treated with inducers of the ARE. A direct activity of PKC on Nrf2 was demonstrated using *in vitro* kinase assays. Both the purified catalytic subunit or immunoprecipitated kinase from cell extracts could phosphorylate recombinant Nrf2. The Nrf2 kinase activity of PKC in the cell immunoprecipitates was shown to increase following exposure of the cultured cells to compounds that stimulate ARE-driven transcription. Therefore, compounds that induce ARE-driven transcription can activate PKC, which is capable of phosphorylating Nrf2 directly.

These studies were followed up by the identification of a serine residue in the N-terminal region of Nrf2 that is a target for PKC phosphorylation.[25]

---

[16] R. Yu, C. Chen, Y.-Y. Mo, V. Hebbar, E. D. Owuor, T.-H. Tan, and A.-N. T. Kong, *J. Biol. Chem.* **275,** 39907 (2000).

[17] L. M. Zipper and R. T. Mulcahy, *Biochem. Biophys. Res. Commun.* **278,** 484 (2000).

[18] J. Alam, C. Wicks, D. Stewart, P. Gong, C. Touchard, S. Otterbein, A. M. K. Choi, M. E. Burrow, and J.-S. Tou, *J. Biol. Chem.* **275,** 27694 (2000).

[19] K. W. Kang, J. H. Ryu, and S. G. Kim, *Mol. Pharm.* **58,** 1017 (2000).

[20] K. W. Kang, M. K. Cho, C. H. Lee, and S. G. Kim, *Mol. Pharm.* **59,** 1147 (2001).

[21] J.-M. Lee, J. M. Hanson, W. A. Chu, and J. A. Johnson, *J. Biol. Chem.* **276,** 20011 (2001).

[22] J. Li, J.-M. Lee, and J. A. Johnson, *J. Biol. Chem.* **277,** 388 (2002).

[23] K. W. Kang, S. J. Lee, J. W. Park, and S. G. Kim, *Mol. Pharm.* **62,** 1001 (2002).

[24] H.-C. Huang, T. Nguyen, and C. B. Pickett, *Proc. Natl. Acad. Sci. USA* **97,** 12475 (2000).

[25] H.-C. Huang, T. Nguyen, and C. B. Pickett, *J. Biol. Chem.* **277,** 42769 (2002).

[26] T. Nguyen, T. H. Rushmore, and C. B. Pickett, *J. Biol. Chem.* **269,** 13656 (1994).

Initially, six potential PKC phosphorylation sites within the rat Nrf2 primary sequence were identified by computer-aided analysis. Short peptides were designed that would mimic these sites and were synthesized for use in competition with purified recombinant Nrf2 in the kinase assay. A peptide corresponding to a potential site around serine 40 was effective at blocking the phosphorylation of Nrf2 by PKC. Mutation of this serine to alanine resulted in a mutant protein that could not be phosphorylated by PKC *in vitro*. The physiological role of this phosphorylation site seems to be in affecting the interaction between Nrf2 and Keap1. Serine 40 resides in a region of Nrf2 designated the Neh2 domain that has been shown to be important for the interaction between Nrf2 and Keap1 repressor.[12] In protein association experiments, active PKC reduced the ability of Nrf2 to interact with Keap1. By contrast, mutant Nrf2 protein, with an alanine at position 40, could still interact with Keap1 even in the presence of active kinase. This mutation did not affect the ability of transcription factors to bind to the ARE or to transactivate the ARE when overexpressed in cells. However, when cotransfected with a construct expressing the Keap1 protein, reporter gene expression was significantly more repressed with the serine mutant than with the wild-type protein. These data suggest that PKC plays an important part in maintaining Nrf2 activity and subsequent ARE-driven transcription by reducing Keap1-mediated repression (Fig. 1).

In addition to promoting the dissociation of Nrf2 from Keap1, inducing agents can increase the stability and therefore steady-state level of the protein in the cell.[27–30] Like many transcription factors, the amount of Nrf2 in the cell is tightly regulated; the protein has a relatively short half-life of less than 20 min. Western blotting experiments demonstrated that treating cells with inducing agents leads to an increase in cellular Nrf2. This increase in Nrf2 was not a result of induced *Nrf2* gene transcription, but was caused by posttranscriptional mechanisms leading to its stabilization.[27]

Protein phosphorylation was implicated as a possible mechanism regulating Nrf2 stability when protein phosphatase inhibitors were used to impose a hyperphosphorylated state in the cell. This gave rise to an increase in the amount of Nrf2 protein. By use of a panel of pharmacological kinase inhibitors, the increase in Nrf2 protein following treatment of cultured cells with *t*BHQ seemed to be mostly dependent on the activity of

[27] T. Nguyen, P. J. Sherratt, H.-C. Huang, C. S. Yang, and C. B. Pickett, *J. Biol. Chem.* **278,** 4536 (2003).

[28] D. Stewart, E. Killeen, R. Naquin, S. Alam, and J. Alam, *J. Biol. Chem.* **278,** 2396 (2003).

[29] M. McMahon, K. Itoh, M. Yamamoto, and J. D. Hayes, *J. Biol. Chem.* **278,** 21592 (2003).

[30] K. Itoh, N. Wakabayashi, Y. Katoh, T. Ishii, T. O'Connor, and M. Yamamoto, *Genes Cells* **8,** 379 (2003).

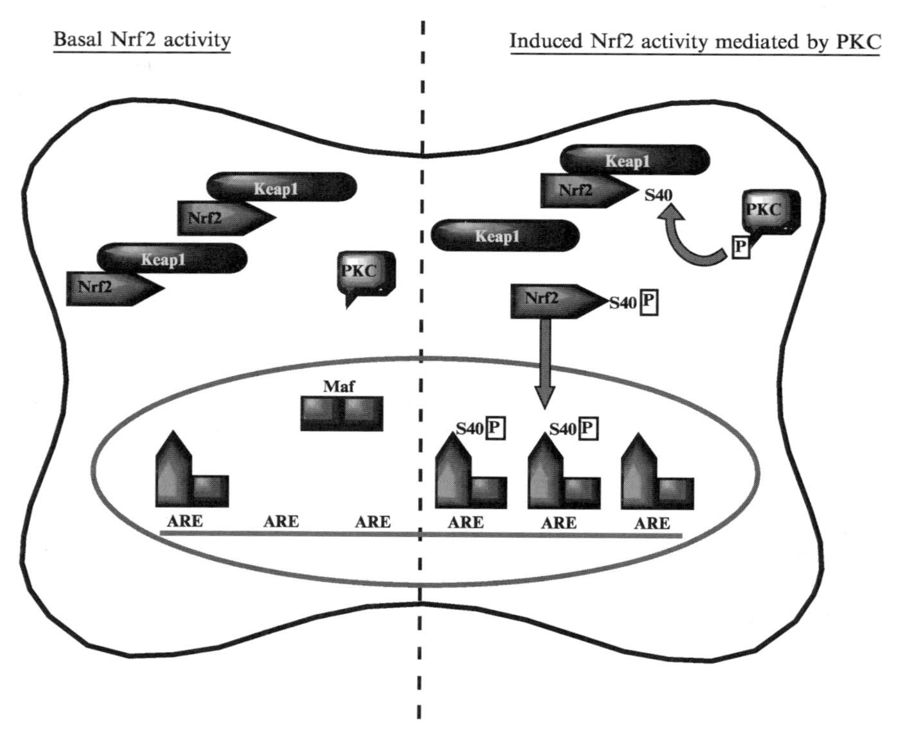

FIG. 1. Under nonstimulated conditions, Nrf2 is repressed by Keap1, which localizes it to the cytoplasm of the cell. There is, however, a low level of basal ARE activity, because not all Nrf2 exists in the repressed state. Following induction by compounds such as *t*BHQ or PMA, PKC isoenzymes are activated and can then directly phosphorylate Nrf2 at serine 40. Once phosphorylated, Keap1 is less able to bind to and repress Nrf2, which subsequently accumulates in the nucleus of the cell to increase transactivation of the ARE. (See color insert.)

the MAPK pathway. This suggests that multiple signal transduction pathways are important for the regulation of Nrf2 activity. The PKC isoenzymes play a critical part in maintaining Nrf2 in an unrepressed state from Keap1 to function in the nucleus, whereas other pathways, possibly the MAPK cascade, contribute by increasing protein stability. Whether the MAPK cascade directly affects the Nrf2 protein has yet to be determined. Together the combined activity of these kinases would lead to a higher level of Nrf2 in the nucleus of the cell to increase ARE-driven transcription. This chapter describes how *in vitro* and cell culture experiments have been used to dissect the role of PKC and implicate the MAPK pathway in the activation of the ARE.

Methods

Use of Selective Activators and Pharmacological Inhibitors of Signaling Pathways in Reporter Gene Assays and Translocation Studies

Cell Culture. The human HepG2 and rat H4IIEC3 hepatoma cell lines were obtained from the American Type Culture Collection. These established cell lines are suitable for work involving regulation of the ARE, because many of the inducible members of the ARE gene battery are highly regulated in the liver. Both cell lines were maintained in Dulbecco's modified Eagle's medium (DMEM) high glucose supplemented with 10% (v/v) fetal calf serum, penicillin/streptomycin, and L-glutamine. All cell culture reagents were obtained from Invitrogen (Carlsbad, CA). The cells were housed at 37° in a humidified Forma Scientific (Marietta, OH) 3250 incubator in the presence of 7% (v/v) CO_2.

Reporter Gene Constructs and Transfection. Reporter gene assays have been instrumental through the course of these studies. The plasmids used in transient transfection assays have been described previously.[2,31] Two plasmids were routinely used that contain the minimal promoter from the rat *GSTA*2 subunit gene fused to the coding sequence for the chloramphenicol acetyl transferase (CAT) enzyme. Upstream of the minimal promoter, the ARE enhancer from either the rat *GSTA*2 subunit gene or the rat quinone oxidoreductase *(QR)* gene has been inserted. These reporter gene constructs are designated *GSTA*2 ARE-CAT and *QR* ARE-CAT, respectively, and allow the expression of CAT under control of the *GSTA*2 minimal promoter to be enhanced by inducers of the ARE. These plasmids were routinely introduced into cells by transient transfection using lipofectamine plus reagent according to the manufacturer's instructions (Life Technologies, Carlsbad, CA). Before transfection, cell media was removed and replaced with Opti-MEM, a low-serum antibiotic-free media that is optimal for liposomal-mediated transfection. Cells were allowed to recover overnight following transfection in fresh Opti-MEM. In addition, some of the studies were also performed using a rat hepatoma H4IIEC3 cell line containing a stably integrated rat *QR* ARE-CAT construct.

Cell Treatment and Reporter Gene Assays. The inducing agents used were *tert*-butylhydroquinone (*t*BHQ) and phorbol 12-myristate 13-acetate (PMA). The redox cycling compound *t*BHQ is a model inducing agent of ARE-enhanced gene expression. The pharmacological inhibitors of PKC phosphorylating activity used in these experiments were staurosporine and Ro-32–0432. Cells were treated with inducers in the absence or

[31] L. V. Favreau and C. B. Pickett, *J. Biol. Chem.* **270,** 24468 (1995).

presence of the PKC inhibitors for 18 h. Typical concentrations of the compounds used in these experiments were *t*BHQ, 50–100 μM; PMA, 10–100 nM; Ro-32–0432, 2 μM; and staurosporine, 15 nM. In control experiments, cells were treated with solvent alone, which unless stated otherwise is DMSO. Following the treatment period, cells were harvested in M-PER Mammalian Protein Extraction Reagent (Pierce, Rockford, IL), and cell lysates were assayed for CAT activity. The CAT assay reaction contains 200 mM TRIS-HCl, pH 7.5, 25 μg of *n*-butyryl or acetyl CoA, 0.075 μCi/2.75 KBq D-threo-[dichloroacetyl-1,2-^{14}C]-chloramphenicol (NEN, Boston, MA), and cell lysate to a final reaction volume of 100 μl. The reactions were incubated for up to 1 h at 37°. Radiolabeled products and remaining substrate were extracted by ethyl acetate followed by separation on thin-layer chromatography (TLC) silica plates using a mixture of chloroform and methanol (19:1). The amount of butylated or acetylated radiolabeled products were quantified by Phosphorimage analysis using a Fujifilm FLA-2000 (Stamford, CT).

A representative reporter gene experiment is shown in Fig. 2, which indicates in conjugation with other such assays that inhibitors of PKC could impair induction of the ARE by both *t*BHQ and phorbol ester. This implicates the importance of PKC in the regulation of the ARE.

Translocation Studies. The Keap1 repressor protein localizes Nrf2 in the cytoplasm of the cell. It was tested whether blocking PKC phosphorylating activity using inhibitors affected the subcellular localization of Nrf2 following exposure to *t*BHQ. The subcellular distribution of Nrf2 was determined by immunocytochemistry and complemented with biochemical fractionation techniques.

Exponentially growing HepG2 or H4IIEC3 cells were trypsinized onto coverslips and treated with *t*BHQ or PMA typically for 4 h, or in cases of testing, the effects of an inhibitor, pretreated with staurosporine or Ro-32–0432 for 1 h before exposure to *t*BHQ. Cells were washed with PBS and fixed with 3% (w/v) paraformaldehyde solution for 10 min. Excess fixative was removed, and remaining reactive aldehyde groups were neutralized with glycine dissolved in PBS. The hepatoma cells were then permeabilized with 0.2% (v/v) Triton X-100 in PBS for 5 min. Following removal of the detergent solution, cells were washed with 0.5% (w/v) BSA in PBS (PBS–BSA) before blocking in this solution for 20 min. Immunocytochemistry was performed using an affinity-purified rabbit polyclonal anti-Nrf2 antibody (sc-722; Santa Cruz Biotechnology, Santa Cruz, CA) diluted in PBS–BSA along with RNase. The primary antibody was detected with an FITC-conjugated anti-rabbit IgG antibody (62–6111; Zymed Laboratories, San Francisco, CA) diluted in PBS–BSA. Propidium iodide (PI) was included in the secondary antibody incubation to counterstain the

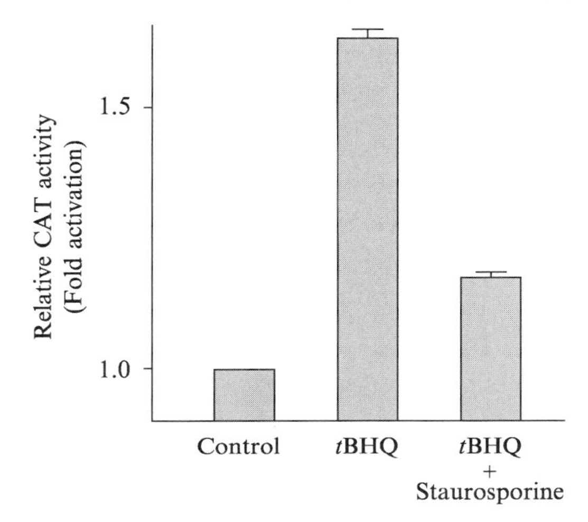

FIG. 2. This is a representative reporter gene experiment. In this instance, a rat hepatoma H4IIEC3 cell line that has been stably transfected with the rat *QR* ARE-CAT was used. Cells were seeded in six-well plates and allowed to recover overnight before being exposed to solvent (DMSO), *t*BHQ alone, or *t*BHQ in the presence of the PKC inhibitor staurosporine, for 18 h. The inducer *t*BHQ was added to a final concentration of 100 μM, and the inhibitor staurosporine was used at 15 n*M*. Experiments were performed in triplicate. Staurosporine reduces the induction of ARE-driven transcription by *t*BHQ.

cells to verify the location and integrity of the nuclei. Fluorescence was monitored with a confocal laser scanning microscope (Leica DM IRBE, Bannockburn, IL).

To confirm the immunocytochemistry data, cells exposed to *t*BHQ or PMA, in the presence or absence of a pharmacological PKC inhibitor, were fractionated into cytosolic and nuclear constituents. Cells were washed once in cold PBS before harvesting by scraping off the plate. The intact cells were then pelleted by centrifugation at 500*g* for 5 min at 4°, the supernatant was discarded, and stored at −80° until required. Cell pellets were resuspended in ice-cold lysis buffer consisting of 50 m*M* TRIS, pH 8.0, 10 m*M* NaCl, 5 m*M* $MgCl_2$, 0.5% (v/v) Nonidet P40, supplemented with a Complete Protease Inhibitor Cocktail (Roche Applied Science, Indianapolis, IN), and mixed gently for 15 min at 4°. Further centrifugation at 800*g* for 10 min at 4° yielded the cytosolic fraction in the supernatant and the nuclear fraction in the pellet. Following removal of the cytosol, the intact nucleus was resuspended and washed once with the lysis buffer described previously. The nuclear extract was prepared by resuspending the nuclear pellet in an extraction buffer containing 20 m*M*

HEPES, pH 7.8, 0.5 *M* NaCl, 1 m*M* EDTA, 1 m*M* DTT, 20% (v/v) glycerol, Complete Protease Inhibitor Cocktail. The suspension was then rotated for 30 min at 4° before centrifugation at 30,000*g* for 1 h at 4°. Protein concentrations in the cytosolic and nuclear extracts were determined by the method of Bradford.

Fractions were resolved by sodiumdodecyl sulfate–polyacrylamide gel-electrophoresis (SDS–PAGE), transferred onto a PVDF membrane, and the relative amounts of Nrf2 protein present were determined by Western blotting using the anti-Nrf2 antibody described previously. Immunoreactive polypeptides were detected following enhanced chemiluminescence (ECL) by autoradiography.

The results showed that pharmacological PKC inhibitors could perturb the nuclear accumulation of Nrf2 in response to *t*BHQ or PMA (Fig. 3). Thus, a mechanism for PKC-mediated activation of the ARE could be its regulation of the cellular localization of Nrf2.

Determining NF-E2–Related Factor 2 as a Phosphoprotein and Identifying Sites of NF-E2–Related Factor 2 Phosphorylation by Protein Kinase C

Determination of the Cellular Phosphorylation State of NF-E2–Related Factor 2. Having implicated PKC is important in the activation of ARE-driven transcription, it is then of interest to determine whether the Nrf2 protein is affected by PKC directly or indirectly. First, the question of whether Nrf2 is indeed a phosphoprotein may be answered by testing whether the endogenous protein is phosphorylated *in vivo*.

Hepatoma cells were cultured for 1 h in phosphate-free MEM supplemented with dialyzed FBS, while in the exponential growth phase, and then metabolically labeled for 3.5 h with [^{32}P]orthophosphate (1 mCi per 60-mm dish). During this labeling period, cells were exposed to *t*BHQ, β-naphthoflavone (βNF), or PMA in time course experiments. In addition, cells were pretreated with a kinase inhibitor for 1 h followed by the addition of *t*BHQ for 30 min. In control experiments, cells were maintained in the radioactive media alone. After washes in ice-cold phosphate-free MEM, cells were lysed in lysis/IP buffer (50 m*M* TRIS, pH 8.0, 150 m*M* NaCl, 1% (v/v) NP-40, 0.5% (w/v) DOC, 0.1% (w/v) SDS, 50 m*M* NaF, 1 m*M* Na_3VO_4, 20 m*M* β-glycerophosphate, 1 μM okadaic acid) for 30 min on ice. Lysates were then cleared by centrifugation at 14,000 rpm for 15 min at 4° in a microcentrifuge. Radiolabeled Nrf2 was immunoprecipitated from cell lysates by incubating it with an anti-Nrf2 antibody overnight at 4°, followed by the addition of Protein A-Trisacryl beads (Pierce, Rockford, IL). The mixture was rotated for 1 h at 4° and then washed at least three times in lysis/IP buffer supplemented with

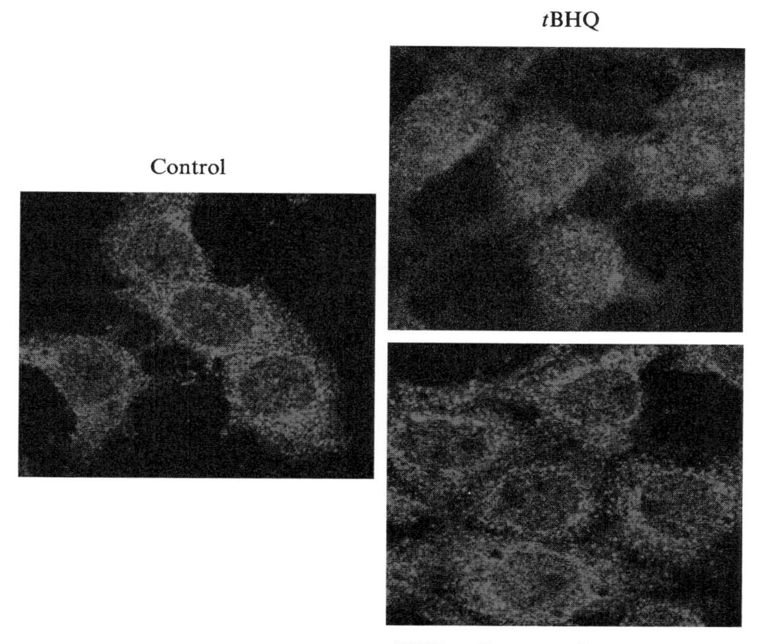

FIG. 3. Immunocytochemistry was used to demonstrate the accumulation of endogenous Nrf2 in the nucleus of cells following exposure to *t*BHQ. This can be impaired by pretreatment of the cells with PKC inhibitors. Rat hepatoma H4IIEC3 cells trypsinized onto coverslips were exposed to solvent (DMSO), *t*BHQ alone for 4 h, or pretreated with staurosporine for 1 h before treatment with *t*BHQ. *t*BHQ was used at a final concentration of 100 μM, and staurosporine was used at 15 n*M*. (See color insert.)

0.4 *M* NaCl. Precipitates were resolved by SDS–PAGE and analyzed by autoradiography.

Phosphorylation of NF-E2–Related Factor 2 by Protein Kinase C. To determine whether PKC may be capable of phosphorylating the Nrf2 protein, *in vitro* kinase assays were performed. Recombinant His–tagged rat Nrf2 was expressed in *Escherichia coli* and purified by nickel-affinity chromatography to use as a substrate in these reactions. The purified protein would then be combined with catalytic subunits of rat brain PKC (Calbiochem, San Diego, CA) in a reaction containing 25 m*M* HEPES (pH 7.5), 10 m*M* $MgCl_2$, 200 μM ATP, and 2 μCi [γ-^{33}P]ATP. Assays were performed at 30° and stopped at various times by the addition of sample buffer for SDS–PAGE analysis. The effect of PKC inhibitor was examined by preincubation of the kinase and inhibitor for 10 min at room temperature before carrying out the assay for 30 min at 30°. The level of radiolabel incorporation into Nrf2 was detected by autoradiography and measured by Phosphorimage analysis.

Phosphorylation of Nrf2 was also assayed using endogenous PKC immunoprecipitated from HepG2 cell lysates. A monoclonal anti-PKC antibody conjugated to agarose (Santa Cruz Biotechnology, Santa Cruz, CA) was used to precipitate PKC from lysates of HepG2 cells treated with *t*BHQ, βNF, or PMA for various times, according to the immunoprecipitation protocol described previously. The anti-PKC immunocomplexes were then incubated with purified recombinant rat Nrf2 and [γ-^{33}P]ATP for 30 min at 30° in a reaction buffer from the GIBCO/BRL Protein Kinase C Assay System (Life Technologies, Carlsbad, CA). Kinase reactions were stopped by the addition of SDS–PAGE sample buffer, the products resolved by SDS–PAGE, and the level of [γ-^{33}P]ATP incorporation determined by a Phosphorimage analysis.

Identification of Serine 40 of Rat NF-E2–Related Factor 2 as a Target for Protein Kinase C Phosphorylation. Taking advantage of the known consensus pattern of PKC phosphorylation sites ([S/T]-X-[R/K], where *X* is any amino acid residue), computer-aided analysis was performed on the primary sequence of rat Nrf2. The 597–amino acid polypeptide contained seven potential sites for direct phosphorylation by PKC at S40, S378, T417, T418, S439, S589, and T594. To determine which of these sites may be important on the native Nrf2 protein, six 8-residue peptides were synthesized corresponding to the potential PKC target sites (one of the peptides corresponded to both the T417 and T418 sites). All the peptides were readily soluble in 20% (v/v) DMSO except peptide aa 36–43, whose solubility was greatly improved when it was extended at both ends by one residue to a 10 aa residue peptide. These synthetic peptides mimicking potential Nrf2 PKC sites were used to compete with the transcription factor in the *in vitro* kinase assay. The peptides were included in the kinase reaction at a final concentration of 5 m*M*. Only one of the peptides, targeting serine 40, was shown to be an effective inhibitor of Nrf2 phosphorylation by the PKC catalytic subunit.

To confirm these peptidomimetics data a site-directed mutagenesis approach was then used. A rat Nrf2 cDNA bearing an AGT->GCT (Ser-to-Ala) mutation at amino acid position 40 (Nrf2-S40A) was expressed as a recombinant His–tagged protein in *E. coli* and purified by nickel-affinity chromatography. Purified Nrf2-S40A was then used in parallel with wild-type Nrf2 as a substrate in the *in vitro* PKC assay described previously. The single amino acid change from serine to alanine abolished the phosphorylation of Nrf2 by PKC. The lack of residual phosphorylation in this mutant indicates that serine 40 seems to be the only PKC site, which was consistent with the peptide competition data.

Determining the Importance of Serine 40 Phosphorylation on the Interaction Between NF-E2–Related Factor 2 and Keap1

In Vitro *Association Assays*. To determine whether the increase of nuclear Nrf2 by PKC-mediated activation of the ARE is due to an effect on the interaction between Nrf2 and Keap1, *in vitro* association assays between the two proteins were performed. The site for PKC phosphorylation is within the Neh2 domain of Nrf2, which has been shown to be the region where Nrf2 interacts with Keap1. Native rat Nrf2 and rat Keap1 protein was produced *in vitro* using the TNT-coupled wheat germ extract system (Promega, Madison, WI). The TNT reaction contained [^{35}S]methionine and was performed according to the manufacturer's instructions. The reaction products were then incubated together for 15 min at 30° before immunoprecipitation by an anti-Nrf2 antibody in an IP buffer containing 50 m*M* TRIS (pH 8.0), 150 m*M* NaCl, 1% (v/v) NP-40, 0.5% (w/v) DOC, 0.1% (w/v) SDS for 4 h at 4°, followed by the addition of Protein A-Trisacryl beads (Pierce, Rockford, IL). The mixture was rotated for 1 h at 4° and washed extensively in IP buffer containing 0.3 *M* NaCl. Precipitates were resolved by SDS–PAGE and detected by autoradiography. The intensity of the ^{35}S-labeled polypeptides was quantified by Phosphorimage analysis.

To determine the effect of PKC phosphorylation on the interaction between Nrf2 and Keap1, an equal amount of labeled wild-type or S40A mutant Nrf2 protein were incubated with PKC and Keap1 in kinase assay buffer for 30 min at 30° before immunoprecipitation. Similar amounts of Keap1 were coprecipitated with both wild-type and S40A mutant Nrf2 in the absence of PKC. In the presence of PKC, the amount of Keap1 associated with wild-type Nrf2 was reduced by approximately 50%. By contrast, the S40A mutant interacted with Keap1 to a similar extent in the absence or presence of PKC. Furthermore, the reduction in the amount of Keap1 coprecipitated with wild-type Nrf2, because the action of PKC was abolished when PKC was preincubated in the presence of 10 n*M* staurosporine. Thus, phosphorylation of Nrf2 by PKC at Ser40 seems to reduce the ability of Nrf2 and Keap1 to interact *in vitro*.

Effect of Phosphorylation on Binding of NF-E2–Related Factor 2/Maf Complexes to the Antioxidant Response Element

Nrf2 can bind to the ARE as a heterodimer with a small Maf protein. Electrophoretic mobility shift assays (EMSA) were used to assess whether interaction of the Nrf2/Maf complex with the ARE is affected when Nrf2 is phosphorylated by PKC. A double-stranded oligonucleotide containing the rat *QR* gene ARE sequence was synthesized to use as a probe. The sequence of the DNA probe was 5′-GATTTCAGTCTAGAGTCACA**GT-GA**C**T**TTG**GC**AAAATCTGAGCCG-3′ (ARE core sequence highlighted in bold). The probe was end labeled with [γ-^{32}P]ATP using T4 polynucleotide kinase. Purified wild-type or S40A mutant Nrf2 protein was preincubated with rat MafK (unless stated otherwise), which was produced in a TNT reaction, for 20 min at 25°. The labeled DNA probe was then added, and the mixture was incubated for 20 min at 30°. DNA–protein complexes were resolved in 6% (w/v) polyacrylamide gels under nondenaturing conditions and then detected by autoradiography. As controls, competition experiments were performed using a 200-fold molar excess of either unlabeled probe or a random 43-base oligonucleotide, which was included in the preincubation mixture at 25° before the addition of the labeled probe. Furthermore, supershift assays were also carried out in which an anti-Nrf2 antibody was introduced following the binding reaction and mixed for 4 h at 4° before electrophoresis. To determine whether the phosphorylation of Nrf2 by PKC may affect its binding to the ARE, wild-type or S40A mutant Nrf2 protein was first incubated in kinase assay buffer for 1 h at 30° in the presence or absence of PKC. An aliquot of these reactions was then incubated with MafK and the labeled probe for EMSA.

In control EMSA experiments, neither wild-type or S40A mutant Nrf2 protein nor MafK could bind to the ARE probe alone. In the presence of MafK, both wild-type and mutant Nrf2 formed complexes with the ARE probe that could be supershifted using the Nrf2 antibody. The intensity and mobility of PKC-phosphorylated wild-type Nrf2 complexes are indistinguishable from those of nonphosphorylated wild-type Nrf2 or the S40A mutant defective for PKC phosphorylation. This suggested that the formation of ARE-binding transcription factor complexes containing Nrf2 and small Maf proteins was not affected by the phosphorylation of Nrf2 by PKC.

Assessment of an In Vivo *Role for Serine 40 of Rat NF-E2–Related Factor2*

Cotransfection reporter gene experiments were used to determine whether the Nrf2-S40A mutant was functional and could transactivate the ARE. The introduction of exogenous Nrf2 into a number of hepatoma cell lines has previously been shown to result in a dose-dependent activation of ARE-enhanced gene expression.

Hepatoma cells were seeded into six-well plates. The *QR*-ARE CAT reporter gene was cotransfected with either a pcDNA3 construct encoding the wild-type or the S40A mutant rat Nrf2 protein by the methods described previously. In separate transfection experiments, a pcDNA3 construct encoding the rat Keap1 protein was also included along with each of the two Nrf2 expression constructs. The plasmids used for each transfection constituted, where appropriate, 1 μg of the CAT reporter DNA, 0.12 μg of the wild-type or mutant Nrf2 expression construct, 0.06 μg of the Keap1 expression construct, and empty pcDNA3 plasmid up to a total of 2 μg. The relative CAT activity was determined by Phosphorimage analysis.

The wild-type Nrf2 protein resulted in a similar level of CAT activity in transfected cell extracts as that seen with the S40A mutant. However, when Keap1 was included in the transfection, the activity of both the wild-type and mutant transcription factors was repressed. Keap1 repressed transactivation of the ARE by the S40A mutant to a much greater extent than the wild-type protein, which may reflect the higher affinity Keap1 has for phosphorylation-deficient Nrf2.

Effects of Phosphorylation on the Stability of the NF-E2–Related Factor 2 Protein

Treatment of Hepatoma Cells with Pharmacological Inhibitor. In addition to promoting the dissociation of Nrf2 from Keap1, inducing agents such as *t*BHQ cause an increase in the amount of Nrf2 protein within the cell. This effect may also be dependent on protein phosphorylation. The protein phosphatase inhibitor, okadaic acid (OA) was used to impose a hyperphosphorylated state within the cells. It was examined whether OA in combination with *t*BHQ would affect the level of Nrf2 in the cell. In addition, similar experiments were performed using a range of pharmacological kinase inhibitors in combination with *t*BHQ to determine which signal transduction pathways may affect the level of Nrf2 protein in the cell. The inhibitors used in these experiments and their working concentration are shown in Table I.

HepG2 cells were seeded in six-well plates and cultured in complete media for 24 h. The media were then replaced with Opti-MEM, and the cells were incubated overnight. The use of low serum culture conditions is important in particular with respect to the MAPK cascade, which is stimulated by serum components. In cotreatment experiments, cells were then exposed to the inhibitors for 1 h before the addition of *t*BHQ to a final concentration of 50 μM in the cell culture medium. The cells were then incubated for a further 4 h in the presence of both compounds. Control

TABLE I
PHARMACOLOGICAL PHOSPHATASE AND KINASE INHIBITORS

Inhibitor	Target	Concentration used
Okadaic acid	Protein phosphatases	50 nM
U0126	MEK1	10 μM
PD 98059	MEK1	50 μM
Staurosporine	PKC	15 nM
Ro 320 432	PKC	2 μM
LY 294 002	PI3K	10 μM
Wortmanin	PI3K	200 nM

experiments were performed when cells were incubated in Opti-MEM alone or in Opti-MEM containing *t*BHQ for 4 h.

Western Blotting Experiments to Determine the Level of Cellular NF-E2–Related Factor 2

Following the treatment period, cells were washed once with ice-cold PBS and then harvested in SDS loading sample buffer. Aliquots of cell lysates were then resolved by SDS–PAGE. The resolved polypeptides were transferred onto PVDF and immunoblotted for Nrf2 and GAPDH polypeptides. The GAPDH protein was used to normalize for protein loading. Antibodies used in this experiment include a polyclonal serum that was raised against a peptide sequence mapping to the N-terminus of Nrf2 and a monoclonal antibody that recognizes GAPDH (Research Diagnostics, Inc., Flanders, NJ). The Nrf2 antibody was used at a dilution of 1:5000, and the GAPDH antibody was used at a dilution of 1:10,000. The primary antibodies were detected with the appropriate secondary horseradish peroxidase–conjugated antibodies and immunoreactive polypeptides were detected by ECL.

A typical autoradiograph of the detectable level of Nrf2 in treated HepG2 cell lysates is shown in Fig. 4. In this experiment, more than one inhibitor of each kinase pathway has been used, because no conclusion can be accurately drawn from the use of a single inhibitor. The immunoblot clearly shows that *t*BHQ increases the amount of cellular Nrf2. This level is further increased by cotreatment with OA. Interestingly, the PKC inhibitors had no effect on the level of Nrf2 protein in HepG2 cells and neither did the PI3K inhibitors. The MEK1 inhibitors, PD 98059 and U0126, however, do impair an increase in Nrf2 protein in response to *t*BHQ, which would suggest that this pathway is important in mediating the stabilization of Nrf2 in response to inducing agents.

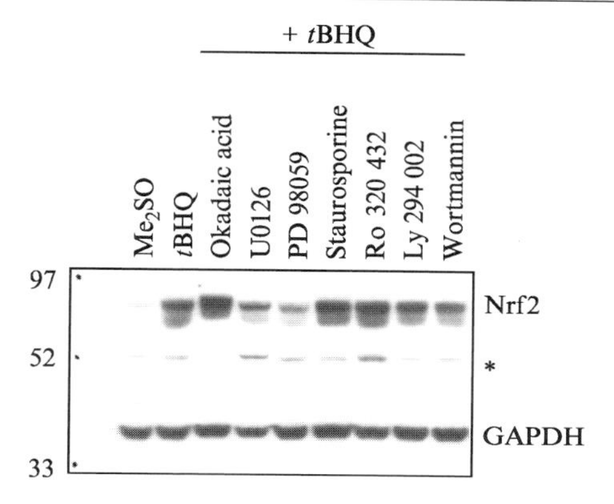

FIG. 4. Western blotting experiments were used to determine which cell signaling pathways are important for *t*BHQ-induced accumulation of Nrf2. HepG2 cells were seeded in six-well plates and exposed to solvent (DMSO), *t*BHQ alone, or *t*BHQ in combination with the inhibitor compounds detailed in Table I for 4 h. Cell lysates were then immunoblotted with anti-Nrf2 and anti-GAPDH antibodies. The asterisk denotes a nonspecific cross-reacting polypeptide. Treatment with *t*BHQ alone increases the level of cellular Nrf2 in cell lysates compared with the solvent control. The use of phosphatase inhibitors potentiates the increase in Nrf2. By contrast, the use of inhibitors of the MAPK cascade suppress the increase of cellular Nrf2 in response to treatment with *t*BHQ. Interestingly, inhibitors of PKC phosphorylating activity have little effect on the induced level of cellular Nrf2 as do inhibitors of PI3K.

Concluding Comments

It is clear that the activity of Nrf2 is regulated by at least two signaling pathways in hepatoma cells. The interaction between Nrf2 and the Keap1 repressor protein is affected by a direct action of PKC on the Nrf2 protein. Phosphorylation of the transcription factor at serine 40 leads to a weaker interaction with the repressor. In addition, stabilization of the Nrf2 protein through a mechanism involving the MAPK cascade is also important for regulating transactivation of the ARE. It is at present not known whether a kinase in this cascade acts on Nrf2 directly or exerts its effects through targeting other components of the ARE pathway.

[23] Analysis of Transcription Factor Remodeling in Phase II Gene Expression with Curcumin

By DALE A. DICKINSON, KAREN E. ILES, AMANDA F. WIGLEY, and HENRY JAY FORMAN

Introduction

Curcumin has been used for centuries in folk medicine and is now being rigorously investigated for use in Western medicine. Routine consumption of curcumin is correlated with the decreased incidence of certain cancers, including prostate and colon. Curcumin has been ascribed diverse roles, including preventing and/or alleviating inflammation; preventing carcinogenesis and mutagenesis; and having tumoricidal, nemicidal, fungicidal, and bactericidal properties, although the underlying mechanism(s) remain poorly defined. Many of the biological properties of curcumin are hypothesized to be manifestations of increased cellular defense capacity to stress. That is, curcumin primes an adaptive response to stress, which, depending on the disease and/or the experimental conditions, manifests itself in different ways. Phase II genes encode proteins that protect the cell from stress through the removal of damaging compounds. We have recently found that curcumin strongly induces glutathione, an important antioxidant with many roles in maintaining cellular health through increased gene expression.[1] The molecular mechanism for the upregulation of these phase II defense genes was due to remodeling of the transcription factor complexes driving expression of these genes in response to curcumin. These changes were correlated with changes in the overall abundance of the transcription factor proteins, their cellular location, and in the case of c-Jun, with activation by phosphorylation. These studies with curcumin revealed the usefulness of investigating the composition of transcription factor complexes in mediating phase II gene expression in addition to considering the overall DNA binding activity.

Biological Properties of Curcumin

Curcumin Basics

Curcumin (diferuloylmethane, CAS 458-37-7) is a low molecular weight polyphenol (F.W. 368.37) that is derived from the rhizomes of *Curcuma* spp., a member of the ginger family *Zingiberaceae*. Curcumin is the major

0076-6879/04 $35.00

component of tumeric, a principal ingredient in curry dishes, where it is used as a flavoring agent in both vegetarian and nonvegetarian food preparations. In this powdered form, curcumin is consumed daily by approximately a quarter of the world's population. Curcumin is also used as common dye C.I. Natural Yellow #3 (C.I. 75300).

Curcumin in Medicine

More than a natural coloring and flavoring agent, curcumin has been routinely used in folk medicines, especially Indian and Chinese, where it has been ascribed many protective and restorative properties. The more common uses for curcumin are for its anti-inflammatory properties,[2,3] as well as tumoricidal, antiprotozoal, and antioxidant effects.[4–7] Routine dietary consumption of curcumin is correlated with decreased incidence of several forms of cancer, including colon cancer.[8] This natural electrophilic product has been demonstrated to have protective effects against oxidative injury, although the mechanisms responsible for this activity remain uncertain. Understanding the molecular mechanisms underlying these biological properties is an area of intense investigation (for a recent review see Araujo and Leon[9]). It is clear that curcumin has complex effects on the cellular redox tone, perhaps because of innate antioxidant properties related to structure but also from the ability of this compound to affect enzyme systems involved in the defense against oxidative stress. Curcumin has two α,β-unsaturated ketone groups that can function as Michael acceptors (Fig. 1). For example, the addition of glutathione (GSH) to one of the carbon–carbon double bonds of curcumin is catalyzed by glutathione *S*-transferase (GST) and functions as a mechanism to remove curcumin from the cell. The effect of curcumin on GSH content is a likely mechanism by which curcumin can function as such a broad adaptogen.

[1] D. A. Dickinson, K. E. Iles, H. Zhang, V. Blank, and H. J. Forman, *FASEB J. 10.1096/fj.02-0566fje* (2003).
[2] A. Mukhopadhyay, N. Basu, N. Ghatak, and P. K. Gujral, *Agents Actions* **12,** 508 (1982).
[3] A. J. Ruby, G. Kuttan, K. D. Babu, K. N. Rajasekharan, and R. Kuttan, *Cancer Lett.* **94,** 79 (1995).
[4] R. C. Srimal and B. N. Dhawan, *J. Pharm. Pharmacol.* **25,** 447 (1973).
[5] M. T. Huang, R. C. Smart, C. Q. Wong, and A. H. Conney, *Cancer Res.* **48,** 5941 (1988).
[6] C. A. Araujo, L. V. Alegrio, D. C. Gomes, M. E. Lima, L. Gomes-Cardoso, and L. L. Leon, *Mem. Inst. Oswaldo Cruz* **94,** 791 (1999).
[7] A. C. Reddy and B. R. Lokesh, *Food Chem. Toxicol.* **32,** 279 (1994).
[8] R. T. Greenlee, T. Murray, S. Bolden, and P. A. Wingo, *CA Cancer J. Clin.* **50,** 7 (2000).
[9] C. C. Araujo and L. L. Leon, *Mem. Inst. Oswaldo Cruz* **96,** 723 (2001).

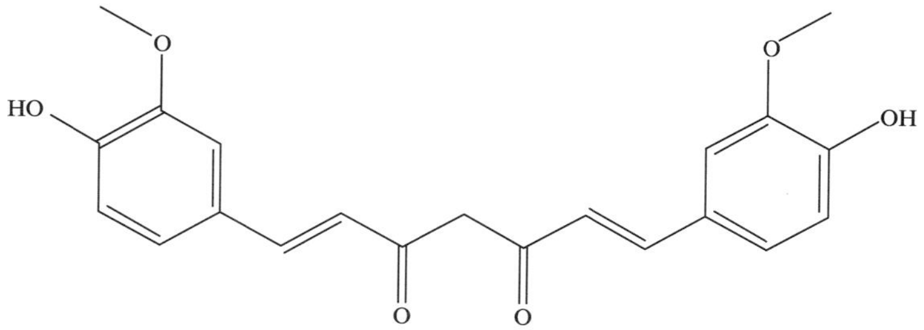

FIG. 1. Structure of curcumin. The phytochemical curcumin, 1,7-bis(4-hydroxy-3-methoxyphenyl)-1,6-heptadien-3,5-dione, has two methoxyphenyl groups and two α,β-unsaturated ketone groups. The methoxyphenyl groups have been hypothesized to be important in mediating anti-inflammatory effects, although there is some evidence suggesting that the α,β-unsaturated ketone moiety is responsible for this activity and for the antiparasitic activity as well. Curcumin, having a double bond in conjugation with a carbonyl, can undergo Michael addition by GSH across the carbon–carbon double bond by means of GST-catalyzed reactions that are partially responsible for their removal from the cell.

Glutathione and Curcumin

Generally found in the millimolar range, GSH is the most abundant nonprotein thiol in the cell. GSH serves as a reducing equivalent, a substrate for several peroxidases that reduce hydroperoxides, and as a substrate for GST-catalyzed conjugation reactions, resulting in the removal of deleterious xenobiotics from the cell (for review, see Dickinson and Forman[10]). The de novo synthesis of GSH is regulated at several levels in the cell. Depletion of GSH by either conjugation reactions, or by oxidized glutathione (GSSG) formation (caused by increased H_2O_2 production) under the catalysis of glutathione peroxidase (GSHPx), will result in an increased production of GSH. Increased synthesis of glutamate-cysteine ligase (GCL) subunits through a combination of increased transcription and mRNA stability is generally believed to be the predominant mechanism for increasing GSH,[11] although to some extent a decrease in GSH will cause a transient increase in the activity of pre-existing GCL by reducing feedback inhibition by GSH.[12] Both mechanisms cause an increase in GSH synthesis. Increased GCL synthesis results in a prolonged GSH increase, whereas the decreased feedback inhibition mechanism produces only a transient increase.

[10] D. A. Dickinson and H. J. Forman, *Biochem. Pharmacol.* **64,** 1019 (2002).
[11] R.-M. Liu, L. Gao, J. Choi, and H. J. Forman, *Am. J. Physiol.* **275,** L861 (1998).
[12] P. G. Richman and A. Meister, *J. Biol. Chem.* **250,** 1422 (1975).

Characteristics of Phase II Genes

Many of the folk uses for curcumin, such as relieving inflammation, involve the treatment of conditions that have been shown to result, at least in some situations, from the deleterious effects of reactive oxygen species (ROS). Under normal physiological conditions, an organism can reduce or eliminate these negative effects of ROS and electrophilic insults with antioxidants such as GSH and thioredoxin and also through the actions of phase II detoxification enzymes such as NAD(P)H:quinone oxidoreductase (NQO1) and members of the GST family. The conditions curcumin is believed to be effective in treating may result from an imbalance between the generation of ROS and the ability of the organism to remove them.

Although early studies focused on the negative effects of ROS, many recent reports demonstrate that some ROS, at relatively low levels, can serve as second messengers in cells, especially hydrogen peroxide (H_2O_2). Low levels of ROS can lead to physiological redox signaling, whereas high or "supraphysiologic" levels of ROS can lead to signaling for apoptosis or can directly cause necrosis. Physiological redox signaling leads to the activation of cellular defense systems that make the cell more resistant to subsequent challenges. This is adaptive signaling. The phase II genes are key components of the adaptive response, and it has been determined through DNA sequencing studies that many phase II genes contain a specific consensus element in their promoters termed the electrophilic response element (EpRE, sometimes called the antioxidant response element, or ARE).[13]

The Electrophilic Response Element

The EpRE element is a common regulatory motif found in the promoters of many phase II genes, including the GSH biosynthetic genes *Gclc*[14] and *Gclm,*[15] and classic phase II genes such as *Nqo1* and *Gst-Ya,* in which it was first described.[16,17] Genes having the EpRE element can be regulated in response to H_2O_2,[18] quinones such as *tert*-butylhydroquinone (*t*BHQ),[19,20] and the classic inducer 12-*O*-tetradecanoylphorbol-13-acetate

[13] A. K. Jaiswal, *Biochem. Pharmacol.* **48,** 439 (1994).
[14] R. T. Mulcahy and J. J. Gipp, *Biochem. Biophys. Res. Comm.* **209,** 227 (1995).
[15] R. T. Mulcahy, M. A. Wartman, H. H. Bailey, and J. J. Gipp, *J. Biol. Chem.* **272,** 7445 (1997).
[16] T. H. Rushmore, R. G. King, K. E. Paulson, and C. B. Pickett, *Proc. Natl. Acad. Sci. USA* **87,** 3826 (1990).
[17] T. H. Rushmore and C. B. Pickett, *J. Biol. Chem.* **265,** 14648 (1990).
[18] I. Rahman, A. Bel, B. Mulier, M. F. Lawson, D. J. Harrison, W. MacNee, and C. A. D. Smith, *Biochem. Biophys. Res. Comm.* **229,** 832 (1996).
[19] J. Li, J. M. Lee, and J. A. Johnson, *J. Biol. Chem.* **277,** 388 (2002).
[20] A. N. Kong, E. Owuor, R. Yu, V. Hebbar, C. Chen, R. Hu, and S. Mandlekar, *Drug Metab. Rev.* **33,** 255 (2001).

(TPA).[21] The precise sequence of the EpRE element itself varies between genes but has the general consensus sequence TGCNNN(C/G)TCA. The architecture of this element is often described as two perfect or imperfect TPA response element (TRE) elements, with various numbers of nucleotides between them. The TRE sequences can be found in any orientation with respect to one another.

How EpRE elements regulate genes is not completely understood. Inducers of EpRE genes also increase the overall EpRE DNA binding activity in the electrophoretic mobility shift assay (EMSA). This implies that stimulation causes the proteins in the EpRE binding complex to be either modified in some manner that increases their binding affinity and/or to move into the nucleus from the cytosol. It is generally accepted that the transcriptional or repressor activity of a transcription factor-*cis* element complex, including the EpRE-binding complex, depends on which component proteins are present in the complex. It is known that Jun family members, Fos family members, Nrf1 and Nrf2, and the small Maf proteins can bind to the EpRE element. The EpRE binding complex is possibly a dimer, similar to the AP-1 complex, although some authors have suggested a ternary complex. The proteins that bind to the EpRE element can either inhibit or induce expression. Many studies have been done using the overexpression of potential transcription factors with reporter constructs. Although this method can demonstrate the capacity of the overexpressed protein to induce expression, the result may actually have little relevance in normal gene expression, where the protein must compete with others for binding. Studies using knockout models for transcription factors in animals may yield potentially more relevant information, but for many of the putative EpRE transcription factors, such models do not yet exist or may be lethal. In general, though, it seems that Jun and Nrf proteins are inducers of EpRE-mediated expression, whereas Fos and the small Maf members are inhibitory.[22,23] Whether additional posttranslational modifications are required for enhanced transcriptional activation in EpRE-driven expression, such as the phosphorylation of c-Jun, remains unknown.

The Glutathione Biosynthetic Genes as Model Phase II Genes

The de novo synthesis of GSH occurs by the sequential action of two adenosine triphosphate (ATP)–dependent enzymes. The first and rate-limiting enzyme, GCL, combines glutamate and cysteine. The dipeptide

[21] T. Nguyen, T. H. Rushmore, and C. B. Pickett, *J. Biol. Chem.* **269,** 13656 (1994).
[22] L. M. Zipper and R. T. Mulcahy, *Biochem. Biophys. Res. Comm.* **278,** 484 (2000).
[23] S. Dhakshinamoorthy and A. K. Jaiswal, *J. Biol. Chem.* **275,** 40134 (2000).

generated by GCL is then combined with glycine by the second enzyme, glutathione synthase (GS), to yield GSH (γ-L-glutamyl-L-cysteinyl-glycine). The phase II enzyme GCL is a heterodimer composed of a catalytic (GCLC) and a modulatory (GCLM) subunit.[24] Each subunit is encoded by a separate gene in humans.[25] Both of these genes, *Gclc* and *Gclm*, contain EpRE-enhancer elements, and many reports have demonstrated EpRE-mediated expression,[15,26,27] although other enhancer elements, notably the AP-1 binding site TRE, can mediate *Gcl* expression.[28] The glutathione *S*-transferases (GSTs) are a large family of phase II enzymes that catalyze the conjugation of GSH to specific compounds. Earlier reports have demonstrated that low-dose curcumin increased the expression of certain GSTs[29] and affected the intracellular content of GSH,[30] although the mechanisms underlying the latter effect were not formally investigated. Recently, we demonstrated that the phase II genes *Gclc* and *Gclm* are expressed in response to low dose curcumin and determined the underlying molecular mechanism to involve remodeling of the transcription factor complex that binds to EpRE.[1] The methods used to determine these findings will be discussed in detail in the rest of this chapter.

Determining Phase II Gene Induction

Curcumin has been shown, in different systems, to possess many characteristics. How this one compound can have so many and varied effects remains uncertain, although it is becoming clear that curcumin functions as an adaptogen. The cells are more robust (i.e., they are more able to withstand stress-induced damage). The phase II genes encode proteins that help protect the cell from stress, and as such seem to be likely candidates for upregulation by curcumin. It is prudent to first determine that the compound of interest is affecting phase II enzymes before performing more detailed experiments to determine the relative composition of the compound-affected transcription factor complex.

[24] R. Sekura and A. Meister, *J. Biol. Chem.* **252,** 2599 (1977).
[25] K. Tsuchiya, R. T. Mulcahy, L. L. Reid, C. M. Disteche, and T. J. Kavanagh, *Genomics* **30,** 630 (1995).
[26] H. R. Moinova and R. T. Mulcahy, *J. Biol. Chem.* **273,** 14683 (1998).
[27] A. C. Wild, J. J. Gipp, and R. T. Mulcahy, *Biochem. J.* **332,** 373 (1998).
[28] D. A. Dickinson, K. E. Iles, N. Watanabe, T. Iwamoto, H. Zhang, D. M. Krzywanski, and H. J. Forman, *Free Radic. Biol. Med.* **33,** 974 (2002).
[29] J. T. Piper, S. S. Singhal, M. S. Salameh, R. T. Torman, Y. C. Awasthi, and S. Awasthi, *J. Biol. Chem.* **30,** 445 (1998).
[30] S. S. Singhal, S. Awasthi, U. Pandya, J. T. Piper, M. K. Saini, J. Z. Cheng, and Y. C. Awasthi, *Toxicol. Lett.* **109,** 87 (1999).

Methods to Study Phase II Expression

Demonstrating an effect on a phase II gene could be done at several levels. The activity of the specific protein could be assayed, the content of the specific protein could be determined, and the mRNA content for that protein could be measured. Each of these end points has both benefits and shortcomings.

Specific activity is certainly the most meaningful indicator in terms of biology: as activity increases, so does the cells capacity for defense, but total activity could be affected by multiple mechanisms, including posttranslational mechanisms. The content of the specific protein can be determined, if an appropriate antibody exists. Protein content is generally reflective of activity, although an increase in protein content may not have biological meaning if that protein requires a posttranslational modification to be active. The increase in protein content can result from several mechanisms, including decreased degradation and/or increased synthesis, further complicating data interpretation. The content of the mRNA for the protein is often a good choice, because techniques to measure mRNA are standard, and determining gene-specific mRNA is relatively straightforward using modern techniques such as real-time polymerase chain reaction (PCR). If signaling for gene expression is to be examined, and especially if inhibitors of signaling are to be used, then mRNA quantification is the most reliable indicator.

Methods to Determine mRNA Content

It deserves to be restated that although mRNA content is often thought to be reflective of gene transcription, the contribution of increased mRNA stability cannot be excluded. Indeed, message stabilization has been demonstrated to play a role in the *Gcl* genes in response to the bioactive lipid 4-hydroxy-2-nonenal.[31] Although nuclear run-on experiments demonstrate gene transcription directly, this method is not routine and requires advanced skill. The total content of a specific mRNA is often biologically meaningful, but with the considerations mentioned previously, posttranscriptional steps may mean that increased mRNA content does not necessarily correspond to increased defensive capacity. Regardless, elevation of mRNA content for phase II genes is a quick and generally reliable method to assess an enhanced defensive capacity of the cell.

Traditionally, mRNA levels were determined using Northern analysis of membrane-bound, fractionated RNA. This method, although reliable,

[31] R. M. Liu, L. Gao, J. Choi, and H. J. Forman, *Am. J. Physiol.* **275,** L861 (1998).

is only semiquantitative and relatively time-consuming. Gene-specific probes are necessary; often a partial cDNA, which may or may not be readily available, is required. PCR-based methods simplify this process. The reactions are fast, do not require fractionation of the RNA, and allow for straightforward gene-specific detection by primer design. Reverse-transcription–PCR (RT–PCR) was the first of the PCR-based methods to quantify mRNA, but a newer method called real-time PCR is more reliable and does not require as many controls as RT–PCR. More importantly, real-time PCR can be truly quantitative, actually reporting the copy number of a message in a cell, whereas RT–PCR remains semiquantitative.

Basics of Real-Time Polymerase Chain Reaction. Real-time PCR quantifies the initial amount of the template specifically, sensitively, and reproducibly and is preferable to other forms of quantitative RT–PCR, which detect the amount of final amplified product. Real-time PCR monitors the fluorescence emitted during the reaction as an indicator of amplicon production during each PCR cycle (i.e., in real time). Other methods typically rely on end-point measurements, when concentrations of reagents can become limiting. In contrast, real-time PCR measurements are taken from the exponential range of the reaction, when component concentrations are not limiting. In addition, this removes post-PCR processing as a potential source of error. The real-time progress of the reaction can be viewed in some systems, such as Cepheid SmartCycler (Cephid, Sunnyvale, CA).

The real-time PCR method is based on the detection and quantification of a fluorescent reporter. This signal increases in direct proportion to the amount of PCR product in a reaction. By recording the amount of fluorescence emission at each PCR cycle, it is possible to monitor the reaction during the exponential phase when the first significant increase in the amount of PCR product correlates to the initial amount of target template. There are two general methods for the quantitative detection of the amplicon: fluorescent probes and DNA-binding agents. Fluorescent probes use the fluorogenic 5′ exonuclease activity of Taq polymerase to measure the amount of target sequences in cDNA samples. This method allows for multiplex reactions but is not as straightforward as the use of DNA binding dyes and is relatively expensive.

The cheaper, more straightforward, and generally reliable alternative is the use of double-stranded DNA binding dye chemistry, which quantifies the amplicon production by the use of a nonsequence specific fluorescent intercalating agent, such as SYBR Green (Applied Biosystems, Bedford, MA) or ethidium bromide. SYBR Green is a minor groove-binding dye, so it only binds to double-stranded DNA. Unfortunately, SYBR Green–based detection cannot distinguish specific amplifications from nonspecific, which may arise from primer–dimer product or promiscuous primer binding. A relatively minor and easily controllable consideration is that longer amplicons create a stronger signal. Obviously SYBR Green can only be used in singleplex reactions.

The threshold cycle, or the C_T value, is the cycle at which the system begins to detect the increase in the signal associated with an exponential growth of PCR product during the log-linear phase. This phase provides the most useful information about the reaction. The C_T value is the most important parameter for quantification. The higher the initial amount of target, the sooner the accumulated product is detected in the PCR process, and the lower the C_T value. Cepheid SmartCycler software determines the C_T value by a mathematical analysis of the growth curve. These values can be used to determine relative levels of gene expression and, with the use of standards, can be used to determine the precise copy number.

RNA Preparation for Real-Time Polymerase Chain Reaction. Total RNA can be collected by any standard method. For laboratories that currently do not analyze RNA, TRIzol reagent (Invitrogen, Carlsbad, CA) is a good choice. Extract total RNA according to the manufacturer's instructions and dissolve in diethyl pyrocarbonate (DEPC)–treated H_2O. Contaminating DNA must then be removed by treatment with DNase. A kit such as DNA-*free* (Ambion, Austin, TX) is economical, reliable, and straightforward to use. We recommend following the manufacturer's instructions precisely, which will remove all contaminating DNA, leaving only RNA. This RNA must be quantified spectrophotometrically at 260 nm.

Reverse Transcription of RNA for Real-Time Polymerase Chain Reaction. Real-time PCR amplifies DNA. As such, it is first necessary to reverse transcribe (RT) the DNA-free RNA. This is easily accomplished using TaqMan random hexamers (Applied Biosystems, Bedford, MA). Following the manufacturer's instructions, DNA-free RNA, RT-PCR stock, and DEPC-H_2O are reverse transcribed in a thermal cycler, using the following times and temperatures: 10 min at 25°, 30 min at 48°, 5 min at 95°, then held at 4°. The RT product from this reaction, which is now a suitable substrate for real-time PCR, can be stored at −20°. We suggest making several reaction volumes of RT product at once, allowing for the quantification of several genes without having to repeat this step.

Real-Time Polymerase Chain Reactions. Real-time PCR must be performed on an instrument capable of detecting real-time fluorescence values that are integrated with appropriate software. The Cepheid SmartCycler 1.2 is a good choice and is packaged with outstanding software. RT products from the preceding step should be thawed on ice while preparing stocks. For each sample to be tested, prepare a cocktail consisting of 12.5 μl SYBR Green PCR mastermix (Applied Biosystems, Bedford, MA), 5.5 μl sterile distilled H_2O, and 1 μl each of the sense and antisense

TABLE I
PRIMER SEQUENCES FOR USE IN REAL-TIME PCR

Gene	Strand	Sequence (from 5′ to 3′)	T_m	Amplicon Size (nt)
Gclc	Sense	ATGGAGGTGCAATTAACAGAC	60°	206
	Anti-sense	ACTGCATTGCCACCTTTGCA	60°	
Gclm	Sense	GCTGTATCAGTGGGCACAG	60°	196
	Anti-sense	CGCTTGAATGTCAGGAATGC	60°	
Nqo1	Sense	AGTTTGCTTACACTTACGCTG	60°	212
	Anti-sense	CCCAATGCTATAAGATGTCAG	60°	
GstA4	Sense	GAGAACCCTGATTGACATGTA	60°	194
	Anti-sense	GCTGATTACCAACAAGAAAGC	60°	
Gapdh	Sense	TGGGTGTGAACCATGAGAAG	60°	199
	Anti-sense	CCATCACGACACAGTTTCC	60°	
β-actin	Sense	CATGGAGTCCTGTGGCATC	60°	195
	Anti-sense	GGAGCAATGATCTTGATCTTC	60°	

primers (Table I). Mix 22.5 μl of this PCR cocktail and 2.5 μl of the appropriate RT product and pipette into a 25-μl SmartCycler tube (Cepheid, Sunnyvale, CA). Tubes should be given a quick spin in a picofuge containing the special attachment for the Cepheid tubes. Samples are cycled for 10 min at 95° (this step is necessary to activate the DNA polymerase in the PCR mix), 20 s at 95° (to separate the double-stranded amplicon DNA), 40 s at 62° (or 2° above your specific annealing temperature), and 40 s at 72°. This is repeated 45 times. When finished, the samples can be held at 4°. If desired, PCR samples can be mixed with DNA loading dye and run on a gel to verify the presence of a single band (the gene-specific amplicon).

Primers for Real-Time Polymerase Chain Reaction. When using DNA binding dyes, it is absolutely essential that the primers used are very specific for the target. Sense and anti-sense primers should be designed so that they have the same melting temperature, T_m, of at least 58°. This helps to ensure specificity of binding. The distance between the primers determines the size of the amplified product (the amplicon). The amplicon size should be consistent among the genes to be measured. Other groups have empirically determined that amplicons in the range of 150–200 work best; amplicons should not be greater than 400 or smaller than 50 nucleotides. Software to help with primer design is recommended to decrease the likelihood of forming primer-dimers and hairpin loops. Other primer considerations, such as maintaining the G + C content at 30–80%, avoiding runs of a single nucleotide, etc., are standard.

Primer sequences for human *Gclc* and *Gclm,* the GSH biosynthetic genes, and for *GstA4* (the specific GST isoforms that adds GSH to carbon–carbon double bonds, as found in curcumin) and *Nqo1,* two typical phase II genes, are listed in Table I. For each gene-specific pair, the primers were designed to have similar T_m and to produce amplicons of similar size, approximately 200 nucleotides.

Glyceraldehyde-3-phosphate dehydrogenase *(Gapdh)* is used as an active reference to determine relative gene expression. By using an endogenous control as an active reference, quantification of an mRNA target can be normalized for minute differences in the amount of total RNA added to each reaction. The most common choices for internal controls are 18S RNA, *Gapdh,* and *β-actin*. The issue of the choice of a normalizer has recently been reviewed by Suzuki *et al.*[32] It has been found that each of the preceding three common choices can vary in levels during growth, development, or with experimental condition. Because the chosen normalizer mRNA species must be proportional to the amount of input RNA, a combination may be necessary. It is possible, however, to validate the chosen normalizer for the target cell or tissue under the experimental conditions. We have found that *Gapdh* is constant in human bronchial epithelial cells (HBE1) exposed to curcumin.[1] Gene-specific primers for both *Gapdh* and *β-actin* are listed in Table I.

In addition to the primary considerations listed previously, there are other factors that may need to be optimized. Several good reviews of real-time PCR are available and should be consulted in addition to the preceding method for both background reading and troubleshooting in incidences of poor results.[33]

Curcumin Induces Phase II Genes in HBE1 Cells. We recently demonstrated that low-dose curcumin induces the mRNA content of the GSH biosynthetic genes *Gclc* and *Gclm* and the classic phase II genes *Nqo1* and *GstA4.*[1] These experiments were performed on a Cepheid SmartCycler 1.2 using the methods and reagents detailed in the preceding section. These expression data correlated well with GCL content and with the intracellular GSH content, demonstrating that curcumin is increasing the redox buffering capacity of the cells, through increased GSH content. Moreover, that increased capacity was further strengthened by increased content of the synthetic enzyme GCL, allowing for rapid replenishment of any GSH consumed in the defense reactions, and further, that the prolonged increase in the *Gcl* mRNA species would allow for rapid production of more GCL. In essence, the low-dose exposure of curcumin led to adaptive responses

[32] T. Suzuki, P. J. Higgins, and D. R. Crawford, *Biotechniques* **29,** 332 (2000).
[33] S. A. Bustin, *J. Mol. Endocrinol.* **29,** 23 (2002).

in these cells, with both an immediate increase in an important antioxidant, GSH, and the ability to rapidly generate more. Using an inhibitor of transcription, actinomycin D, we demonstrated that the increase in *Gcl* mRNA was due primarily to increased transcription and not message stabilization. This finding provided a basis to investigate the DNA-binding activity of potential transcription factors and, more importantly, to determine the composition of the complexes.

Transcription Complex Analysis

The electrophoretic mobility shift assay (EMSA) was first described in a seminal article in *Cell* in 1986.[34] It is also referred to in the literature as the gel shift, band shift, gel mobility shift, and DNA retardation assay. Although numerous modifications of the original method have been reported since then, the basic premise remains the same. Gel shift is a powerful *in vitro* method to draw inferences about which transcription complexes are bound to consensus sequences in the promoter region under conditions of interest *in vivo*. The gel shift is used to determine whether binding to a particular site (TRE, κB, SRE, EpRE, etc.) is increased under a given set of conditions. Because many transcription factor complexes, such as AP-1, have multiple potential constituent proteins, the super-shift identifies which proteins are present. It can also be used to determine whether the constituents of the transcription factor complex change when the cell is activated, stressed, etc. This is also important because some AP-1 complexes (i.e., c-Fos/c-Jun) generally stimulate transcription, whereas others (i.e., Fra2/JunD) may inhibit transcription.[35] However, this can also vary between cell type and needs to be determined individually for each system.[36]

Electrophoretic Mobility Shift Assay

The EMSA is an *in vitro* method that models transcription factor binding and in many cases transcriptional activation within the cell. A high-salt extraction method is used to prepare a nuclear lysate or extract. This extract is incubated with a labeled oligonucleotide, typically around 20 nucleotides in length, in a binding buffer that is optimized for that oligo. Traditionally, radiolabels (^{32}P) have been used, but now fluorescent and

[34] R. Sen and D. Baltimore, *Cell* **46,** 705 (1986).

[35] L. A. Kobierski, H. M. Chu, Y. Tan, and M. J. Comb, *Proc. Natl. Acad. Sci. USA* **88,** 10222 (1991).

[36] J. J. Andreucci, D. Grant, D. M. Cox, L. K. Tomc, R. Prywes, D. J. Goldhamer, N. Rodrigues, P. A. Bedard, and J. C. McDermott, *J. Biol. Chem.* **277,** 16426 (2002).

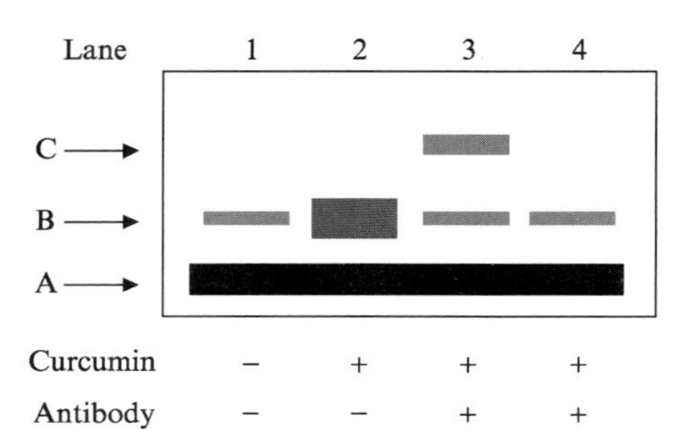

FIG. 2. Diagrammatic representation of EMSA and immunodepletion result. This cartoon exemplifies the results that can be obtained with EMSA and immunodepletion experiments. Band A represents the unbound, labeled oligo. Band B represents oligo that is bound to a transcription factor complex. Band C represents a classic super-shifted band, resulting from the coincubation of nuclear extract, labeled oligo, and specific antibody. In practice, however, as reported by us and many others, Band C is often not observed. Regardless, a reduction in the intensity of band B is observed when coincubated with antibody, whether an additional band is observed or not. This can be seen by comparing lane 4 with lane 2. It should be noted that for many consensus elements, there is always some basal activity (lane 1). If treatment increases the overall binding activity, an increase in intensity should be observed compared with control (lane 2 versus lane 1).

chemiluminescent labels or probes are also available. The samples are electrophoresed, and the bands are visualized through exposure of photographic film or with an electronic emission counting device such as an InstantImager (Perkin Elmer, Boston, MA) for radiolabeled probes or an AlphaImager (Alpha Infotech, San Leandro, CA) for fluorescently labeled probes. If transcription factors are present in the sample that bind to the promoter region of the gene of interest *in vivo,* the synthetic oligo is designed so that they will bind to it as well. The bound oligo-transcription factor complex migrates slower through the gel and separates from the free, unbound probe. This is seen on the EMSA as two distinct bands: one that has been retarded or shifted, and one that has migrated freely through the gel (Fig. 2, bands B and A, respectively). If the shifted band increases with intensity with a given treatment, this is consistent with increased binding to the consensus sequence in vivo (Fig. 2, band B, lane 2 versus lane 1).

Preparation of Nuclear Extracts. Cells should be treated according to the experimental system and collected as appropriate. Collected cells should be centrifuged at 500*g* for 5 min at 4° to generate a cell pellet. The buffer should be carefully and quickly aspirated from the tube to avoid disturbing the cell pellet. Nuclei can be extracted from the whole cell pellet and lysed using the NE-PER kit following the manufacturer's instructions

(Pierce Chemical Co., Rockford, IL). A 1:100 dilution of a stock protease inhibitor cocktail (Sigma, St. Louis, MO; P-8340) and a 1:500 dilution of a stock phosphatase inhibitor cocktail (Sigma P-5726) should be added to the supplied NE-PER lysis buffers. Aliquot the lysate into several prechilled tubes to avoid multiple freeze–thaw cycles. Reserve one aliquot for protein analysis, quick-freeze the others in liquid nitrogen, then store at −80°. Determine protein concentration using any standard method.

DNA Sequences for Electrophoretic Mobility Shift Assay. The TRE oligo (5′-CGCTTGA*TGAGTCA*GCCGGAA, sense) is available commercially from Promega (Madison, WI; E2301). This oligo differs from the classical TRE consensus sequence by a point mutation. Fortuitously, however, it is the identical sequence as the TRE-like element at −263 in the human *Gclc* promoter. The EpRE oligo (5′-GCGCTGAGTCAC, sense) is identical to the "EpRE 4" in the human *Gclc* promoter.[15] Because this is not commercially available, single-stranded oligos for both the sense and antisense strand can be custom made and annealed to form a double-stranded oligo on site. These double-stranded oligos need to be labeled in some manner so that they are detectable in the assay system. We recommend end-labeling with radioactivity, using T4 polynucleotide kinase, following the manufacturer's directions. A reaction mixture of 10 μl H_2O, 4 μl of 10× forward buffer (Invitrogen, Calsbad, CA), 2 μl of T4 poly kinase (Invitrogen), and 2 μl of γ-^{32}P-ATP (approximately 4000 Ci/mmol) are incubated at 37° for 15 min. The reaction is stopped with the addition of 1 μl of 0.5 *M* EDTA, pH 8.0, and the oligo mixture is chilled on ice for 2 min. The unincorporated radioactivity needs to be removed, so probe purification is necessary. We recommend using size-exclusion columns, such as NucTrap (Stratagene, La Jolla, CA). Follow the specific manufacturer's directions. Percent incorporation can be determined by measuring the radioactivity in a small aliquot of probe using a scintillation counter. Generally, at least 50% incorporation should be expected.

Incubation, Gel Electrophoresis, and Imaging. The gel shift assay is performed as previously described.[28,37] Two micrograms of nuclear protein is preincubated for 15 min at 37° with 5× gel shift binding buffer (Promega E 3581). Following this preincubation step, 3 μl of ^{32}P-labeled TRE or EpRE oligo (10^6 cpm/pmol) is added, and samples are incubated for an additional 20 min at room temperature. HeLa extract (Promega E3621) should be used as a positive control. A 10-fold excess of cold (unlabeled) probe should be added to duplicate samples to ensure that the shifted band actually represents specific binding. The samples are electrophoresed on native

[37] K. E. Iles, D. A. Dickinson, N. Watanabe, T. Iwamoto, and H. J. Forman, *Free Radic. Biol. Med.* **32,** 1304 (2002).

5% polyacrylamide/TBE gels (Invitrogen) at 250 V for 30 min, transferred to blotting paper (VWR 28303–102), and dried for 1 h. Dried membranes can either be exposed to film and densitometry used to estimate radioactivity, or radioactive decay can be measured directly by electronic autoradiography, using, for example, a Packard InstantImager.

Other Considerations. Values obtained from the InstantImager (raw counts) need to be normalized to the control group (100%) and then expressed as relative means ± the relative error of the mean (REM). Because replicate immunodepletion assays produce raw counts that cannot be compared directly, statistical analysis needs to be performed comparing the percent increase in binding versus control. Paired *t* tests should be used when comparing two treatment groups. When comparing multiple groups, ANOVA can be performed with a Tukey's test done post hoc.

Curcumin Increases Electrophilic Response Element and AP-1 DNA Binding Activities in HBE1 Cells. Curcumin has been used by us as a model to investigate the induction of phase II gene expression.[1] In EMSA studies with curcumin, we found that AP-1 binding activity in HBE1 cells exposed to 10 μM curcumin increased with time, peaked 30 min after incubation, and remained elevated for at least 3 h. Accompanied by this increase in AP-1 binding activity was increased EpRE DNA binding activity. EpRE binding activity increased early, 15 min after incubation, and remained so to 3 h. These data illustrate several of the useful applications of the gel shift assay in dissecting the signaling pathways leading to phase II gene expression. Time-course experiments coupled with EMSA analysis reveal the timing of the activation of specific transcription factor binding in a system. They are in fact critical, because binding to different enhancer elements may occur at different time points in the same system. As well, increased binding is often an early event in the signaling process but may take hours in some systems, so it is essential to rigorously examine a wide range of time points before concluding that a particular enhancer element is not activated in a given system. In our system, although both EpRE and AP-1 are activated, the timing is sequential (EpRE precedes AP-1) but later coordinates, because EpRE binding remains active after AP-1 binding is upregulated. This illustrates the usefulness of EMSA in defining the complexities of signaling data.

Immunodepletion

Traditionally, the "super-shift" has been described as an adaptation of the gel shift aimed at identifying the transcription factors bound to the oligonucleotide. All of the steps of a standard gel shift are performed, but antibodies to the proteins suspected to be in the transcription factor complex are added to the reaction mixture. In theory, this should produce three bands: the quickly migrating free probe band (Fig. 2, band A), the retarded protein–oligo complex (Fig. 2, band B), and the "super-shifted" antibody–protein–oligo complex (Fig. 2, band C). This third band migrates more slowly through the gel, because its movement is impeded by the additional weight of the bound antibody. In this scenario, if a third band (Fig. 2, band C) is detected, then this is taken as evidence that the suspected protein is indeed bound to the oligo, and present in the transcription factor complex.

The super-shifting of the band occurs when the antibody does not interfere with binding of the complex to the DNA sequence. But, because many enzymes lose activity when bound to an antibody,[38] it is not surprising that the binding of an antibody to a transcription factor causes it to lose affinity for DNA.[39] In this case, which was our experience with the AP-1 complexes in HBE1 cells, there is no third band. Nonetheless, what always occurs with antibody binding to a transcription factor, regardless of whether there is a shifted band or not, is a decrease in the intensity of the original complex (Fig. 2, band B, lane 2 versus lanes 3 and 4). We have termed this antibody-mediated decrease in DNA binding activity "immunodepletion." This decrease in intensity is evidence that the transcription factor of interest was present in the original complex that was bound to the labeled oligonucleotide. Furthermore, these decreases can be measured and used to draw inferences about the relative amounts of specific transcription factors in a given binding complex.

Immunodepletion Method. To determine the constituent proteins of the AP-1 and EpRE binding complexes, samples were prepared as described previously, except that 4 μg of sample was used, and subsequent to all the previous incubation steps, the samples were incubated for an additional 2 h at room temperature with 4 μg of antibody to the various AP-1 and EpRE complex proteins (all antibodies were obtained from Santa Cruz, except for the anti-sera to the small Maf proteins, which was a generous gift from Volker Blank, McGill University, Montréal, Canada). Duplicate samples were run that were incubated with a nonspecific antibody (to p65) to demonstrate that the decrease in binding was not a result of nonspecific interactions.

Curcumin Remodels the Electrophilic Response Element and AP-1 Complexes in HBE1 Cells. Immunodepletion experiments were performed in our curcumin model to further dissect the curcumin-mediated changes leading to increased *Gcl* mRNA. The AP-1 transcription factor complex usually consists of members of the Jun and Fos families. We found that

[38] C. Esposito, F. Paparo, I. Caputo, M. Rossi, M. Maglio, D. Sblattero, T. Not, R. Porta, S. Auricchio, R. Marzari, and R. Troncone, *Gut* **51,** 177 (2002).

[39] V. V. Tyulmenkov and C. M. Klinge, *Steroids* **65,** 505 (2000).

the basal AP-1 binding complexes contained a mixture of c-Fos, Fra1, JunB, and JunD, but no c-Jun was detectable, whereas the AP-1 binding complex in cells exposed to curcumin contained a significant amount of c-Jun. In the basal EpRE binding complex, a significant amount of c-Jun, Nrf2, and MafG/MafK was detected. Exposure to curcumin led to a distinct appearance of JunD in the EpRE binding complex. These data illustrate some of the useful applications of the immunodepletion assay. First, they can delineate differences between basal and activated complexes. This can be critical, because the composition of the binding complex can dictate whether it is activating or repressing transcription. It also illustrates the complexities of transcriptional activation, because some transcription factors (for example, c-Jun) may form part of the binding complex to more than one consensus sequence. This introduces the possibility of competition for transcription factors when multiple pathways are "turned on" at once.

Conclusions

Phase II gene products help to protect cells from damaging environments. Curcumin functions as an adaptogen and is responsible for many beneficial health effects. We have shown that curcumin can induce phase II gene expression in HBE1 cells. How this occurs was demonstrated at the level of the transcription factor complexes that bind to the enhancer elements common to phase II genes. Curcumin increased the overall DNA binding activity of AP-1 and EpRE complexes. In addition to this increase, the composition of the complexes was changed in response to curcumin. These changes, although not discussed in detail here, correlated with total cellular content, nuclear localization, and posttranslational activation.[1] The induction of phase II genes in response to a dietary component such as curcumin provides a basis for the continued study of "functional foods" in preventive medicine. The tools outlined in this chapter should help investigators appreciate the various levels of analysis that can be pursued in delineating the expression of this important class of genes.

Acknowledgments

This work was supported by a Research Development Award from the Office of Postdoctoral Education, University of Alabama at Birmingham to DAD, and by grant ESO5511 from the National Institutes of Health to HJF.

[24] Quinones and Glutathione Metabolism

By NOBUO WATANABE, DALE A. DICKINSON, RUI-MING LIU, and HENRY JAY FORMAN

Introduction

The metabolism of quinones and of glutathione (GSH) is remarkably intertwined. Most quinones can be conjugated to GSH as their major route to elimination. When quinones are redox cycled, H_2O_2 is produced, the elimination of which depends on the use of GSH by glutathione peroxidase (GSHPx). Quinones are excellent inducers of the enzyme glutamate cysteine ligase (GCL), which catalyzes the rate-limiting step in GSH synthesis. Recent studies of redox signaling pathways have indicated that GSH plays a major role in the regulation of such pathways. Xenobiotics, including quinones, are able to activate transcription factors that bind to the xenobiotic response element and electrophile response element (also called the antioxidant response element). GSH not only is involved in the removal of xenobiotics but also modulates the signaling pathways leading to the activation of transcription factors. Thus, induction by xenobiotics and oxidants of enzymes involved in the metabolism of quinones, such as NQO1, are modulated by GSH. In this chapter, we will review the methods used for measurement of quinone redox cycling, GSH content, GCL mRNA content, and γ-glutamyl transpeptidase (GGT), a second GSH-metabolizing enzyme that is also induced by H_2O_2.

Enzymatic Reduction of Quinones

Quinone compounds (Q) can be reduced by cellular reductases. The reduction can be either one- or two-electron (Fig. 1). One-electron reduction yields a semiquinone radical ($Q^{\bullet -}$). Several intracellular flavoenzymes, including NADPH-cytochrome P450 reductase and NADH-cytochrome b_5 reductase, can mediate the one-electron reduction of quinones. Recently, all the three isoforms of nitric oxide synthetase (NOS), whose reductase domains have a high sequence homology with P450 reductase, have also been demonstrated as being capable of one-electron reduction of quinones.[1,2] Quinone reductases have been identified in the plasma membrane,

[1] A. P. Garner, M. J. Paine, I. Rodriguez-Crespo, E. C. Chinje, P. Ortiz De Montellano, I. J. Stratford, D. G. Tew, and C. R. Wolf, *Cancer Res.* **59,** 1929 (1999).
[2] H. Matsuda, S. Kimura, and T. Iyanagi, *Biochim. Biophys. Acta* **1459,** 106 (2000).

0076-6879/04 $35.00

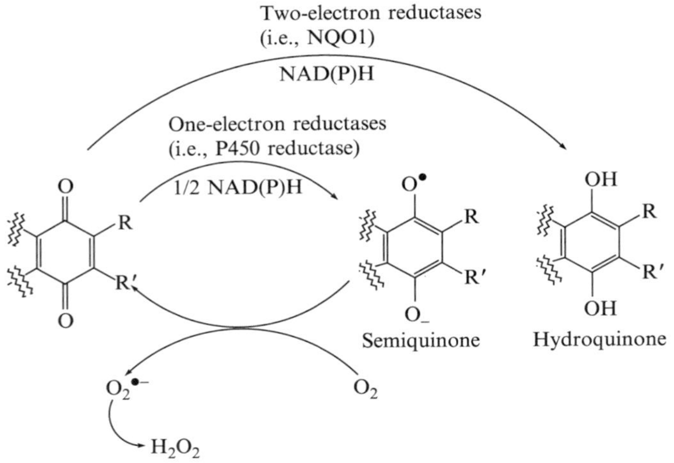

FIG. 1. Quinone reduction by cellular reductases. Quinones can be reduced by one-electron or two-electron oxidoreductases. Semiquinones can readily react with O_2, whereas the hydroquinones generally do not react directly with O_2.

although their precise identities remain unresolved.[3–5] The product of one-electron reduction, $Q^{\bullet -}$, reacts rapidly with O_2 to form superoxide anion, $O_2^{\bullet -}$, regenerating Q. $O_2^{\bullet -}$ can dismute either spontaneously or rapidly through catalysis by superoxide dismutase (SOD) to H_2O_2 and O_2, although this is in competition with the diffusion-limited reaction of $O_2^{\bullet -}$ with nitric oxide (NO) to yield peroxynitrite ($ONOO^-$).

Two-electron reduction of a quinone yields the corresponding hydroquinone, QH_2. Several two-electron quinone reductases, including NQO2 (for review see Long and Jaiswal[6]), have been identified.[7,8] Among them, NQO1 (NAD(P)H quinone oxidoreductase 1; DT-diaphorase), the prime cytosolic quinone reductase, has been well characterized.[9–11] This obligatory

[3] V. E. Kagan, A. Arroyo, V. A. Tyurin, Y. Y. Tyurina, J. M. Villalba, and P. Navas, *FEBS Lett.* **428,** 43 (1998).
[4] S. Yamashoji, *Biochem. Mol. Biol. Int.* **44,** 555 (1998).
[5] C. Kim, F. L. Crane, W. P. Faulk, and D. J. Morre, *J. Biol. Chem.* **277,** 16441 (2002).
[6] D. J. Long, 2nd and A. K. Jaiswal, *Chem-Biol. Inter.* **129,** 99 (2000).
[7] A. K. Jaiswal, *Arch. Biochem. Biophys.* **375,** 62 (2000).
[8] T. Kishi, T. Takahashi, S. Mizobuchi, K. Mori, and T. Okamoto, *Free Radic. Res.* **36,** 413 (2002).
[9] C. Lind, E. Cadenas, P. Hochstein, and L. Ernster, *Methods Enzymol.* **186,** 287 (1990).
[10] E. Cadenas, *Biochem. Pharmacol.* **49,** 127 (1995).
[11] D. Ross, J. K. Kepa, S. L. Winski, H. D. Beall, A. Anwar, and D. Siegel, *Chem-Biol. Inter.* **129,** 77 (2000).

two-electron reducing flavoenzyme can reduce a wide spectrum of quinones and thereby compete with the one-electron reductases for the quinone substrate. Because QH_2 is usually less reactive than $Q^{\bullet -}$, inhibition of NQO1 action by a chemical inhibitor (i.e., dicumarol) is generally expected to elevate reactive oxygen species (ROS) production in quinone-exposed cells by favoring one-electron reductase-mediated redox cycling. Evidence for this phenomenon has been well demonstrated in hepatocytes.[12] It is important to note, however, that this purported antioxidative scheme of NQO1 is applicable when the resultant QH_2 is resistant to autoxidation,[10] or cells may further metabolize the QH_2, for example, by the glucuronide conjugation system, to eliminate quinones. If QH_2 is not removed, however, it can go through autoxidation and therefore generate ROS (see later). Thus, NQO1 can contribute to increased ROS production.[13–15]

Location of Quinone Reduction Is Important

The two naphthoquinones, menadione (2-methyl-1,4-naphthoquione; MQ) and 2,3-dimethoxy-1,4-naphthoquinone (DMNQ), have been extensively used as oxidative stressors in various studies (Fig. 2). MQ can both redox cycle and arylate thiolates in proteins and amines in DNA, whereas DMNQ can only redox cycle.[16] Although ROS production by these quinones is widely believed to occur intracellularly through one-electron reductase-mediated redox cycling, $O_2^{\bullet -}$, $Q^{\bullet -}$, and QH_2 have been detected extracellulary by many investigators in numerous cell types, including hepatyocytes.[12,13,16–21] Because $O_2^{\bullet -}$ and $Q^{\bullet -}$ cannot cross the plasma membrane because of their innate reactivity and charge, their presence in the extracellular environment implies their generation at or outside the plasma membrane. One potential mechanism for extracellular ROS generation is

[12] H. Thor, M. T. Smith, P. Hartzell, G. Bellomo, S. A. Jewell, and S. Orrenius, *J. Biol. Chem.* **257,** 12419 (1982).

[13] L. M. Nutter, E. O. Ngo, G. R. Fisher, and P. L. Gutierrez, *J. Biol. Chem.* **267,** 2474 (1992).

[14] P. R. Gardner, *Arch. Biochem. Biophys.* **333,** 267 (1996).

[15] J. J. Pink, S. M. Planchon, C. Tagliarino, M. E. Varnes, D. Siegel, and D. A. Boothman, *J. Biol. Chem.* **275,** 5416 (2000).

[16] T. W. Gant, D. N. Rao, R. P. Mason, and G. M. Cohen, *Chem-Biol. Inter.* **65,** 157 (1988).

[17] G. Powis, B. A. Svingen, and P. Appel, *Mol. Pharmacol.* **20,** 387 (1981).

[18] G. M. Rosen and B. A. Freeman, *Proc. Natl. Acad. Sci. USA* **81,** 7269 (1984).

[19] R. M. Liu, D. W. Nebert, and H. G. Shertzer, *Toxicol. Appl. Pharmacol.* **122,** 101 (1993).

[20] C. Bayol-Denizot, J. L. Daval, P. Netter, and A. Minn, *Biochim. Biophys. Acta* **1497,** 115 (2000).

[21] J. L. Lee, O. N. Bae, S. M. Chung, M. Y. Lee, and J. H. Chung, *Chem-Biol. Inter.* **137,** 169 (2001).

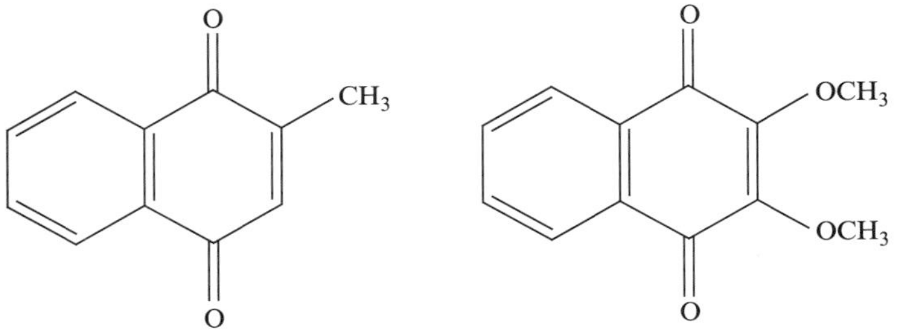

MQ
(2-methyl-1,4-naphthoquinone)

DMNQ
(2,3-dimethoxy-1,4-naphthoquinone)

FIG. 2. Structures of MQ and DMNQ. The methoxy residues of DMNQ block conjugation.

the autoxidation of QH_2 derived from the action of NQO1. Indeed, QH_2 from both MQ and DMNQ is not stable at physiological pH and will undergo nonenzymatic reactions with Q and $O_2^{\bullet -}$ and yield H_2O_2 and thus appears as an overall QH_2 autoxidation:

$$QH_2 + O_2 \rightarrow Q + H_2O_2 \tag{1}$$

although QH_2 does not significantly react directly with O_2[22,23] (Fig. 3).

In some nonhepatocyte cells, inhibition of NQO1 has been shown to result in decreased ROS production by MQ, indicating that NQO1 participates in ROS production.[13,18] Recently, we have also demonstrated that in lung adenocarcinoma A549-S cells, QH_2 generated from NQO1 reduction of MQ and DMNQ, become the predominant (>90%) source of H_2O_2 in the extracellular space.[24] This study provides further evidence that the process of QH_2 autoxidation actually consists of several elementary steps (Fig. 3), an $O_2^{\bullet -}$-driven oxidation:

$$QH_2 + O_2 \rightarrow Q + H_2O_2 \tag{2}$$

and a parallel comproportionation reaction:

$$QH_2 + Q \rightarrow 2Q^{\bullet -} + 2H^+ \tag{3}$$

both of which are followed by autoxidation of the semiquinone, resulting in the generation of superoxide, which then oxidizes QH_2 or dismutes into H_2O_2:

[22] K. Ollinger, G. D. Buffinton, L. Ernster, and E. Cadenas, *Chem-Biol. Inter.* **73,** 53 (1990).
[23] R. Munday, *Free Radic. Res.* **32,** 245 (2000).
[24] N. Watanabe and H. J. Forman, *Arch. Biochem. Biophys.* **411,** 145 (2003).

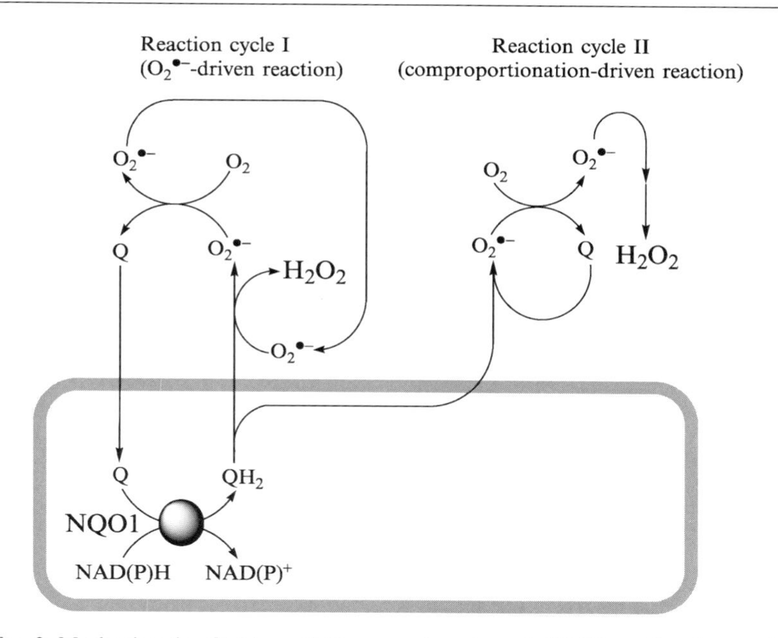

FIG. 3. Mechanism for Q-driven H_2O_2 generation through NQO1-mediated transplasma membrane redox cycling of naphthoquinone in A549-S cells. The illustration is from Watanabe and Forman,[24] with slight modifications under the copyright transfer agreement with Elsevier.

$$Q^{\bullet -} + O_2 \rightarrow Q + O_2^{\bullet -} \quad (4)$$

$$2O_2^{\bullet -} + 2H^+ \rightarrow O_2 + H_2O_2 \quad (5)$$

Clearly, MQ and DMNQ exposure can lead to the generation of ROS both intracellularly by means of one-electron reductases and extracellularly by means of NQO1, as suggested in the pioneering studies.[12,13,18] Given this perspective, it is important to use at least these two quinones in experiments. One also should keep in the mind that the relative contributions of intracellular and extracellular ROS-generating pathways can depend on the cell type.

ROS, especially H_2O_2, are important signaling molecules regulating a wide range of cellular responses. Depending on the amount and the site of generation, ROS may be involved in processes such as apoptosis/necrosis, cellular adaptation to an oxidative environment, the induction of antioxidant genes, and signal transduction.[25] It is therefore of paramount

[25] D. A. Dickinson and H. J. Forman, *Ann. N. Y. Acad. Sci.* **973,** 488 (2002).

importance to know how and where Q-driven ROS production takes place when investigating a Q-induced cellular response and how to quantify this production. In this chapter, we will describe the method used to determine the site of the naphthoquinone-driven H_2O_2 production, the NQO1 assay, and how cells increase GSH biosynthesis when exposed to Q.

Methods to Investigate Quinone Metabolism

Quantitation of Quinone-Derived H_2O_2

Several methods have been used in the measurement of H_2O_2 derived from quinone-exposed cells. Estimation from oxygen consumption is one reliable method but is limited by the number of samples that can be handled in one experiment. Horseradish peroxidase (HRP)–mediated oxidation of a redox probe has also been frequently used. However, in addition to the oxidation of the redox probe, HRP can also oxidize various QH_2 in a H_2O_2-dependent manner:

$$QH_2 + H_2O_2 \rightarrow Q + 2H_2O \quad (6)$$

or an O_2-dependent manner[26]:

$$QH_2 + 1/2O_2 \rightarrow Q + H_2O \quad (7)$$

Therefore, this method needs to be applied cautiously.

We used the FOX assay (ferrous iron oxidation in xylenol orange; for detail see Nourooz-Zadeh[27] in this series) for the measurement of H_2O_2 in the quinone-exposed cells. This method relies on chelation by xylenol orange of ferric iron that results from the Fenton reaction (Fig. 4). The hydroxyl radical ($^{\bullet}OH$), a by-product of this reaction, is removed by butylated hydroxytoluene (4.4 m*M*). There are several advantages with this method over other methods. First, the specificity is relatively high. This is probably due to its acidic, less aqueous reaction condition (25 m*M* H_2SO_4 in 90% methanol) where QH_2 is stable. Second, with the combination of multichannel pipettes and 96-well plates, a high-throughput assay is possible.

Mechanistic Study for Quinone-Derived Reactive Oxygen Species Production. Because quinone-exposed cells not only generate H_2O_2 but also remove it, it is essential to prevent H_2O_2 consumption without affecting other cellular functions. The procedure described here was used in a study for H_2O_2 production in A549-S cells exposed to MQ and

[26] M. H. Klapper and D. P. Hackett, *J. Biol. Chem.* **238,** 3736 (1963).
[27] J. Nourooz-Zadeh, *Methods Enzymol.* **300,** 58 (1999).

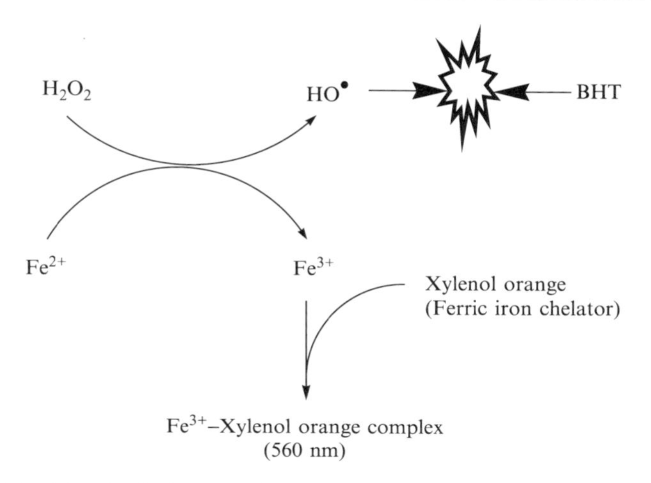

FIG. 4. Principle of FOX assay. Fenton chemistry yields hydroxyl radical, which is scavenged by a large excess of butylated hydroxytoluene (BHT) and Fe^{3+} that reacts with xylenol orange.

DMNQ.[24] To this end, GSH/GPx-dependent H_2O_2 scavenging activity is abrogated by GSH depletion by buthionine-[R,S]-sulfoximine (BSO) pretreatment, and catalase is subsequently inactivated by 3-amino-1,2,4-triazole (ATZ) pretreatment. This treatment, in addition to preserving H_2O_2, can also decrease the loss of NADPH that would occur through the need to reduce glutathione disulfide (GSSG) and thereby preserve NADPH as a source of electrons for quinone reduction. This protocol therefore provides a means to investigate the mechanism of quinone generation of H_2O_2 in the absence of H_2O_2 removal systems that would confound what is already a complex system. In some cell lines, however, such as rat L2 cells, severe GSH depletion by BSO may cause cell death.

Buthionine-[R,S]-sulfoximine and 3-amino-1,2,4-triazole Pretreatments. Cells are seeded initially at 0.7×10^4/well in 96-well plates, incubated for 12–18 h, and then treated with 100 μl of BSO (250 μM) in culture medium (F12K/10% FCS) for 24 to 27 h. Under these conditions, 95% of GSH can be depleted by 24 h. In some cell types, maximum GSH depletion occurs much more rapidly. The medium is removed by manifold aspirator, and cells are subsequently incubated with 25 mM ATZ in the culture medium for 1 to 2 h before the experiment. Endogenous catalase is completely inactivated by 30 min. The presence of ATZ has no effect on the viability of A549-S cells at least up to 24 h, and because catalase activity is not essential for viability in the absence of oxidative stress, this should apply to other cells as well. In A549-S cells, pretreatment with BSO or ATZ had only marginal effects on either GSH content or antioxidant enzyme activities other than the respective targets,[24] and the combination of BSO plus ATZ pretreatment completely abolished the H_2O_2 consumption capacity of the cells (Fig. 5A), which resulted in the apparent H_2O_2 production by either Q (Fig. 6).

Measuring Quinone-Derived H_2O_2 by FOX Assay. BSO plus ATZ pretreated cells are washed with PBS (200 μl × 2 times) before quinone exposure, and the cells are incubated at 37° with 50 μl of the quinone to be tested in KRP (pH 7.4) containing 5 m*M* glucose. After 30 min, 20 μl of the

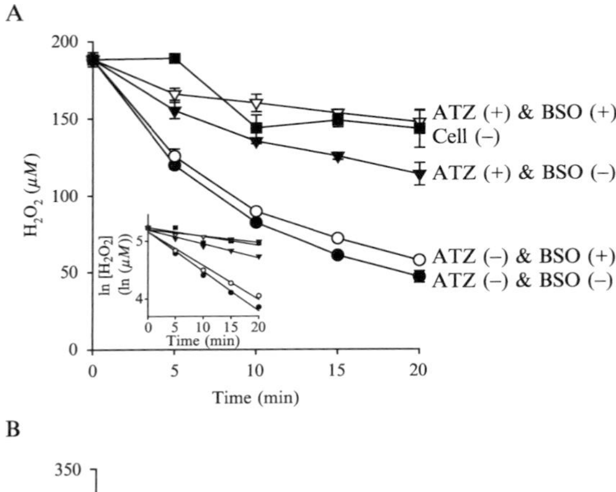

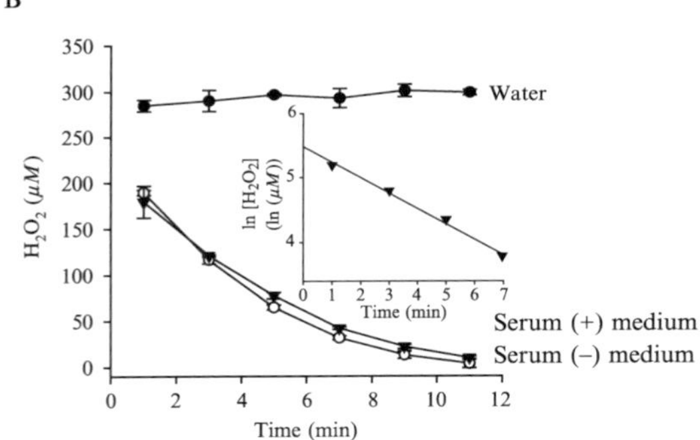

FIG. 5. (A) Time course of disappearance of H_2O_2 in F12K medium in the presence or absence of 10% serum. Inset, First-order nature of the disappearance reaction of H_2O_2 in the serum-containing medium. (B) H_2O_2 consumption ability of A549-S cells following treatment with ATZ, BSO, and the combination of both. *Note*: BSO plus ATZ treatment can completely abolish the cell's capacity to consume H_2O_2 in KRP. The data are from Watanabe and Forman[24] under the copyright transfer agreement with Elsevier.

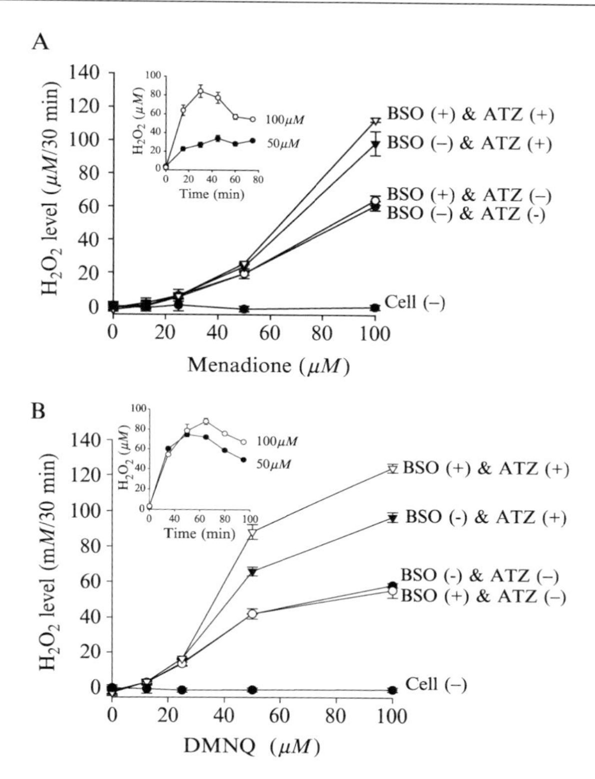

FIG. 6. MQ (A)- or DMNQ (B)-driven H_2O_2 production in A549-S cells following treatment with ATZ, BSO, or the combination of both. Inset, Time course of quinone-induced H_2O_2 production in BSO plus ATZ-treated A549-S cells. The data are from Watanabe and Forman[24] under the copyright transfer agreement with Elsevier.

buffer is removed and mixed in a 96-well plate with 180 μl of FOX reagent (100 μM xylenol orange, 4.4 mM butylated hydroxytoluene, 250 μM $Fe(NH_4)_2(SO_4)_2$, 25 mM H_2SO_4 in 90% [v/v] methanol), sealed with an adhesive sheet to avoid evaporation of methanol, and incubated for 20 min at room temperature. The absorbance at 560 nm is read in a plate reader. Because the iron complex tends to layer in the bottom, the plates should be shaken briefly before measurement. The H_2O_2 concentration is calculated from a standard curve made with authentic H_2O_2. With the A549-S cell line, there were no substantial differences in H_2O_2 levels between the total culture (containing cells) and in the buffer alone. Therefore, H_2O_2 levels in the buffer from Q-exposed cells could be used.

Quinone-induced H_2O_2 accumulated linearly during the first 30 min after the exposure to each quinone, reaching around 100 μM, and then the accumulated H_2O_2 disappeared with time (Fig. 6, insets). A similar transient accumulation and disappearance of ROS production by MQ has also been reported in other cell lines.[4,20] The transient nature of H_2O_2 production is possibly due to oxidative damage to the cells and/or a depletion of O_2 in the buffer, because maximum O_2 concentration in an aqueous solution under air at atmospheric pressure is about 140 μM at 37°. Therefore, the H_2O_2 levels were measured at 30 min after the exposure to each quinone.

Validation of FOX Assay. The FOX assay, in principle, can detect lipid hydroperoxides as well as other reducing agents that may include QH_2. To evaluate the authenticity of the H_2O_2 signal detected, catalase (20~100 μg/ml) should be added to cells during Q exposure. In A549-S cells, catalase completely eliminated measurable H_2O_2 generated by MQ or DMNQ (Fig. 7). When pharmacological inhibitors, such as dicumarol, are added in the culture during quinone exposure, the potential of each additional reagent to affect the H_2O_2 measurement should be estimated. In our evaluation, none of the following agents, MQ (100 μM), DMNQ (100 μM), 2-deoxyglucose (30 mM), dephenyleneiodonium (20 μM), dicumarol (20 μM), or superoxide dismutase (SOD) (100 μg/ml), influenced the detection of authentic H_2O_2.

Interpretation of Data. In A549-S cells, the potent NQO1 inhibitor dicumarol ($Ki = 0.5$ nM, with respect to NADH[28]; inhibited H_2O_2 generation 90% with MQ and 100% with DMNQ [Fig. 4]). In fact, as low as even 0.8 μM dicumarol exhibited the maximum inhibition for both quinones.[24] This suggests that H_2O_2 produced during exposure to either quinone was predominantly derived from autoxidation of QH_2 that was generated by the action of NQO1.

Superoxide dismutase also lowered H_2O_2 generation throughout the MQ concentration range, but the inhibitory effect was partial and decreased with increasing MQ concentration; 70% inhibition at 50 μM and 50% at 100 μM (Fig. 7A). SOD also partially prevented DMNQ-induced H_2O_2 generation. However, the inhibition was more prominent at low quinone concentration; 50% inhibition at <50 μM and less than 20% at 100 μM (Fig. 7B). Because SOD does not get into cells, these data indicate the involvement of two mechanisms in extracellular H_2O_2 generation through QH_2 autoxidation (Fig. 3). At low quinone concentrations, QH_2

[28] S. Chen, K. Wu, D. Zhang, M. Sherman, R. Knox, and C. S. Yang, *Mol. Pharmacol.* **56,** 272 (1999).

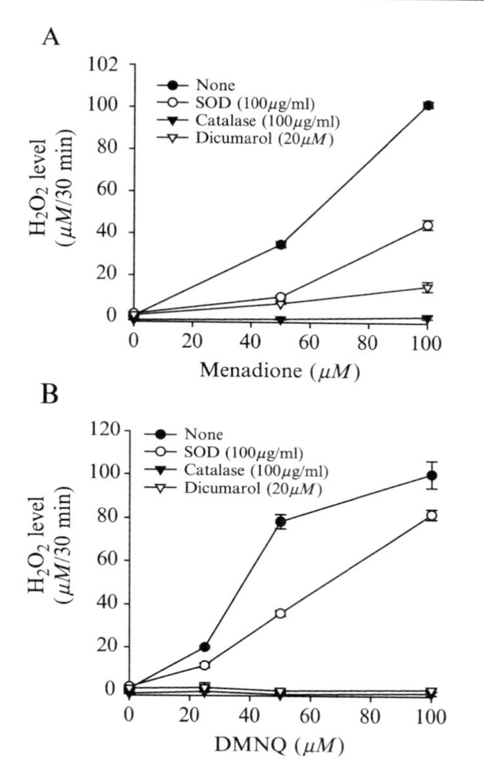

FIG. 7. Effect of SOD, catalase, and dicumarol on MQ- (A) or DMNQ-induced H_2O_2 production in BSO plus ATZ-treated A549-S cells. Each agent was added simultaneously with Q.

generated by NQO1 diffuses out of the plasma membrane where it undergoes $O_2^{\bullet-}$-driven QH_2 autoxidation[22,23,29,30]:

$$QH_2 + O_2^{\bullet-} \rightarrow Q^{\bullet-} + H_2O_2 \tag{8}$$

SOD thereby can inhibit this process. It is important to realize that if H_2O_2 is generated extracellularly through one-electron reductase-mediated redox cycling, that is, by means of $Q^{\bullet-}$ formation, SOD cannot inhibit H_2O_2 generation. The resulting $Q^{\bullet-}$ could then reduce O_2 to $O_2^{\bullet-}$, which, in turn, could oxidize another QH_2 that diffuses out of the cell

[29] T. Ishii and I. Fridovich, *Free Radic. Biol. Med.* **8,** 21 (1990).
[30] R. Munday, *Free Radic. Biol. Med.* **22,** 689 (1997).

with the resulting parental quinone re-entering this cycle. The initial $O_2^{\bullet-}$ may come from either a plasma membrane NAD(P)H oxidase of a non-phagocyte-type, such as NOX1,[31] or from oxidation of the $Q^{\bullet-}$ formed in the comproportionation reaction. Because of an abundant presence of CuZnSOD in the cytosol (in A549-S cells, it is 80 μg/ml assuming a cytosol volume of 10 μl/10^6 cells), this $O_2^{\bullet-}$-driven QH_2 autoxidation is possible only outside the cells. On the other hand, at high Q concentration, the SOD-uninhibited H_2O_2 generation, particularly in DMNQ exposure, became prevalent (Fig. 7). This can be attributable to the predominance of comproportionation reaction[23]:

$$QH_2 + Q \rightarrow 2Q^{\bullet-} + 2H^+ \tag{9}$$

over the $O_2^{\bullet-}$-mediated propagation reaction in the QH_2 autoxidation process (Fig. 3). We have shown evidence using a cytotoxicity assay that the comproportionation-dependent H_2O_2 generation also occurred in the extracellular space to a greater extent than in the cytosol and that this NQO1-mediated H_2O_2 generation is in fact a consequence of a defense against the arylation toxicity of quinones.[24]

NQO1 Activity Assay

Overview of NQO1 Assay. A high-throughput NQO1 assay using a 96-well plate and a plate reader with a kinetic assay function is described here. It is based on the well-established measurement of NQO1 activity using dicoumarol-inhibitable reduction of 2,6-dichlorophenolindophenol (DCIP; Fig. 8) introduced in a previous volume in this series.[9] A plate reader such as the SpectraMax Plus (Molecular Devices, Sunnyvale, CA) can be used to conduct kinetic assays in the ultraviolet (UV) and visible ranges.

Measuring NQO1 Activity. Suitably diluted four-fold final concentration of cell lysates ($4 \times 10 \sim 4 \times 50$ μg/ml in 0.1% Triton X-100/0.1 *M* NaPi, pH 7.4) are mixed with $4 \times$ v/1000 of either dicumarol (10 m*M* stock in 0.1 N NaOH; to 4×10 μM) or vehicle in test tubes designed for a 96-well format tube rack (available from BioRad, Hercules, CA or Fisher, Pittsburgh, PA). With a multichannel pipette, 50 μl of the cell lysates are plated into wells of a 96-well plate and 50 μl of 0.1 *M* NaPi (pH 7.4) are layered. Any air bubbles should be ruptured at this stage by touching the surface with a new pipette tip. The reaction is initiated by adding 100 μl of $2 \times$ reaction mixture (2×100 μM DCIP, 2×200 μM NADPH, 2×5 μM FAD in 0.1 *M* NaPi [pH 7.4]), followed by quick shaking on the plate reader. Change in absorbance at 600 nm is monitored at 22°, and the initial

[31] K. K. Griendling, D. Sorescu, and M. Ushio-Fukai, *Circ. Res.* **86,** 494 (2000).

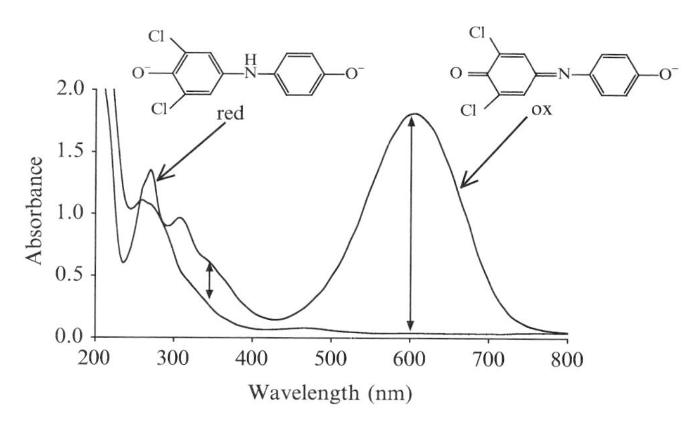

FIG. 8. Change in the absorbance spectrum of 2,6-dichloroindophenol (100 μM) before and after reduction with an equimolar amount of DTT at pH 7.4. The absorbances at 340 nm and 600 nm are noted.

velocity (dA_{600}/min) is obtained from the linear range. The rates are then converted into the rates of DCIP reduction using a system-specific extinction coefficient (12.42 mM^{-1} at 200 μl/well) that was calculated from an external DCIP standard in the same buffer (see later). Dicumarol-inhibitable DCIP reduction is calculated, and 1 unit of the enzyme activity is defined as the amount of enzyme required for the reduction of 1 μM of DCIP per min.

Validation of NQO1 Activity. Because DCIP can be reduced not only by NQO1 but also potentially by other enzymes in the lysate, it is essential to evaluate the authenticity of the NQO1 activity with DCIP by measuring the stoichiometry. If the dicumarol-inhibitable DCIP reduction is mediated solely by NQO1, the dicumarol-inhibitable rate of DCIP reduction and that of NADPH oxidation could be the same. The rate of NADPH oxidation can be monitored in the identical reaction system as previously on the plate reader at 340 nm. However, it should be noted that the spectrum change in DCIP reduction is associated with not only the visible region but also the UV region, which encompass 340 nm (Fig. 8). Therefore, when calculating the rate of NADPH oxidation, the concentration of NADPH corresponding to the change in $DCIP_{340\ nm}$ should be subtracted from the apparent NADPH oxidation. In this system, the ratio of Δ $DCIP_{340\ nm}$ to Δ $DCIP_{600\ nm}$ following complete reduction by DTT is 0.190 (Fig. 8). Therefore, the rate of DCIP reduction (d[DCIP]/dt) and the genuine rate of NADPH oxidation (d[NADPH]/dt) are given as the following equations:

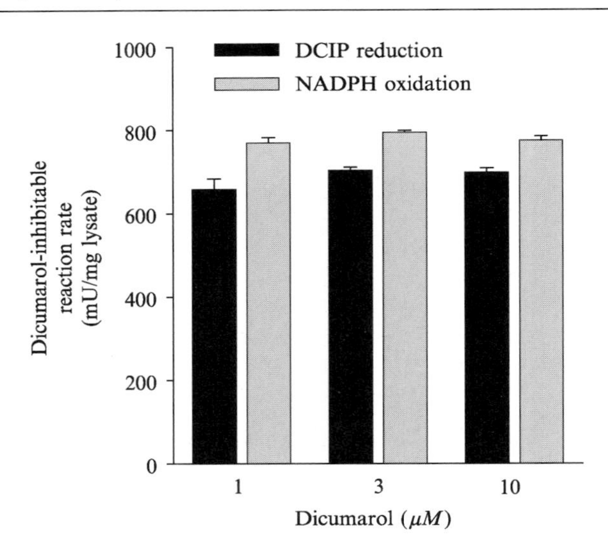

FIG. 9. Relationship between the dicumarol-inhibitable rate of DCIP reduction and that of NADPH oxidation at various dicumarol concentrations in A549 cells. Measurements are made using the correction of the contribution of DCIP at 340 nm as described in the text.

$$d[\mathrm{DCIP}]/dt = 1/\varepsilon_1 \times d\mathrm{A}_{600}/dt$$

$$\varepsilon_1 = 12.4\ \mathrm{m}M^{-1}\text{at } 200\ \mu\mathrm{l/well}$$

$$d[\mathrm{NADPH}]/dt = 1/\varepsilon_2 \times d\mathrm{A}_{340}/dt - 1/\varepsilon_2 \times 0.190 \times d\mathrm{A}_{600}/dt$$

$$\varepsilon_2 = 3.30\ \mathrm{m}M^{-1}\text{ at } 200\ \mu\mathrm{l/well}$$

where, ε_1 and ε_2 are the extinction coefficients of DCIP at 600 nm and NADPH at 340 nm, respectively, in the same buffer at 200 μl/well. Figure 9 shows the almost 1:1 stoichiometry in DCIP reduction and NADPH oxidation in the presence of various concentrations of dicumarol in A549-S cell lysate, leading to the conclusion that the dicumarol-inhibitable DCIP reduction in this cell line is almost completely catalyzed by NQO1.

Special Considerations for the FOX and NQO1 Assays

Basal Cell Culture Media. For the measurement of H_2O_2, as well as other experiments with quinones, it is important to remember that the cell culture medium used in the experiment can affect the outcome. Several commercially available culture media contain millimolar concentrations of pyruvate, a potent H_2O_2 scavenger.[32] For example, F12K medium (Invitrogen, Carlsbad, CA), which is routinely used to maintain A549-S cells, contains 2 m*M* pyruvate, and thereby the half-life of H_2O_2 in the medium at 37° is ~3 min, regardless of serum content (Fig. 5B). Pyruvate in this basal media prevents significant accumulation of quinone-derived H_2O_2 in the medium. Therefore, if the measurement of H_2O_2 is to be carried out in cell culture medium, the proper selection of medium is essential. A good alternative for many experiments is to use glucose (5 m*M*)-containing Krebs-Ringer phosphate buffer (KRP; pH 7.4) for H_2O_2 measurement.

Effect of Serum Proteins on Dicumarol Action. Dicumarol avidly binds to serum albumin.[33] Therefore, the effect of dicumarol can be strongly antagonized by the presence of BSA or serum in the culture medium.[34] In our experiments, although the simultaneous addition of $<1\ \mu M$ dicumarol could completely inhibit DMNQ-driven H_2O_2 production in KRP, the cytoprotective effect of even 20 μM dicumarol (without preincubation) was dramatically diminished against H_2O_2-dependent cell killing in serum-containing medium.[24]

Effect of Confluency of Cells. NQO1 activity, as well as some other antioxidant enzyme activities, can be affected by the confluency of the cells.[35] In the case of HeLa cells, NQO1 activity is reported to increase more than 10-fold as they reach confluency.[34] Moreover, the presence of pyruvate, which removes H_2O_2 in the culture medium, suppresses the basal expression of NQO1.[34] Thus, culture condition should be designed carefully in experiments with quinones.

Effect of Cell Origin. NQO1 activity in cancer cells is often elevated compared with normal counterparts.[11] It should be noted that in human liver cells, in contrast to nonhepatocyte human cells, and even rat liver cells, NQO1 expression is quite low.[36] NQO1 has at least one genetic polymorphism. The most prevalent NQO1 mutation is a Pro-187-Ser substitution designated as NQO1*2.[11] The NQO1*2/*2 phenotype, which accounts for 4% and 22% of Caucasian and Chinese populations, respectively, virtually lacks NQO1 activity because of rapid proteasome-mediated degradation.[11] Therefore, in any model using cells of human origin for an experiment with quinones, NQO1 activity should be determined in advance.

[32] S. Desagher, J. Glowinski, and J. Premont, *J. Neurosci.* **17,** 9060 (1997).
[33] S. Garten and W. D. Wosilait, *Biochem. Pharmacol.* **20,** 1661 (1971).
[34] R. I. Bello, C. Gomez-Diaz, F. Navarro, F. J. Alcain, and J. M. Villalba, *J. Biol. Chem.* **276,** 44379 (2001).
[35] N. Watanabe, D. A. Dickinson, D. M. Krzywanski, K. E. Iles, H. Zhang, C. J. Venglarik, and H. J. Forman, *Am. J. Physiol. Lung Cell. Mol. Physiol.* **283,** L726 (2002).
[36] D. Siegel and D. Ross, *Free Radic. Biol. Med.* **29,** 246 (2000).

Quinone Metabolism and Glutathione Content

Glutathione Content as a Physiological Marker of Quinone Metabolism

The physiological response of the cell following quinone exposure can be complex. We have already discussed the induction of NQO1, a phase II quinone-metabolizing enzyme. There are many phase II enzymes, which function to protect the cell from stress. The metabolism of quinones such as MQ and DMNQ often results in the production of ROS, which can have subsequent deleterious effects. Several other phase II enzymes, including glutathione reductase (GR) and the glutathione *S*-transferases (GST), help maintain the intracellular environment in a reduced state, while simultaneously marking compounds for export, respectively. GSH is used as a substrate in these reactions. GSH, a tripeptide (γ-L-glutamyl-L-cysteine-glycine), is the most abundant nonprotein thiol in the cell. The content of GSH responds to changes in the environment and is exquisitely sensitive to changes in the redox state of the cell (see next section). As such, GSH content and oxidation state are often used as physiological markers of quinone-induced stress.

Glutathione Content in Response to Quinones

Under physiological conditions, GSSG content in the total cellular GSH pool is maintained at less than 1% because of the action of GR and an ample supply of the reductant NADPH, generated primarily from the pentose phosphate pathway. Quinone exposure, however, has a profound effect on the redox status of the cells in terms of the GSSG/GSH ratio. This is because quinone redox cycling establishes a futile circuit with respect to NADPH; both the quinone-driven H_2O_2 generation and the elimination of H_2O_2 by means of the GSH/GPx system consume NADPH, which leads to a rise in the GSSG/GSH ratio. This redox shift favors the formation of protein mixed-disulfides by reaction of their cysteine residues with GSSG. Cells, however, manage to circumvent mixed-disulfide formation initially by decreasing GSSG accumulation through excretion by means of poorly identified mechanisms, followed by an increase in GSH content through de novo synthesis. For this reason, in general, exposure to a quinone often results in a biphasic response in the total GSH content; an initial drop in GSH content, which is associated with a transient increase in GSSG, followed later by a pronounced increase in GSH through biosynthesis. For MQ, the initial depletion of GSH is also due to the GST-mediated conjugation of GSH to MQ, marking it for export from the cell. Therefore, generally, MQ causes a more severe GSH depletion than does

DMNQ.[16,37] Depending on cell type and MQ concentration, an increase in GSSG may occur transiently.[16,38,39] DMNQ, on the other hand, typically causes a much smaller decrease in GSH content, although a transient increase in GSSG content has been reported resulting from GSH oxidation in response to H_2O_2 production.[40] Nonetheless, both MQ and DMNQ show a similar increase in GSH content at later time points (typically >2 h after initial exposure).

Measuring Glutathione Content

The most useful method for measuring the intracellular content of GSH with simultaneous measurement of GSSG is high-performance liquid chromatography (HPLC) separation. A variation of the method of Fariss and Reed[41] has been used extensively by us and others to accurately and reproducibly measure GSH and derivatives. Although this is not a high-throughput method, it has a major advantage in that, in addition to being able to measure GSH, the content of GSSG can be simultaneously measured. Together these determinations can be used to calculate total glutathione (GSH + GSSG). Total glutathione can also be measured by an enzymatic recycling assay by first eliminating GSH and then removing the excess agent used to accomplish that feat.[42,43] Although this method works well, it is far more cumbersome than HPLC. Because measurements of both GSH and GSSG are relevant to quinone metabolism, however, we will focus on the HPLC method.

The procedure of Fariss and Reed, with modification, is straightforward. First, cell pellets from a six-well cluster dish are acidified with 1 ml of 10% perchloric acid containing 2 m*M* EDTA and 7.5 nmol of γ-glutamyl-glutamic acid (Bachem, King of Prussia, PA) as an internal standard used to calculate recovery. After centrifugation at 13,000*g* for 5 min, 450 μl of the supernatant is transferred to a fresh microfuge tube, 45 μl of 10 m*M* iodoacetic acid in 0.2 m*M* m-cresol purple is added to the supernatant, and the pH is adjusted to 8–9 using sequential additions of sodium bicarbonate. This is incubated in the dark at room temperature for 15 min, then 450 μl of a 1% dinitrofluorobenzene solution is added, and the reaction mixture is incubated at 4° overnight. The following day, 45 μl of 1 *M*

[37] G. Bellomo, H. Thor, and S. Orrenius, *Methods Enzymol.* **186,** 627 (1990).
[38] P. C. Brown, D. M. Dulik, and T. W. Jones, *Arch. Biochem. Biophys.* **285,** 187 (1991).
[39] T. J. Chiou and W. F. Tzeng, *Toxicology* **154,** 75 (2000).
[40] M. Shi, E. Gozal, H. A. Choy, and H. J. Forman, *Free Radic. Biol. Med.* **15,** 57 (1993).
[41] M. Fariss and D. J. Reed, *Methods Enzymol.* **143,** 101 (1987).
[42] T. P. M. Akerboom and H. Sies, *Methods Enzymol.* **77,** 373 (1981).
[43] M. E. Anderson, *Methods Enzymol.* **113,** 548 (1985).

L-lysine is added to react with the excess dinitrofluorobenzene; the salt precipitate is removed by centrifugation, and HPLC analysis is carried out with a system coupled to a spectrophotometric detector. Elution solvents are 80% methanol, 20% water *(solvent A)*, and 0.5 *M* sodium acetate in 64% methanol *(solvent B)*. After a 100-μl injection of the derivatized sample, the mobile phase is maintained at 70% *solvent A* and 30% *solvent B* for 5 min, followed by a 15-min linear gradient to 100% *solvent B* at a flow rate of 1.5 ml/min. The mobile phase is held at 100% *solvent B* for 40 min. Dinitrofluorobenzene derivatives are detected at 365 nm. Standards using commercially available purified GSH and GSSG are run under the same conditions, and either the peak areas of the peak heights are measured. GSH and GSSG contents are calculated on the basis of GSH or GSSG standards.

Transcriptional Regulation of Glutathione Biosynthesis

Overview of Transcriptional and Posttranscriptional Components

The de novo synthesis of GSH is a well-established component of adaptation to stress from oxidants and xenobiotics. The increase in GSH content results from a combination of increased protein content and activity of the synthetic enzymes glutamatecysteine ligase (GCL, E.C. 6.3.2.2), glutathione synthase (GS, E.C. 6.3.2.3), and γ-glutamyl transpeptidase (GGT, E.C. 2.3.2.2). GCL and GS, working sequentially, synthesize GSH from the component amino acids. GCL is generally considered to be rate-limiting. The regulation of GCL (sometimes called γ-glutamylcysteine synthetase, or GCS) has been extensively studied, especially in response to quinones and ROS production, and will be discussed further in the next section. The regulation of GS remains relatively unstudied, although a few research reports along with the clinical literature suggest that GS is important, and changes in its activity and/or content can be deleterious.[44,45] GGT, unlike the synthetic GCL and GS enzymes, breaks down GSH, but in so doing plays the important role of returning cysteine to the cell; cysteine is often sufficiently low in concentration that it is rate-limiting for GCL activity. Similar to GCL, the regulation of GGT has been reported in response to quinone exposure and ROS production and is discussed next.

[44] R. Njalsson, K. Carlsson, B. Olin, B. Carlsson, L. Whitbread, G. Polekhina, M. W. Parker, S. Norgren, B. Mannervik, P. G. Board, and A. Larsson, *Biochem. J.* **349,** 275 (2000).

[45] S. W. Brusilow and A. L. Horwich, *In* "The Metabolic and Molecular Basis of Inherited Disease," pp. 1187–1232. McGraw-Hill, New York (1995).

Reactive Oxygen Species–Induced Expression of Glutamayl Cysteine Ligase

By use of Northern analysis of total RNA, several groups have reported that the mRNAs for the GCL genes, *Gclc* and *Gclm*, and that for GGT, *Ggt,* increase with exposure to ROS or compounds that produce ROS, in addition to other species of physiological and toxicological interest.[46,47] *Gclc* and *Gclm* are separate, single-copy genes, which each produce one of the two subunits of GCL, both of which are required for physiological functioning. *Gclc* encodes the catalytic, or heavy, subunit. GCLC is about 73 kDa in size and possesses the catalytic activity of GCL. GCLM, encoded by *Gclm,* is a 31-kDa modulatory subunit, which regulates the activity of the holoenzyme; it affects this by altering the K_m for glutamate and the K_i for GSH. Both of these genes have several different potential enhancer elements in their promoters, most notably TRE or TRE-like elements, and EpRE elements. These elements are binding sites for transcription factor complexes that are known to respond to cellular stress. The components of these complexes contain redox-sensitive transcription factors, such as the Jun family. Hence, genes driven by TRE or EpRE elements often respond to stress conditions that impinge on the cell, making them good candidates as end points in stress studies. For example, *Gclc* and *Gclm* message content has been used extensively as end points in studying the cellular response to quinones. It should be noted here however, that mRNA content is not a true measure of transcription, because a change in mRNA stability can also contribute to an increased amount of message through decreased degradation. For example, the lipid peroxidation end product 4-hydroxy-2-nonenal (4HNE) has been shown to not only increase the transcription of *Gclc* and *Gclm* but also affects the mRNA turnover rate.[48] Although this complicates analysis of the response to 4HNE, it does not seriously negate the use of mRNA content as a valuable end point. By use of MQ and DMNQ, we have demonstrated that exposure to either of these quinones results in increased content of both *Gcl* mRNA species.[49,50]

[46] A. C. Wild and R. T. Mulcahy, *Free Radic. Res.* **32,** 281 (2000).

[47] R.-M. Liu, M. M. Shi, C. Giulivi, and H. J. Forman, *Am. J. Respir. Cell Mol. Biol.* **274,** L330 (1998).

[48] R.-M. Liu, L. Gao, J. Choi, and H. J. Forman, *Am. J. Physiol.* **275,** L861 (1998).

[49] M. M. Shi, T. Iwamoto, and H. J. Forman, *Am. J. Physiol.* **267,** L414 (1994).

[50] D. A. Dickinson, D. R. Moellering, K. E. Iles, R. P. Patel, A.-L. Levonen, A. Wigley, V. M. Darley-Usmar, and H. J. Forman, *Biol. Chem. Hoppe-Seyler* **384,** 527 (2003).

Reactive Oxygen Species–Induced Expression of γ-Glutamyl Transpeptidase

The regulation of GGT is more complex than GCL. In humans, there are several GGT-encoding genes. Although only one GGT gene is found in rodents, more than five GGT transcripts have been identified so far. Nevertheless, using a rat model, we previously demonstrated that in airway epithelial cells, exposure to ROS-producing quinones led to an increase in GGT mRNA content in addition to GCL mRNAs. As with the *Gcl* genes, both transcription and mRNA stability can contribute to increased *Ggt* mRNA content in response to quinone exposure. An exciting, but complicating, observation though is that the relative contribution of these two mechanisms leading to increased mRNA content depends on the nature of the quinone, that is, whether it works primarily through conjugation (*t*BHQ) or through redox cycling (DMNQ). The conjugation of GSH to *t*BHQ resulted in increased GGT mRNA through decreased degradation, whereas the redox-cycling quinone DMNQ increased GGT mRNA through activating gene transcription.[51,52] Taken together, these data demonstrate that the response to a quinone, in terms of regulation of a gene expression, can occur at transcriptional and/or posttranscriptional levels, and one must carefully examine all these potential changes to build a comprehensive picture of response.

Measuring Reactive Oxygen Species–Induced Expression of Glutamylcysteine Ligase and γ-Glutamyl Transpeptidase

For most of the studies reported to date in the literature, traditional Northern analysis has been used. For the *Gcl* genes, this has been done with cDNA probes; *Ggt* is more troublesome, and the use of cRNA probes has been necessitated in some cases when the mRNA content is low. A real-time quantitative PCR-based method has recently been developed by us to measure *Gcl* mRNA content and is described in detail in an accompanying chapter in this volume.[53] Standard Northern analysis using radiolabeled probes will not be described in detail here, except for specifics necessary for GCL and GGT quantification.

To detect *Gclc* and *Gclm* mRNA, cDNA probes for each are generated using primers based on rat kidney sequences. In our hands, these probes

[51] R. M. Liu, M. M. Shi, C. Giulivi, and H. J. Forman, *Am. J. Physiol. Lung Cell. Mol. Physiol.* **274,** L330 (1998).

[52] R. M. Liu, H. Hu, T. W. Robison, and H. J. Forman, *Am. J. Respir. Cell Mol. Biol.* **14,** 186 (1996).

[53] D. A. Dickinson, K. E. Iles, A. Wigley, and H. J. Forman, *Methods Enzymol.* **378,** 302 (2004).

hybridize with human, rat, mouse, and bovine mRNA. Total RNA from rat tissue (we have used the rat lung epithelial cell line L2) is reverse transcribed using standard methods. Gene-specific sequences are as follows:

$$Gclc5'\text{sense} : 5' - \text{AGACACGGCATCCTCCAGTT} - 3'$$

$$Gclc3'\text{antisense} : 5' - \text{CTGACACGTAGCTCGGTAA} - 3'$$

$$Gclm5'\text{sense} : 5' - \text{AGACCGGGAACCTGCTCAAC} - 3'$$

$$Gclm5'\text{antisense} : 3' - \text{CATCACCCTGATGCCTAAGC} - 3'$$

The cDNAs generated from these can then be radiolabeled with [^{32}P]dCTP using a random-primed labeling system. Use approximately 25–50 ng of purified cDNA for each reaction, which will probe one membrane. The probe for *Gclm* should be used before *Gclc,* because the binding affinity for *Gclm* is relatively poor: use caution during the wash steps for *Gclm,* regularly checking the membrane for radioactivity.

To detect message for GGT in rat lung epithelial cells, a cRNA probe, not a cDNA probe, is required. Using an in vitro transcription system, radiolabel the GGT DNA fragment resulting from *Not*I digestion of pBluescript CGT plasmid (available on request) with [α-^{32}P]CTP according to the manufacturer's instructions to generate the cRNA probe. This radiolabeled probe should be purified using standard phenolchloroform extraction before use in hybridization. The probe, once labeled, should be stored for no more than 3 days at $-80°$.

Conclusion

The metabolic fates of quinones and GSH are intricately linked. In this chapter, we have described that linkage and how to measure several of the relevant parameters of those interactions. We used recent studies from our laboratory as examples. The reader should be aware that differences in the expression of one- and two-electron reductases, as well as differences in the chemical and biological properties among quinones, can produce markedly different results. Nonetheless, the assays described herein and the precautions in method and interpretation are generally applicable, regardless of these differences. The measurement of H_2O_2 is one of the critical issues, and it should be noted that the FOX assay described here is not the only assay that may be used; measurement of superoxide in biological systems, which was not addressed here, remains a difficult problem in quantitative analysis. Finally, the biological manifestation of quinone exposure often involves changes in GSH content with accompanying gene expression for increased GSH synthesis. Methods describing standard measurement of these common end points are described.

Acknowledgments

The authors acknowledge Elsevier for granting us to reproduce figures from Watanabe and Forman, 2003.[24] The valuable contributions of many former members of our laboratory are noted. This work was supported by grant ES05511 from the National Institutes of Health.

[25] Doxorubicin Cardiotoxicity and the Control of Iron Metabolism: Quinone-Dependent and Independent Mechanisms

By GIORGIO MINOTTI, STEFANIA RECALCATI, PIERANTONIO MENNA, EMANUELA SALVATORELLI, GIANFRANCA CORNA, and GAETANO CAIRO

Introduction

Doxorubicin (DOX) is the leading compound of a broad family of extractive or pharmaceutically engineered anticancer anthracyclines. Since its introduction in several investigational and approved chemotherapy regimens, DOX has contributed to improved life expectancy of countless patients affected by carcinomas, sarcomas, or lymphomas.[1] The activity of DOX against tumors is nonetheless accompanied by acute and chronic toxicities to the heart. The acute toxicity develops immediately after initiation of DOX treatment and consists of arrhythmias or hypotensive episodes, which do not represent an indication to discontinue the anthracycline regimen. In contrast, the chronic cardiotoxicity develops late in the course of therapy or any time after completion of cumulative dose regimens and causes congestive heart failure, often refractory to standard medications.[2] Iron plays a pivotal role in mediating cardiotoxicity induced by DOX. This became evident in the mid 1970s, when a series of nonpolar derivatives of ethylenediamine tetraacetic acid (EDTA), including a *bis*-ketopiperazine code-named ICRF-187 and subsequently given the nonproprietary name of dexrazoxane, was shown to afford protection in most common animal models of anthracycline-induced cardiotoxicity (reviewed by Minotti

[1] R. B. Weiss, *Semin. Oncol.* **19,** 670 (1992).
[2] P. K. Singal and N. Iliskovic, *N. Engl. J. Med.* **339,** 900 (1998).

0076-6879/04 $35.00

et al.[3]). Structure-activity studies then revealed that dexrazoxane acted by entering cardiomyocytes, and by undergoing stepwise hydrolysis of the two piperazine rings, yielding a diacid diamide able to chelate low molecular weight iron.[1–3] A role for iron has been maintained ever since and has been confirmed by the fact that dexrazoxane decreased the incidence or severity of anthracycline-induced cardiotoxicity in several clinical studies, often allowing patients to receive significantly greater cumulative doses of DOX without increasing its potential to cause congestive heart failure.[3]

The protective efficacy of dexrazoxane against DOX-induced cardiotoxicity offers a good example of how iron can be both good and bad to the cell. As a matter of fact, iron is essential for such important functions like oxygen transport and use and DNA synthesis or detoxification processes. To execute such multiple tasks the cell can count on a low molecular weight iron pool (often referred to as the labile iron pool, LIP), which constantly equilibrates with the sites of deposition or metabolic use. At the same time, a pathologic expansion of the LIP exposes the cell to potentially damaging reactions that iron can cause directly or by converting normal byproducts of cell respiration, like superoxide anion ($O_2^{\bullet -}$) and hydrogen peroxide (H_2O_2), into highly damaging hydroxyl radicals or equally aggressive ferryl ions or oxygen-bridged Fe(II)/Fe(III) complexes.[4,5] The fact that dexrazoxane protects against DOX-induced cardiotoxicity anticipates that anthracyclines expand LIP inside cardiomyocytes. How precisely this occurs is nonetheless a matter of controversy. Some believe that the cellular LIP expands in response to a mobilization of iron induced by DOX in extracellular fluids[6]; here we review experimental evidence suggesting that DOX and other anthracyclines can act at an intracellular level, perhaps by altering the function of iron regulatory proteins (IRP) that serve to maintain LIP within physiologic concentrations.

The Iron Regulatory Proteins

Functions and Structure-Activity Considerations

The size of cellular LIP, and hence the maintenance of iron-mediated homeostatic processes, depends on the concerted regulation of the expression of transferrin receptor (TfR) and ferritin (Ft), which mediate iron

[3] G. Minotti, G. Cairo, and E. Monti, *FASEB J.* **13,** 199 (1999).

[4] G. Cairo and A. Pietrangelo, *Biochem. J.* **352,** 241 (2000).

[5] G. Cairo, S. Recalcati, A. Pietrangelo, and G. Minotti, *Free Radic. Biol. Med.* **32,** 1237 (2002).

[6] L. Gille, M. Kleiter, M. Willmann, and H. Nohl, *Biochem. Pharmacol.* **64,** 1737 (2002).

uptake by internalizing iron-laden transferrin or sequester iron in a catalytically inactive form, respectively. Mechanisms for the regulation of TfR or Ft subunits at a transcriptional level, mediated by changes in iron availability, have been described in detail,[5,7,8] but a major mechanism is operated at a posttranscriptional level by the so-called IRP-1 and IRP-2. Both these proteins are located in the cytoplasm and bind with high affinity to conserved iron responsive elements (IRE) in the untranslated regions of Ft and TfR mRNAs.[9] When cellular iron is low, IRP-1 and IRP-2 bind to IRE and stabilize the mRNA for TfR while also decreasing translation of mRNA for Ft, favoring iron uptake over sequestration and expanding LIP. The opposite occurs when cellular iron is low. Under the latter conditions, the affinity of IRP-1 or IRP-2 for IRE diminishes, causing decreased stability of TfR mRNA and enhanced translation of Ft mRNA. The mechanisms through which IRP-1 can sense iron levels and modulate its affinity for IRE involve reversible assembly–disassembly of a [4Fe-4S] cluster. In iron-replete cells, the cluster is assembled, and IRP-1 displays aconitase activity, similar to that of the mitochondrial enzyme, which reversibly isomerizes citrate to isocitrate through a *cis*-aconitate intermediate in the tricarboxylic acid cycle; in iron-depleted cells, the cluster is lacking, and IRP-1 functions as an RNA-binding protein.[9] The last few years have witnessed improved knowledge of the mechanisms leading to partial or complete disassembly of the Fe-S cluster of cytoplasmic aconitase. The fourth iron atom of the cluster (needed for catalytic activity and referred to as Fe_a) is solvent-exposed and easily removable from the cluster, resulting in formation of a [3Fe-4S] protein devoid of aconitase activity; however, this protein would also lack the ability to recognize IRE.[10] A complete switch of aconitase to IRP-1 only occurs when nitrogen-centered radicals or certain anthracycline metabolites extend the process of Fe removal to the remaining iron centers of the cluster (referred to as Fe_{b1-3}).[4,5,11,12] IRP-2 is highly homologous to IRP-1 (79% at the amino acid level) but lacks aconitase activity because of its inability to assemble a [4Fe-4S] cluster and is characterized by the presence of a 73-amino acid insertion in the N-terminus.[9,13] The mechanisms through which IRP-2 can sense iron levels and modulate its affinity

[7] L. Bianchi, L. Tacchini, and G. Cairo, *Nucleic Acids Res.* **27,** 4223 (1999).

[8] C. N. Lok and P. Ponka, *J. Biol. Chem.* **274,** 24147 (1999).

[9] M. W. Hentze and L. Kuhn, *Proc. Natl. Acad. Sci. USA* **93,** 8175 (1996).

[10] X. Brazzolotto, J. Gaillard, K. Pantopoulos, M. W. Hentze, and J. M. Moulis, *J. Biol. Chem.* **274,** 21625 (1999).

[11] G. Minotti, R. Ronchi, E. Salvatorelli, P. Menna, and G. Cairo, *Cancer Res.* **61,** 8422 (2001).

[12] G. Cairo, R. Ronchi, S. Recalcati, A. Campanella, and G. Minotti, *Biochemistry* **41,** 7435 (2002).

[13] B. R. Henderson, *BioEssays* **18,** 739 (1996).

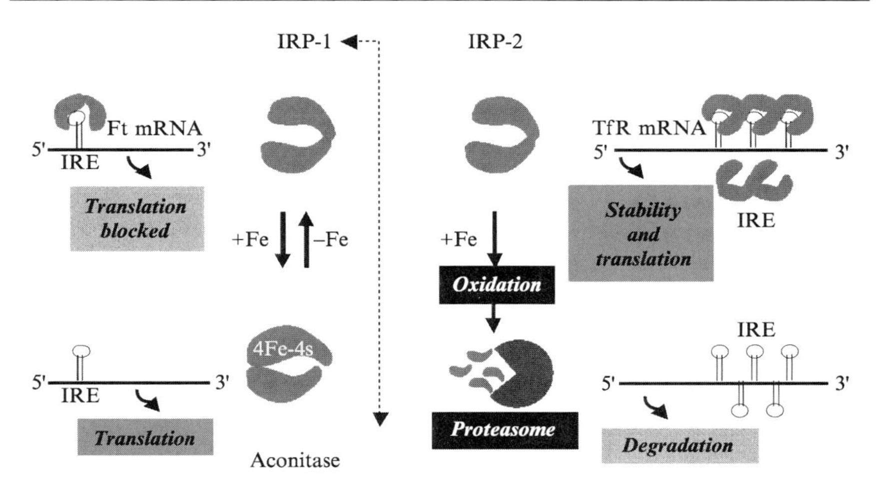

FIG. 1. Simplified representation of iron-mediated switch between aconitase and IRP-1 mechanisms underlying Fe-dependent cluster assembly/disassesmbly, leading to aconitase↔ IRP-1 switches, or Fe-induced IRP-2 degradation. (See color insert.)

for IRE are quite different from those described for aconitase/IRP-1. When the LIP is too large, some iron binds to the 73-amino acid insertion, promoting site-specific formation of free radicals, which oxidatively modify IRP-2 and prime it to proteasome-mediated degradation.[4,14] When the LIP is too small, IRP-2 is protected from degradation and/or undergoes de novo synthesis.[15] The mechanisms underlying IRP-2 degradation or aconitase↔ IRP-1 switches are sketched in Fig. 1.

Methodological Aspects

IRP-1 activity is measured by RNA electromobility shift assays (RE-MSA) that detect formation of RNA–protein complexes. After the initial report of two IRE-binding complexes in rat liver extracts,[16] the method has remained essentially unchanged. Synthetic RNAs encompassing IRE motifs are transcribed in vitro and incubated with cytoplasmic extracts, followed by resolution of the RNA–protein complexes on native polyacrylamide gels. Treatment of samples with ribonuclease T1, which degrades unprotected RNA probe, and heparin, which displaces nonspecifically

[14] K. Iwai, S. K. Drake, N. B. Wehr, A. M. Weissman, T. LaVaute, N. Minato, R. Klausner, R. L. Levine, and T. A. Rouault, *Proc. Natl. Acad. Sci. USA* **95,** 4924 (1998).

[15] B. R. Henderson and L. C. Kuhn, *J. Biol. Chem.* **270,** 20509 (1995).

[16] E. A. Leibold and H. N. Munro, *Proc. Natl. Acad. Sci. USA* **85,** 2171 (1988).

bound proteins, ensures specific detection of IRE–IRP complexes.[17] Ultraviolet (UV) cross-linking of these complexes and analysis by denaturing gel electrophoresis was initially used to characterize the proteins involved. Analysis of cytoplasmic extracts by REMSA detects IRP-1, which is active in IRE binding, (i.e., the so-called spontaneous activity); treatment of samples with reducing agents (e.g., 2-mercaptoethanol [2-ME]) before REMSA usually increases IRP-1 activity, allowing for detection of the so-called total activity. The latter procedure has become a common practice, because the ratio between spontaneous and total IRP is thought to offer information on the relative abundance of aconitase vs. IRP-1, and hence on the size of LIP, and can also serve to ensure that equal amounts of proteins have been properly loaded on the gel. 2-Mercaptoethanol increases IRP-1 activity by reducing disulfide bonds between cys-437, an important determinant of IRP-IRE interactions, and cys-503 or cys-506.[18] What has remained uncertain is whether 2-ME actually serves to optimize detection of total IRP-1. 2-Mercaptoethanol does not abolish aconitase activity,[18] a finding suggesting that it does not extrude Fe-S clusters. Moreover, there are conditions under which the imunodetectable levels of aconitase/IRP-1 remain unchanged throughout an experiment, but the RNA-binding activity decreases regardless of the presence of 2-ME in the assay,[11] a finding suggesting that 2-ME does not always measure total IRP-1 in a sample. In an attempt to measure total IRP-1 more accurately, some adopt the technique of disassembling Fe-S clusters with ferricyanide.[19,20] Because ferricyanide acts as an oxidant, this procedure may be confounded by the possible oxidation of –SH residues in the apoprotein; hence, ferricyanide treatment must be followed by addition of 2-ME to ensure complete reduction and activation of the apoprotein.[19,20] As accurate as this procedure can be, the ratio of ferricyanide to clusters remains highly critical, and irreversible oxidation of the apoprotein should always be taken into account. We suggest that a safer and less speculative procedure would be that of using 2-ME to evaluate redox changes that have occurred at cys-437 during sample manipulation and storage or after exposure to known oxidants during the course of an experiment. Thus, activation of IRP-1 by 2-ME may offer evidence for the presence of some disulfide bridges in the apoprotein, whereas lack of activation may offer evidence that oxidation(s) has proceeded beyond disulfide bridges and has formed sulfur species,

[17] G. Cairo and A. Pietrangelo, *J. Biol. Chem.* **269,** 6405 (1994).

[18] H. Hirling, B. R. Henderson, and L. C. Kuhn, *EMBO J.* **13,** 453 (1994).

[19] D. J. Haile, T. A. Rouault, J. Harford, M. C. Kennedy, G. A. Blondin, H. Beinert, and R. D. Klausner, *Proc. Natl. Acad. Sci. USA* **89,** 11735 (1992).

[20] J. C. Kwok and D. R. Richardson, *Mol. Pharmacol.* **62,** 888 (2002).

which cannot be reverted to –SH by 2-ME.[11,12,21,22] An alternative approach to quantifying the ratio between the clusterless apoprotein and the cluster-containing holoprotein might be that of measuring both IRP-1 activity and aconitase activity (whether through the consumption of *cis*-aconitate at 240 nm [$\varepsilon = 3.6$ mM^{-1} cm^{-1}] or the assay that couples citrate to isocitrate isomerization with isocitrate dehydrogenase–dependent reduction of NAD^+ to $NADH+H^+$ [$\varepsilon_{340} = 6.22$ mM^{-1} cm^{-1}]).[19,22] This approach is both biochemically and biologically appropriate, because cells almost certainly contain a mixed pool of aconitase and IRP-1 within which variations in enzymatic activity would be mirrored by opposite changes in IRE-binding activity. Potential drawbacks are nonetheless introduced by the aforementioned existence of a (3Fe-4S) species that lacks both enzymatic and RNA-binding activity[10]; it is also worth noting that the levels of silenced aconitase–IRP-1 may vary from one sample to another, depending on the phosphorylation status of the protein and consequent modifications in Fe-S cluster stability and turnover.[23] An accurate determination of apo/holo protein ratios may be further limited by the possible presence of extensively oxidized IRP-1, which does not respond to activation by 2-ME and leads to an erroneous underestimation of total RNA-binding activity.

IRP-2 is similar to IRP-1 in exposing –SH residues that mediate IRE recognition and RNA binding, but its expression is regulated quite differently; for example, inflammatory agents activate IRP-1 but repress IRP-2 in a mouse macrophage cell line.[24] This may be highly relevant for the expression of IRP-controlled mRNAs, because different subsets of IRE-containing mRNAs are bound preferentially by IRP-1 or IRP-2.[13] These factors need to be taken into account when analyzing the pattern of activities of the two IRPs. From a technical viewpoint, one should also consider that IRE complexes with murine IRP-1 and IRP-2 can easily be distinguished with common REMSA because of their different mobilities, but human IRP-1 and IRP-2 comigrate and must be separated through the use of specific antibodies. To avoid cross-reactivity, anti-IRP-2 antibodies are raised against the 73-amino acid sequence specific to this IRP. Anti-IRP-1 or IRP-2 antibodies can thus be used in gel "supershift" assays to ascertain whether the RNA-binding activity of a cytoplasmic extract is contributed by IRP-1 or IRP-2. The fact that IRP-2 is regulated at the level of protein degradation also rationalizes assays on the basis of immunoblotting

[21] Y. M. Torchinsky, *in* "Sulfur in Proteins," p. 48. Pergamon Press, New York, 1981.

[22] G. Minotti, S. Recalcati, A. Mordente, G. Liberi, A. M. Calafiore, P. Preziosi, and G. Cairo, *FASEB J.* **12,** 541 (1998).

[23] N. M. Brown, M. C. Kennedy, W. E. Antholine, R. S. Eisenstein, and W. E. Walden, *J. Biol. Chem.* **277,** 7246 (2002).

[24] S. Recalcati, D. Taramelli, D. Conte, and G. Cairo, *Blood* **91,** 1059 (1998).

rather than on gel supershift; this techniques actually offers a more accurate quantitation of IRP-2 levels. At the same time, the presence of the 73-amino acid degradation may favor artifactual degradation of IRP-2 during sample preparation; this problem can be obviated by including protease inhibitors in homogenization or sonication buffers.

The Anthracyclines

Metabolism and Structure-Activity Considerations

Doxorubicin-induced chronic cardiotoxicity is always seen with dilated cardiomyopathy and subcellular lesions varying from myofibrillar loss to cytoplasmic vacuolization (mostly caused by dilation of the sarcoplasmic reticulum). This morphologic pattern is observed in both humans and laboratory animals and reflects the action of metabolites formed after uptake and biotransformation of DOX inside cardiomyocytes. As shown in Fig. 2, DOX is composed of aglyconic and sugar moieties. The aglycone (called doxorubicinone) consists of a tetracyclic ring with adjacent quinone-hydroquinone moieties in rings C-B and of a short side chain containing a carbonyl group at C-13 and a primary alcohol at C-14; the sugar (called daunosamine) is attached by a glycosidic bond to C-7 in the tetracyclic ring and consists of a 3-amino-2,3,6-trideoxy-L-fucosyl moiety.

The structure of DOX is complex enough to allow for formation of several metabolites, making it difficult to anticipate whether the acute and

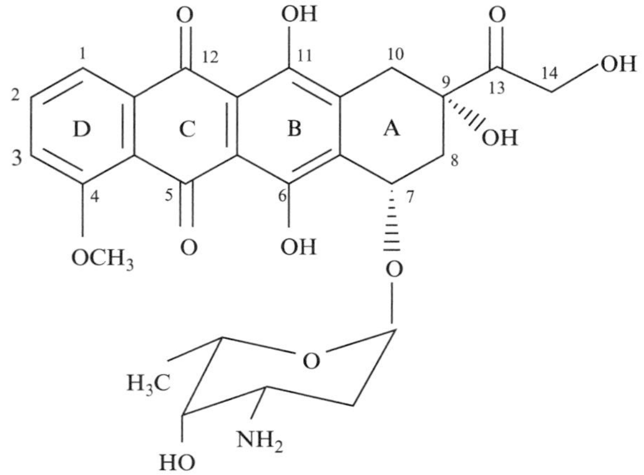

FIG. 2. Structure of DOX.

chronic toxicities were mediated by different levels of a given metabolite or by formation of structurally distinct metabolites. Moreover, studies in laboratory animals are heavily influenced by species-related and strain-related differences in drug metabolism; therefore, some DOX metabolites might be detected in the heart of a given animal species or strain but not in the heart of others, making it difficult to predict which particular metabolite(s) would form and cause cardiotoxicity in humans. On the basis of several recent studies in our laboratories, performed in myocardial samples obtained from patients undergoing open-chest cardiac surgery,[22,25–28] we suggested that the acute and chronic cardiotoxicities induced by DOX were mediated by structurally distinct metabolites or by-products of the anthracycline molecule. In particular, we proposed that the acute phase is mediated by one-electron reduction of the quinone moiety in ring C (hereafter referred to as Q), resulting in formation of a semiquinone ($Q^{\bullet -}$) that readily regenerates its parent Q by reducing oxygen to $O_2^{\bullet -}$ and H_2O_2 (Fig. 3). This reaction is mediated by a number of oxidoreductases endowed with one-electron Q-reductase activity (microsomal NAD[P]H-cytochrome P450 or b_5 reductases, nuclear cytochrome b_5 reductase(s), mitochondrial NADH dehydrogenase, cytoplasmic NADH-dependent xanthine dehydrogenase, nitric oxide synthase).[29,30] Under appropriate conditions $Q^{\bullet -}$ can regenerate Q by reductively cleaving the glycosidic bond between daunosamine and ring A of doxorubicinone, yielding 7-deoxydoxorubicinone (Fig. 3). It is uncertain whether such reaction occurs by intramolecular or intermolecular electron transfer; in either case, reductive deglycosidation would be accompanied by increased generation of reactive oxygen species (ROS), because 7-deoxydoxorubicinone is much more lipophilic than DOX and can orientate Q in the closest proximity to reductases embedded in membrane bylayers.

A broad panel of anthracycline metabolites can nonetheless be formed independent of electron addition to Q. In particular, we have shown that human cardiac cytosol is equipped with reductase-type glycosidases that

[25] S. Licata, A. Saponiero, A. Mordente, and G. Minotti, *Chem. Res. Toxicol.* **13,** 414 (2000).
[26] G. Minotti, A. F. Cavaliere, A. Mordente, M. Rossi, R. Schiavello, R. Zamparelli, and G. F. Possati, *J. Clin. Invest.* **95,** 1595 (1995).
[27] G. Minotti, S. Licata, A. Saponiero, P. Menna, A. M. Calafiore, G. Di Giammarco, G. Liberi, F. Animati, A. Cipollone, S. Manzini, and C. A. Maggi, *Chem. Res. Toxicol.* **13,** 414 (2000).
[28] G. Minotti, A. Saponiero, S. Licata, P. Menna, A. M. Calafiore, G. Teodori, and L. Gianni, *Clin. Cancer Res.* **7,** 1511 (2001).
[29] G. Powis, *Free Radic. Biol. Med.* **6,** 63 (1989).
[30] J. Vasquez-Vivar, P. Martasek, N. Hogg, B. S. Masters, K. A. Pritchard, Jr., and B. Kalyanaraman, *Biochemistry* **36,** 11293 (1997).

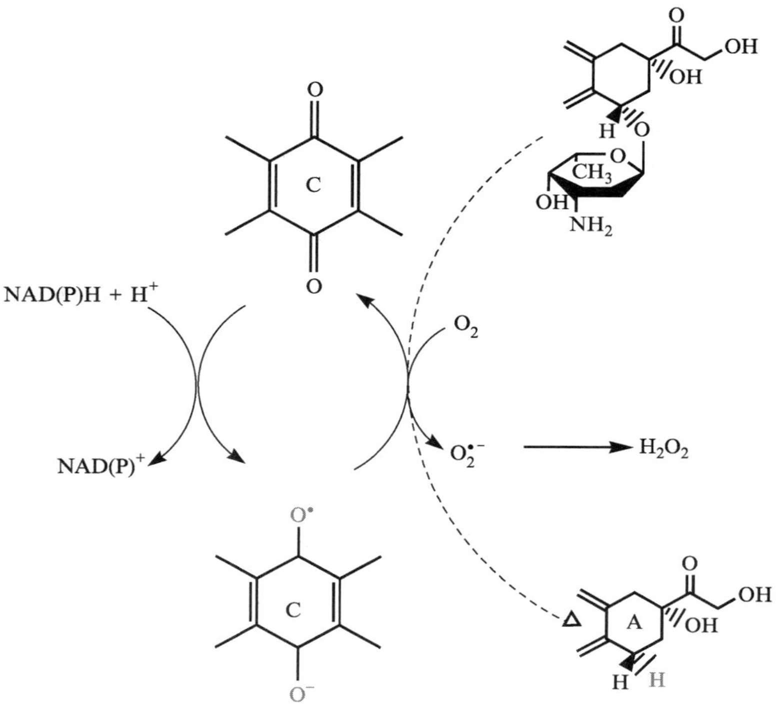

FIG. 3. Quinone-dependent DOX metabolism. Simplified representation of NAD(P)H-dependent $Q^{\bullet-}$ formation, leading to $O_2^{\bullet-}$ and H_2O_2 or reductive deglycosidation. (See color insert.)

directly convert DOX to 7-deoxydoxorubicinone (Fig. 4I). The pathophysiological relevance of this particular mechanism of reductive deglycosidation probably is superior to that mediated by $Q^{\bullet-}$, because it occurs at concentrations of DOX ~10–20 times lower than those required to see deglycosidation during redox cycling of Q; moreover, Q-independent formation of 7-deoxydoxorubicinone by cytoplasmic glycosidases is seen at physiologic pO_2, whereas that mediated by Q only occurs when pO_2 is sufficiently low to favor competition of the glycosidic bond for $Q^{\bullet-}$.[25,29,31] Human cardiac cytosol is also equipped with hydrolase-type glycosidases that convert DOX to doxorubicinone (Fig. 4II), possibly similar to 7-deoxydoxorubicinone in exhibiting improved partitioning in cellular organelles. Whether doxorubicinone contributes to the formation of ROS by

[31] L. Gille and H. Nohl, *Free Radic. Biol. Med.* **23,** 775 (1997).

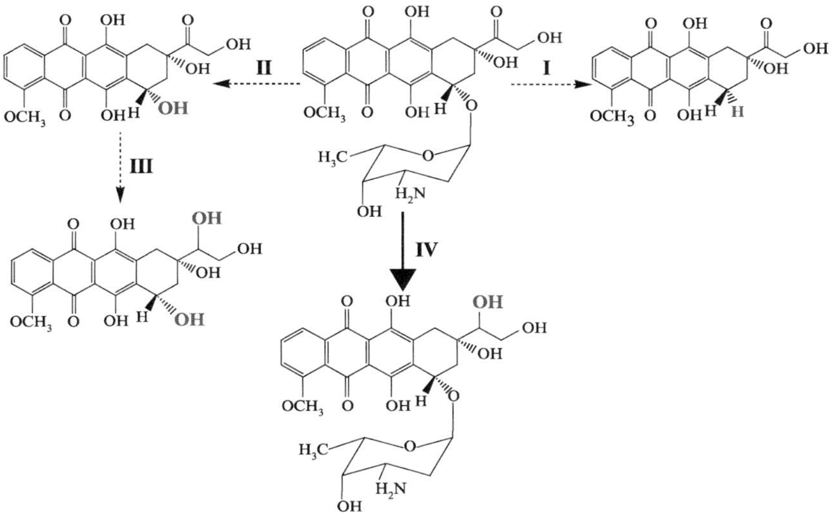

FIG. 4. Quinone-independent DOX metabolism. Pathways of Q-independent reductive deglycosidation, leading to 7-deoxydoxorubicinone (I); hydrolase-type deglycosidation followed by carbonyl reduction, leading to doxorubicinone (II) and doxorubicinolone (III); direct carbonyl reduction, leading to DOXol (IV). Based on Licata *et al.*[25] (See color insert.)

membrane-bound Q reductases is nonetheless uncertain. In fact, the conversion of DOX to doxorubicinone is tightly coupled with two-electron reduction of the side chain C-13 carbonyl group by cytoplasmic aldo/keto- or carbonyl reductases,[25] yielding a secondary alcohol metabolite (doxorubicinolone), which is relatively more polar than its parent carbonyl aglycone (Fig. 4III). Moreover, the presence of a secondary alcohol moiety in the side chain has long been known to increase the K_m of Q for one-electron reductases, limiting formation of $Q^{\bullet -}$ and its redox coupling with oxygen to give ROS.[32] On balance, the acute phase of cardiotoxicity seems to be characterized by a prevailing formation of Q-derived ROS, generated by DOX or 7-deoxydoxorubicinone. This metabolic pattern changes during the chronic phase of cardiotoxicity, characterized by preferred reduction of DOX by aldo/keto- or carbonyl reductases and consequent formation of its side chain secondary alcohol metabolite doxorubicinol (DOXol, Fig. 4IV). Preferred conversion of DOX to DOXol, coupled with increased

[32] P. G. Gervasi, M. R. Agrillo, L. Citti, R. Danesi, and M. Del Tacca, *Anticancer Res.* **6,** 1231 (1986).

polarity and reduced affinity of the anthracycline molecule for one-electron Q reductases, should diminish formation of ROS during the chronic phase of cardiotoxicity.

Methodological Aspects

There are established methods for detecting $Q^{\bullet-}$ and ROS in reconstituted chemical systems or anthracycline-perfused isolated heart models using electron spin resonance/spin trapping; similarly, there are several high-performance liquid chromatography (HPLC), thin-layer chromatography (TLC), or mass spectrometry protocols for analyzing metabolites formed through Q-independent reactions. These techniques will not be described here; the reader is addressed to refs. 25, 27, 33–37, and related bibliographic links for an extensive coverage. Here it is worth noting that in biologic systems, DOXol exists as S or R diastereoisomers, depending on which particular reductase isoform is present in a given tissue, whereas DOXol prepared by chemical procedures (e.g., $NaBH_4$ reduction of the C-13 carbonyl group) always consists of a mixture of (S) and (R) diastereoisomers. Chirality does not introduce technical problems when measuring DOXol, because biologic (S) or (R) isomers cochromatograph with (S)+(R) standards in most common TLC or HPLC conditions. Problems could arise if (S) or (R) isomers exhibited different activities toward biological targets possibly involved in cardiotoxicity. In this regard, we found that synthetic DOXol and enzymatically produced DOXol were equally effective at delocalizing Fe(II) in human cardiac cytosol,[22,26] a reaction that we will soon describe to contribute to inducing chronic cardiotoxicity. Whether (S) or (R) or (S)+(R) DOXol shared reactivity in other putative mechanisms of toxicity (e.g., inhibition of Ca^{2+}-Mg^{2+} ATPase in sarcoplasmic reticulum, f_0–f_1 proton pump in mitochondria, and Na^+-K^+ ATPase and Na^+-Ca^{2+} exchanger in sarcolemma) was not established.[38]

[33] B. Kalyanaraman, K. M. Morehouse, and R. P. Mason, *Arch. Biochem. Biophys.* **286,** 164 (1991).

[34] S. Rajagopalan, P. M. Politi, B. K. Sinha, and C. E. Myers, *Cancer Res.* **48,** 4766 (1988).

[35] S. Fogli S, R. Danesi, F. Innocenti, A. Di Paolo, G. Bocci, C. Barbara, and M. Del Tacca, *Ther. Drug Monit.* **21,** 367 (1999).

[36] W. J. van der Vijgh, P. A. Maessen, and H. M. Pinedo, *Cancer Chemother. Pharmacol.* **26,** 9 (1990).

[37] J. Bloom, P. Lehman, M. Israel, O. Rosario, and W. A. Korfmacher, *J. Anal. Toxicol.* **16,** 223 (1992).

[38] R. J. Boucek, Jr., R. D. Olson, D. E. Brenner, M. E. Ogumbumni, M. Inui, and S. Fleischer, *J. Biol. Chem.* **262,** 15851 (1987).

Mechanisms of Iron-Mediated Anthracycline-Dependent Cardiotoxicity

Role of Quinone-Derived Reactive Oxygen Species and Iron in Doxorubicin-Induced Cardiomyocyte Apoptosis: General Considerations

Until a few years ago apoptosis was disregarded as a mechanism of DOX-induced cardiotoxicity, but this picture has now changed. Apoptosis seems to mediate the acute cardiac effects of DOX both in cultured cells[39] and in vivo[40,41]; however, apoptosis seems to be less pronounced in the setting of chronic cardiomyopathy, during which it may involve cell types other than cardiomyocytes (e.g., cells of the conducting system).[42] Studies in cultured bovine aortic endothelial cells and adult rat cardiomyocytes have shown that DOX induces NF-kB activation and nuclear translocation, followed by development of apoptosis.[39] All such processes require iron uptake in the cell, and its reaction with DOX-derived H_2O_2, as major indices of apoptosis (e.g., caspase 3 activation and cytochrome *c* release) were inhibited by interventions against iron (anti-TfR antibodies, dexrazoxane), ROS (glutathione peroxidase overexpression, cell permeable scavengers), or both (overexpression of methallothioneins, which scavenge ROS and chelate iron).[39,43,44] These results attest to the importance of DOX or 7-deoxydoxorubicinone-derived $Q^{\bullet -}$ during the course of acute cardiotoxicity, leading to concomitant amplification of LIP and formation of ROS in excess of the low levels of ROS-detoxifying enzymes that characterize cardiomyocytes compared with other cell types.[3] Molecular insights into the mechanisms of DOX-induced apoptotic signaling will be provided by others in a separate chapter.[45] Here it is important to emphasize that NF-kB activation is antiapoptotic in cancer cells and actually confers resistance to DOX in some human tumors.[46] The opposite consequences

[39] S. Wang, S. Kotamraju, E. Konorev, S. Kalivendi, J. Joseph, and B. Kalyanaraman, *Biochem. J.* **367,** 729 (2002).

[40] O. J. Arola, A. Saraste, K. Pulkki, M. Kallajoki, M. Parvinen, and L. M. Voipio-Pulkki, *Cancer Res.* **60,** 1789 (2000).

[41] N. P. Dowd, M. Scully, S. R. Adderley, A. J. Cunningham, and D. J. Fitzgerald, *J. Clin. Invest.* **108,** 585 (2001).

[42] D. Kumar, L. A. Kirshenbaum, T. Li, I. Danelisen, and P. K. Singal, *Antioxid. Redox Signal.* **3,** 135 (2001).

[43] S. Kotamraju, C. R. Chitambar, S. V. Kalivendi, J. Joseph, and B. Kalyanaraman, *J. Biol. Chem.* **277,** 17179 (2002).

[44] G. W. Wang, J. B. Klein, and Y. J. Kang, *J. Pharmacol. Exp. Ther.* **298,** 461 (2001).

[45] S. Kotamraju, S. V. Kalivendi, E. Konorev, C. R. Chitambar, J. Joseph, and B. Kalyanaraman, *Methods Enzymol.* **378,** 362 (2004).

[46] A. Arlt, J. Vorndamm, M. Breitenbroich, U. R. Folsch, H. Kalthoff, W. E. Schmidt, and H. Schafer, *Oncogene* **20,** 859 (2001).

of NF-kB activation in cardiac vs. cancer cells might be exploited to design pharmacological interventions that prevent apoptosis in the heart but not in tumors, improving the therapeutic index of DOX.

Quinone-Dependent Dysregulation of Iron Homeostasis during Acute Cardiotoxicity: Role of Ferritin and IRP-2 and Analysis of Pathophysiologic Consequences

There are at least two mechanisms through which Q dysregulates iron homeostasis while also producing ROS in the acute phase of anthracycline-induced cardiotoxicity. One mechanism involves the ability of $Q^{\bullet -}$ ($E^{o\prime}$ −0.4 V) to reduce and release iron that is safely stored in the polynuclear ferric oxohydroxide core of ferritin ($E^{o\prime}$ −0.23 V). Ferritin iron is released by $Q^{\bullet -}$ directly, that is, by means of electron-tunneling relayed by redox sites on the surface and transprotein channels of ferritin, or indirectly, that is, by means of the intermediacy of $O_2^{\bullet -}$; the latter is small enough to penetrate the transprotein channels of ferritin and has a reduction potential lower than that of ferritin-bound iron ($E^{o\prime}$ −0.33 V). An extensive coverage of DOX-dependent ferritin iron release can be found in ref. 47. A second mechanism involves changes in IRP activities. IRP-2 readily undergoes degradation in H9c2 rat embryo cardiomyocytes[11] or adult rat cardiomyocytes[20] exposed to DOX. This is seen at concentrations of DOX both lower[11] and higher[20] than the plasma peak obtained in patients after standard infusions of the anthracycline (~7.5–10 μM) and reflects the unique susceptibility of IRP-2 to ROS-induced degradation; in fact, IRP-2 is inactivated by any other anthracycline forming ROS (e.g., daunorubicin) but not by analogues in which $Q^{\bullet -}$ formation is precluded by chemical substitution(s) of Q (e.g., 5-iminodaunorubicin).[11] The effects of acute anthracycline treatment on aconitase and/or IRP-1 are less clear or controversial. The exposure of cell lysates to $O_2^{\bullet -}$ and H_2O_2 does not switch aconitase to IRP-1, not even when ROS are produced at rates (~ 1.8 nmol $O_2^{\bullet -}$/min)[48] that exceed those produced by an intracellular pool of DOX (0.06 nmol $O_2^{\bullet -}$/min) (this value is approximated on the basis of determination of DOX uptake in isolated cardiomyocytes and its redox cycling in microsomes enriched in NADPH-cytochrome P450 reductase).[11] If anything, sustained fluxes of ROS might have the effect of reducing IRP-1 by attacking a pre-existing pool of the clusterless apoprotein, but RNA binding still would be rescued by 2-ME.[48] The apparent lack of reactivity of $O_2^{\bullet -}$ and H_2O_2 toward aconitase raised the possibility that $Q^{\bullet -}$-derived

[47] G. Minotti, *Chem. Res. Toxicol.* **6,** 134 (1993).

[48] G. Cairo, E. Castrusini, G. Minotti, and A. Bernelli-Zazzera, *FASEB J.* **10,** 1326 (1996).

ROS might acutely damage cardiomyocytes without perturbing homeostatic processes linked to [4Fe-4S] clusters. Supportive evidence was obtained by exposing isolated cardiomyocytes to mitoxantrone, an analogue that lacks sugar and carbonyl residues and thus serves as a good model to focus on the formation and reactivity of ROS while also avoiding interpretation problems caused by concomitant processes of deglycosidation and carbonyl reduction. Mitoxantrone did induce cardiomyocyte damage (evidenced by lactate dehydrogenase release) and predictably induced an extensive loss of ROS-sensitive IRP-2, but neither event was accompanied by changes in aconitase or IRP-1 activities.[11] Aconitase inactivation, apparently induced by ROS, did occur in two studies in which adult rat cardiomyocytes were exposed to DOX; however, the concentrations of DOX adopted in these studies were quite high (often in the range of 20–40 μM), and bicarbonate was used in selected experiments to further enhance DOX uptake and $O_2^{\bullet -}$ reactivity.[49,50] In another study, a converse decrease of aconitase and an increase of IRP-1 were seen after exposing bovine aortic endothelial cells to rather low concentrations of DOX (e.g., 0.5 μM).[43] These changes were mediated by ROS and formed the basis to suggest that ROS expanded LIP by switching aconitase to IRP-1.[43] Important factors to be kept in mind when interpreting this latter study pertain to the precise mechanisms of ROS–cluster interactions and the cellular site(s) at which ROS are produced. In regard to the first point, one should consider that, as already mentioned, aconitase switches to IRP-1 only when its Fe-S cluster has undergone complete disassembly. Neither $O_2^{\bullet -}$ nor H_2O_2 would be able to do so, not even if they were formed high enough to approach the cluster; in fact, there is solid evidence that $O_2^{\bullet -}$ and H_2O_2 remove Fe_a but not Fe_{b1-3},[10,51] an effect sufficient to inactivate aconitase but not to activate IRP-1. In regard to the second point, one should consider that nitric oxide synthase (NOS) serves a major or prevailing site of $O^{\bullet -}$ formation in endothelial cells.[30,52] Because endothelial NOS is located primarily in plasma membrane caveolae, a major fraction of H_2O_2 would be produced by DOX within the plasmalemma environment, from which type I/IIa protein phosphatases are known to transduce signals that activate IRP-1 independent of ROS–cluster interactions.[53] It is therefore quite

[49] E. A. Konorev, M. C. Kennedy, and B. Kalyanaraman, *Arch. Biochem. Biophys.* **368,** 421 (1999).

[50] E. A. Konorev, H. Zhang, J. Joseph, M. C. Kennedy, and B. Kalyanaraman, *Am. J. Physiol. Heart Circ. Physiol.* **279,** H2424 (2000).

[51] P. R. Gardner, I. Ranieri, L. B. Epstein, and C. R. White, *J. Biol. Chem.* **270,** 13339 (1995).

[52] S. V. Kalivendi, S. Kotamraju, H. Zhao, J. Joseph, and B. Kalyanaraman, *J. Biol. Chem.* **276,** 47266 (2001).

[53] P. Pantopoulos and M. W. Hentze, *EMBO J.* **14,** 2917 (1995).

possible that DOX-dependent IRP-1 activation in endothelial cells occurred through membrane-transduced signals rather than through direct ROS–cluster interactions. Whether similar mechanisms operate in cardiomyocytes is unknown at present, but detailed investigations of constitutive or inducible NOS expression in embryonic or postbirth cardiomyocytes show that either isoform undergoes remarkable downregulation at birth.[54] We therefore believe that, in adult cardiomyocytes, $Q^{\bullet -}$ would form primarily in intracellular organelles from which ROS eventually diffuse to inactivate IRP-2 while leaving IRP-1 unchanged or reversibly inactivated at worst. On the basis of these premises, the effects of DOX on iron homeostasis during the acute phase of cardiotoxicity should be confined to an expansion of LIP, induced by $Q^{\bullet -}$ and $O_2^{\bullet -}$ through a release of iron from ferritin and partially counteracted by ROS-dependent IRP-2 degradation, aimed at reducing TfR-mediated iron uptake and consequent aggravation of oxidative stress. This framework accommodates the protective efficacy of anti-TfR antibodies, because iron uptake from extracellular fluids would contribute to expanding LIP inside cardiomyocytes. It also accommodates the potential benefit of cell-permeable ROS scavengers or any other antioxidant that prevents or mitigates the consequences of reactions between ROS and excess LIP.

Quinone-Independent Dysregulation of Iron Homeostasis during Chronic Cardiotoxicity: Role of IRP-1 and Analysis of Pathophysiologic Consequences

A dysregulation of iron homeostasis, occurring during the chronic phase of anthracycline-induced cardiotoxicity and involving alterations of IRP, must be appraised in light of the preferred conversion of DOX to DOXol, which characterizes this phase. That DOXol mediates chronic cadiotoxicity is indicated by the following lines of evidence: (1) accelerated development of cardiomyopathy in transgenic mice bearing cardiac-restricted overexpression of carbonyl reductases converting DOX to DOXol[55]; (2) accelerated development of congestive heart failure in breast cancer patients who receive DOX plus other chemotherapeutics that facilitate its reduction to DOXol[28]; (3) milder and/or less progressive cardiomyopathy in rats exposed to investigational anthracyclines forming less alcohol metabolite than DOX[56,57]; (4) preferred accumulation of DOXol over

[54] W. Bloch, B. K. Fleischmann, D. E. Lorke, C. Andressen, B. Hops, J. Hescheler, and K. Addicks, *Cardiovasc. Res.* **43,** 675 (1999).

[55] G. L. Forrest, B. Gonzalez, W. Tseng, X. Li, and J. Mann, *Cancer Res.* **60,** 5158 (2000).

[56] G. Minotti, M. Parlani, E. Salvatorelli, P. Menna, A. Cipollone, F. Animati, C. A. Maggi, and S. Manzini, *Br. J. Pharmacol.* **134,** 1271 (2001).

other metabolites in postmortem myocardial samples from patients exposed to cumulative doses of the anthracyline.[58] Doxorubicinol has no direct reactivity toward IRP-2,[11] nor would it be a major source of IRP-2–degrading ROS because of its reduced affinity for Q reductases. The cardiotoxic potential of DOXol is more directly attributable to its ability to convert cytoplasmic aconitase into a "null protein" lacking enzymatic and IRP-1 activities. This has been observed after exposing human cardiac lysates to a bolus of purified DOXol[22] or after delivering pharmacokinetic concentrations of DOX to isolated cardiomyocytes, which converted DOX to DOXol within the experiment time.[11] The basic foundations of the action of DOXol have been confirmed by some[59] but not by others[20]; such discrepancies are explained by experimental factors that need to be carefully considered.

Studies in Cell-Free Systems: Methodological Aspects and Biochemical Determinants. In human cardiac lysates, purified DOXol reacts with the [4Fe-4S] cluster of cytoplasmic aconitase and releases Fe(II), which is chelated and detected spectrally with ferrozine.[22] Ferrous iron release exceeds the amount of Fe_a titrated by inducing limited cluster disassembly with ammonium persulfate, suggesting that the action of DOXol extends beyond Fe_a and involves Fe_{b1-3}.[22] Iron release best occurs in the presence of the aconitase substrate-intermediate *cis*-aconitate and proceeds through oxidation of DOXol back to DOX according to an apparent stoichiometry of one DOXol oxidized per two Fe released (corresponding to a 1:1 stoichiometry if one considers that the secondary alcohol moiety is a two-equivalent reduced carbonyl). The requirement for *cis*-aconitate is pathophysiologically relevant, because [4Fe-4S] clusters would almost certainly contain substrates in vivo[18] and probably reflects sterical and redox events that facilitate DOXol–cluster interactions (*cis*-aconitate ligation of Fe_a, reversible intensification of the Fe[II] character of Fe_a, and 180° rotation of *cis*-aconitate around Fe_a).[22] The fact that DOXol oxidation is stoichiometric with the release of four iron atoms also indicates that the metabolite must reduce Fe_{b1-3}, exhibiting Fe(II)/Fe(III) character, to an iron species with net Fe(II) character.[22] The importance of substrate-mediated sterical and redox events is confirmed by the fact that fluorocitrate, converted by aconitase into by-products that displace *cis*-aconitate from Fe_a, suppresses

[57] G. Sacco, R. Giampietro, E. Salvatorelli, P. Menna, N. Bertani, G. Graiani, F. Animati, C. Goso, C. A. Maggi, S. Manzini, and G. Minotti, *Br. J. Pharmacol.* **139,** 641 (2003).

[58] D. J. Stewart, D. Grewaal, R. M. Green, N. Mikhael, R. Goel, V. A. Montpetit, and M. D. Redmond, *Anticancer Res.* **13,** 1945 (1993).

[59] X. Brazzolotto, M. Andriollo, P. Guiraud, A. Favier, and J.-M. Moulis, *Biochim. Biophys. Acta* **1593,** 209 (2003).

DOXol oxidation and Fe(II) release.[22] The concentrations of DOXol and *cis*-aconitate are critical to see redox coupling of DOXol with Fe-S clusters and to draw conclusions of biological relevance. Anthracycline-induced biological effects are, by definition, pharmacokinetic events; therefore, DOXol should be used in amounts that reproduce those formed by the the aldo/keto- or carbonyl-reductases of the cardiac samples under investigation (e.g., human cardiac lysates were shown to form 3.5–4 nmol DOXol/mg prot./4h; hence, 1 mg of cytosolic protein was reconstituted with no greater than 4 nmol of purified DOXol/mg prot.).[22] On the other hand, the concentration of *cis*-aconitate should be sufficiently high to ensure cluster saturation throughout the experiment (in 1-h studies with 1 mg protein of human cardiac cytosol [containing approximately 2 mU aconitase], DOXol-dependent iron release best occurred when *cis*-aconitate was $\sim$100 μM, an amount that would saturate intact Fe-S clusters for no less than $\sim$50 min[22]). Another factor to be taken in consideration pertains to the presence of electron acceptors that compete with aconitase for DOXol. This is not the case with oxygen and ferritin, because DOXol neither reduces oxygen to $O_2^{\bullet -}$ nor redox couples with ferritin under nonenzymatic conditions.[26] Nevertheless, DOXol, like any other anthracycline molecule with adjacent quinone/hydroquinone moieties, might chelate iron present in a low molecular weight form, giving anthracycline-iron complexes that confound an interpretation of direct reactions between DOXol and clusters. The amount of anthracycline-chelatable iron clearly increases during sample preparation because of the disruptive actions of homogenization and centrifugation procedures; at the same time, these procedures would also cause partial oxidation or disassembly of Fe-S clusters. To avoid all such problems, we found it convenient to adopt a procedure through which [4Fe-4S] clusters were reconstituted under reducing conditions by incubating cytosol with cysteine and ferrous ammonium sulfate at an optimized ratio of 1000 or 50 nmol:mg prot., respectively; unreacted cysteine and ferrous ammonium sulfate, and any other adventitious iron present in the samples, were then removed by gel filtration on Sepharose 6B minicolumns, followed by 65% ammonium sulfate precipitation, and consecutive dialyses against decreasing concentrations of *cis*-aconitate (to protect clusters against oxidative inactivation) and EDTA (to remove low molecular weight iron but not Fe atoms coordinated within Fe-S clusters).[22] One last factor to be considered pertains to the poor stability of anthracyclines, and especially of its reduced metabolites, on prolonged storage in buffers. The purity of DOXol must therefore be checked by HPLC or two-dimensional TLC just before the experiments, and experiments should be done at near-physiological pH to avoid ionization of amino and phenolic residues that influence the redox behavior of anthracyclines.[60] Under such defined conditions, DOXol-dependent cluster disassembly causes a loss of aconitase activity, which is not accompanied by an increase of IRP-1 activity but actually results in a significant decrease of RNA binding, insensitive to 2-ME and accompanied by an inability of the apoprotein to reassemble a cluster and recover enzyme activity. These responses suggest that reduced RNA binding and impeded reassembly of Fe-S structures may be due to oxidation of apoprotein –SH residues beyond formation of disulfide bridges. The apoprotein transiently formed after cluster disassembly would not be damaged by either residual DOXol or newly formed DOX; this can be demonstrated by incubating DOXol or DOX with cytosol that has previously been subjected to cluster disassembly and IRP-1 activation with dithiothreitol at pH 8.9.[22] However, RNA binding decreases if IRP-1 is exposed to stoichiometric amounts of the products of reactions between DOXol and clusters (i.e., ~2 nmol DOX and ~4 nmol low mol wt Fe(II) in experiments conducted with 4 nmol DOXol and 1 mg protein of human cardiac cytosol).[22] Irreversible modifications of the clusterless apoprotein seem to involve formation of DOX-Fe(II) complexes and their subsequent oxidation to Fe(II)-Fe(III) species with prooxidant activity.[22,61] Interestingly, neither superoxide dismutase (SOD) nor catalase (CAT) protected the apoprotein/IRP-1 from the damaging action of DOX-Fe(II), nor could either enzyme prevent conversion of aconitase into a null protein in whole cytosol exposed to DOXol.[22] While indicating that DOXol introduces reducing equivalents into the cluster by direct electron transfer mechanisms rather than by oxygen activation, these results also demonstrate that $O_2^{\bullet -}$ and H_2O_2 do not mediate –SH oxidation by DOX-Fe(II) or do so in a protein environment that is sterically precluded to SOD and CAT. Accordingly, DOX-Fe(II) has long been known to promote SOD-insensitive and CAT-insensitive oxidation of other cell constituents (e.g., polyunsaturated fatty acids).[62] These findings have been largely confirmed by Brazzolotto *et al.*[59] after reconstituting HPLC-grade DOXol with recombinant human cytoplasmic aconitase/IRP-1. In these studies, DOXol was far superior to DOX at abolishing aconitase activity. The process of iron removal seemed to be limited to Fe_a, as evidenced by electron spin resonance detection of a [3Fe-4S] protein, but this is explained by at least two factors: (1) incubations were carried out at pH 7.9, high enough to alter the redox behavior of

[60] P. Menna, E. Salvatorelli, R. Giampietro, G. Liberi, G. Teodori, A. M. Calafiore, and G. Minotti, *Chem. Res. Toxicol.* **15,** 1179 (2002).

[61] G. Minotti, C. Mancuso, A. Frustaci, A. Mordente, S. A. Santini, A. M. Calafiore, G. Liberi, and N. Gentiloni, *J. Clin. Invest.* **98,** 650 (1996).

[62] G. Minotti, *Arch. Biochem. Biophys.* **268,** 398 (1990).

anthracyclines; (2) *cis*-aconitate was used at concentrations probably too low to ensure cluster saturation throughout the experiment time (10 μM *cis*-aconitate vs. 10 μM aconitase/IRP-1). Nevertheless, Brazzolotto *et al.*[59] did obtain convincing evidence for formation of a null protein, because they found that removal of Fe_a (which should neither increase nor decrease RNA binding by mixed aconitase/IRP-1 pools) actually caused 2-ME insensitive loss of IRP-1 activity. The effect was most evident in the presence of *cis*-aconitate, as one would expect if the substrate were needed to promote DOXol-Fe_a interactions and consequent formation of DOX-Fe(II) complexes that then attacked the clusterless IRP-1. Similar supportive evidence was obtained by reconstituting DOXol with lysates, although the latter were derived from tumor cells rather than from cardiomyocytes. In the case of lysates from DOX-resistant small cell lung carcinoma lines, constitutively characterized by a complete switch of aconitase into IRP-1, there was no effect of either DOX or DOXol on spontaneous or total IRP-1. This was consistent with the absence of anthracycline-labile Fe-S clusters in the samples. Different results were obtained in DOX-sensitive lines characterized by the presence of a mixed pool of aconitase/IRP-1. In the latter samples, anthracyclines irreversibly and concentration dependently decreased IRP-1 activity by mechanisms that were consistent with DOXol attack to Fe-S clusters and conversion of aconitase into a null protein. In fact, DOXol was more effective than DOX at inactivating IRP-1; moreover, anthracyclines did not affect the clusterless pool with spontaneous IRP-1 activity but selectively reduced the RNA-binding pool activatable by 2-ME, as one would expect if a cluster-containing protein had been converted by DOXol into a null protein bearing irreversible –SH oxidations.[59] The results obtained by us[22] and Brazzolotto *et al.*[59] have not been replicated by Kwok and Richardson[20] when reconstituting DOXol with lysates from either cardiomyocytes or iron-loaded human melanoma SK-Mel-28 cells. In these latter studies, DOX or DOXol did not inactivate aconitase or decrease IRP-1 activity; this is explained by experimental factors, which were neglected by the authors. First, there is no information on whether and how DOXol stability and purity were checked before experiments. Second, incubations were carried out at pH 8.0, and no attempt was made to prevent reactions of the anthracycline molecule with adventitious iron likely present in the lysates. Finally, there may have been unrecognized problems with the concentration of anthracyclines and *cis*-aconitate and their critical ratios to [4Fe-4S] clusters. An interpretation of aconitate/cluster ratios is precluded by the lack of any precise information on the final volume adopted in preparing the incubation mixtures, making it difficult to conclude whether the substrate was too low to promote anthracycline/cluster collisions or perhaps too high (a condition having the opposite effect of protecting clusters from chemical modifications[18,19]). Regarding DOX and DOXol, there is no information on whether they were used at concentrations reflecting anthracycline uptake in intact cells or reproducing the aldo/keto- or carbonyl-reductase activities of the cell lines used. In some experiments, anthracyclines were used at 20 μM, a concentration of no pharmacokinetic value. Kwok and Richardson[20] did obtain evidence for a reversible inactivation of IRP-1 by DOX-Fe complexes in lysates of rat cardiomyocytes or SK-Mel-28 cells, similar to what was observed by us in human cardiac cytosol[22]; however, Kwok and Richardson chose to work with the ferric form of DOX-Fe complexes and adopted concentrations of DOX and Fe(III) several times higher than those possibly formed after DOXol oxidation and Fe(II) release from the cluster. Again, these are quite artifical conditions that should be called into question before drawing conclusions of biological relevance. Similar reservations were raised long ago by Gelvan and Samuni[63] when considering the possible role of DOX-Fe(III) complexes in vivo. In conclusion, there is solid evidence for a preferred reaction of aconitase/IRP-1 with DOXol rather than with DOX in cell-free systems, but this reaction has not always been studied under proper conditions.

Studies in Cellular Systems. The effects of DOX on IRP activities have been studied in H9c2 rat embryo cardiomyocytes exposed to pharmacokinetic concentrations of DOX (1–10 μM). Under such defined conditions, DOX is metabolized to form both $Q^{\bullet -}$ and DOXol and still converts aconitase into a null protein. The reaction pattern and the relative contribution of DOXol vs. $Q^{\bullet -}$-derived ROS have been dissected by comparing DOX to analogues, such as the already mentioned 5-iminodaunorubicin and mitoxantrone, which only form secondary alcohol metabolite or ROS, respectively. When using these carefully designed settings, one can see that DOX concentration-dependently converts aconitase to IRP-1 through the formation of an intracellular pool of DOXol that induces cluster disassembly without oxidizing cys-437. Thus, aconitase activity declines but can be rescued with cysteine and ferrous ammonium sulfate; at the same time, IRP-1 activity increases regardless of whether it is assayed in absence or presence of 2-ME. $Q^{\bullet -}$-derived ROS intervene at a later stage of the reaction and seem to synergize DOX-Fe(II) complexes, formed after DOXol dependent cluster iron release, in converting the transiently formed IRP-1 into a null protein. Thus, aconitase activity can no longer be rescued by cysteine/ferrous ammonium sulfate, and IRP-1 activity decreases both in

[63] D. Gelvan and A. Samuni, *Cancer Res.* **48,** 5645 (1988).

the absence and the presence of 2-ME. The mechanisms of DOX-induced conversion of aconitase/IRP-1 into a null protein are shown in Fig. 5.

IRP-1 inactivation has been observed also by Brazzolotto *et al.*[59] after 24 h exposure of DOX-sensitive tumor cells to as little as 0.5 μM anthracycline. Such inactivation was reversible (i.e., IRP-1 was rescued by 2-ME); it is quite possible that slightly higher but still pharmacologically relevant concentrations of DOX could have caused irreversible inactivation as that observed in H9c2 cardiomyocytes. IRP-1 activity did not change at all in resistant cells, not even when DOX was increased to 12.5 μM to overcome resistance-related phenomena like anthracycline efflux or vesiculation.[59] Again, this latter finding is well explained by the fact that resistant cells had little or no Fe-S clusters that could be attacked by DOXol. It is worth noting that DOX-induced conversion of aconitase/IRP-1 into a null protein occurred without appreciable changes in the levels of immunodetectable protein in H9c2 cardiomyocytes.[11] A decrease of immunoreactive protein only occurred when Kwok and Richardson[20] exposed DOX-sensitive SK-Mel-28 cells to as high as 20 μM DOX. In the same study, a decrease of IRP-1 activity was observed after a 6-h exposure of adult rat

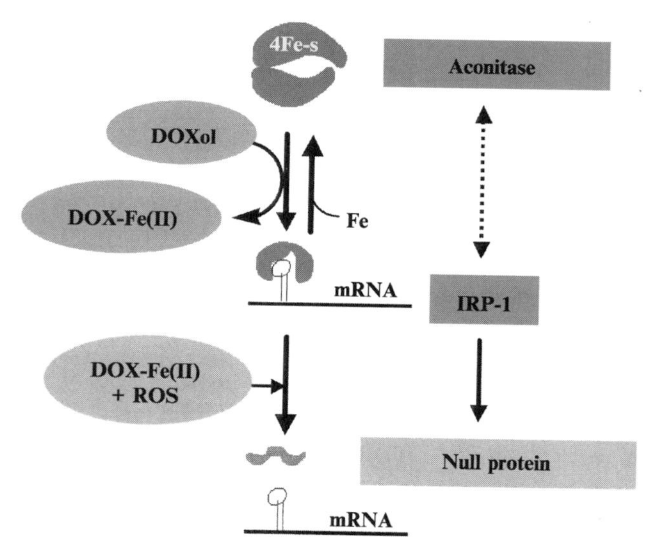

FIG. 5. Proposed mechanisms for DOX-dependent conversion of aconitase/IRP-1 into a null protein during chronic cardiotoxicity. Simplified representation of (1) DOXol oxidation with Fe-S clusters, leading to reversible conversion of aconitase to IRP-1 and concomitant formation of DOX-Fe(II); (2) synergistic interaction of DOX-Fe(II) with ROS, leading to IRP-1 inactivation and formation of a null protein. (See color insert.)

cardiomyocytes to 1–25 μM DOX, but IRP-1 returned to initial levels after 24-h incubations.[20] An interpretation of these findings is limited by the fact that the authors neither monitored DOX metabolism nor assessed whether fluctuations in IRP-1 activity were accompanied by opposite changes in aconitase activity during the course of their experiments. Such a lack of information precludes adequate evaluation of the biochemical events induced by DOX in these studies.

Pathophysiological Consequences and Final Considerations. The conversion of aconitase/IRP-1 into a null protein may have a prevailing impact in inducing chronic cardiomyopathy. In fact, the presence of a null protein makes cells unable to sense iron levels and to coordinate TfR and ferritin levels in a manner suited to a correct equilibration of iron between the distinct sites of deposition or metabolic use[64]; it may also result in a misplacement of iron ions at cellular sites that govern the contraction–relaxation cycle of the heart but loose their function after sterical occupation by this metal (e.g., the calcium release channel, ryanodine receptor 2 of the sarcoplasmic reticulum).[3] These processes may contribute to chronic impairment of cardiac metabolism and contractility, regardless of whether cardiomyocytes eventually undergo apoptosis. Formation and cluster reactivity of DOXol during the chronic phase of cardiotoxicity, perturbing cellular iron trafficking, rationalize the protective efficacy of chelation therapy. Moreover, the reduced activity of DOXol in ROS formation, its ability to attack Fe-S clusters by ROS-independent mechanisms, and the interaction of ROS with DOX-Fe(II) in a sterically hindered protein environment all rationalize the reduced efficacy of antioxidants against the development of chronic cardiomyopathy in patients or laboratory animals that best reproduce human pharmacokinetics (e.g., dogs).[3] The unique reactivity of DOXol toward aconitase/IRP-1 and the precise mechanisms of null protein formation clearly call for validation in clinical settings or proper animal models of in vivo cardiotoxicity. Likewise, preliminary observations that DOX-sensitive or DOX-resistant tumor cells exhibit different ratios of aconitase to IRP-1 might anticipate a possibility to target iron trafficking in sensitive cells while not affecting iron homeostasis in the heart. Research efforts are now going in this direction.

Acknowledgments

Work in the authors' laboratories was supported by MURST FIRB 2002 ("Post-genomics of iron diseases"), MURST Cofin 2001 and 2002, Associazione Italiana Ricerca sul Cancro, and MURST Center of Excellence on Aging at the University of Chieti.

[64] N. H. Gehring, M. W. Hentze, and K. Pantopoulos, *J. Biol. Chem.* **274,** 6219 (1999).

[26] Oxidant-Induced Iron Signaling in Doxorubicin-Mediated Apoptosis

By SRIGIRIDHAR KOTAMRAJU, SHASI V. KALIVENDI, EUGENE KONOREV, CHRISTOPHER R. CHITAMBAR, JOY JOSEPH, and B. KALYANARAMAN

Introduction

Doxorubicin (DOX) or adriamycin, a quinone-containing anthracycline antibiotic, is a widely used chemotherapeutic drug for treating leukemia, breast cancer, Hodgkin's disease, or sarcomas.[1,2] The clinical efficacy of this drug is greatly restricted because of the development of a severe form of cardiomyopathy or congestive heart failure in cancer patients treated with this drug.[3,4] Children treated for leukemia with DOX developed heart problems years after cessation of DOX chemotherapy.[1–3] Despite cardiotoxicity, DOX is still included in most chemotherapeutic regimens because of its broad-spectrum antitumor activity and therapeutic efficacy. Current evidence indicates that DOX-mediated cardiotoxicity may be caused by increased generation of reactive oxygen species (ROS) such as superoxide ($O_2^{\bullet -}$), hydrogen peroxide (H_2O_2), or hydroxyl radicals ($^{\bullet}OH$) through redox-activation of DOX.[5–15]

[1] M. R. Bristow, M. E. Billingham, J. W. Mason, and J. R. Daniels, *Cancer Treat. Rep.* **62,** 873 (1978).

[2] R. A. Minow, R. S. Benjamin, and J. A. Gottlieb, *Cancer Chemother. Rep.* **6,** 198 (1975).

[3] P. K. Singal, *N. Eng. J. Med.* **339,** 900 (1998).

[4] P. K. Singal, C. M. R. Deally, and L. E. Weinberg, *J. Mol. Cell. Cardiol.* **19,** 817 (1987).

[5] D. B. Sawyer, R. Fukazava, M. A. Arstall, and R. A. Kelly, *Circ. Res.* **84,** 257 (1999).

[6] S. Kotamraju, E. A. Konorev, J. Joseph, and B. Kalyanaraman, *J. Biol. Chem.* **275,** 33585 (2000).

[7] L. Wang, W. Ma, R. Marcovich, J.-W. Chen, and P. H. Wang, *Circ. Res.* **83,** 516 (1998).

[8] J. Narula, N. Haider, R. Virmani, T. G. Disalvo, F. D. Kolodgie, R. J. Hijjar, U. Schmidt, M. J. Semigran, G. W. Dec, and B. A. Khaw, *N. Engl. J. Med.* **335,** 1182 (1996).

[9] S. Kalivendi, S. Kotamraju, H. Zhao, J. Joseph, and B. Kalyanaraman, *J. Biol. Chem.* **276,** 47266 (2001).

[10] V. Gouaze, M. E. Mirault, S. Carpentier, R. Salvayre, T. Levade, and N. Andrieu-Abadie, *Mol. Pharmacol.* **60,** 488 (2001).

[11] K. A. Kampbell, M. R. Lashley, J. K. Wyatt, M. H. Nantz, and R. D. Britt, *J. Am. Chem. Soc.* **123,** 5710 (2001).

[12] K. A. Campbell, E. Yikilmaz, C. V. Grant, W. Gregor, A-F. Miller, and R. D. Britt, *J. Am. Chem. Soc.* **121,** 4714 (1999).

[13] G. Minotti, G. Cairo, and E. Monti, *FASEB J.* **13,** 199 (1999).

[14] B. Kalyanaraman, E. Perez-Reyes, and R. P. Mason, *Biochim. Biophys. Acta* **630,** 119 (1980).

[15] J. Goodman and P. Hochstein, *Biochem. Biophys. Res. Commun.* **77,** 797 (1977).

0076-6879/04 $35.00

Emerging literature indicates that myocardial impairment caused by DOX may involve myocyte and endothelial apoptosis.[5,6] Apoptosis in myocardium ultimately leads to cardiomyopathy through a systematic reduction in the number of cardiomyocytes. From a therapeutic point of view, DOX-induced cardiomyocyte apoptosis has broader implications in mitigating cardiotoxicity. Currently, DOX cardiotoxicity is treated with antioxidant and iron chelation therapy.[6,16,17] Further understanding of the oxidative mechanisms leading to DOX-induced apoptosis may result in an effective antioxidative and antiapoptotic therapy for mitigating DOX cardiotoxicity. Recent data from our and other laboratories suggest that exposure of adult and neonatal rat cardiomyocytes to submicromolar levels of DOX induces a significant amount of apoptosis.[6,18]

The proapoptotic effect of DOX in myocytes and endothelial cells has been attributed to intracellular iron and H_2O_2 formation.[6,9] Recently, we and others showed that intracellular iron plays a critical role in initiating oxidant-induced apoptosis through upregulation of transferrin receptor (TfR).[19–21] Transferrin receptor synthesis is regulated by interaction of the iron regulatory protein (IRP) with the iron-responsive element (IRE) present on the 3′-untranslated region of TfR mRNA. Iron-responsive elements serve as sensors of cellular iron.[20–24] Mitochondrial toxins that stimulate $O_2^{\bullet -}$ and H_2O_2 formation cause excessive accumulation of cellular iron by means of activation of IRP-1.[25] Antioxidants inhibited quinone-induced TfR overexpression and the associated iron uptake, implicating a role for oxidant-induced iron signaling mechanism.[19] In this chapter, we discuss the methodological aspects linking DOX-induced intracellular oxidative stress, iron signaling, and apoptosis. Pertinent redox parameters

[16] D. Kumar, L. A. Kirshenbaum, T. Li, I. Danelisen, and P. K. Singal, *Antioxid. Redox Signal* **3,** 135 (2001).

[17] J. L. Speyer, M. D. Green, E. Kramer, M. Rey, J. Sanger, C. Ward, N. Dubin, V. Ferrans, P. Stecy, A. Zeleniuch-Jacquotte *et al., N. Engl. J. Med.* **319,** 745 (1988).

[18] J. Nitobe, S. Yamaguchi, M. Okuyama, N. Nozaki, M. Sata, T. Miyamoto, Y. Takeishi, I. Kubota, and H. Tomoike, *Cardiovasc. Res.* **57,** 119 (2003).

[19] S. Kotamraju, C. R. Chitambar, S. V. Kalivendi, J. Joseph, and B. Kalyanaraman, *J. Biol. Chem.* **277,** 17179 (2002).

[20] R. D. Klausner, G. Ashwell, J. Van Renswounde, J. B. Harford, and K. R. Bridges, *Proc. Natl. Acad. Sci. USA* **80,** 2263 (1983).

[21] R. D. Klausner, J. Van Renswounde, G. Ashwell, C. Kempf, A. N. Schechter, A. Dean, and K. R. Bridges, *J. Biol. Chem.* **258,** 4715 (1983).

[22] N. H. Gehring, M. W. Hentze, and K. Pantopoulos, *J. Biol. Chem.* **274,** 6219 (1999).

[23] K. Pantopoulous and M. W. Hentze, *EMBO J.* **14,** 2917 (1995).

[24] K. Pantopoulos, S. Mueller, A. Atzberger, W. Ansorge, W. Stremmel, and M. W. Hentze, *J. Biol. Chem.* **272,** 9802 (1997).

[25] U. Testa, E. Pelosi, and C. Peschle, *Crit. Rev. Oncogenesis* **4,** 241 (1993).

measured are the following: intracellular glutathione (GSH) levels, aconitase activity, IRP–IRE interaction, TfR expression, cellular iron uptake, DCFH oxidation to DCF, and apoptosis.

Measurement of Intracellular Oxidative Stress

Reduced Glutathione. The cellular redox signaling involves the post-transcriptional modification of proteins involving the GSH redox cycle and thioredoxin. Glutathione, a substrate of GSH peroxidase, detoxifies intracellular peroxides. Thus, the measurement of intracellular GSH is a critical redox factor in doxorubicin-induced apoptosis.

Glutathione was measured by high-performance liquid chromatography (HPLC) as the *o*-phthalaldehyde (OPA) adduct at pH 8.0.[26] *o*-Phthalaldehyde (36.4 m*M*) was dissolved in ethanol and then diluted 1:10 (vol/vol) with borate buffer (75 m*M*, pH 8.0). Following treatment of bovine aortic endothelial cells (BAEC) with DOX in the presence or absence of antioxidants, cells were washed twice with Dulbecco's phosphate-buffered saline (DPBS), suspended in 250 μl of PBS, and lysed by sonication (1 cycle, 12% power for 30 s). After centrifugation at 10,000 rpm for 2 min, 200 μl of the clear supernatant was mixed with OPA (50 μl) and incubated at room temperature for 30 min. Samples (100 μl) were injected onto a Kromasil C-18 column and eluted isocratically with a mobile phase consisting of 150 m*M* sodium acetate/methanol (91.5:8.5). The OPA–GSH adduct was monitored using a fluorescence detector operating at excitation and emission wavelengths at 250 and 410 nm, respectively. Quantification was performed using GSH as a standard, and the values were expressed as nmoles/mg protein. DOX treatment caused a time-dependent decrease in intracellular GSH levels.[6]

Measurement of Aconitase Activity. The release of the "solvent-exposed" iron and oxidation of the $[4Fe\text{-}4S]^{2+}$ cluster accompanies loss of catalytic activity of this enzyme.[27] Rapid reactivation is achieved by iron–sulfur cluster reduction and Fe^{2+} insertion. The balance between the inactive and active form of aconitase provides a sensitive measure of the changes in the steady-state $O_2^{\bullet -}$ levels occurring in cells and mitochondria during redox activation.[27] Intracellular aconitase activity is a critical redox parameter in DOX-induced endothelial apoptosis.

Bovine aortic endothelial cells were grown to subconfluence and treated with 0.5–1 μM doxorubicin. After the treatment, cells were washed twice with ice-cold PBS, scraped, and centrifuged at 3000 rpm for 10 min to pellet the cells, and homogenized using a 25-gauge syringe needle (10

[26] F. Tietze, *Anal. Biochem.* **27,** 502 (1969).

[27] P. R. Gardner, *Biosci. Rep.* **17,** 33 (1997).

strokes) with 200 μl of lysis buffer (0.2% Triton X-100, 100 μM DTPA, and 5 mM citrate in PBS). The lysate was centrifuged at 3000 rpm for 10 min to pellet out the cellular debris and unlysed cells. The supernatant was further centrifuged at 12,000 rpm, aconitase activity was measured in the supernatant, and protein was estimated by the Lowry method. The activity of aconitase (100 μg protein in 100 μl) was measured in 100 mM TRIS-HCl, pH 8.0 (800 μl), containing 20 mM DL-trisodium isocitrate (100 μl). The rate of change of absorbance was followed for 2 min at 240 nm in a ultraviolet (UV) spectrophotometer. An extinction coefficient for *cis*-aconitate of 3.6 mM^{-1} cm^{-1} at 240 nm was used to calculate the enzyme activity, and the values were expressed as U/mg protein.[28]

Inactivation of aconitase is often used as a physiologically relevant indicator of the intracellular oxidant formation. DOX treatment inhibited 40% of the total aconitase activity in endothelial cells within 2–3 h.

Measurement of Complex I Activity. The NADH/ubiquinone oxidoreductase or complex I is the largest mitochondrial respiratory chain complex. The inhibition of complex I activity has been shown to generate $O_2^{\bullet-}$.[29] Thus, the measurement of complex I activity represents a viable indicator of cellular redox environment.

Endothelial cells were grown in 100-mm diameter culture dishes to subconfluency and treated with doxorubicin. After the treatment, cells were washed twice with 5 ml of ice-cold PBS. Monolayers were scraped into 5 ml of PBS and placed in 15-ml Falcon tubes. Cells were pelleted at 500g at 4° for 10 min and resuspended in 2.5 ml of TES buffer (0.25 M sucrose, 1 mM EGTA, and 10 mM triethanolamine-acetate at pH 7.0). Cells were homogenized with a Dounce homogenizer with 15 strokes. The resulting material was transferred to a 2-ml tube and centrifuged at 1500g. The postnuclear supernatant was centrifuged at 10,000g at 4° for 10 min to obtain a mitochondria-enriched pellet. The pellet was washed twice with 1 ml of homogenization buffer. The protein content of the pellet was assayed by the Bradford method. After isolation, the mitochondria-enriched pellet can be stored at −80° for <2 wk before measurements of the complex I activity. The mitochondrial pellet was subjected to "freeze-thawing" three times. Twenty microliters (0.3 mg protein) of mitochondrial homogenate was mixed with 930 μl of 10 mM potassium phosphate buffer, pH 8.0, in a 1-ml cuvette containing 50 μl of 100 μM NADH. The rate of NADH oxidation was monitored at 340 nm for 2 min in a

[28] E. A. Konorev, M. C. Kennedy, and B. Kalyanaraman, *Arch. Biochem. Biophys.* **368,** 421 (1999).

[29] Y. Hsiu-Chuan, T. D. Oberley, C. G. Gairola, L. I. Szweda, and D. K. St. Clair, *Arch. Biochem. Biophys.* **362,** 59 (1999).

UV spectrophotometer. Then 5 μl of 10 mM ubiquinone-1 was added, and the stimulated rate of NADH oxidation was measured as the complex I activity, using an extinction coefficient of 6.81 mM^{-1} cm^{-1} at 340 nm.[29] To further confirm the role of complex I, the reaction was carried out in the presence of 5 μl of 2 mM rotenone, a well-known complex I inhibitor. Finally, the rotenone-inhibitable complex I activity was expressed as μM min^{-1} mg^{-1} protein. DOX treatment (0.5 μM) caused a 50% inhibition in complex-I activity within 2–3 h in BAEC.

Iron Regulatory Protein–Iron Responsive Elements Measurements by Electrophoretic Mobility Shift Assay (EMSA). The posttranscriptional regulation of cellular iron metabolism in higher eukaryotes is controlled by the binding interaction between the IRP-1 and IRP-2 with IREs. Increased binding of IRPs to IREs stabilizes transferrin receptor mRNA, leading to increased iron uptake. In most mammalian cells, IRP-1 is more abundant than IRP-2 and can therefore be considered as the dominant posttranscriptional regulator of cellular iron metabolism. Recently, we and others have demonstrated that IRP-1 (but not IRP-2) is rapidly activated by extracellular H_2O_2 and increases their binding to IREs.[19] Thus, there is a distinct regulatory connection between iron metabolism and oxidative stress.

The IRP/IRE binding was measured by EMSA. ^{32}P-labeled IRE mRNA for the RNA band-shift assay was prepared using as a template, a 1000-base pair rat L ferritin pseudogene that contains the conserved IRE sequence. The plasmid (*p66-L* gene) containing this insert (which was generously provided by Dr. Elizabeth Leibold, University of Utah, Salt Lake City, Utah) was linearized with *Sma*I (Life Technologies, Inc.) and used for *in vitro* transcription of IRE mRNA. Transcription was carried out with Sp6 RNA polymerase using a Riboprobe transcription system from Promega (Madison, WI).

Preparation and Purification of ^{32}P-labeled mRNA Probe. This probe was prepared using a riboprobe *in vitro* transcription kit from Promega using a Sp6 RNA polymerase. The final reaction mixture (25 μl) contained the following components: 5× Sp6 RNA polymerase buffer 5 μl (final concentration is 1×); 2.5 μl of 100 mM DTT (10 mM final); 1 μl of 40,000 U/ml RNasin ribonuclease inhibitor (20 U final); 2.5 μl of 1 mM GTP (100 μM final); master mix of 2.5 mM each of ATP, CTP, UTP (0.5 mM each final); 5 μl of α^{32}P GTP (50 μCi final, 1 Ci = 0.037 MBq); 0.5–1 μg (1.5 μl) of linearized p66 plasmid (DNA template); and finally the volume was adjusted to 25 μl with RNase-free water. The reaction mixture was incubated for 2–3 h at 37°. After incubation, 25 μl of formamide was added to the sample and boiled for 2 min to denature RNA.

The reagents for purification of labeled transcript on a 5% polyacrylamide/urea gel were urea (15 g), 5 ml of acrylamide (30% stock, 19:1 [19 g acrylamide, 1 g *bis* acrylamide in 66.67 ml water]), 6 ml of 5× TRIS-borate EDTA buffer (5× TBE = TRIS base 54 g, boric acid 27.5 g, 20 ml of 0.5 *M* EDTA, pH 8.0, and finally made up to a total volume of 1000 ml with water), 20 μl of TEMED, 180 μl of 10% ammonium persulfate (APS). (*Note*: Mix urea and acrylamide in 5 ml of water in a conical flask using a stir bar. Because it takes time to get this into solution, this step must be started early.) The experimental protocols are as follows: Measure the volume in a graduated cylinder and calculate the amount of additional water needed to make up 30 ml, including 0.2 ml of the TEMED and APS, which should be added just before pouring the gel. Pour a large, thin gel for this (20 × 20 cm × 0.75 mm). Use a long thin comb for deep wells. Before loading samples in wells, flush out wells with buffer, and load the sample. Simultaneously, run bromophenol blue (BPB)/xylene cyanol sample dye (0.25% BPB/0.25% xylene cyanol/30% glycerol) in a separate lane to monitor the run. Run the gel at 30 mA in 1 × TBE buffer until top dye is half way down the gel. IRE-RNA runs with the top dye (i.e., xylene cyanol).

Removal of ^{32}P mRNA from Gel. Lay the gel flat and remove the upper glass plate. Keep gel wet with TBE buffer and cover it with plastic wrap and mark it with a radioactive/fluorescent marker to orient position. Place an x-ray film over the gel and expose it at room temperature in the darkroom for 1–2 min using an intensifying screen and develop the film. Align the film with gel to locate the position of the hot band. Do this by placing the film flat on the view box and laying the gel (on the glass plate) over it. Cut out the hot band from the gel by cutting directly through the plastic wrap with a scalpel. Transfer the gel band to a microfuge tube (without the plastic wrap) and crush the gel with a P200 pipette tip. Add about 0.5–1 ml of 0.5 *M* ammonium acetate to tube and elute RNA from the crushed gel slice by tumbling tube overnight on a rotator at room temperature. The next day, centrifuge the sample for 10 min at 5000*g* to pellet the gel to the bottom of the tube and remove the supernatant. To increase the yield, add 100 μl of ammonium acetate to the gel pellet and centrifuge again for about 30 min. Concentrate the sample to about 100 μl in a speed vacuum centrifuge or alternately, the sample can be concentrated by using the nucleic acid extraction kit supplied from Quiagen (Valencia, CA) and elute the RNA in 100 μl of RNase-free water or TRIS buffer, pH 8.0. Precipitate the RNA by adding 10 μl of 3 *M* sodium acetate and 200 μl of ice-cold 70% ethanol. Incubate the mixture at -20° for a few hours or overnight to precipitate RNA. All the subsequent steps should be done at 4°. Centrifuge the sample at high speed for 10 min to pellet out the

RNA. Wash the pellet with ice-cold 70% ethanol and centrifuge at 7500 rpm for 10 min. Check the washes with a Geiger counter to make sure that radioactivity (nonspecific) is no longer present in the supernatant from the pellet. Air dry the pellet and dissolve the labeled RNA in about 100 μl of water. Count 2 μl and use the cpm for reference. Store the labeled RNA at -20° until use. This should be good for 2–3 weeks until the cpm decreases. Use 10,000–40,000 cpm (0.2–0.5 ng) for band-shift assay.

Preparation of Cell Lysates for the Band Shift Assay. Bovine aortic endothelial cells were grown to subconfluence in 60-mm culture dishes and treated with doxorubicin. After the treatment, cells were washed with DPBS and scraped, and the cell pellet was resuspended in 200 μl of buffer A (20 m*M* HEPES, pH 7.5, 5% glycerol, 0.5 m*M* EDTA, 25 m*M* KCl, 1 m*M* DTT, 1% nonidet P-40) containing a cocktail of protease inhibitors (one tablet of protease inhibitor cocktail supplied from Amersham (Piscataway, NJ) per 10 ml of buffer A). DTT and protease inhibitor cocktail should be added freshly. Briefly lyse the cells by pipetting up and down and keep the tubes on ice for 15 min and vortex occasionally to ensure cell lysing. Centrifuge at 14,000 rpm for 30 min at 4°. Collect the supernatant and estimate the protein by the Bradford method and take 20–40 μg protein for the band-shift assay.

Band-Shift Assay (RNA–Protein Binding Reaction). Before the actual binding reaction is initiated, make a large nondenaturing 5% polyacrylamide gel (7.5 ml of 30% acrylamide [60:1], 9 ml of 5× TBE, 28.2 ml of water, 20 μl of TEMED, 270 μl of 10% APS) 1.5-mm thick and allow to polymerize. Pre-run the gel for 30–60 min at 30 mA. This can be done while the binding reaction is in progress. The gel must be run with a circulating water cooler.

Binding Reaction. Set up the binding reaction in small Eppendorf tubes or in 96-well plates behind a radioactive shield. This assay has a narrow range of sensitivity, so be meticulous about accurate pipetting and protein concentrations. The binding reaction includes 2 μl of 10× band shift buffer (100 m*M* HEPES, pH 7.6, 30 m*M* $MgCl_2$, 400 m*M* KCl, 50% glycerol, and 1 m*M* DTT [DTT should be added freshly each time]), 16 μl of 20–40 μg protein (lysate). If sample is to be reduced, add 0.42 μl of β-mercaptoethanol (final concentration is 2%) to the mixture and incubate 10 min at room temperature (RT) before adding hot RNA. Then add 2 μl of ^{32}P RNA (25,000–30,000 cpm/μl) and incubate at RT for 30 min. Then add 2 μl of RNase T1 (1 U/μl) to remove the unbound labeled RNA and incubate for 10 min at RT. Finally, add 2 μl of heparin (50 mg/ml) and further incubate for 10 min at RT. Before loading, mix samples with 3 μl of 6× DNA loading dye, the samples are run at 30–50 mA (do not run the gel too far, because the free hot RNA will run into the buffer. Stop the gel when

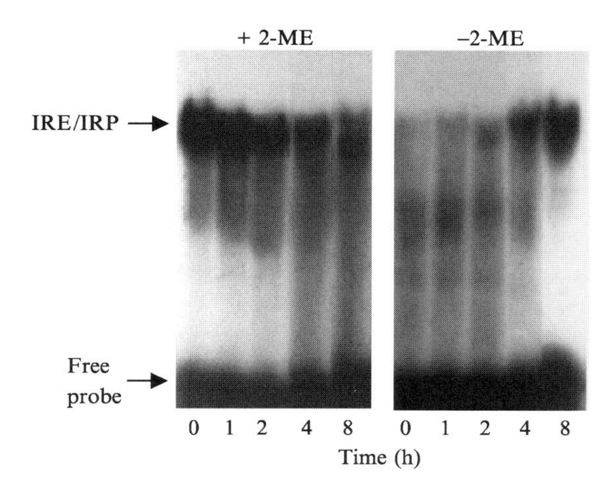

FIG. 1. Doxorubicin-induced increase in IRE/IRP binding as a function of time in BAEC.

bromophenol blue is about 2 cm from the bottom. Place the gel flat and remove top glass plate only. Place a sheet of 1 *M* Whatman paper over the gel and carefully peel back, lifting gel along with the paper, cover the gel with plastic wrap, dry the gel, keep it in an x-ray cassette along with the film (should be done in the dark), keep the cassette overnight at -80°, and develop the film to visualize the bands. DOX treatment caused an increase in IRE-IRP binding in BAEC after 2–4 h (Fig. 1).

Determination of Transferrin Receptor Levels

The iron transport in the plasma is carried out by transferrin, which transfers iron to cells through its interaction with a specific membrane receptor. Recently, we and others[19,30] have shown that intracellularly generated H_2O_2 from DOX metabolism or extracellular addition of glucose/glucose oxidase to endothelial cells resulted in the increased expression of TfR (Fig. 2). This suggests that peroxide-induced oxidative stress increases the TfR–dependent iron uptake, which is blocked in the presence of a monoclonal IgA class of TfR antibody (to show the specific role of TfR) or different antioxidants like Fe (III) tetrakis (4-benzoic acid) porphyrin (FeTBAP) that scavenges both intracellular superoxide and H_2O_2. Transferrin receptor levels in cells can be measured by Western analysis as follows.

[30] C. Ziemann, A. Burkle, G. F. Kahl, and K. I. Hirsch-Ernst, *Carcinogenesis* **20,** 407 (1999).

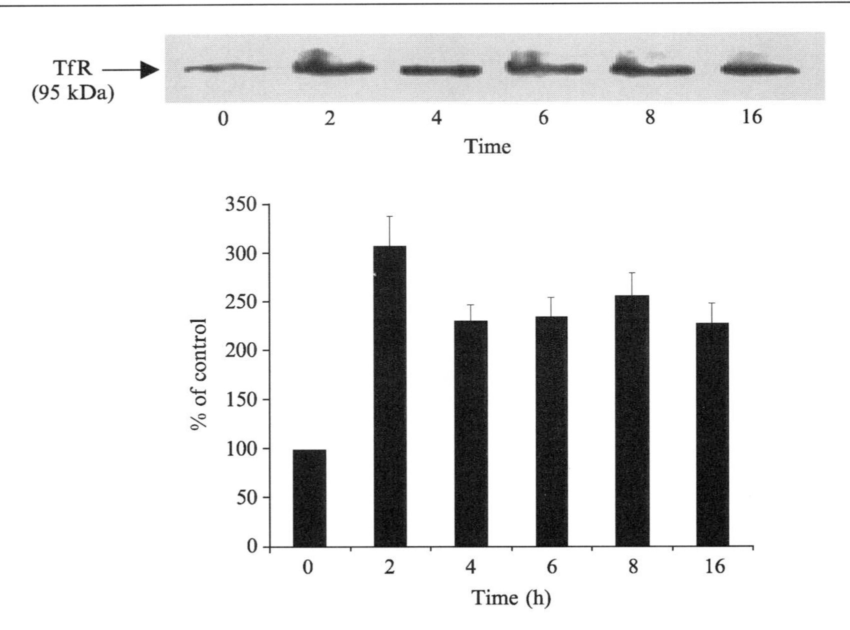

FIG. 2. Effect of doxorubicin on transferrin receptor levels in endothelial cells. BAEC were treated with 0.5 μM DOX for different time periods as indicated and transferrin receptor (TfR) levels were determined by Western analysis using the anti-TfR antibody. Note that following a 2-h incubation with DOX, TfR levels were increased and remained elevated for 16 h.

Bovine aortic endothelial cells were grown to subconfluence and treated with doxorubicin for different time points. After the treatment, cells were washed once with DPBS and resuspended in 200 μl of RIPA buffer (20 mM TRIS–HCl, pH 7.4, 2.5 mM EDTA, 1% Triton-X 100, 1% sodium deoxycholate, 1% SDS, 100 mM NaCl, 100 mM sodium fluoride). To a 10-ml solution of the preceding, the following agents were added: 1 mM sodium vanadate, 10 μg/ml aprotinin, 10 μg/ml leupeptin, and 10 μg/ml pepstatin inhibitors. Cells were homogenized by passing the suspension through a 25-gauge needle (10 strokes). The lysate was centrifuged at 750*g* for 10 min at 4° to pellet out the nuclei. The remaining supernatant was centrifuged for 30 min at 12,000 rpm. Protein was determined by the method of Lowry, and 20 μg was used for the Western blot analysis. Proteins were resolved on 8% SDS–polyacrylamide gels and blotted onto nitrocellulose membranes. Membranes were washed with TBS (140 mM NaCl, 50 mM TRIS–HCl, pH 7.2) containing 0.1% Tween-20 (TBST) and 5% skim milk to block the nonspecific protein binding. Membranes were incubated with mouse anti-human TfR monoclonal antibody (Zymed, San Francisco, CA), 1 μg/ml in TBST for 2 h at room temperature, washed five times, and then incubated with horseradish peroxidase–conjugated rabbit anti-mouse IgG (1:5000) for 1.5 h at room temperature. The TfR band was detected with the ECL method (Amersham, Buckinghamshire, UK).[31,32]

Measurement of ^{55}Fe Uptake in Endothelial Cells

Bovine aortic endothelial cells were grown in DMEM containing 10% FBS until subconfluence. On the day of the treatment with doxorubicin, the medium was replaced with DMEM containing 2% FBS, and the cells were allowed to adjust to the medium conditions; 0.1 μCi of ^{55}Fe supplied as ferric chloride (Perkin Elmer, Boston, MA, stock solution [1 mCi was diluted in 0.1 *N* HCl]) was added to the medium for various time points, and its levels were measured as a function of time. Cells were washed twice (2 $\times$ 5 ml) with DPBS and lysed with PBS containing 0.1% Triton X-100 (300 μl) by briefly pippetting up and down and incubating on ice for 15 min min with occasional vortexing. The lysate was centrifuged at 3000 rpm for 10 min, and the supernatant was collected, and 275 μl was taken into radioactive vials containing 10 ml of scintillation fluid and counted in a beta counter.[33] Simultaneously, 2 μl of the supernatant was taken to estimate the total protein. Radioactive counts in different samples were normalized to their respective protein concentrations, and the values were expressed as cpm/mg protein. Results indicate that concomitant with an increase in TfR expression, there was at least a twofold increase in ^{55}Fe uptake in DOX-treated cells that was antagonized by anti-TfR antibody treatment.

Detection of Intracellular Oxidants

The oxidation of 2′,7′-dichlorodihydrofluorescein (DCFH), a nonfluorescent probe, to a green fluorescent product, dichlorofluorescein (DCF), has been used to measure intracellular H_2O_2 by numerous investigators.[34,35] This assay was originally used to monitor intracellular oxidant produced during oxidative stress or apoptosis.[36,37] The cell-permeable

[31] R. Taetle, J. Castagnola, and J. Mendelsohn, *Cancer Res.* **46,** 1759 (1986).

[32] K. A. Pritchard, Jr., M. K. O'Banion, J. M. Miano, N. Vlasic, U. G. Bhatia, D. A. Young, and M. B. Stemerman, *J. Biol. Chem.* **269,** 8504 (1994).

[33] C. R. Chitambar and P. A. Seligman, *J. Clin. Invest.* **78,** 1538 (1986).

[34] R. Cathcart, E. Schwiers, and B. N. Ames, *Anal. Biochem.* **134,** 111 (1983).

[35] J. Duranteau, N. S. Chandel, A. Kulisz, Z. Shao, and P. T. Schumacker, *J. Biol. Chem.* **273,** 11619 (1998).

[36] C. P. LeBel, H. Ischiropoulos, and S. C. Bondy, *Chem. Res. Toxicol.* **5,** 227 (1992).

nonfluorescent probe, DCFH-diacetate (DCFH-DA), is hydrolyzed by intracellular esterases to form the active probe, DCFH. Previous studies implicated a role for redox-active iron in cellular oxidation of DCFH to DCF,[19,36] although the origin of cellular iron was not known. However, reports also indicate that stimulation of vascular smooth muscle cells with mitogenic growth factors induce a large increase in DCF fluorescence minutes after ligand stimulation.[38] In these studies, DCFH oxidation is unlikely to be mediated by oxidant-induced cellular iron signaling. Recent data suggest that DCFH oxidation in cells may be caused by oxidants formed from the interaction between heme proteins (i.e., myoglobin, cytochrome *c*) and H_2O_2.[39] Therefore, to investigate the specific role of iron in DCFH oxidation we have used an IgA class of TfR antibody (42/6), which specifically blocks the TfR-dependent iron uptake in BAECs wherein we showed that the iron uptake is predominantly TfR dependent. Our results indicate that anti-TfR antibody (42/6) almost completely blocked the doxorubicin-induced DCFH oxidation by means of an iron-mediated pathway (Fig. 3). DCF green fluorescence in endothelial cells is measured as follows.

Bovine aortic endothelial cells were grown to subconfluence and were treated with doxorubicin and other antioxidants (pretreated for 1 h before the addition of doxorubicin). After the treatment, the medium was aspirated, and cells were washed with 4 ml of DPBS twice and incubated in 2 ml of fresh culture medium without FBS. DCFH-DA was added at a final concentration of 10 μM and incubated for 20 min. Note that DCFH-DA was added after DOX treatment. The cells were then washed twice with DPBS and maintained in 1 ml culture medium. Fluorescence was monitored using a Nikon fluorescence microscope (excitation 488 nm, emission 610 nm) equipped with a FITC filter. The intensity values were calculated using the Metamorph image analysis software (Universal Imaging Corporation, Downingtown, PA).

Hydroethidine (Dihydroethidium) Staining

One of the popular assays for detecting superoxide in cells and tissues involves the use of fluorescence-based techniques.[40,41] Generally, the red fluorescence arising from oxidation of hydroethidine (HE) (also

[37] D. J. Kane, T. A. Sarafian, R. Anton, H. Hahn, E. B. Gralla, J. S. Valentine, T. Ord, and D. E. Bredesen, *Science* **262,** 1274 (1993).

[38] M. Sundaresan, Z. X. Yu, V. J. Ferrans, K. Irani, and T. Finkel, *Science* **270,** 296 (1995).

[39] T. Ohashi, A. Mizutani, A. Murakami, S. Kojo, T. Ishii, and S. Taketani, *FEBS Lett.* **511,** 21 (2002).

[40] G. Rothe and G. Valet, *J. Leukoc. Biol.* **47,** 440 (1990).

[41] W. Carter, P. K. Narayanan, and P. Robinson, *J. Leukoc. Biol.* **55,** 253 (1994).

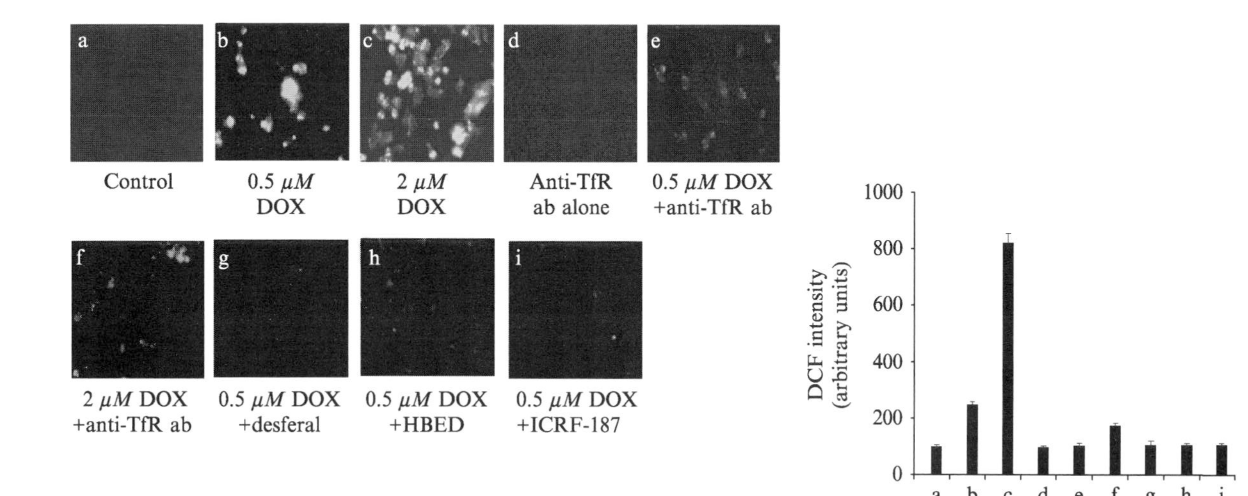

FIG. 3. Effect of anti-TfR antibody and iron chelators on DOX-induced oxidative stress, as measured by DCF staining. BAEC were treated with 0.5 μM and 2 μM DOX alone or in the presence of anti-TfR ab (12 μg/mL, IgA class) or iron chelators as indicated for 4 h. The medium was then aspirated, and cells were washed twice with DPBS and subsequently incubated with 10 μM DCF-DA for 20 min. The cells were then washed with DPBS and maintained in 1 ml of the culture medium. The green fluorescence, characteristic of DCF, was measured using FITC and rhodamine filters, respectively. Data shown are representative of three separate experiments.

dihyroethidium) is detected.[42,43] The redox-sensitive fluorophore hydroethidine (dihydroethidium) has been used to monitor intracellular oxidative stress. The procedure to measure the superoxide levels described here is similar to that of the DCFH oxidation protocol. Following pretreatment of BAEC with antiapoptotic antioxidants and doxorubicin, culture medium was aspirated and cells were washed twice with DPBS and incubated in fresh culture medium without FBS. Hydroethidine (10 μM) was added to the cells, and the incubation was continued for 20 min during which hydroethidine was oxidized to ethidium. Fluorescence images were obtained using a Nikon fluorescence microscope equipped with a rhodamine filter. The fluorescence intensity values from three different fields of view were calculated using the Metamorph software, and the average values were represented.

Intracellular Iron Chelators

ICRF-187 has been used to prevent DOX-induced cardiotoxicity in cancer patients.[17] ICRF-187 itself does not chelate iron; however, following intracellular esterase-dependent hydrolysis, ICRF-187 acquires a bidentate EDTA-like structure, which can chelate intracellular iron released from iron stores. HBED is currently being used as an intracellular iron chelator in several cellular studies.[44] DOX-induced apoptosis, as measured by caspase-3 activation, was dramatically lowered in cells preincubated with ICRF-187, desferral, or HBED. If cells were not preincubated with the iron chelators, they were not effective in inhibiting DOX-induced apoptosis.[19]

Measurement of Apoptosis

The morphological characteristics of apoptosis or programmed cell death include cell shrinkage, membrane blebbing, release of apoptotic bodies, and the loss of structural organization of the nucleus.[45,46] The condensation of nuclear chromatin is accompanied by the activation of an endonuclease that initially cleaves the chromosomal DNA into 50- to 300-kb fragments. In most cell systems, the onset of fragmentation leads to the formation of internucleosomal fragments of 180-bp length. Detection of DNA fragments is the biochemical basis of most methods that are currently used in the quantitation of apoptosis.

[42] F. Miller, D. Gutterman, D. Rios, D. Heistad, and B. Davidson, *Circ. Res.* **82,** 1298 (1998).
[43] T. Paravicini, L. Gulluyan, G. Dusting, and G. Drummond, *Circ. Res.* **91,** 54 (2002).
[44] R. J. Bergeron, J. Wiegand, and G. M. Brittenham, *Blood* **93,** 370 (1999).
[45] F. L. Kiechle and X. Zhang, *J. Clin. Ligand Assay* **21,** 58 (1998).
[46] F. L. Kiechle and X. Zhang, *Clin. Chim. Acta* **326,** 27 (2002).

DNA Laddering

DNA laddering by agarose gel electrophoresis has been used as a semi-quantitative method of detection of apoptosis induced by DOX (Fig. 4). Endothelial cells were grown to subconfluence in 100-mm cell culture dishes. After DOX treatment, the culture medium was removed and centrifuged at 3000*g* for 5 min to collect detached cells. Adherent cells were lysed with 1 ml of lysis buffer (10 m*M* TRIS–HCl, pH 8.0) containing EDTA (10 m*M*) and Triton X-100 (0.5%), and then pooled with pellets made of detached cells. RNA was digested using 10 μl of RNase (1 mg/ml) and incubated at 37° for 1 h followed by the addition of 20 μg/ml proteinase K treatment for 2 h at 50°. DNA was extracted by adding a 200-μl mixture of phenol, chloroform, and isoamyl alcohol (25:24:1) and vortexed for 15 s by inverting the Eppendorf tube up and down and centrifuged at 10,000*g* for 10 min. The upper aqueous phase that contains the DNA was carefully removed, transferred into a new Eppendorf tube, and the extraction step was repeated to completely remove the contaminants. To the supernatant, an equal volume of isopropanol was added to precipitate the DNA and stored overnight at −20° (overnight incubation is preferred so as to precipitate out the small fragments). The sample was then centrifuged at 12,000*g* for 20 min at 4°. The pellet was washed once with 0.5 ml of 70% ethanol to remove the precipitated salt and centrifuged again at 12,000*g* for 10 min. The resulting pellet was air-dried for 5–10 min (avoid overdrying, which will make it difficult for DNA to dissolve) and dissolved in an adequate volume (20–40 μl) of autoclaved water or

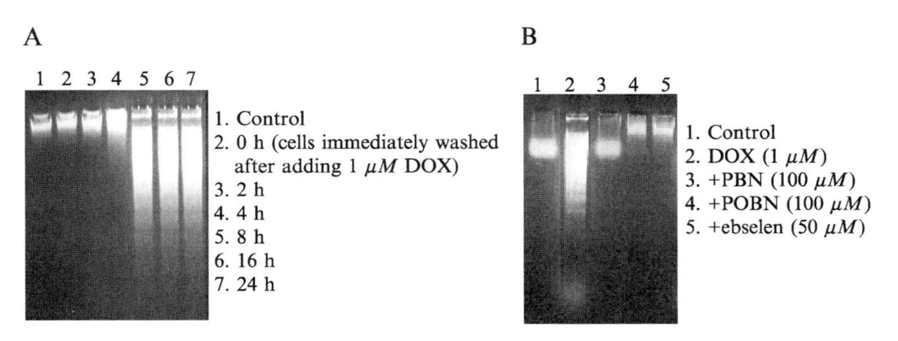

FIG. 4. Doxorubicin-induced apoptosis was inhibited by nitrones and ebselen. (A) BAEC were cultured in Dulbecco's modified Eagle's medium containing 10% FBS. After reaching subconfluence, cells were treated with DOX alone for different times to monitor the time course of DNA fragmentation. (B) Same as A but in the presence of either 100 μM *N-tert*-butyl-α-phenylnitrone (PBN), 100 μM α-(4-pyridyl-1-oxide)-*N-tert*-butylnitrone (POBN), or 50 μM ebselen for 16 h. DNA was extracted and electrophoresed on a 2% agarose gel. Cells were pretreated with agents 1 h before the addition of DOX.

TRIS buffer, pH 8.0, containing EDTA. Avoid pipetting the DNA solution up and down to promote resuspension, because this may cause shearing of the intact DNA. DNA dissolves best under slightly alkaline conditions. DNA concentration is measured at 260 nm in a UV spectrophotometer. Approximately 5–10 μg of the DNA was taken, and an equal volume of 2× sample buffer (0.25% bromphenol blue, 30% glycerol) was added and electrophoretically separated on a 2% agarose gel containing 1 μg/ml ethidium bromide and visualized under UV transillumination.

Disadvantages

The observation of oligonucleosomal DNA fragments by DNA laddering is one of the most widely used assays for the detection of apoptosis. However, this assay has numerous disadvantages: lack of sensitivity and specificity and lengthy preparation time, requiring a high level of expertise and inability to discern specific cells in a population that are undergoing apoptosis. Apoptotic endonucleases not only affect cellular DNA by producing the classical DNA ladder but also generate free 3′-OH groups at the ends of these DNA fragments.

The Terminal Deoxynucleotidyltransferase–Mediated Nick-End Labeling Assay

The terminal deoxynucleotidyltransferase-mediated nick-end labeling (TUNEL) assay is used for microscopic detection of apoptosis.[47] This assay is based on labeling of 3′-free hydroxyl ends of the fragmented DNA with fluorescein-dUTP catalyzed by terminal deoxynucleotidyltransferase (TdT). Procedures are followed according to the commercially available kit (ApoAlert) from Clontech (Palo Alto, CA).

This protocol is given for the adherent endothelial cells, although it can also be used for suspension cells and tissue section with minor modifications. The cells were grown to subconfluence on Lab-Tek chamber slides or on poly-L-lysine–coated slides prepared in the laboratory, and apoptosis was induced by adding DOX. Following induction, slides were washed by dipping twice in a coplin jar containing phosphate-buffered saline (PBS). *Always use a fresh jar each time.* The slides were fixed by immersing in fresh 4% formaldehyde prepared in PBS at 4° for 30 min (at this stage fixed slides can be stored for up to 2 weeks in 70% ethanol at −20°). After fixation, the slides were immersed in a coplin jar containing fresh PBS for 5 min × 2 at room temperature. Cells were then permeabilized by immersing the slides in a coplin jar containing prechilled 0.2% Triton-X

[47] Y. Gavrieli, Y. Sherman, and S. A. Ben-Sasson, *J. Cell. Biol.* **119,** 493 (1993).

100 prepared in PBS and incubated for 5–10 min. *In our laboratory, we incubate for 7 min.* Slides were washed in a coplin jar containing PBS for 5 min × 2 at room temperature. Slides were removed from PBS and tapped gently to remove excess liquid and covered with 100 μl of equilibration buffer (200 mM potassium cacodylate, pH 6.6, 25 mM TRIS–HCl, pH 6.6, 0.2 mM DTT, 0.25 mg/ml BSA, 2.5 mM cobalt chloride). By use of forceps, a piece of plastic coverslip was gently placed on top of the cells to evenly spread the buffer and equilibrated at room temperature for 10 min. min. In the meantime, the TdT reaction mixture consisting of 45 μl of equilibration mixture, 5 μl of fluorescein-tagged dUTP, and 1 μl of terminal deoxynucleotidyl transferase enzyme (0.5 μg protein/μl) was prepared. By use of forceps, the plastic coverslip was removed, the slide was gently tapped to remove excess liquid, the edges were carefully blotted and dried with tissue paper, and 50 μl of TdT reaction mixture was gently placed on the cells on a ~5-cm^2 area. *At this stage, do not allow the cells to dry out.* From this point, samples were protected from light by wrapping the coplin jar in aluminum foil or keeping the jar in a covered box. By use of forceps, a piece of plastic coverslip was gently placed on top of the cells to evenly spread the liquid. To perform the tailing reaction, the slide was placed in a dark humidified 37° incubator for 1 h. After the reaction, the plastic coverslip was removed, and the reaction was terminated by immersing the slide in a coplin jar containing 2× SSC (0.3 M sodium chloride and 30 mM trisodim citrate) buffer for 15 min. The cells were washed with PBS by transferring the slide to a fresh coplin jar for 5 min × 2, and the cells were stained with propidium iodide (10 ng/ml final concentration made in PBS) for 5–10 min. The cells were then washed by transferring the slide to a fresh coplin jar filled with deionized water for 5 min × 2. At this stage, one can either immediately visualize the fluorescence or can add a drop of antifade solution onto the slide and cover it with a glass coverslip to keep the fluorescence for a longer time. In either case, it is better to view the slide as soon as possible.

Apoptotic cells exhibit a strong nuclear green fluorescence that can be detected using a standard fluorescein filter (520 nm). All cells stained with propidium iodide exhibit a strong red cytoplasmic fluorescence at 620 nm using a rhodamine filter that indicates all cells. The areas of apoptotic cells can be detected by fluorescence microscopy–equipped with fluorescein isothiocyanate (FITC) filter where they appear green (TUNEL positive). The quantification of apoptosis was performed using a Metamorph image analysis package.

Figure 5 shows that treatment of BAEC with 0.5–1 μM of DOX enhanced the fraction of TUNEL-positive BAEC from 2–65%. Pretreatment with iron chelators or anti-TfR antibody inhibited DOX-induced apoptosis.

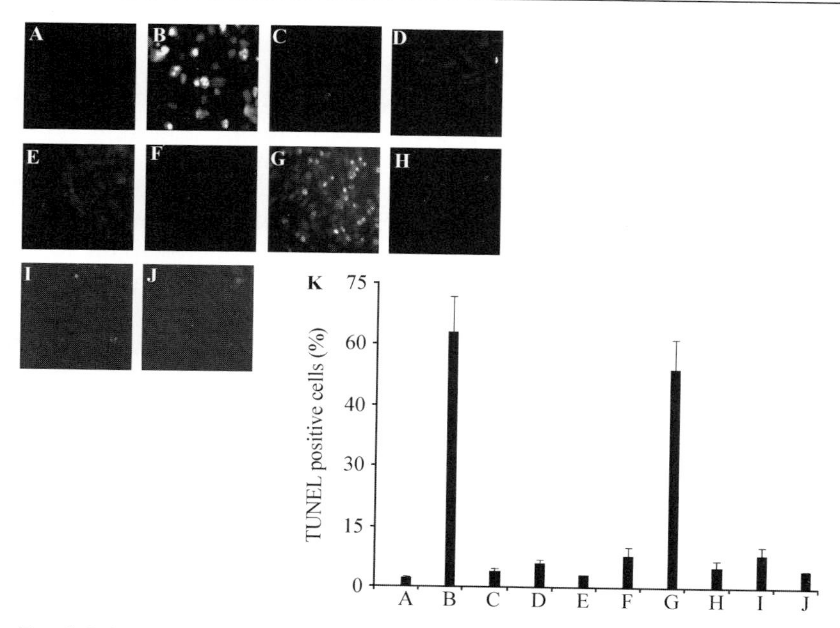

FIG. 5. Inhibition of DOX-induced apoptosis in BAEC treated with anti-TfR antibody, iron chelators, or antioxidants. BAEC were treated with DOX for 8 h and other agents as shown below. Cells were then harvested, stained for TUNEL positive cells, and examined by fluorescence microscopy (A) control cells, (B) cells treated with 0.5 μM DOX, (C) cells were pretreated (2 h) with anti-TfR antibody before the addition of DOX, (D) the same as above, except that DOX and anti-TfR antibody were added at the same time, (E) cells were treated with anti-TfR antibody alone, (F) cells were pretreated with ICRF-187 iron chelator for 2 h before the addition of 0.5 μM DOX, (G) cells were treated with 0.5 μM DOX and ICRF-187 without pretreatment, (H) cells were treated with 100 μM PBN for 2 h before the addition of 0.5 μM DOX, (I) cells were treated with 50 μM ebselen for 2 h before the addition of 0.5 μM DOX, (J) cells were treated with 20 μM FeTBAP for 2 h before the addition of 0.5 μM DOX, and (K) percent apoptosis in (A–J) calculated using Metamorph image analysis software. Data shown are the representative of three separate experiments.

Caspase-3 Activity

Caspases are a family of proteases that mediate apoptotic cell death. Caspase 3 (also referred to as CPP32, Yama, and apopain) is a member of the CED-3 subfamily of caspases and is one of the critical enzymes of apoptosis. The activated caspase 3 can specifically cleave most of the caspase-related substrates, including many key proteins such as the nuclear enzyme poly(ADP-ribose) polymerase (PARP), the inhibitor of caspase-activated deoxyribonuclease (ICAD), gelsolin, and fodrin, which are proteins involved in apoptosis regulation.[48,49]

The caspase 3 colorimetric assay is based on the hydrolysis of the peptide substrate, acetyl-Asp-Glu-Val-Asp *p*-nitroanilide (Ac-DEVD-pNA), by caspase 3, resulting in the release of the *p*-nitroaniline (pNA) moiety. The concentration of the pNA released from the substrate is calculated from the absorbance values at 405 nm or from a calibration curve prepared with defined pNA solutions. The caspase 3 activity in cells was measured as follows: Endothelial cells were grown to subconfluence and treated with DOX (0.5–1 μM). Following treatment with DOX and other antioxidants, cells were washed with Dulbecco's PBS and lysed with 100 μl of 1× lysis buffer diluted from 5× stock (250 m*M* HEPES, pH 7.4, 25 m*M* CHAPS, 25 m*M* DTT) (5× stock solution can be stored at −20° for at least 2 weeks) and passed through a 25-gauge needle for 10 times to disrupt the cell membranes; samples were incubated on ice for 10 min and then centrifuged in a microcentrifuge at 12,000*g* for 3 min at 4° to precipitate out the cellular debris. Then 75 μl of the lysate was added to an equal volume of 2× assay buffer, which was diluted from 10× stock (200 m*M* HEPES, pH 7.4, 1% CHAPS, 50 m*M* DTT, 20 m*M* EDTA). Finally, 5 μl of 4 m*M* of substrate (DEVD-pNA purchased from Sigma Chemicals (St. Louis, MO), and prepared in DMSO) was added and incubated in a 37° water bath for 1 h, and absorbance was read at 405 nm in a spectrophotometer. Protein was estimated by the Bradford method using 2 μl of the lysate, and the caspase-3 activity was calculated as nmol pNA/mg protein. The absorbance is usually read at zero time (to eliminate any nonspecific absorbance) and after 60 min. The difference in absorbance was used to calculate the activity. Figure 6A shows the time course of caspase-3 proteolytic activation in DOX-treated endothelial cells.

Mitochondrial Cytochrome c *Release*

Mitochondria play an important role in the apoptotic signaling pathway. Several studies have shown that cellular stress can induce the release of cytochrome *c* (cyt *c*) from the intermembrane space (IMS) of mitochondria into the cytosol and has a major role in further cascading the programmed cell death by means of activating the executioners of cell death.[50,51] In mammals, cyt *c* triggers the assembly of the apoptosome.

[48] H. Sakahira, M. Enari, and S. Nagata, *Nature* **391,** 96 (1998).

[49] S. Kamada, H. Kusano, H. Fujita, M. Ohtsu, R. C. Koya, N. Kuzumaki, and Y. Tsujimoto, *Proc. Natl. Acad. Sci. USA* **95,** 8532 (1998).

[50] K. F. Ferri and G. Kroemer, *Natl. Cell. Biol.* **3,** E255 (2001).

[51] D. R. Green and J. C. Reed, *Science* **281,** 1309 (1998).

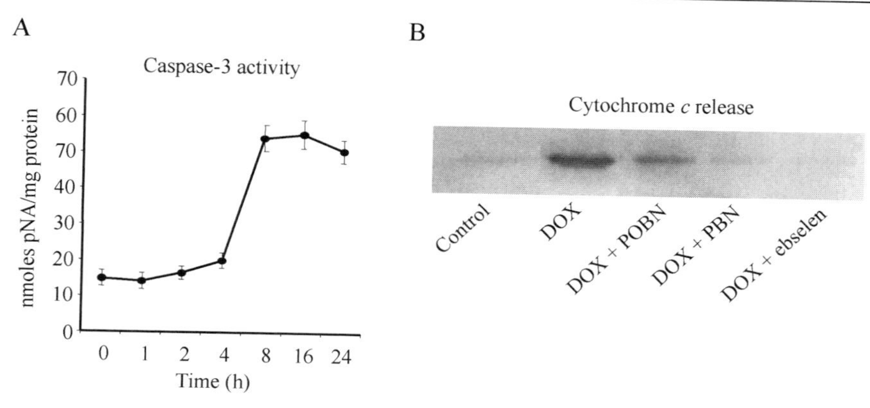

FIG. 6. Doxorubicin-induced mitochondrial cytochrome *c* release and kinetics of caspase-3 activation. (A) BAEC were treated with 1 μM DOX for different time periods; caspase-3 activity was measured by monitoring the release of para-nitroaniline at 405 nm. (B) Cells were treated with 1 μM DOX for 16 h in the presence (pretreatment for 1 h) or absence of 100 μM POBN, 100 μM PBN, or 50 μM ebselen. The lysate (25 μg of protein) was subjected to 14% SDS–PAGE followed by Western analysis using antibodies specific for cytochrome *c*.

The apoptosome is a complex composed of cyt *c*, apoptotic protease activating factor-1 (Apaf-1), and deoxy adenosine triphosphate (dATP). Successive binding of cyt *c* and dATP converts Apaf-1 from a closed menomeric configuration to an open heptameric platform for procaspase 9 assembly, leading to its activation, which in turn would activate caspase 3.[52]

To investigate the role of mitochondria in doxorubicin-induced apoptosis, the release of mitochondrial cyt *c* into the cytosol was measured (Fig. 6B). Endothelial cells were grown to subconfluency and treated with doxorubicin. After the treatments, cells were washed with DPBS and suspended in 150 μl of lysis buffer (2 mM NaH_2PO_4, 16 mM Na_2HPO_4, 150 mM NaCl, and 40 μg/ml saponin). Cells were homogenized by passing the suspension through a 25-gauge needle (10 strokes). The lysate was centrifuged at 750*g* for 10 min at 4° to pellet out the nuclei. The remaining supernatant was centrifuged for 15 min at 10,000*g*. The pellet was used as the mitochondrial fraction and the supernatant as the cytosolic fraction. After determining the protein concentration, 10 μg of the supernatant protein mixed with an equal volume of 2× Laemmli sample buffer and boiled for 5 min and loaded on to a 14% SDS–polyacrylamide gel to resolve the proteins and transferred onto nitrocellulose membrane at a constant

[52] M. Van Gurp, N. Festjens, G. Van Loo, X. Saelens, and P. Vandenabeele, *Biochem. Biophys. Res. Commun.* **304,** 487 (2003).

voltage of 32 V (0.5 V/cm^2) for 16 h at 4°. The membrane was washed once with TRIS-buffered saline (TBS) (140 m*M* NaCl, 50 m*M* TRIS, pH 7.2) containing 0.1% Tween-20 before blocking nonspecific binding with TBST containing 5% skim milk (one can also use 5% fetal calf serum or 5% bovine serum albumin to block the nonspecific binding) for 1 h. The nitrocellulose membrane was then incubated with the mouse anti-cyt *c* antibody (clone 7H8.C12, PharMingen, San Diego, CA) at 1 μg/ml in TBST containing 2% skim milk for 1–2 h at room temperature, washed 4 × 5 min with TBST, and further incubated with an anti-mouse IgG-horseradish peroxidase antibody (1:5000) for 1–2 h. The membrane was washed 4 × 5 min again with TBST and a final wash with TBS for 5 min. Finally, the bands were visualized by enhanced chemiluminescence method (Amersham Pharmacia Biotech, Piscataway, NJ).

Summary

We propose that TfR is an effective "gatekeeper" and modulator of DOX-induced apoptosis in endothelial cells. The proposed model for oxidant-induced cellular uptake of iron is shown in Fig. 7. DOX induces

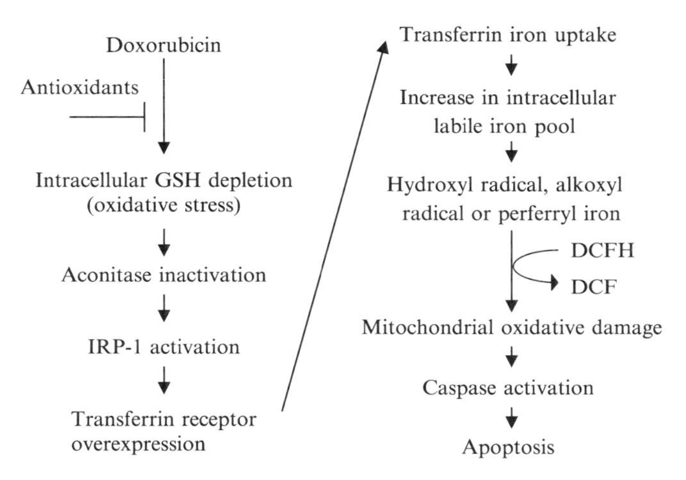

FIG. 7. A proposed model for oxidant-induced cellular influx of iron, DCFH oxidation, and apoptosis. The "oxidant stimulus" collectively refers to experimental conditions involving exposure of cells to hydroperoxides, peroxynitrite, or other preformed oxidants. ROS and RNS are also generated intracellularly following activation of cells by cytokines, UV radiation, or chemotherapeutic agents. Peroxide-induced DCFH oxidation is catalyzed by iron transported through TfR in endothelial cells. Antioxidants are postulated to inhibit TfR expression and iron uptake.

intracellular GSH depletion and inactivates the aconitase enzyme. This, in turn, increases the levels of noniron-bound IRP-1, enabling it to bind to the 3′-untranslated region of TfR mRNA, thereby enhancing the stability and level of expression of TfR. Elevation of TfR expression enhances the cellular uptake of iron, exacerbates oxidative stress, and stimulates apoptosis. Antioxidants and cell-permeable iron chelators diminish DOX-induced intracellular ROS and inhibit the accumulation of iron-free IRP-1, thereby blocking the induction of TfR expression and iron uptake. The oxidant-induced iron signaling mechanism is a new perspective that should be more fully explored in DOX cardiotoxicity.

Acknowledgment

This work was supported by the National Institutes of Health Grant CA77822.

Author Index

Numbers in parentheses are footnote reference numbers and indicate that an author's work is referred to although the name is not cited in the text.

A

Abdelmohsen, K., 258, 259, 260(7), 268(7), 269(7), 272(7)
Åberg, F., 5, 9, 15, 17, 146, 157, 185
Ackrell, B. A., 147
Adachi, O., 18
Adair, W. L., 153
Adams, J., 276
Adderley, S. R., 351
Addicks, K., 354
Adesnik, M., 223
Agarwal, S., 141, 142, 142(9), 150
Agner, E., 171
Agrillo, M. R., 349
Agrup, G., 101
Ahmad, K., 276
Ahn, C., 259
Ahn, Y. I., 238(12), 239, 273
Aithal, H. N., 10
Akerboom, T. P. M., 335
Akman, S., 259
Alam, I., 112(19), 113
Alam, J., 223, 229, 232(48), 233(48), 237(48), 238(12; 13), 239, 255, 273, 275, 276, 287(18), 288, 289
Alam, S., 8, 255, 276, 289
Alam, Z., 112(18), 113
Albagli, O., 234, 276
Albert, R. H., 32(8), 33
Alcain, F. J., 181, 197, 210, 211(14), 212(20), 213, 213(14), 214(16), 215(14), 333
Ale-Agha, N., 259, 260(7), 268(7), 269(7), 272(7)
Alegrio, L. V., 303
Alesnik, M., 186
Alessi, D. R., 254
Alho, H., 15
Allen, P. L., 207
Alleva, R., 170, 171(3)
Alò, F. P., 158
Alonso, A., 270
Altenbach, H. J., 206
Alzari, P. M., 32
Amano, T., 33
Ames, B. N., 64, 132, 371
Amzel, L. M., 187
Anderson, G. L., 110
Anderson, M. E., 335
Anderson, S. O., 95
Andersson, B., 124, 125, 126(14), 127, 127(14), 128(14), 130(14; 24)
Andersson, M., 4, 5, 9, 15(28)
Andres, D. A., 4
Andressen, C., 354
Andreucci, J. J., 313
Andrews, G. K., 241
Andrews, N. C., 229, 233, 274
Andrieu-Abadie, N., 200, 206, 362
Andriollo, M., 355, 357(59), 358(59), 360(59)
Angel, P., 228
Angelin, B., 163(15), 164
Animati, F., 347, 350(27), 354
Anlar, N., 113, 114(20), 119(20)
Ansorge, W., 363
Antholine, W. E., 345
Anthony, C., 18
Anton, R., 371(37), 372
Anwar, A., 320, 333(11)
Aoki, Y., 274
Appel, P., 321
Appelkvist, E. L., 5, 6, 9, 10, 14(10), 15, 16, 126, 157, 170, 185
Araujo, C. A., 303
Araujo, C. C., 303
Archakov, A. I., 77, 78(28)

Arenz, C., 206
Ariese, F., 34
Arin, M., 223, 224(30)
Arlt, A., 351
Armeni, T., 159
Arnér, E. S. J., 134
Arnold, R. S., 183
Arola, O. J., 351
Arroyo, A., 181, 184, 186(22; 44), 187, 201, 206(6), 207, 208, 210, 211(14), 213(14), 214(10; 11), 215(14), 320
Arstall, M. A., 362, 363(5)
Asard, H., 183
Ashida, K., 233
Ashwell, G., 363
Askonas, L. J., 32
Assaf, A. R., 110
Asunis, I., 229, 242, 274
Atalla, A., 33
Atzberger, A., 363
Augé, N., 200
Auricchio, S., 317
Avelange-Macherel, M. H., 125, 128(18)
Averboukh, L., 67
Aviles, R. J., 163
Awasthi, S., 307
Awasthi, Y. C., 307
Ayaki, H., 238(14), 239
Azzi, A., 52, 84

B

Babu, K. D., 303
Bach, T. J., 130
Backhaus, R. A., 131
Bae, O. N., 321
Baggiolini, M., 221
Bagnaresi, P., 182, 188(30; 32)
Bailey, H. H., 273, 305, 307(15), 315(15)
Baillie, T. A., 114, 118(23)
Bale, W. F., 5
Balendran, A., 254
Balmain, A., 221
Baltimore, D., 313
Bamba, T., 125
Bamberger, S., 213
Bannai, S., 273, 280(6), 287
Barbara, C., 350
Barch, D. H., 238(10), 239
Bargossi, A. M., 157, 171, 176(13)
Barr, R., 179, 181, 183, 184(4), 185(4), 190(4), 191, 192(4), 194(83), 196(83), 198(4)
Barrette, T. R., 124
Barrios, R., 223, 224(30)
Barron, E. S. G., 180
Barroso, M. P., 201, 201(17), 202, 205(7), 210, 212(17), 214(17)
Barry, S., 121
Basso, B., 182, 188(32)
Basu, N., 303
Battino, M., 16, 149, 156, 156(7; 11), 157, 158, 159, 170, 171, 176(13)
Bauer, C. E., 222
Baxter, A., 11
Bayol-Denizot, C., 321, 328(20)
Beal, M. F., 16
Beall, H. D., 320, 333(11)
Beaudet, A. L., 336
Beck, W. T., 67
Becker, G. W., 181
Bedard, P. A., 313
Begu, D., 131
Beinert, H., 344, 359(19)
Bel, A., 305
Beland, F. A., 113, 117, 123(21)
Belinsky, M., 228
Bello, R. I., 181, 186(22), 187, 201, 206(6), 208, 214(10), 333
Bellomo, G., 321, 323(12), 335
Benjamin, R. S., 362
Bennett, M. J., 31, 32, 32(8), 33
Bensasson, R. V., 102
Ben-Sasson, S. A., 376
Bensimon, S., 273, 274(2)
Bentinger, M., 4, 10, 11(34), 13, 14(44), 15(34)
Berchtold, G. A., 41
Bérczi, A., 182, 184, 188(31)
Beresford, S. A., 110
Bergeron, R. J., 374
Berglund, L., 163(15), 164
Berlin, J. A., 64, 66
Bernelli-Zazzera, A., 352
Bernstein, L., 110
Berry, M. N., 5
Beyer, R. E., 132, 156, 187
Bezombes, C., 200
Bhagavan, H. N., 15
Bhatia, U. G., 371
Bianchet, M. A., 187

Bianchi, G. P., 157
Bianchi, L., 342
Billingham, M. E., 362
Bilsen, T., 181
Biondi, R. M., 254
Bird, T. H., 222
Biscardi, A. M., 67
Biswal, S., 248, 278
Bizarri, M., 188
Björnstedt, M., 131, 133, 134(14; 15), 135(17), 136
Black, L. J., 112(17), 113
Blackburn, M., 112(16), 113
Blair, I. A., 33, 38(14), 46(14), 48(14), 53(14), 54(14), 59, 61, 62, 63
Blake, D. G., 239
Blank, V., 302(1), 303, 312(1)
Blatt, A. H., 55
Bleczinski, W., 56
Blizzard, T. A., 114, 118(23)
Bloch, B., 102
Bloch, W., 354
Block, O., 206
Blond-Elguindi, S., 112(19), 113
Blondin, G. A., 344, 359(19)
Bloom, D., 238(11), 239
Bloom, J., 350
Bly, C. G., 5
Boak, A., 27
Board, P. G., 336
Bocci, G., 350
Bodell, W. J., 116
Bohm, W., 153
Böhmer, F. D., 265
Bohren, K. M., 32
Boinapally, S., 229, 232(48), 233(48), 237(48), 238(13), 239
Bolden, S., 303
Bollinger, J. A., 19, 24(17)
Bolton, E. M., 187
Bolton, J. L., 110, 112, 112(18; 19), 113, 114(20), 116, 118(14; 15), 119, 119(20), 120, 121, 121(51), 122, 122(11), 123
Bompadre, S., 156, 158, 159, 164(19), 165, 166(19), 169(19), 170, 171, 174
Bondy, S. C., 371, 372(36)
Bonneu, M., 131
Bonora, P., 186, 188, 188(59)
Booth, J., 45
Boothman, D. A., 67(13; 14), 68, 69(14), 321
Borovansky, J., 88(4), 89, 108
Borre, M., 18
Borrelli, M. J., 276
Bossa, F., 193(96), 194
Böttger, M., 180, 183, 185(42)
Boucek, R. J., Jr., 350
Bouillaud, F., 124
Bouvier, F., 131
Bovina, C., 16
Bowry, V. W., 15, 156, 163, 170
Boyland, J. E., 45
Bozak, R. E., 277
Bradford, M. M., 226
Brand, K., 179
Brazzolotto, X., 342, 345(10), 353(10), 355, 357(59), 358(59), 360(59)
Bredesen, D. E., 371(37), 372
Breen, A. P., 63, 221
Breimer, L. H., 221
Breitenbroich, M., 351
Brennan, M. L., 163
Brenner, D. E., 350
Brewer, K. A., 254
Bridge, A., 179, 184(4), 185(4), 190(4), 192(4), 196, 198(4)
Bridges, K. R., 363
Brightman, A. O., 183, 191, 194, 194(83), 196(83)
Brightwell, J., 354(57), 355
Brismar, K., 16
Bristow, M. R., 362
Britigan, B. E., 221
Britt, R. D., 362
Brittenham, G. M., 374
Briviba, K., 266
Brock, B. J., 186, 188(61)
Bronmann, W. G., 67(13), 68
Brooks, D. E., 213
Brouchet, A., 200
Brown, A. J., 163
Brown, D. E., 27, 28
Brown, K., 117, 221
Brown, M. S., 4, 8
Brown, N. M., 345
Brown, P. C., 335
Brunk, U. T., 5, 148
Bruno, A. P., 200
Bruno, M., 194
Brusa, M. A., 68, 72, 72(16), 73
Brzovic, P., 25, 26(21)

Buchczyk, D. P., 259, 260(7), 268(7), 269(7), 272(7)
Buckhout, T., 181, 184, 185(47)
Buettner, G. R., 210
Buffinton, G. D., 322, 329(22)
Bundgaard, J. R., 18
Bungert, J., 276
Burczynski, M. E., 33, 35, 36(13), 38, 38(13), 39(13; 30), 45(13; 30), 46(13), 47, 48, 48(13; 30), 50(30), 53(13), 54, 61
Burdeleski, M., 171
Burgess, J. R., 187, 201, 206(6), 208, 214(10)
Burke, V., 163
Burkle, A., 369
Burlingame, A. L., 18
Burnett, P., 186, 223
Burón, M. I., 181, 197, 201, 205(7), 208, 210, 214(11; 16)
Burr, J. A., 187
Burrow, M. E., 287(18), 288
Bustin, S. A., 312
Butler, J., 195, 196(108)
Buttle, M. J., 276
Buxton, G. V., 103
Buytaert, P., 121, 123(46)
Buzdar, A. U., 117
Bychkov, R., 14
Byun, J., 120

C

Cadenas, E., 68, 87(21), 132, 133(13), 151, 187, 320, 322, 329(22), 330(9)
Cai, D. Y., 19, 20(16), 23(15), 28, 29
Cai, X., 114, 118(23)
Cairo, G., 340, 341, 342, 342(4; 5), 343(4), 344, 345, 347(22), 350(22), 351(3), 352, 355(22), 356(22), 357(22), 358(22), 359(22), 361(3)
Calabrese, L., 193(96), 194
Calafiore, A. M., 345, 347, 347(22), 350(22; 27), 354(28), 355(22), 356(22), 357, 357(22), 358(22), 359(22)
Caldwell, S., 194
Calkins, M. J., 243, 251(39), 278
Calzado, M. A., 201
Camara, B., 131
Campanella, A., 342, 345(12)
Cantin, A. M., 121
Cao, A., 229, 233, 242, 274
Caputo, I., 317
Carew, T. E., 163
Carlinin, P., 193(96), 194
Carlsson, K., 336
Carpentier, S., 362
Carpentier, Y. A., 6, 7(18), 200
Carr, B. I., 259
Carter, C., 207
Carter, W., 372
Carthew, P., 116
Casamayor, A., 254
Cass, A. E., 207
Castagnola, J., 371
Castedo, M., 201
Castelli, G. P., 16, 149
Castillo, F. J., 181, 183(25), 191(25)
Castrusini, E., 352
Casula, L., 233
Cathcart, R., 371
Caubergs, R., 183
Cavalcanti, M. da S. B., 67
Cavalieri, A. F., 347, 350(26), 356(26)
Cavalieri, E. L., 34, 45, 120, 120(43), 121
Cavazzoni, M., 187
Cavin, C., 244, 287
Ceccarelli, R., 67
Cerny, R. L., 120(43), 121
Chan, J. Y., 229, 231, 232, 238, 238(15), 239, 274, 278
Chan, K., 229, 231, 233, 242, 243, 244, 251(39), 274, 278
Chanas, S. A., 244, 287
Chandel, N. S., 371
Chang, J. C., 229, 231
Chang, M., 112(19), 113
Chapman, J., 14
Chau, Y.-P., 67
Chavin, W., 100
Chedekel, M. R., 105
Chen, C., 240, 287(16), 288, 305
Chen, H., 134
Chen, J.-W., 362
Chen, K., 163
Chen, L., 34
Chen, Q., 114, 118(23)
Chen, S., 187, 223, 328
Chen, Y., 100, 121, 122
Chen, Z. W., 18
Cheng, A. L., 259

Cheng, G., 183
Cheng, J. Z., 307
Cheung, P. C., 254
Chiba, T., 232, 233(63), 273, 280(6), 286
Chinje, E. C., 319
Chiou, T. J., 335
Chisolm, G. M., 163
Chitambar, C. R., 351, 353(43), 362, 363, 366(19), 369(19), 371, 372(19), 374(19)
Cho, M. K., 254, 288
Cho, N. M., 191, 192(89), 194, 194(89), 195(89), 198(89)
Choi, A. M., 223, 238(12; 13), 239
Choi, A. M. K., 273, 287(18), 288
Choi, J., 304, 308, 337
Choi, M. K., 229, 232(48), 233(48), 237(48)
Chojnacki, T., 3, 8, 10, 13, 14(44), 126, 152
Chopra, R. K., 15
Chow, C. K., 15
Chow, W., 259
Choy, H. A., 335
Chretien, D., 16
Chu, W. A., 252, 257(50), 288
Chueh, J., 189, 190(79), 191(79)
Chueh, P.-J., 181, 189(28), 190(28), 191, 191(28), 192(89), 194, 194(89), 195, 195(89; 97), 196(104), 197, 198, 198(89; 105)
Chung, J. H., 321
Chung, S. M., 321
Cipollone, A., 347, 350(27), 354
Cirillo, R., 354(57), 355
Citti, L., 349
Clapper, M. L., 234
Clark, M. G., 180, 184(13)
Clarke, C. F., 15, 216
Clarkson, P., 164
Clemens, J. A., 112(17), 113
Cobb, C. E., 211
Cohen, G. M., 321, 335(16)
Cohen, M. S., 221
Colditz, G. A., 110
Cole, R. N., 237, 238, 243(1), 277, 287
Coleman, J. E., 101
Coleman, W. B., 261
Collakova, E., 125
Coluzzi, P., 259
Comb, M. J., 313
Comeau, E., 116
Commentz, J., 171
Con, M.-J., 67
Conney, A. H., 35, 303
Conte, D., 345
Convert, O., 121
Cook, C., 207
Cook, J. L., 229, 232(48), 233(48), 237(48), 238(13), 239
Cooksey, C. J., 99, 100, 101(27), 108, 109
Cooley, L., 275, 276
Cordoba, F., 181(29), 182, 184, 186(44), 187(29), 188(29), 197, 198(117), 216
Corna, G., 340
Cornforth, R. H., 153
Corsaro, C., 107
Cortez, C., 45, 55
Cotter, T. G., 196
Cox, D. M., 313
Coyle, J. T., 243
Cradenas, F., 51
Cramer, F., 153
Crane, F. L., 179, 180, 181, 182, 183, 184, 184(13), 185(45; 46), 186, 186(44), 191, 194(83), 196(83), 197, 201, 208, 210, 211(14), 212(17; 20), 213, 213(14), 214(11; 17), 215(14), 216, 320
Crawford, D. R., 312
Cremonesi, P., 120
Crick, D. C., 4
Croft, K. D., 163
Cundari, E., 67
Cundy, K. C., 64
Cunningham, A. J., 351
Curreli, N., 186, 188(57)
Curtin, P. T., 229
Cuvillier, O., 200

D

D'Albuquerque, I. L., 67
Dale Poulter, C., 155
Dallner, G., 3, 4, 5, 6, 7, 8, 9, 10, 10(17), 11, 11(19; 34), 12, 13, 14, 14(10; 44), 15, 15(28; 34), 16, 17, 125, 126, 126(14), 127, 127(14), 128(14), 129, 130, 130(14; 24), 131, 131(30), 139, 146, 156(5), 157, 163, 163(15), 164, 170, 185
Dalton, T. P., 239
Danelisen, I., 351, 363
Danesi, R., 349, 350
Daniel, V., 236, 240

Daniels, J. R., 362
Danile, V., 273, 274(2)
Danilov, L. L., 126, 128(23)
Danson, M. J., 133(21), 134
Darley-Usmar, V. M., 337
Datta, S., 18
Daugas, E., 201
D'Aurelio, M., 16
Daval, J. L., 321, 328(20)
Davidson, B., 374
Davidson, M. B., 229, 274
Davis, R. J., 259
Dbaibo, G. S., 206
Deak, M., 254
Dean, A., 363
Dean, G., 207
Dean, R. T., 163
de Araujo, M. do C., 67
Debrauwer, L., 121
Dec, G. W., 362
de Cabo, D., 191, 201, 205(15), 206(15)
de Cabo, R., 187, 194, 201, 206(6), 208, 214(10)
Decker, H., 90, 97
Degen, J., 259
Degrassi, F., 67
De Greef, J. A., 183
deGrey, A. D. N. J., 195, 197(11)
DeHahn, T., 181, 189(28), 190(28), 191(28)
Dehal, S. S., 118
del Castillo-Olivares, A., 195, 196(104)
DellaPenna, D., 124, 125
De Long, M. J., 235, 243
Del Tacca, M., 349, 350
DeLuca, I., 188
Demarinis, R. M., 41
De Martinis, M., 158
De Matteis, F., 116
Deng, P. K. S., 223
DePierre, J. W., 10
Der, C. J., 287
Dernburg, A. F., 276
Desagher, S., 333
De Salvia, A., 67
Deutsch, U., 259
Devanesan, P. D., 34, 120, 120(43), 121
Devreese, B., 18
Dewald, B., 221
Deweindt, C., 234, 276
Deyashiki, Y., 33
Dhakshinamoorthy, S., 222, 224(16), 228(16), 229(16), 233, 234, 236(16), 237(16), 238(11), 239, 306
Dhawan, B. N., 303
Dhordain, P., 234, 276
Diab, K., 206
Dialameh, G. H., 258
Diamond, A. M., 67
Di Bernardo, S., 187
Dickinson, D. A., 302, 302(1), 303, 304, 307, 312(1), 315, 315(28), 319, 323, 333, 337, 338
Dieter, M. Z., 239
Di Giammarco, G., 347, 350(27)
Diliberto, E. J., Jr., 207
Dillinger, R., 90
DiNinno, F. P., 114, 118(23)
Dinkova-Kostova, A. T., 237, 238, 243(1), 277, 287
Di Paolo, A., 350
Dipple, A., 43
Disalvo, T. G., 362
Disch, 130
Disteche, C. M., 307
Dixit, V. M., 200
Do, T. Q., 15
Dodge, J. A., 111
Doi-Yoshioka, H., 206
Dolan, M. E., 67
Dolan, P. M., 244, 274, 287
Domann, F. E., 276
Dongen, W. V., 121, 122, 123(46)
Dooley, D. M., 18, 19, 24(17), 27, 28, 29
Döring, O., 183, 185, 185(42)
Doss, G. A., 114, 118(23)
Dove, J. E., 28, 29, 30
Dowd, N. P., 351
Drake, S. K., 343
Driscoll, T. A., 206
Drummond, G., 374
Druzhinina, T. N., 126, 128(23)
Dryer, J. L., 196
Dryhurst, G., 121(51), 122
Dubin, M., 67, 68, 70(15), 72, 72(16), 73, 75(15), 80(15)
Dubin, N., 363, 374(17)
DuBois, J. L., 17
Duckworth, H. W., 101
Duff, G. A., 88(2), 89
Duffy, S., 243

Duine, J. A., 18, 28
Dulik, D. M., 335
Dunlap, J. C., 199
Dupont, I. E., 6, 7(18)
Duranteau, J., 371
Durst, M., 270
Dusting, G., 374
Dwivedy, I., 120, 120(43), 121

E

Echtay, K. S., 124
Eckardt, D., 259
Eckhart, W., 259, 268(11)
Edens, W. A., 183
Edery, P., 16
Edholm, L. E., 101
Edlund, C., 4, 5, 9, 15(28), 170
Edlund, P. O., 171
Edmunds, L. N., Jr., 199
Egner, P. A., 244
Eiberger, J., 259
Eicken, C., 97
Eisenstein, R. S., 345
El-Assad, W., 206
Elcombe, C. R., 287
Eleouet, J. F., 229
Ellem, K. A., 196
Ellis, E. M., 33, 238
Elmberger, P. G., 4, 5, 14(10), 163(15), 164
Elmore, B. O., 19, 24(17)
El-Sabban, M., 206
Elsen, S., 222
Elthon, T. E., 187, 188(75)
Em, O., 191
Embrechts, J., 121, 122, 123(46)
Enari, M., 379
Engel, C. K., 276
Engel, J. D., 175(17), 234, 242, 243, 274, 275, 276, 276(17; 21), 282(21), 287, 289(12)
Enochs, W. S., 88(3), 89
Epstein, L. B., 353
Erb, H., 251
Erdjument-Bromage, H., 229, 274
Ericsson, J., 4, 8, 11
Eriksson, L. C., 133, 134(14; 15)
Eriksson, M., 157, 163(15), 164
Ernster, L., 14, 15, 16, 68, 70(17), 124, 125, 126(14), 127(14), 128(14), 130(14), 132, 133(13), 139, 146, 156(5), 157, 163, 185, 186, 320, 322, 329(22), 330(9)
Esch, G. L., 261
Esmans, E. L., 121, 122, 123(46)
Esposito, C., 317
Essigmann, J. M., 63
Estabrook, R. W., 68
Estévez, A., 208, 214(11)
Evans, D. C., 114, 118(23)
Exton, J. H., 254

F

Fabisiak, J. P., 156(9), 157
Fahl, W. E., 229, 242, 287
Faig, M., 187
Falcone, J. F., 112(17), 113
Fan, P., 112, 116, 118(14)
Faris, R. A., 261
Fariss, M., 335
Farmer, P. B., 117
Farr, A. L., 159
Fassy, F., 201
Fato, R., 16, 149, 187
Fattorini, D., 164(19), 165, 166(19), 169(19), 170, 171, 174
Faulk, P., 182
Faulk, W. P., 197, 320
Favier, A., 355, 357(59), 358(59), 360(59)
Favreau, L. V., 236, 291
Feigelson, H. S., 110
Feild, T. S., 124
Feir, B., 132
Fenoll, L. G., 99
Ferguson, S. J., 146
Fernández-Ayala, D. J. M., 201, 201(17), 202, 205(15), 206(15)
Fernandez Villamil, S. H., 67, 68, 70(15), 72, 72(16), 73, 75(15), 80(15)
Ferrans, V., 363, 372, 374(17)
Ferrante, L., 158
Ferri, E., 156
Ferri, K. F., 379
Festing, M., 116
Festjens, N., 380
Fickling, M. M., 264
Field, J. M., 64, 66
Fieser, L. F., 55
Finckh, B., 171
Finkel, T., 372

Fiore, M., 67
Fiorella, P. L., 157, 171, 176(13)
Fiorentini, D., 132, 187
Fiorini, R. M., 170
Fischer, A., 264
Fisher, G. R., 321, 322(13), 323(13)
Fitzgerald, B. J., 11
Fitzgerald, D. J., 351
Fleischer, S., 350
Fleischmann, B. K., 354
Fleury, C., 124
Fliesler, S. J., 4
Flower, L., 56, 58(63), 62(63)
Flowers-Geary, L., 35, 45(48), 46, 53, 54(53), 56
Fluckiger, R., 27
Flynn, T. G., 31
Fogli, S., 350
Folkers, K., 16, 156(7), 157, 164
Folsch, U. R., 351
Forman, H. J., 302, 302(1), 303, 304, 307, 308, 312(1), 315, 315(28), 319, 322, 323, 323(24), 325(24), 326(24), 327(24), 328(24), 330(24), 333, 333(24), 335, 337, 338
Formiggini, G., 16
Fornari, F. A., 201, 206(12)
Forrest, G. L., 354
Forsmark, P., 146
Forsmark-Andree, P., 14, 124, 185
Forster, M. J., 139, 141(3), 149
Forthoffer, N., 201, 204(14), 205(14), 206(14)
Foster, A. B., 112(16), 113
Foster, C. H., 41
Foster, N., 88(2), 89
Fournier, F., 121
Frantz, C., 179, 180
Frasch, M., 276
Fredlund, B., 182, 188(31)
Fredlund, K. M., 184, 187, 188(76)
Freeman, B. A., 321, 322(18), 323(18)
Freeman, H. C., 18
Freeman, M. L., 255, 276
Frei, B., 156, 163
Freshwater, S., 239
Fridovich, I., 74, 80, 329
Friling, R. S., 273, 274(2)
Frustaci, A., 357
Frydman, B., 67
Fu, P. P., 45, 55
Fu, X., 163
Fujisaki, T., 125
Fujita, H., 379
Fujita, K., 89
Fujiwara, K. T., 233
Fujiwara, T., 274
Fukazava, R., 362, 363(5)
Fukui, T., 29
Fukunaga, Y., 157, 171
Fukusaki, E. I., 125
Fukuyoshi, Y., 133
Furumura, M., 105

G

Gabbay, K. H., 32
Gaikwad, A., 223, 224(30)
Gaillard, J., 342, 345(10), 353(10)
Gairola, C. G., 365, 366(29)
Galli, M. C., 132, 187
Gallop, P. M., 27
Galloway, D. C., 239
Gant, T. W., 321, 335(16)
Ganther, H. E., 134
Gao, L., 304, 308, 337
Garbiras, B. J., 67
García-Borrón, J. C., 90, 95, 98(17), 100
Garcia-Canero, R., 180
García-Cánovas, F., 99
Garcia-Hendugo, G., 197
García-Ruiz, P. A., 99
Gardner, P. R., 321, 353, 364
Garner, A. P., 319
Garratt, P. J., 99, 100, 101(27)
Garten, S., 333
Gascoyne, P. R., 211
Gaskell, M., 117
Gauthier, S., 112
Gavrieli, Y., 376
Gay, G. A., 162
Gederaas, O., 181
Gehring, N. H., 361, 363
Geilin, C., 194
Gelboin, H. V., 35, 221, 223(6)
Gelei, S., 201
Gelvan, D., 359
Genova, M. L., 16
Gentiloni, N., 357
Gerber, P. A., 259, 260(7), 268(7), 269(7), 272(7)

Gerdemann, C., 97
Geremia, E., 107
Geromel, V., 16
Gervasi, P. G., 349
Gewirtz, D. A., 201, 206(12)
Ghatak, N., 303
Giachetti, A., 354(57), 355
Giampietro, R., 357
Gianni, L., 347, 354(28)
Giannis, A., 206
Gibson, J., 195
Gilbert, H. F., 191
Gill, H. S., 45
Gille, L., 341, 348
Gimeno, C. J., 64
Gimino Martins, D., 67
Gipp, J. J., 239, 273, 305, 307, 307(15), 315(15)
Girotti, S., 156
Giuliano, G., 16
Giulivi, C., 337, 338
Glowinski, J., 333
Goel, R., 355
Gohil, K., 158
Goijman, S., 67
Gold, M. H., 186, 188(61)
Goldenberg, H., 180, 183
Goldhamer, D. J., 313
Goldstein, J. L., 4, 8
Goldzheher, H. W., 121
Gomes, D. C., 303
Gomes-Cardoso, L., 303
Gómez-Díaz, C., 181, 186(22), 200, 201, 201(17), 202, 204(18), 205(7; 18), 206(18), 210, 211(14), 212(17), 213(14), 214(17), 215(14), 333
Gomez-Rey, M. L., 191
Gong, P., 287(18), 288
Gonzales-Reyes, J. A., 181(29), 182, 187(29), 188(29), 197, 198(117)
Gonzalez, B., 354
Gonzalez, F. J., 7, 11(19), 223
Gooden, J. K., 120(43), 121
Goodenough, D. A., 259
Goodman, J., 112(16), 113, 362
Gopishetty, S. R., 61
Gorin, M. B., 118
Gorini, A., 158
Gorman, A., 196
Goto, G., 14
Goto, Y., 25
Gottlieb, J. A., 362
Gouazé, E., 200, 206
Gouaze, V., 362
Gozal, E., 335
Gralla, E. B., 371(37), 372
Grant, C. V., 362
Grant, D., 313
Grant, S., 201, 206(12)
Grebing, C., 180, 181, 184(13), 197
Green, D. R., 379
Green, E. L., 29
Green, M. D., 363, 374(17)
Green, R. M., 355
Greenlee, R. T., 303
Greenstock, C. L., 103
Gregor, W., 362
Greppin, H., 199
Grese, T. A., 111
Grewaal, D., 355
Griendling, K. K., 330
Grifi, D., 186, 188(57)
Grimshaw, C. E., 32
Grisham, J. W., 261
Grisham, M. B., 221
Grnler, J., 3
Grollman, A. P., 117
Groopman, J. D., 33, 244
Gross, M. L., 120, 120(43), 121
Grossi, G., 157, 171, 176(13)
Grünler, J., 4
Gu, C., 123
Guarente, L., 226
Guerrini, F., 188
Guingerich, F. P., 221
Guiraud, P., 355, 357(59), 358(59), 360(59)
Gujral, P. K., 303
Guldenagel, M., 259
Gulluyan, L., 374
Gunther, H., 41
Gupta, S. C., 56
Guss, J. M., 18
Gutierrez, P. L., 321, 322(13), 323(13)
Gutterman, D., 374

H

Hacisalihoglu, A., 18
Hahn, H., 371(37), 372
Haider, N., 362
Haile, D. J., 344, 359(19)

Hainaut, P., 61
Haines, T. H., 124
Haller, H., 14
Hamamura, K., 12, 13(43), 14(43)
Hamasaki, N., 132
Hamilton, R. L., 5
Hammond, M. L., 114, 118(23)
Han, X., 229
Han, X. D., 242
Han, X. L., 274
Handa, S., 201
Hannun, Y. A., 200, 201, 206, 206(1)
Hanson, J. M., 241, 252, 257(50), 288
Hara, A., 33
Harford, J. B., 344, 359(19), 363
Harrer, J. M., 239
Harris, H., 101
Harrison, D. J., 305
Hartmann, C., 25, 26(21)
Hartzell, P., 321, 323(12)
Harvey, I., 18
Harvey, R. G., 33, 35, 36(13), 38, 38(13; 14), 39(13; 30), 43, 45, 45(13; 23; 30; 47; 48), 46, 46(13; 14; 23), 47, 48, 48(13; 14; 23; 30), 49(24), 50, 50(30; 50), 51, 52, 52(50), 53, 53(50; 13; 14), 54(14; 53), 55, 56, 61
Hashizume, N., 132
Hatayama, I., 232, 233(63), 273, 280(6), 286
Hatefi, Y., 165, 172, 179, 185(2)
Hayakawa, T., 133
Hayashi, N., 230, 232, 233(63), 273, 274, 275, 276, 278(13), 280(6), 284(13), 286
Hayashi, T., 187
Hayes, J. D., 33, 238, 238(5; 7), 239, 244, 256, 287, 289
Hayes, M. K., 187, 188(75)
Hazen, S. L., 163
He, L., 239
Hearing, V. J., 105
Hebbar, V., 240, 287(16), 288, 305
Heide, L., 125
Heistad, D., 374
Helman, W. P., 103
Hemmerlin, A., 130
Henderson, B. E., 110
Henderson, B. R., 342, 343, 344, 345(13), 355(18), 359(18)
Henderson, C. J., 244, 287
Hendrix, S. L., 110
Hentze, M. W., 342, 345(10), 353, 353(10), 361, 363
Herrlich, P., 265
Hertel, R., 188
Hescheler, J., 354
Heuer, S., 183, 185(42)
Hicks, R. J., 277
Hidaka, M., 191
Hidalgo, A., 197
Higginbotham, S., 34, 45, 120(43), 121
Higgins, P. J., 312
Hijjar, R. J., 362
Hirashima, M., 133
Hirling, H., 344, 355(18), 359(18)
Hirooka, K., 125
Hirota, S., 26
Hirsch-Ernst, K. I., 369
Hixson, D. C., 261
Hobeck, J., 191
Hochstein, P., 68, 132, 133(13), 320, 330(9), 362
Hodgson, J. M., 163
Hoffman, L. A., 180, 256
Hofmann, K., 200
Hogdall, E. V. S., 18
Hogg, N., 347, 353(30)
Hohmura, K. I., 276
Holloway, P. W., 152
Holm, P., 15
Holmgren, A., 134, 136
Holtzclaw, W. D., 223, 229, 237, 238, 243(1), 277, 287
Hoog, S. S., 32
Hops, B., 354
Hortobagyi, G. N., 117
Hoshino, H., 274, 275, 276
Hou, Y.-t., 54
Houen, G., 18
Howard, B. V., 110
Hrubec, T. C., 181
Hsieh, R. K., 259
Hsiu-Chuan, Y., 365, 366(29)
Hu, H., 338
Hu, R., 240, 305
Huang, C. M., 179
Huang, H.-C., 229, 232(50), 233(50), 237(50), 242, 252, 255, 256(56), 276, 286, 288, 289
Huang, M. T., 303
Huang, R. C. C., 236
Huang, Z., 112(18), 113

Hubner, C., 171
Huhtala, H., 15
Huizinga, M., 36
Hundall, T., 124
Hung, C.-F., 35, 54, 61
Hung, H. L., 259
Hutson, J. L., 11

I

Iba, H., 233
Ibrahim, W. G., 15
Ida, Y., 157, 171
Idriss, N. Z., 206
Igarashi, K., 229, 230, 232, 233(63), 234, 243, 273, 274, 275, 276, 276(21), 278(13), 280(6), 282(21), 284(13), 286, 287, 289(12)
Ikeda, T., 18
Iles, K. E., 302, 302(1), 303, 307, 312(1), 315, 315(28), 333, 337, 338
Iliskovic, N., 340
Imada, I., 12, 14
Innocenti, F., 350
Inui, M., 350
Ioannidis, G., 158
Irani, K., 372
Isaksson, M., 197
Ischiropoulos, H., 371, 372(36)
Ishida, T., 33
Ishii, T., 80, 232, 233(63), 234, 243, 273, 275, 276, 276(21), 280(6), 282(21), 286, 287, 289, 289(12), 329, 372
Ishikawa, M., 252, 254(49)
Iskander, K., 223, 224(30)
Islam, N. B., 56
Isler, O., 258
Israel, M., 350
Ito, E., 230, 274, 278(13), 284(13)
Ito, N., 276, 282(30)
Ito, S., 89, 91, 92, 105
Itoh, K., 232, 233(63), 234, 238, 243, 243(1), 244, 248, 256, 273, 274, 275, 276, 276(21), 277, 278, 280(6), 282(21), 286, 287, 289, 289(12)
Itoh, S., 121
Iverson, S. L., 113, 114(20), 119(20)
Ivins, J. K., 41, 42(33), 54(33)
Iwabuki, H., 18
Iwai, K., 343
Iwamoto, T., 26, 307, 315, 315(28), 337
Iyanagi, T., 185, 319

J

Jackson, R. D., 110
Jacobs, E., 191
Jaenicke, E., 97
Jaffrézou, J. P., 200
Jaiswal, A. K., 68, 186, 187, 221, 222, 223, 223(33; 34), 224, 224(16; 30), 226(36), 227, 228, 228(15; 16), 229, 229(15; 16; 36), 231(47), 232(47), 233, 233(47), 234, 234(47), 236(15; 16), 237(16; 47), 238(8; 11), 239, 242(8), 286, 305, 306, 320
Jakobsson-Borin, Å., 17
James, R. F., 238
Janes, S. M., 18, 28, 28(2)
Jankowiak, R., 34
Jarman, M., 112(16), 113, 116
Jarvis, W. D., 201, 206(12)
Jayaraman, J., 8
Jay-Gerin, J. P., 121
Jeffrey, A. M., 41
Jemiola-Rzeminska, M., 124
Jeong, H., 34
Jergil, B., 101
Jerina, D. M., 41, 43, 45, 56
Jernstrom, B., 238
Jevnikar, M. G., 112(17), 113
Jewell, S. A., 321, 323(12)
Jez, J. M., 31, 32, 32(8), 33
Jiang, L., 251
Jiang, Q., 287
Jo Davisson, V., 155
Johansson, S. L., 120(43), 121
Johnson, C. I., 108, 109
Johnson, D. A., 238, 241, 251
Johnson, J., 111
Johnson, J. A., 238, 239, 241, 243, 244, 245, 245(38), 248, 251, 251(39), 252, 257(38), 257(50), 278, 288, 305
Johnson, K. C., 110
Jones, B. N., 110
Jones, C. D., 112(17), 113
Jones, D. P., 144
Jones, T. W., 335
Jongejan, J. A., 18

Jordan, A. M., 109
Jordan, V. C., 118
Joseph, J., 351, 353, 353(43), 362, 363, 363(6; 9), 366(19), 369(19), 372(19), 374(19)
Joseph, P., 68, 187, 222, 223, 228(15), 229(15), 236(15)
Joshi, V. C., 8, 10
Joyard, J., 125, 128(18)
Juda, G. A., 19
Judah, D. J., 33
Jurkiewicz, B. A., 210

K

Kadhom, N., 16
Kagan, H. M., 27
Kagan, V. E., 147, 156(6; 9), 157, 208, 320
Kahl, G. F., 369
Kalén, A., 5, 6, 14(10), 126, 163(15), 164, 170, 185
Kalinchuk, N. A., 126, 128(23)
Kalivendi, S. V., 351, 353, 353(43), 362, 363, 363(9), 366(19), 369(19), 372(19), 374(19)
Kallajoki, M., 351
Kalthoff, H., 351
Kalyanaraman, B., 347, 350, 351, 353, 353(30; 43), 362, 363, 363(6; 9), 365, 366(19), 369(19), 372(19), 374(19)
Kamada, S., 379
Kampbell, K. A., 362
Kamzalov, S., 138
Kan, Y. W., 229, 231, 232, 233, 242, 243, 244, 251(39), 274, 278
Kane, D. J., 371(37), 372
Kanemitsu, M. Y., 259, 268(11)
Kang, D., 132
Kang, K. W., 254, 255, 287(19), 288
Kang, Y. J., 351
Kano, K., 18
Kansler, T. W., 248, 278
Kanzaki, T., 133
Kar, S., 259
Karas, R. H., 164
Karin, M., 228
Karuzina, I. I., 77, 78(28)
Kasprzak, K. S., 221
Kataoka, K., 229, 233, 274
Katoh, K., 232, 233(63)
Katoh, Y., 234, 237, 238, 243, 243(1), 273, 275, 276, 276(21), 277, 280(6), 282(21), 286, 287, 289, 289(12)
Katsuoka, F., 175(17), 242, 274, 276(17)
Kaur, S., 18
Kavanagh, T. J., 307
Kawaguchi, K., 18
Kawai, S., 233
Kay, G. F., 196
Keaney, J. F., Jr., 163
Keenan, R. W., 126
Keenan, T. W., 199
Keller, R. K., 3, 4, 11, 126
Kelley, P. M., 207
Kelly, R. A., 362, 363(5)
Kelso, R., 276
Kempf, C., 363
Kennedy, M. C., 344, 345, 353, 359(19), 365
Kensler, T. W., 33, 244, 248, 273, 274, 278, 287
Kepa, J. K., 320, 333(11)
Kettle, A. J., 163
Khan, T. H., 109
Khaw, B. A., 362
Khoo, J. C., 163
Khorasanizadeh, S., 11
Khrapova, N. G., 146(7), 147
Kiechle, F. L., 374
Killeen, E., 255, 276, 289
Kim, C., 181, 182, 191, 192(89), 194, 194(89), 195(89), 198(89), 320
Kim, I. F., 236
Kim, J. K., 18
Kim, M. C., 132
Kim, M. J., 233
Kim, S. G., 254, 255, 287(19), 288
Kimura, S., 223, 319
King, R. A., 90
King, R. G., 305
Kirshenbaum, L. A., 351, 363
Kishi, T., 132, 157, 171, 179, 184(3), 185, 185(3), 186, 192(3), 198(3), 320
Kishishita, S., 26
Kitamura, T., 133
Klapper, M. A. H., 324
Klausner, R. D., 343, 344, 359(19), 363
Klein, J. B., 351
Klein, M., 124
Kleining, H., 125, 128(12)
Klein-Sazanto, A. J. P., 223

Kleiter, M., 341
Klinge, C. M., 317
Klingenberg, M., 124
Klinger, W., 223
Klinman, J. P., 17, 18, 19, 20(16), 23(15), 24, 25, 26, 26(21; 22), 27, 28, 28(2; 27), 29, 30
Klotz, L.-O., 258, 259, 260(7), 266, 268(7), 269(7), 272(7)
Knebel, A., 265
Knight, L. P., 33
Knoth, J., 207
Knowles, P. F., 27, 276, 282(30)
Knox, R., 223, 224(30), 328
Kobayashi, A., 125, 230, 273, 274, 278(13), 284(13)
Kobierski, L. A., 313
Kockx, M., 121, 123(46)
Kocsis-Bedard, S., 121
Kogame, K., 230, 274, 278(13), 284(13)
Kohlschutter, A., 171
Kohsaka, H., 201
Koike, K., 133
Koike, M., 133
Koike, R., 201
Kojo, S., 372
Kole, P. L., 45
Kolesnick, R. N., 201, 206(12)
Kolodgie, F. D., 362
Kong, A.-N., 240, 287, 287(16), 288, 305
Konorev, E. A., 351, 353, 362, 363(6), 365
Kontush, A., 171
Kooperberg, C., 110
Koppenol, W. H., 195, 196(108)
Korfmacher, W. A., 350
Kornberg, T. G., 276
Kostetskii, E. Y., 155
Kotamraju, S., 351, 353, 353(43), 362, 363, 363(6; 9), 366(19), 369(19), 372(19), 374(19)
Kotchen, J. M., 110
Koya, R. C., 379
Kraft, A. D., 251
Kramer, E., 363, 374(17)
Krasasakis, K., 194
Krebs, B., 97
Kretzschmar, M., 223
Kreutzer, D. A., 63
Krikorian, A., 206
Kristensson, K., 4, 170
Kritharides, L., 163
Kroemer, G., 201, 379
Krol, E., 112(18), 113
Kruczek, M., 126
Kruger, S., 180
Krzywanski, D. M., 307, 315(28), 333
Ku, H.-H., 148
Kubota, I., 363
Kuhn, L. C., 342, 343, 344, 355(18), 359(18)
Kulisz, A., 371
Kumagai, T., 240
Kumakura, S., 206
Kumar, D., 351, 363
Kumar, V., 18
Kunihisa, M., 125
Kuo, M.-L., 67
Kupfer, D., 118
Kurata, W. E., 259, 268(11)
Kuroda, S., 18
Kusano, H., 379
Kushi, Y., 201
Kuttan, G., 303
Kuttan, R., 303
Kuzumaki, N., 379
Kwak, M.-K., 244, 248, 274, 278, 287
Kwok, J. C., 344, 352(20), 355(20), 358(20), 359(20), 360(20), 361(20)
Kwong, L. K., 139, 140, 141(5), 147, 148(12)
Kwong, M., 231, 232, 238, 278

L

Laaksonen, A., 124
Labrie, F., 112
LaCroix, A. Z., 110
Laemmli, U. K., 204
Lagendijk, J., 171
Lagerstedt, A., 15
Lai, C. J., 32
Lake, B. G., 77
Laloi, M., 124
Lambert, C., 103
Lambeth, J. D., 183
LaMont, J. T., 69
Lampe, P. D., 259, 268(11)
Land, E. J., 88, 94, 96(12), 98, 99, 100, 101(27), 102, 103, 105, 105(12), 108, 109
Landi, L., 132, 187
Lang, J. K., 157, 158, 171
Langsjoen, P., 164
Larm, J. A., 179, 201

Larsson, A., 336
Larsson, L. I., 18
Lash, L. H., 144
Lashley, M. R., 362
Lass, A., 139, 141, 141(3; 4), 142, 142(9), 143, 147, 148(11; 12), 149, 149(11), 150, 151, 151(17)
Last, J. A., 221
Latter, A. J. M., 108, 109
Lau, A. F., 259, 268(11)
Laudenschlager, W. G., 113
Laurent, G., 200
LaVaute, T., 343
Lawen, A., 179, 201
Lawler, J., 191, 199
Lawrence, J., 194, 197
Lawson, M. F., 305
Learmonth, B. A., 195
LeBel, C. P., 371, 372(36)
Lebideau, M., 16
Leclercq, G., 112(16), 113
Lecocq, G., 234, 276
Lee, C. H., 254, 288
Lee, C. P., 14
Lee, H. H., 45
Lee, I. Y., 16
Lee, J. L., 321
Lee, J.-M., 238, 241, 243, 245(38), 251(39), 252, 257(38; 50), 278, 288, 305
Lee, M. Y., 321
Lee, P. J., 238(12), 239, 273
Lee, S.-H., 33, 38(14), 46(14), 48(14), 53(14), 54(14), 59, 61
Lee, S. J., 255, 288
Lee, Y. H., 223
Legault, C., 110
Legros, N., 112(16), 113
Lehman, P., 350
Lehr, R. E., 45
Lei, W., 287
Leibold, E. A., 343
Leichtweis, S. B., 163
Lemay, R., 121
Lemiere, F., 121, 122, 123(46)
Lenaz, G., 16, 149, 156, 187, 195
Lenormand, P., 266
Leon, L. L., 303
Leone, L., 156, 158
Leong, L., 259
Leprince, D., 234, 276
Leth, A., 171
Letters, J., 163
Levade, T., 200, 206, 362
Levine, R. L., 343
Levonen, A.-L., 337
Lewis, M., 31, 32, 32(8), 33
Leykauf, K., 270
Li, C. J., 67, 69
Li, J., 238, 243, 244, 245, 245(38), 248, 257(38), 288, 305
Li, K.-M., 45
Li, R., 187
Li, R. B., 18
Li, S. M., 125
Li, T., 351, 363
Li, W., 119
Li, X., 354
Li, Y., 69, 224, 226(36), 229(36), 234
Li, Y. Z., 67
Liberi, G., 345, 347, 347(22), 350(22; 27), 355(22), 356(22), 357, 357(22), 358(22), 359(22)
Licata, S., 347, 348(25), 349(25), 350(25; 27), 354(28)
Lichtenthaler, H. K., 130
Liebler, D. C., 187
Liehr, J. G., 110
Liem, H. H., 51
Lim, C. K., 116
Lima, M. E., 303
Lind, C., 68, 132, 133(13), 320, 330(9)
Lind, J., 151
Lindbladh, C., 101
Lindgren, A., 184, 185(45), 197
Lindsey, C. C., 200
Linnane, A. W., 179
Lins Lacerda, A., 67
Lippa, S., 170
Littarru, G. P., 16, 156, 156(7; 11), 157, 164(19), 165, 166(19), 169(19), 170, 171, 171(3), 174
Liu, A. A., 67
Liu, B., 206
Liu, D., 229
Liu, L. F., 67
Liu, R.-M., 304, 308, 319, 321, 337, 338
Liu, T. W., 259
Liu, W., 229
Liu, X., 112, 118(14), 123
Ljones, T., 209

Lok, C. N., 342
Lokesh, B. R., 303
Lombini, A., 188
Long, D. J. II, 222, 223, 223(33; 34), 224, 224(16; 30), 228(16), 229(16), 236(16), 237(16), 320
Lönnrot, K., 15
Loo, L. W., 259, 268(11)
López-Lluch, G., 201, 201(17), 202, 205(7; 15), 206(15)
Lorke, D. E., 354
Löw, H., 179, 180, 181, 184, 184(13), 185(45), 197, 212(20), 213
Löw, P., 9, 15(28)
Lowery, M. D., 97
Lowry, O. H., 159
Lu, A. Y. H., 222
Lu, R., 231
Luethy, M. H., 187, 188(75)
Luft, F. C., 14
Lund, E., 163(15), 164
Lundahl, J., 16
Luster, D. G., 181, 184, 185(47)
Lüthen, H., 183, 185(42)
Lüthje, S., 183, 185, 185(42)
Lutke-Brinkhaus, F., 125, 128(12)
Lyles, M. M., 191

M

Ma, H., 32, 32(8), 33, 36
Ma, W., 362
Mabon, N., 117
Machida, K., 14
Machnik, G., 223
Macho, A., 201
Maciel, M. C. N., 67
MacKellar, W. C., 186, 197, 208
MacNee, W., 305
Maessen, P. A., 350
Maggi, C. A., 347, 350(27), 354
Maglio, M., 317
Maguire, J. J., 147
Mai, K. H., 248, 278
Mailhot, J., 112
Makar, A., 121, 123(46)
Malkin, H., 109
Malkinson, A. M., 113
Maltby, D., 18
Maltsev, S. S., 126, 128(23)
Mancuso, C., 357
Mandlekar, S., 240, 287, 305
Mann, B. R., 264
Mann, J., 354
Mannervik, B., 336
Manson, M. M., 116
Manzini, S., 347, 350(27), 354, 354(57), 355
Mao, Y., 67
Marcovich, R., 362
Margoliash, E., 195, 196(108)
Margolin, K., 259
Marini, M. G., 233
Marques, M. M., 113, 117, 123(21)
Marsh, J. B., 5
Martasek, P., 347, 353(30)
Martin, E. A., 116, 117
Martín, S. A., 201(17), 202
Martín, S. F., 201, 202, 204(14; 18), 205(14; 15; 18), 206(14; 15; 18)
Martínez-Ortíz, F., 99
Martinus, R. D., 179
Marton, L. J., 67
Marzari, R., 317
Masala, S., 186, 188(57)
Masaln, H., 191
Maser, E., 33
Mashima, R., 163
Mason, H. S., 95, 101(16)
Mason, J. W., 362
Mason, R. P., 321, 335(16), 350, 362
Massiah, M. A., 277
Masters, B. S., 347, 353(30)
Mastovich, S. L., 113
Masuda-Inoue, S., 206
Mathews, F. S., 18
Matschiner, J. T., 258
Matsuda, H., 319
Matsumoto, N., 12
Matsunami, H., 18
Matsuura, K., 33
Matsuzaki, R., 29
Matthews, R. T., 141
May, J. M., 207, 211
Mayer, H., 258
Mazel, P., 75
Mazzanti, L., 170
McBride, O. W., 186, 223
McCague, R., 112(16), 113, 116
McCahon, C. D., 19
McCarron, D. A., 14

McCarthy, A. D., 11
McCay, P. B., 146
McCord, J. M., 221
McCoull, K. D., 35, 54, 61, 62, 63
McCullough, K. D., 261
McDermott, J. C., 313
McDonaugh, K. T., 229
McGowan, A., 196
McGuirl, M. A., 18, 19, 28, 29
McIntire, W. S., 18
McKeown, K. A., 19
McLaughlin, J. A., 211
McLellan, L. I., 239, 244, 287
McLeod, R., 238
McLuckie, K. I., 117
McMahon, M., 244, 256, 287, 289
McMullen, G. L., 179
McPherson, M. J., 27
McWalter, G. K., 287
Mead, E. W., 113
Medeiros Maciel, G., 67
Medin, J. A., 200
Medina, M. A., 181, 197
Meister, A., 304, 307
Melchheier, I., 258
Mellors, A., 146
Melzer, M., 125
Mendelsohn, J., 371
Meneghini, R., 221
Menna, P., 340, 342, 344(11), 345(11), 347, 350(27), 352(11), 353(11), 354, 354(28), 355(11), 357, 360(11)
Meredith, M. J., 276
Merenyi, G., 151
Miano, J. M., 371
Mignotte, V., 229
Mikhael, N., 355
Milbradt, R., 157, 158(19), 171
Miller, A.-F., 362
Miller, F., 374
Miller, K. I., 97
Miller, L. L., 5
Mills, S. A., 19, 24
Minakami, P., 146
Minato, N., 343
Minicucci, L. A., 134
Minn, A., 321, 328(20)
Minotti, G., 340, 341, 342, 342(5), 344(11), 345, 345(11; 12), 347, 347(22), 348(25), 349(25), 350(22; 25–27), 351(3), 352, 352(11), 353(11), 354, 354(28), 355(11; 22), 356(22; 26), 357, 357(22), 358(22), 359(22), 360(11), 361(3), 362
Minow, R. A., 362
Mirault, M. E., 362
Mirejovsky, P., 108
Mishizawa, M., 233
Misra, H. P., 74
Misra, P. C., 184
Miu, R. K., 191, 194(83), 196(83)
Miyabe, Y., 33
Miyajima, N., 233
Miyamoto, M., 240, 243
Miyamoto, T., 363
Miyasaka, N., 201
Mizobachi, S., 186
Mizobuchi, S., 320
Mizushima, N., 201, 279
Mizutani, A., 372
Mo, Y.-Y., 287(16), 288
Moehlenkamp, J. D., 239, 241
Moellering, D. R., 337
Moenne-Loccoz, P., 28
Moffat, G. J., 287
Mohammadi, E., 236
Mohr, D., 15, 164, 165(18), 169, 169(18), 170
Moi, P., 229, 233, 242, 274
Moinova, H. R., 229, 232(49), 233(49), 237(49), 239, 307
Molina Portela, M. P., 67, 68, 72, 72(16), 73
Molines, H., 121
Møller, I. M., 182, 184, 187, 188(31; 76)
Monks, T. J., 121(51), 122
Montecucco, C., 52, 84
Montgomery, C. A., 223
Monti, E., 341, 351(3), 361(3)
Montpetit, V. A., 355
Moore, D. J., 320
Moore, P. H., 121
Moorthy, B., 116, 117
Mordente, A., 345, 347, 347(22), 348(25), 349(25), 350(22; 25; 26), 355(22), 356(22; 26), 357, 357(22), 358(22), 359(22)
Morehouse, K. M., 350
Morgan, R. J., Jr., 259
Mori, K., 186, 320
Mori, T. A., 163
Mori, Y., 18
Morimoto, H., 12, 14

Morré, D. J., 179, 180, 181, 182, 183, 183(3–5; 25), 184, 184(3–5), 185(3–5; 45), 188(35; 38), 189, 189(81), 190(28; 79), 191, 191(25; 28; 38; 79), 192, 192(3; 4; 81; 89), 193, 193(91; 94), 194, 194(35; 80; 83; 84; 89), 195, 195(89; 97), 196, 196(5; 35; 80; 83; 95; 104; 106; 107), 197, 197(35; 95), 198, 198(3–5; 89; 91–95; 105), 199, 199(91; 93–95), 202, 203(19), 213
Morré, D. M., 179, 184(3; 5), 185(3; 5), 189, 190(4; 79), 191, 191(79), 192, 192(3; 89), 193, 193(91; 94), 194, 194(80; 89), 195, 195(89; 97), 196, 196(5; 80; 95; 104; 106; 107), 197, 197(95), 198, 198(3; 5; 89; 91–95; 105), 199, 199(91; 93–95)
Morrow, C. S., 238(14), 239
Mortensen, S. A., 171
Morton, M. R., 228(43), 229, 239, 273, 274(1), 286, 291(2)
Mosca, F., 164(19), 165, 166(19), 169(19), 170, 171, 174
Mosialou, E., 238
Motohashi, H., 175(17), 242, 248, 274, 276(17), 278
Motteram, J. M., 11
Moulis, J.-M., 342, 345(10), 353(10), 355, 357(59), 358(59), 360(59)
Moutsoulas, P., 201
Mu, D., 18, 28
Mueller, S., 363
Muhlenweg, 125
Mukhopadhyay, A., 303
Mulcahy, R. T., 229, 232(49), 233(49), 237, 237(49), 239, 273, 287(17), 288, 305, 306, 307, 307(15), 315(15), 337
Mulier, B., 305
Muller-Eberhard, U., 51
Muller-Rober, B., 124
Multani, A., 223
Munday, R., 80, 322, 329, 329(23), 330(23)
Munnich, A., 16
Muñoz, E., 201
Munro, H. N., 343
Murakami, A., 372
Mure, M., 18, 24, 27, 28, 28(27)
Murphy, J. A., 221
Murphy, J. P., 63
Murphy, T. H., 243, 251
Murray, J. M., 27
Murray, T., 303
Murty, V. S., 56, 59
Muto, A., 275, 276
Myers, C. E., 350
Mysliwa-Kurdziel, B., 124

N

Nabeshima, Y., 232, 233(63), 273, 280(6), 286
Nabuiri, A. M. D., 8
Nagasawa, T., 276
Nagata, S., 279, 379
Nagata-Kuno, K., 132
Nagley, P., 179
Naish-Byfield, S., 108
Naito, Y., 240
Nakajima, O., 275
Nakamura, M., 187
Nakamura, N., 28, 29
Nakamura, T., 12, 13(43), 14(43)
Nakamura, Y., 240
Nakanishi, H., 254
Nakauchi, H., 276
Nakayama, T., 33
Nantz, M. H., 362
Naquin, R., 255, 276, 289
Nara, F., 206
Narayanan, P. K., 372
Narula, J., 362
Nash, T., 75
Naumov, V. V., 146(7), 147
Navarro, F., 181, 184, 186, 186(22; 44), 187, 201, 204(14), 205(14), 206(6; 14), 208, 210, 211(14), 212(17), 213(14), 214(10; 11; 17), 215(14), 333
Navas, P., 15, 180, 181, 181(29), 182, 183(25), 184, 186, 186(22; 44), 187, 187(29), 188(29), 191(25), 197, 200, 201, 201(17), 202, 203(19), 204(14; 18), 205(7; 14; 15; 18), 206(6; 14; 15; 18), 207, 208, 210, 211(14), 212(17; 20), 213, 213(14), 214(10; 11; 16; 17), 215(14), 216, 320
Nayfield, S. G., 118
Neal, G. E., 33, 238
Nebert, D. W., 223, 239, 321
Nedbal, L., 124
Neder, K., 67
Nelson, K. G., 261
Netter, P., 321, 328(20)
Ney, P. A., 229

Ngo, E. O., 259, 321, 322(13), 323(13)
Ngui, J. S., 114, 118(23)
Nguyen, T., 229, 232(50), 233(50), 237(50), 242, 252, 255, 256(56), 276, 286, 287(1), 288, 289, 306
Nicholls, D. E., 146
Nienhuis, A. W., 229
Niimura, Y., 191
Niki, E., 146, 156(8), 157
Nikolic, D., 119, 122
Nilges, M. J., 88(3), 89
Nishikawa, S., 276
Nishimura, N., 274
Nishina, H., 233
Nishino, T., 191
Nishizawa, M., 27, 229, 233, 274, 275
Nitobe, J., 363
Niu, X., 163
Njalsson, R., 336
Njus, D., 207
Noda, M., 233, 274
Nohl, H., 156(6), 157, 341, 348
Nordenbrand, K., 146
Nordgren, H., 67
Nordman, T., 131, 133, 135(17)
Norgren, S., 336
Norling, B., 16, 126, 146, 185
Norris, S. R., 124
Not, T., 317
Nourooz-Zadeh, J., 324
Nowack, D. D., 202, 203(19), 213
Nozaki, N., 363
Numazawa, S., 252, 254(49)
Nunerz de Castro, I., 181
Nutter, L. M., 259, 321, 322(13), 323(13)

O

O'Banion, M. K., 371
Oberley, T. D., 365, 366(29)
Ockene, J., 110
O'Connor, T., 175(17), 242, 274, 276, 276(17), 289
Oetting, W. S., 90
Ogita, T., 206
Ogumbumni, M. E., 350
Ohad, I. I., 124
Ohashi, T., 372
Ohigashi, H., 240
Ohnishi, K., 191
Ohnishi, S. T., 50(50), 51, 52, 52(50), 53(50), 56, 58(63), 62(63)
Ohnishi, T., 50(50), 51, 52, 52(50), 53(50)
Ohno, T., 12, 13(43), 14(43)
Ohshima, H., 121
Ohta, T., 273
Ohtsu, M., 379
Okajima, T., 18, 26
Okamoto, T., 132, 157, 171, 185, 186, 320
Okuyama, M., 363
Olin, B., 336
Olivares, C., 95, 98(17)
Ollinger, K., 322, 329(22)
Olson, R. D., 350
Olson, R. E., 258
Olsson, J. M., 8, 131, 133, 134(14; 15), 135(17)
Ono, Y., 206
Or, T., 371(37), 372
Oradei, A., 170
Orfanos, C. E., 194
Orkin, S. H., 229, 274
Orrenius, S., 67, 68, 321, 323(12), 335
Ort, D. R., 124
Ortiz de Montellano, P., 319
Osawa, T., 240
Osborn, H. M. I., 109
Osipov, A. N., 147
Osowska-Rogers, S., 127, 130(24)
Otani, S., 14
Otterbein, S., 287(18), 288
Owuor, E., 240, 287(16), 288, 305
Oyake, T., 232, 233(63), 273, 274, 280(6), 286

P

Packer, J., 264
Packer, L., 147, 157, 158, 158(19), 171
Padilla, S., 15, 216
Paine, M. J., 319
Palackal, N. T., 33, 35, 38, 38(14), 39(30), 45(30), 46(14), 48, 48(14; 30), 50(30), 53(14), 54, 54(14), 61
Palcic, M. M., 28
Palumbo, A., 92
Pandya, U., 307
Pankratz, M., 244
Pantopoulos, K., 342, 345(10), 353(10), 361, 363
Pantopoulos, P., 353
Paolucci, U., 16

Paparo, F., 317
Paquette, B., 121
Paravicini, T., 374
Pardee, A. B., 67, 69
Paris, A., 121
Park, J.-W., 64, 255, 288
Parker, M. W., 336
Parlani, M., 354
Parmryd, I., 7, 131
Parthasarathy, S., 163
Parvinen, M., 351
Patak, P., 258, 259, 260(7), 268(7), 269(7), 272(7)
Pataki, J., 45
Patel, R. P., 337
Pathak, D. N., 116
Pathak, S., 223
Patil, K. D., 120(43), 121
Patrick, E. J., 180
Patten, G. S., 180
Paul, D. L., 259
Paul, S. M., 121
Paulino, M., 67
Paulson, K. E., 305
Pauss, N., 112(19), 113
Pavel, S., 99, 101(27), 108
Pawlowski, J. E., 32, 36
Paz, M. A., 27
Pearl, M. L., 117
Pelosi, E., 363
Peñalver, M. J., 99
Penel, C., 181, 183(25), 189(28), 190(28), 191(25; 28), 199
Penn, M. S., 163
Penning, T. M., 31, 32, 32(8), 33, 35, 36, 36(13), 38, 38(13; 14), 39(13; 30), 41, 42(33), 45(13; 23; 30; 47; 48), 46, 46(13; 14; 23), 47, 48, 48(13; 14; 23; 30), 49(24), 50, 50(30; 50), 51, 52, 52(50), 53, 53(13; 14; 50), 54, 54(14; 33; 53), 56, 58(63), 59, 61, 62, 62(63), 63, 64, 66, 121(51), 122
Penninger, J. M., 264
Perez-Reyes, E., 362
Pérez-Vicente, R., 201
Perissinotti, L. J., 68, 72, 72(16), 73
Perlaky, L., 223(33), 224
Perry, D. K., 206
Persson, B., 16
Peschle, C., 363
Peter, M. G., 91, 94(9), 95
Peters, J. M., 7, 11(19)
Peters, M. K., 112(17), 113
Peterson, E., 4, 11, 67
Pethig, R., 211
Pettersson, K., 163
Pettus, T. R. R., 200
Pezzuto, J. M., 112(18), 113
Pfeifer, G. P., 61
Pfeiffer, G. R., 234
Phillips, D. H., 111, 116
Phillips, S. E. V., 27, 276, 282(30)
Photiou, A., 109
Piazzi, S., 157, 171, 176(13)
Piccolo, D. E., 41
Pich, M. M., 16
Pickart, C. M., 256
Pickett, C. B., 222, 228(43; 44), 229, 232(50), 233(50), 236, 237(50), 238(6), 239, 242, 252, 255, 256(56), 273, 274(1), 276, 286, 287(1), 288, 289, 291, 291(2), 305, 306
Pieri, C., 159
Pietrangelo, A., 341, 342(4; 5), 343(4), 344
Pietsch, E. C., 238(15), 239
Pike, A. J., 112(17), 113
Pillay, N. S., 238(10), 239
Pinedo, H. M., 350
Pink, J. J., 67(13; 14), 68, 69(14), 321
Pinkus, R., 236, 240
Piper, J. T., 307
Pisha, E., 112, 112(18), 113, 121, 122(11)
Planchon, S. M., 67(13; 14), 68, 69(14), 321
Plastino, J., 29
Playford, D. A., 163
Pletcher, J., 198
Podda, M., 157, 158(19), 171
Pogue, R., 179, 183(5), 184(5), 185(5), 192, 193(94), 194, 196(5), 198, 198(5; 93; 94), 199(93; 94)
Poh-Fitzpatrick, M. B., 51
Polekhina, G., 336
Politi, P. M., 350
Polticelli, F., 193(96), 194
Pomerantz, S. H., 101
Ponka, P., 342
Popjak, G., 152, 153
Popov, V. N., 124
Porcu, M. C., 186, 188(57)
Porta, R., 317
Possati, G. F., 347, 350(26), 356(26)

Potter, G. A., 116
Potterf, S. B., 105
Pou, S., 221
Pouyssegur, J., 266
Powis, G., 321, 347, 348(29)
Premont, J., 333
Prentice, R. L., 110
Prestera, T., 229, 238, 238(12), 239, 240, 240(2), 273
Preziosi, P., 345, 347(22), 350(22), 355(22), 356(22), 357(22), 358(22), 359(22)
Primiano, T., 33, 273
Principato, G., 159
Pritchard, K. A., Jr., 347, 353(30), 371
Prive, G. G., 276
Proger, V., 188
Prota, G., 88, 91, 91(1), 92, 93(1), 97(1), 107(1)
Prywes, R., 313
Pulford, D. J., 238, 238(5; 7), 239
Pulkki, K., 351
Pupillo, P., 182, 186, 188, 188(30; 32; 59)
Purdy, R. H., 121
Purvis, A. C., 124

Q

Qiu, S., 112, 122(11)
Qu, Z.-C., 211
Quiles, J. L., 159
Quinn, P. J., 147, 156(6), 157

R

Radda, G. K., 207
Radjendirane, V., 223, 224, 227, 234
Rahman, I., 305
Raich, N., 229
Rajagopalan, S., 350
Rajasekharan, K. N., 303
RamaKrishna, N. V. S., 34
Ramamathan, R., 120(43), 121
Ramasarma, T., 8, 10
Ramos, C. L., 221
Ramos-Gomez, M., 244, 274, 287
Ramsden, C. A., 88, 98, 99, 100, 101(27), 108, 109
Randall, R. J., 159
Randerath, K., 116, 117
Ranieri, I., 353
Rao, D. N., 321, 335(16)
Raper, H. S., 95
Rapp, S. R., 110
Raschko, J., 259
Rasmusson, A. G., 187, 188(76)
Rastinejad, F., 11
Rathahao, E., 121
Ratnam, K., 32
Read, D., 276
Rebrin, I., 138
Recalcati, S., 340, 341, 342, 342(5), 345, 345(12), 347(22), 350(22), 355(22), 356(22), 357(22), 358(22), 359(22)
Rechsteiner, M., 256
Reddy, A. C., 303
Reddy, C. C., 134
Reddy, J. K., 9(31), 10
Redmond, M. D., 355
Reed, D. J., 335
Reed, J. C., 201, 379
Reid, L. L., 307
Reihnér, E., 163(15), 164
Reinhardt, D., 223
Reiss, Y., 8
Renders, D., 183
Rescigno, A., 186, 188(57)
Rey, M., 363, 374(17)
Rich, K. J., 116
Rich, P. R., 124
Richardson, D. R., 344, 352(20), 355(20), 358(20), 359(20), 360(20), 361(20)
Richardson, P., 164
Richman, P. G., 304
Richter, C., 52, 84
Ricigliano, J. W., 32
Ricordy, R., 67
Ricquier, D., 124
Rieble, S., 186, 188(61)
Riesmeier, J. W., 124
Riley, P. A., 88, 94, 96(12), 98, 99, 100, 101(27), 103, 105, 105(12), 106, 107, 108, 109
Rimke, T., 97
Rinaldi, A. C., 186, 188(57)
Rindgen, D., 62, 63
Rios, D., 374
Roberts, G. C., 117
Roberts, J. E., 88(2), 89
Robertson, K. A., 67(13), 68

Robertson, S., 153
Robins, A. H., 107
Robinson, P., 372
Robison, T. W., 338
Rocque, P. A., 207
Rodenhius, S., 61
Rodrigues, N., 313
Rodríguez-Aguilera, J. C., 201, 201(17), 202, 205(15), 206(15), 207, 208, 210, 212(17), 214(11; 17)
Rodriguez-Crespo, I., 319
Rodríguez-López, J. N., 99
Roepstorff, P., 95
Rogan, E. G., 34, 45, 120, 120(43), 121
Rohde, M., 171
Rohmer, M., 130
Romeo, P. H., 229
Romslo, I., 181
Romualdi, A., 259
Ronchi, R., 342, 344(11), 345(11; 12), 352(11), 353(11), 355(11), 360(11)
Roop, D. R., 223, 223(33; 34), 224, 224(30)
Rorsman, H., 101
Rosario, O., 350
Rosebrough, N. J., 159
Rosen, G. M., 221, 321, 322(18), 323(18)
Rosengren, E., 101
Ross, A. B., 103
Ross, B. C., 11
Ross, D., 187, 320, 333, 333(11)
Ross, R., 110
Rossi, M., 317, 347, 350(26), 356(26)
Rossouw, J. E., 110
Rothe, G., 372
Rötig, A., 16
Rouault, T. A., 343, 344, 359(19)
Roullet, C. M., 14
Roullet, J. B., 14
Routledge, M. N., 117
Ruby, A. J., 303
Rudney, H., 8
Ruggiero, C. E., 18
Ruiyin, C., 9(31), 10
Rundhaugen, L. M., 238(10), 239
Rushmore, T. H., 228(43; 44), 229, 238(6), 239, 273, 274(1), 286, 288, 291(2), 305, 306
Rushton, F. A. P., 108
Rustin, P., 16
Ryan, M., 97
Ryu, J. H., 287(19), 288

S

Sacco, G., 354(57), 355
Saelens, X., 380
Safa, A. R., 67
Sagher, D., 62
Saini, M. K., 307
Sakahira, H., 379
Sakurai, Y., 133
Salameh, M. S., 307
Salvatorelli, E., 340, 342, 344(11), 345(11), 352(11), 353(11), 354, 355(11), 357, 360(11)
Salvayre, R., 200, 206, 362
Samali, A., 67
Samuni, A., 359
Sanchez-Jimenez, F., 181
Sandelius, A. S., 181
Sanders, M. M., 67
Sanders-Loehr, J., 28, 29
Sanger, J., 363, 374(17)
Sanjust, E., 186, 188(57)
Santamaria, A. B., 235
Santini, S. A., 357
Santis, C., 105
Santos-Ocaña, C., 15, 207, 216
Saponiero, A., 347, 348(25), 349(25), 350(25; 27), 354(28)
Sarafian, T. A., 371(37), 372
Saraste, A., 351
Sasaki, T., 264
Sasaki-Irie, J., 264
Sastre, A., 243
Sata, M., 363
Sato, F., 125
Sato, H., 273, 274, 280(6), 287
Sato, K., 33, 222
Sato, T., 12, 13(43), 14(43)
Satoh, K., 232, 233(63), 273, 280(6), 286
Sattler, W., 164, 165(18), 169(18)
Sawada, H., 33
Sawyer, D. B., 362, 363(5)
Saysell, C. G., 27
Sblattero, D., 317
Scagliarini, S., 186, 188(59)
Scalia, M., 107
Schaefer-Ridder, M., 45

Schafer, H., 351
Schaffner-Sabba, K., 70
Schechter, A. N., 363
Schedin, S., 7, 11(19), 12
Schiavello, R., 347, 350(26), 356(26)
Schinina, M. E., 193(96), 194
Schirrmeister, W., 223
Schlechtweg, J., 223
Schlegel, B. P., 31, 32, 33
Schmidt, U., 362
Schmidt, W. E., 351
Schmidt-Ruppin, K. H., 70
Schnaar, R. L., 243
Scholz, R. W., 134
Schramke, K. C., 191
Schuerch, A. R., 70
Schuler, A. R., 67
Schullek, K. M., 113
Schumacker, P. T., 371
Schwartz, B., 18, 29, 30
Schweigener, L., 197
Schwiers, E., 371
Sciuto, S., 107
Scriver, C. R., 336
Scully, M., 351
Sedlak, D., 189, 194(80), 195, 196(80; 104)
Ségui, B., 200
Segura, J. A., 181
Segura-Aguilar, J., 132, 187
Seidegard, J., 238
Sekhar, K. R., 255, 276
Sekura, R., 307
Seligman, P. A., 371
Semigran, M. J., 362
Sen, R., 313
Senthilmohan, R., 163
Seoud, M. A., 111
Serbinova, E., 147
Serrano, A., 181(29), 182, 184, 186(44), 187(29), 188(29)
Service, R. F., 111
Setti, M., 187
Severinghaus, E. M., 5
Shao, Z., 371
Sharp, K. A., 213
Shen, L., 112, 112(18; 19), 113, 114(20), 118(14), 119(20), 120, 121, 122
Shepherd, A. G., 239
Sheppard, A. J., 143
Sherman, M., 328
Sherman, Y., 376
Sherratt, P. J., 255, 256(56), 276, 286, 287(1), 289
Shertzer, H. G., 239, 321
Shi, M. M., 335, 337, 338
Shiah, S.-G., 67
Shibahara, S., 275
Shibata, S., 259
Shibayev, V. N., 126, 128(23)
Shibutani, S., 117, 121
Shigemura, T., 132
Shigenaga, M. K., 64
Shih, A. Y., 251
Shishehbor, M. H., 163
Shitashige, M., 132
Shivdasani, R., 229
Shou, M., 61
Shumaker, S. A., 110
Sichel, G., 107
Siegel, D., 67(14), 68, 69(14), 187, 320, 321, 333, 333(11)
Siegenthaler, P. A., 130
Sies, H., 258, 259, 260(7), 266, 268(7), 269(7), 272(7), 335
Sindelar, P. J., 3, 10, 16, 139, 163
Singal, P. K., 340, 351, 362, 363
Singhal, S. S., 307
Sinha, B. K., 350
Sjöberg, M., 16
Skotland, T., 209
Skulachev, V. P., 124
Sly, W. S., 336
Small, G. J., 34
Smart, R. C., 303
Smit, N. P. M., 99, 101(27), 108
Smith, A. J., 18
Smith, C. A. D., 305
Smith, G. J., 261
Smith, J. D., 261
Smith, L. L., 116
Smith, M. T., 321, 323(12)
Smithgall, T. E., 35, 45(23; 47), 46, 46(23), 48(23), 49(24), 50
Snowden, M. A., 11
So, A., 243
Soderberg, M., 170
Soderhall, J. A., 124
Sohal, R. S., 138, 139, 141, 141(3; 4), 142, 142(9), 143, 146, 147, 148, 148(11; 12), 149, 149(11), 150, 151, 151(17)

Söhl, G., 259
Sohlenius-Sternbeck, A. K., 10
Solano, F., 90, 95, 98(17), 100
Solenghi, M. D., 170, 171(3)
Solis, W. A., 239
Soll, J., 125, 128(11)
Sollai, F., 186, 188(57)
Solomon, E. I., 97
Soltaninassab, R. R., 276
Sorescu, D., 330
Sorrentino, B. P., 229
Soussi, T., 61
Spara, M., 11
Sparla, F., 186, 188(58)
Spencer, S. R., 240
Speyer, J. L., 363, 374(17)
Sprecher, D. L., 163
Sridhar, G., 59
Srimal, R. C., 303
Sriram, P., 116, 117
Srisuma, S., 248, 278
St. Clair, D. K., 365, 366(29)
Stack, D. E., 120, 120(43), 121
Stadler, I., 81
Stearns, R. A., 114, 118(23)
Stecy, P., 363, 374(17)
Stefanick, M. L., 110
Steinberg, D., 163
Steinebach, V., 28
Stemerman, M. B., 371
Stewart, D., 229, 232(48), 233(48), 237(48), 238(13), 239, 255, 276, 287(18), 288, 289, 355
Stocker, R., 15, 156, 156(10), 157, 162, 163, 164, 165(18), 169, 169(18), 170
Stoppani, A. O. M., 67, 68, 70(15), 72, 72(16), 73, 75(15), 80(15)
Storz, G., 222
Stoyanovsky, D. A., 147
Stratford, I. J., 319
Strauss, B., 62
Stremmel, W., 363
Stringer, J. L., 223
Strzalka, K., 124
Su, Q. J., 25, 26(22)
Subbarao, K. V., 105
Suematsu, T., 133
Sugarman, S. M., 117
Sugumaran, M., 95
Suire, 131
Sukuki-Konagai, K., 206
Sukumaran, K. B., 45, 55
Sun, E. E., 182, 184, 185(45; 46), 194, 197
Sun, I. L., 180, 181, 182, 184, 184(13), 185(45; 46), 191, 194(83), 196(83), 197
Sun, J. S., 67, 275
Sun, P., 192, 198(92)
Sun, W. M., 221
Sun, X. Z., 223
Sundaresan, M., 372
Susín, S. A., 201
Sutherland, M. W., 195
Sutter, T. R., 33, 273, 278
Suwabe, N., 276
Suzuki, H., 191, 275
Suzuki, N., 117, 121
Suzuki, S., 29
Suzuki, T., 312
Svingen, B. A., 321
Swanson, S. M., 123
Swartz, H. M., 88(3), 89
Sweeting, M., 191, 195, 196(104)
Swiezewska, E., 4, 8, 10, 13, 14(44), 124, 125, 126(14), 127, 127(14), 128(14), 129, 130, 130(14; 24), 131, 131(30)
Szent-Györgyi, A., 211
Szweda, L. I., 365, 366(29)

T

Tabet, J. C., 121
Tacchini, L., 342
Taetle, R., 371
Taggart, W. V., 258
Tagliarino, C., 67(14), 68, 69(14), 321
Takagi, K., 18
Takahashi, S., 230, 232, 233(63), 273, 274, 275, 278(13), 280(6), 284(13), 286, 287
Takahashi, T., 132, 185, 186, 320
Takaku, K., 275
Takayanagi, R., 146
Takeishi, Y., 363
Takeshige, K., 132, 146
Taketani, S., 372
Taketo, M. M., 275
Takeyasu, K., 276
Talalay, J., 273, 277, 287
Talalay, P., 187, 223, 229, 235, 237, 238, 238(12), 239, 240, 240(2), 243(1), 274
Tamada, Y., 33

Tamagawa, H., 164
Tan, T.-H., 287, 287(16), 288
Tan, Y., 313
Tanaka, M., 206
Tanaka, S., 252, 254(49)
Tanaka, T., 14
Tang, W., 114, 118(23)
Tang, X., 198
Taniguchi, H., 33
Taniguchi, M., 14
Tanizawa, K., 18, 26, 29
Tappel, A. L., 134, 146
Taramelli, D., 345
Tashiro, S., 275
Tatehata, H., 106
Tavassoli, M., 51
Teclebrhan, H., 5, 8, 9
Tedeschi, G., 186, 188(58)
Tempst, P., 229, 274
Teodori, G., 347, 354(28), 357
Terashima, I., 117, 121
Ternes, P., 198
Terwilliger, N., 97
Testa, U., 363
Tetef, M., 259
Tew, D. G., 319
Thal, L., 110
Thatcher, G. R. J., 110
Thelen, M., 221
Thelin, A., 8, 11, 12
Theurkauf, W. E., 275
Thimmulappa, R. K., 248, 278
Thomas, S. R., 156(10), 157, 163
Thompson, A. R., 105, 112(17), 113
Thompson, C. B., 67
Thompson, J. A., 113
Thompson, P. D., 164
Thor, H., 321, 323(12), 335
Thorstensen, K., 181
Thrower, J. S., 256
Thutewohl, M., 206
Tian, G., 26
Tietze, F., 364
Todorovic, R., 34, 120(43), 121
Toki, T., 230, 274, 278(13), 284(13)
Tokumasa, F., 276
Tomasetti, M., 170, 171(3)
Tomc, L. K., 313
Tomoike, H., 363
Torchinsky, Y. M., 345
Torman, R. T., 307
Torti, F. M., 238(14; 15), 239
Torti, S. V., 238(14; 15), 239
Tou, J.-S., 287(18), 288
Touchard, C., 229, 232(48), 233(48), 237(48), 238(13), 239, 287(18), 288
Traber, M. G., 157, 158(19), 171
Traylor, R. S., 201, 206(12)
Troncone, R., 317
Trost, P., 186, 188, 188(58; 59)
Truscott, T. G., 102, 103, 105
Trush, M. A., 121(51), 122
Tsao, M. S., 261
Tseng, W., 354
Tsuchida, S., 222
Tsuchiya, K., 307
Tsuji, Y., 238(14), 239
Tsujimoto, Y., 379
Tsuruda, L. S., 35, 54, 61
Tuczek, F., 90, 97
Tudela, J., 99
Turner, E. E., 45
Turowski, P. N., 29
Turunen, M., 6, 7, 10, 10(17), 11(19; 34), 15(34), 16
Tyulmenkov, V. V., 317
Tyurin, V. A., 208
Tyurina, Y. Y., 156(9), 157, 208, 320
Tzeng, W. F., 335

U

Ubbink, J. B., 171
Uchida, K., 240
Uozumi, T., 191
Upston, J. M., 163
Uro-Coste, E., 200
Ushio-Fukai, M., 330
Usui, A., 132

V

Vadhanavikit, S., 134
Valenti, V., 188
Valentine, J. S., 371(37), 372
Valet, G., 372
Valle, D., 336
Valliant, F., 179
Van Aerden, C., 121

Van Alsine, J. M., 213
van Beeumen, J., 18
van Breemen, R. B., 112, 112(18; 19), 113, 118(14; 15), 119, 121, 122, 123
Vandenabeele, P., 380
Vandenberghe, I., 18
van der Vijgh, W. J., 350
Van Gurp, M., 380
Van Holde, K. E., 97
van Kann, P. J., 124
Van Loo, G., 380
van Marck, E., 121, 123(46)
Vanni, A., 67
Van Renswounde, J., 363
Varenitsa, A. I., 77, 78(28)
Varnes, M. E., 67(14), 68, 69(14), 321
Varón, R., 99
Vasendin, I. M., 155
Vaskovskii, V. E., 155
Vasquez-Vivar, J., 347, 353(30)
Vaughan, J., 264
Vener, V., 124
Venglarik, C. J., 333
Ventura, B., 16
Ventura Pinto, A., 67
Venturella, S., 354(57), 355
Venugopal, R., 222, 228(15), 229, 229(15), 231(47), 232(47), 233, 233(47), 234(47), 236(15), 237(47), 238(8), 239, 242(8), 286
Vermaak, W. J., 171
Vigil, E. L., 179, 180
Villa, R. F., 158
Villalba, J. M., 181, 181(29), 182, 184, 186, 186(44), 187, 187(29), 188(29), 197, 200, 201, 201(17), 202, 204(14; 18), 205(7; 14; 15; 18), 206(6; 14; 15; 18), 207, 208, 210, 211(14), 212(17; 20), 213, 213(14), 214(10; 11; 16; 17), 215(14), 320, 333
Villalba, T. M., 181, 186(22)
Virmani, R., 362
Virodor, V., 105
Vitale, L., 188
Vlasic, N., 371
Vogel, E., 41
Voipio-Pulkki, L. M., 351
Volmat, V., 266
von Montfort, C., 258, 259, 260(7), 268(7), 269(7), 272(7)
Vorndamm, J., 351
Vukomanovic, V., 116
Vuust, J., 18

W

Wactawski-Wende, J., 110
Waechter, C. J., 4
Wagner, A. M., 124
Waikel, R. L., 223(33; 34), 224
Wakabayashi, N., 234, 237, 238, 243, 243(1), 248, 275, 276, 276(21), 277, 278, 280(37), 282(21), 287, 289, 289(12)
Wakamatsu, K., 105
Wakefield, L. M., 207
Walden, W. E., 345
Waldmann, H., 206
Wallace, R. B., 110
Wan, J., 191, 194(83), 196(83)
Wan, Y. J. Y., 10, 11(34), 15(34)
Wang, A., 238(9), 239
Wang, B., 231
Wang, C., 67
Wang, D., 63
Wang, G. W., 351
Wang, H., 163
Wang, H.-G., 201
Wang, H.-M., 67
Wang, J., 183
Wang, L., 362
Wang, M. D., 113
Wang, P. H., 362
Wang, R. W., 114, 118(23)
Wang, S., 192, 193(94), 194, 195, 196(104), 198(94), 199, 199(94), 351
Wang, S. X., 27, 28(27)
Wang, X., 223(33; 34), 224
Wang, Z., 259
Wanke, M., 129, 130, 131, 131(30)
Ward, C., 363, 374(17)
Ward, J. F., 221
Warn-Cramer, B. J., 259, 268(11)
Wartman, M. A., 273, 305, 307(15), 315(15)
Wasley, J. W. F., 70
Wasserman, W. W., 229, 242
Wassertheil-Smoller, S., 110
Watanabe, N., 307, 315, 315(28), 319, 322, 323(24), 325(24), 326(24),

327(24), 328(24), 330(24), 333, 333(24)
Watenabe, M., 12
Watson, M. L., 5
Watson, N. S., 11
Watts, G. F., 163
Weber, C., 157, 158(19), 171
Weber, M. J., 287
Weber, T. J., 213
Weed, J. C., Jr., 111
Wehlin, L., 16
Wehr, N. B., 343
Wehrli, W., 70
Weiner, L. M., 236, 240
Weiss, R. B., 340, 341(1)
Weissman, A. M., 343
Werderitsh, D., 194
Whalen, D. L., 56
Whitbread, L., 336
White, C. R., 353
White, N. H., 116
Whitman, S. P., 238(14), 239
Whitmarsh, A. J., 259
Wicks, C., 287(18), 288
Wiegand, J., 374
Wigley, A. F., 302, 337, 338
Wilce, M. C. J., 18
Wilczak, C. A., 240
Wild, A. C., 229, 232(49), 233(49), 237(49), 307, 337
Wiley, J. C., 45
Wilkinson, F. E., 195
Wilkinson, J. IV, 234
Willecke, K., 259
Williams, C. H. J., 133, 134(18)
Williams, M. C., 5, 121
Williams, N. K., 18, 19, 20(16), 29
Williams, R. J., 11
Williamson, G., 238(9), 239
Willis, R., 164
Willmann, M., 341
Wilmot, C. M., 27
Winder, A. J., 101
Wingo, P. A., 303
Winkler, E., 124
Winski, S. L., 187, 320, 333(11)
Witiak, D. T., 67
Witschi, H., 221
Witting, P. K., 163, 169
Witztum, J. L., 163
Wolf, C. R., 244, 287, 319
Wolvetang, E. J., 179, 201
Wong, C. Q., 303
Wong, G., 251
Woodside, A. B., 155
Wosilait, W. D., 333
Wright, C., 11
Wu, H.-Y., 67
Wu, J., 287
Wu, K., 328
Wu, K. B., 223
Wu, L.-Y., 194, 196, 198
Wuerzberger-Davis, S. M., 67(13), 68
Wyatt, J. K., 362

X

Xia, L., 133, 134(14–16)
Xia, L. J., 164
Xie, T., 68, 228
Xiong, Y., 121
Xu, D. P., 207
Xu, W., 241
Xu, Y., 68, 228
Xu, Z. Q., 276
Xue, F., 275
Xue, H., 14

Y

Yadav, K. D. S., 276, 282(30)
Yagi, H., 43, 56
Yagita, H., 201
Yamaguchi, H., 18
Yamaguchi, K., 29
Yamaguchi, S., 363
Yamamoto, H., 106, 276, 280(37)
Yamamoto, M., 175(17), 229, 230, 232, 233(63), 234, 237, 238, 242, 243, 243(1), 244, 248, 256, 273, 274, 275, 276, 276(17; 21), 277, 278, 278(13), 280(6), 282(21), 284(13), 286, 287, 289, 289(12)
Yamamoto, Y., 171
Yamasato, Y., 206
Yamashita, S., 171
Yamashoji, S., 320, 328(4)
Yamauchi, O., 26
Yamazaki, I., 185
Yan, L.-J., 147, 148(12)

Yan, X. X., 255
Yanagawa, T., 273, 280(6), 287
Yang, C. S., 255, 256(56), 276, 289, 328
Yang, X. L., 223
Yang, Y., 112, 118(15), 123
Yang, Y. F., 207
Yano, Y., 14
Yantiri, E., 195, 196(104)
Yao, D., 112, 118(15)
Yazaki, K., 125
Ye, C. Q., 164
Yeh, H. J., 41
Yekundi, K. G., 258
Yen, T. S., 231
Yikilmaz, E., 362
Yoshida, A., 252, 254(49)
Yoshida, C., 240, 276
Yoshida, T., 252, 254(49)
Yoshie, Y., 121
Yoshizawa, I., 121
Young, D. A., 371
Yu, D., 64, 66
Yu, L., 110, 112, 118(15), 119
Yu, R., 240, 287, 287(16), 288, 305
Yu, Z. X., 372

Z

Zamparelli, R., 347, 350(26), 356(26)
Zamzami, N., 201
Zappacosta, F., 193(96), 194
Zeleniuch-Jacquotte, A., 363, 374(17)
Zhang, D., 223, 328
Zhang, F., 112, 112(19), 113, 116, 118(14; 15), 119, 121, 122, 122(11), 123
Zhang, H., 302(1), 303, 307, 312(1), 315(28), 333, 353
Zhang, X., 374
Zhang, X. X., 10, 11(34), 15(34)
Zhang, Y., 9, 15, 141, 157, 229, 240
Zhao, H., 353, 362, 363(9)
Zhao, Q., 223
Zheng, M., 222
Zhivotovsky, B., 67
Zhong, L., 134
Ziemann, C., 369
Zipper, L. M., 237, 287(17), 288, 306
Zubak, V. M., 18
Zurbriggen, R., 196

Subject Index

A

Acetate, radiolabeling precursor for coenzyme Q labeling, 3
Aconitase
  doxorubicin response
    assay, 364–365
  iron regulatory protein-1 activity, 342
Adriamycin, *see* Doxorubicin
Aldo-keto reductases
  crystal structures, 32–33
  dihydrodiol dehydrogenase activity
    assays
      benzene dihydrodiol oxidation, 41–42
      polyaromatic hydrocarbon *trans*-dihydrodiol oxidation, 43, 45–47
    detection in cell culture, 53–54
    polyaromatic hydrocarbon *trans*-dihydrodiol specificity, 47–48
    types of enzymes, 35–36
  functional overview, 31, 33
  oxygen metabolism studies during *trans*-dihydrodiol oxidation
    Clark electrode measurement of oxygen consumption, 51
    electron paramagnetic resonance measurement of reactive oxygen species, 52–53
    hydrogen peroxide production, 51
    superoxide production, 52
  polyaromatic hydrocarbon *o*-quinone products
    characterization by trapping with β-mercaptoethanol, 49–50
    classification, 56
    DNA lesions
      covalent adducts, 61–62
      oxidative damage, 62–64
      p53 mutagenesis assay, 64–65
    electron paramagnetic resonance of *o*-semiquinone anion radicals, 58
    reactivity with amino acids and proteins, 58–59
    structures and organic synthesis, 55–56
  polycyclic aromatic hydrocarbon activation, 33–35
  purification of recombinant human enzymes from *Escherichia coli*
    cell growth and lysis, 38–39
    chromatography, 39
    cloning, 36, 38
    purification table, 40
  relationship between rat and human family members, 37
  stable transfection systems for DNA lesion and p53 mutation studies, 65–66
  types, 31–32
Antioxidant response element
  antioxidant response signaling of detoxifying enzyme synthesis, 235–238, 243–244
  binding proteins, *see* Nrfs
  consensus sequence, 286
  DNA microarray analysis of *tert*-butylhydroquinone induction of genes, 244–248
  gene distribution, 238–239, 305–306
  mutagenesis studies, 226–229
  Nrf binding assay, 229–234
  phosphatidylinositol 3-kinase in activation
    apoptosis inhibition, 257
    inhibitor studies, 253–254
    Nrf2 studies
      degradation inhibition, 255–257
      nuclear translocation, 254–255, 258
    overview, 252–253
  quinone activation
    Nrf2 role, 242–243
    overview, 238–239
    oxidative stress-dependent mechanisms, 239–241
    oxidative stress-independent mechanisms, 241
Apoptosis
  antioxidant response element activation and apoptosis inhibition, 257

Apoptosis *(cont.)*
doxorubicin cardiotoxicity
apoptosis induction, 351–352, 363
assays
caspase-3 activity, 378–379
cytochrome *c* release, 379–381
DNA laddering, 375–376
terminal deoxynucleotidyl transferase-mediated nick-end labeling assay, 376–377
ARE, *see* Antioxidant response element
Ascorbic acid
coenzyme Q-mediated stabilization studies
ascorbate free radical
electron paramagnetic resonance, 210–211
generation, 209–210
genetic analysis of trans-plasma membrane redox system, 216–217
plasma membrane fractions, 212–215
stabilization by cell lines, 210–212
oxidation, 207
reduction mechanisms, 207–208

B

Bach proteins, phase II enzyme repression, 275

C

CD45 protein tyrosine phosphatase, menadione inhibition
assay, 264
epidermal growth factor receptor activation studies, 264–266
Ceramide signaling, *see* Sphingomyelinase
Cholesterol, coenzyme Q_{10} determination with cholesterol and cholesteryl esters in plasma
applications, 169
extraction, 164–165
high-performance liquid chromatography, 165
interferences, 168–169
linearity, 165
materials, 164
rationale, 162–164
sensitivity, 165, 169
standards, 165
validation, 166, 168
Coenzyme Q
ascorbate stabilization studies
ascorbate free radicals
electron paramagnetic resonance, 210–211
generation, 209–210
genetic analysis of trans-plasma membrane redox system, 216–217
plasma membrane fractions, 212–215
stabilization by cell lines, 210–212
assays
overview, 143
coenzyme Q_{10} high-performance liquid chromatography determination with cholesterol and cholesteryl esters in plasma
applications, 169
chromatography, 165
extraction, 164–165
interferences, 168–169
linearity, 165
materials, 164
rationale, 162–164
sensitivity, 165, 169
standards, 165
validation, 166, 168
extraction, 144
high-performance liquid chromatography with colorimetric detection of coenzyme Q_{10} in plasma
accuracy and precision, 174–175
advantages, 175–176
calibration curves, 174
chromatography, 173
electrochemical detection comparison, 173–175
materials, 172
principles, 171
sample preparation, 172–173
stability of samples, 175
standards, 172
high-performance liquid chromatography with electrochemical detection of coenzyme Q forms and vitamin E
advantages, 161–162

chromatography and detection, 159–160
extraction, 159
materials, 158
mitochondrial sample preparation and treatment, 158–159
overview, 145, 156–158
standards, 158
materials, 143–144
sample preparation and mitochondria isolation, 144
standards, 145
functions, 139, 146, 163
half-life, 138
metabolism analysis
biological systems, 4–7
biosynthesis
activators, 9–11
inhibitors, 8–9
rate determination, 7
diet preparation, 7–8
half-life analysis, 7, 12–14
metabolite identification, 12–14
nuclear receptor regulation of metabolism, 11–12
radiolabeling precursors, 3–4
uptake studies, 14–16
mitochondrial distribution of coenzyme Q_9 and coenzyme Q_{10} by species, 141–142
pentane extraction, 16–17
plasma levels, factors affecting, 170
redox states, 146
sphingomyelinase regulation
assay, 204–205
inhibition of neutral magnesium-dependent enzyme, 201–202
statin effects on levels, 8–9, 171
structure, 138
superoxide generation relationship studies, 142–143
tissue and mitochondrial content in mouse and rat, 139–141
uptake of supplemental coenzyme Q_{10}, 141
vitamin E interactions
inner mitochondrial membrane, 147–149
reduction of α-tocopheryl radicals, 147
superoxide generation studies, 149–151
Connexins, menadione phosphorylation induction
immunohistochemistry analysis, 270–272
overview, 259, 268–269
Western blot analysis, 269–270
Copper amine oxidase
half-reaction assays, 24–26
kinetic mechanism, 24
purification of *Hansenula polymorpha* enzyme
bacterial expression system
cell growth and lysis, 20–22
chromatography, 22–23
recombinant protein expression systems, 19–21, 24
yeast expression system
cell growth, 23
chromatography, 23–24
topaquinone studies
absorbance spectriscopy, 28
biogenesis studies, 29–31
electron paramagnetic resonance, 29
infrared spectroscopy, 28–29
redox staining, 27–28
Curcumin
dietary applications, 302–303
electrophoretic mobility shift assay of antioxidant response element and AP-1 binding activities
curcumin effects, 316
gel electrophoresis and imaging, 315–316
immunodepletion studies for binding protein identification, 316–318
incubation conditions, 315
nuclear extract preparation, 314–315
oligonucleotides, 315
principles, 313–314
glutathione enzyme modulation, 303–304, 306–307
medicinal properties, 302–303
phase II gene induction studies with real-time polymerase chain reaction
curcumin-induced genes, 312–313
overview, 308–310
primers, 311–312
reverse transcription, 310–311
RNA preparation, 310
stress defense, 302
structure, 304

Cytochrome P450 reductase, β-lapachone redox cycling
  aminopyrine N-demethylase assay of quinone effects, 75–77
  aniline hydroxylase assay of quinone effects, 77–79
  inhibitory activity of β-lapachone and derivatives, 79–80
  one-electron reduction, 68
  semiquinone radical formation, 71–73

D

2,3-Dimethoxy-1,4-naphthoquinone, reactive oxygen species production, 321–324
DMNQ, *see* 2,3-Dimethoxy-1, 4-naphthoquinone
DNA microarray, Nrf2 studies
  Keap1 binding of Nrf2, DNA microarray analysis using knockout mice, 283
  knockout and transgenic mouse studies of *tert*-butylhydroquinone induction of genes
    cortical culture, 251–252
    liver, 245, 248, 251
    males versus females, 249–250
Doxorubicin
  anti-tumor activity, 340, 362
  apoptosis assays
    caspase-3 activity, 378–379
    cytochrome *c* release, 379–381
    DNA laddering, 375–376
    terminal deoxynucleotidyltransferase-mediated nick-end labeling assay, 376–377
  cardiotoxicity
    apoptosis induction, 351–352, 363
    iron chelation prevention, 340–341, 374
    lesion morphology, 346
    metabolite toxicity, 347–350
    overview, 340, 362
    quinone-dependent dysregulation of iron homeostasis, ferritin and iron regulatory protein-2 roles, 352–354
    quinone-independent dysregulation of iron homeostasis, iron regulatory protein-1 role
      cardiomyocyte studies, 359–361
      cell-free system studies, 354–359
      pathophysiological consequences, 361
  hydrogen peroxide induction and detection, 371–372
  iron uptake effects in endothelial cells, 371
  metabolism, 347–350
  metabolite detection, 350
  oxidative stress assays
    aconitase, 364–365
    DT-diaphorase, 365–366
    glutathione, 364
    iron regulatory protein electrophoretic mobility shift assay, 366–369
  structure, 346–347
  superoxide induction and hydroethidine detection, 372, 374
  transferrin receptor response, 369–371, 381–382
DT-diaphorase
  antioxidant response signaling of detoxifying enzyme synthesis, 235–238
  assay of activity and special considerations, 330–333
  doxorubicin response assay, 365–366
  expression analysis
    antioxidant treatment, 225
    calcium phosphate transfection, 225
    chloramphenicol acetyltransferase reporter assay, 225–226
    human hepatoblastoma cell culture, 224
    Northern blot, 224
    promoter elements and mutagenesiss, 226–229
    xenobiotic treatment, 225
  inhibition and reactive oxygen species production, 322
  knockout mouse, 223–224
  β-lapachone redox cycling
    semiquinone radical formation, 80–81
    two-electron reduction, 68–69
  Nrf mediation of gene expression
    luciferase reporter assay, 232–234
    plasmids, 231
    transfection, 231–232
  quinone reductase activity, 132–133, 320–321

tissue distribution, 223
transplasma membrane electron transport chain, 185–186

E

ECTO-NOX proteins, *see* External NADH oxidase proteins
Electron paramagnetic resonance
aldo-keto reductase
trans-dihydrodiol oxidation reactive oxygen species, 52–53
o-semiquinone anion radicals, 58
ascorbate free radical, 210–211
β-lapachone redox cycling
spectra acquisition, 71
semiquinone radical formation by cytochrome P450 reductase, 71–73
semiquinone radical formation by DT-diaphorase, 80–81
topaquinone studies, 29
Electrophile response element, *see* Antioxidant response element
Electrophoretic mobility shift assay
curcumin studies of antioxidant response element and AP-1 binding activities
curcumin effects, 316
gel electrophoresis and imaging, 315–316
immunodepletion studies for binding protein identification, 316–318
incubation conditions, 315
nuclear extract preparation, 314–315
oligonucleotides, 315
principles, 313–314
iron regulatory protein response to doxorubicin, RNA electrophoretic mobility shift assay, 343–344, 366–369
Nrf2 binding to antioxidant response element, 297–298
EMSA, *see* Electromophoretic mobility shift assay
Epidermal growth factor receptor tyrosine kinase, menadione activation, 260–266
Estrogens
antiestrogens, *see* Selective estrogen receptor modulators
carcinogenesis mechanisms, 111–113, 123
DNA adduct formation with quinoids
deoxynucleoside reaction with *o*-quinones, 114–115
equine estrogen adducts, 121–123
estradiol adducts, 119–121
estrone adducts, 119–121
quinoid preparation
quinone methides, 113
o-quinones, 113
replacement therapy side effects, 110
External NADH oxidase proteins
age-related protein, 195–196
clock function, 198–199
constitutive proteins, 194, 199
growth functions, 196–199
oscillatory patterns of activity, 192–194
properties, 188–189
protein disulfide thiol interchange, 190–191
substrate specificity, 189–190
terminal oxidases in transplasma membrane electron transport, 188
transplasma membrane electron transport role, 183–185, 196–198
tumor-associated proteins, 194–195, 199

F

Farnesyl pyrophosphate, radiolabeling precursor for coenzyme Q labeling, 4
Ferricyanide reductase, transplasma membrane electron transport, 181–182
Ferritin, quinone-dependent dysregulation of iron homeostasis and role in doxorubicin cardotoxicity, 352–354
FOX assay, quinone-derived hydrogen peroxide
cell system, 324–325
incubation conditions and absorbance measurement, 326–328
interpretation, 328–330
pretreatment of cells, 325–326
special considerations, 333
validation, 328

G

GCL, *see* Glutamylcysteine ligase
  curcumin interactions, 304
  subunits, 307
Glutamylcysteine ligase
  curcumin interactions, 304
  reactive oxygen species-induced expression, 337–339
  subunits, 307
γ-Glutamyl transpeptidase, reactive oxygen species-induced expression, 338–339
Glutathione
  assay, 335–336
  curcumin modulation, 303–304, 306–307
  doxorubicin response assay, 364
  quinone conjugation, 319
  quinone metabolism marker, 334
  quinone response, 334–335
Glutathione reductase, ubiquinone reduction studies
  extraction and high-performance liquid chromatography of ubiquinone-6, ubiquinone-10, and ubiquinol-10, 137–138
  materials, 135–136
  overview, 133–135
  reaction conditions, 136
Glutathione *S*-transferase, curcumin interactions, 307
GST, *see* Glutathione *S*-transferase

H

High-performance liquid chromatography
  coenzyme Q_{10} determination with cholesterol and cholesteryl esters in plasma
    applications, 169
    chromatography, 165
    extraction, 164–165
    interferences, 168–169
    linearity, 165
    materials, 164
    rationale, 162–164
    sensitivity, 165, 169
    standards, 165
    validation, 166, 168
  coenzyme Q_{10} determination with colorimetric detection in plasma
    accuracy and precision, 174–175
    advantages, 175–176
    calibration curves, 174
    chromatography, 173
    electrochemical detection comparison, 173–175
    materials, 172
    principles, 171
    sample preparation, 172–173
    stability of samples, 175
    standards, 172
  electrochemical detection and separation of coenzyme Q forms and vitamin E
    advantages, 161–162
    chromatography and detection, 159–160
    extraction, 159
    materials, 158
    mitochondrial sample preparation and treatment, 158–159
    overview, 145, 156–158
    standards, 158
  glutathione assay, 335–336, 364
  ubiquinone-6, ubiquinone-10, and ubiquinol-10 separation, 137–138
HMG-CoA reductase, *see* 3-Hydroxy-3-methylglutaryl coenzyme A reductase
HPLC, *see* High-performance liquid chromatography
Hydrogen peroxide
  doxorubicin induction and detection, 371–372
  quinone-derived peroxide
    FOX assay
      cell system, 324–325
      incubation conditions and absorbance measurement, 326–328
      interpretation, 328–330
      pretreatment of cells, 325–326
      special considerations, 333
      validation, 328
    horseradish peroxidase detection, 324
3-Hydroxy-3-methylglutaryl coenzyme A reductase, inhibitor effects on coenzyme Q synthesis, 8–9, 171

I

Iron regulatory protein-1
  aconitase activity, 342
  iron sensing mechanism, 342
  quinone-independent dysregulation of iron homeostasis, role in doxorubicin cardotoxicity
    cardiomyocyte studies, 359–361
    cell-free system studies, 354–359
    pathophysiological consequences, 361
  reducing agent enhancement of activity, 344–345
  RNA electrophoretic mobility shift assay, 343–344, 366–369
Iron regulatory protein-2
  iron sensing mechanism, 342–343
  quinone-dependent dysregulation of iron homeostasis, role in doxorubicin cardotoxicity, 352–354
  RNA electrophoretic mobility shift assay, 345, 366–369
  Western blot, 345–346
IRP-1, *see* Iron regulatory protein-1
IRP-2, *see* Iron regulatory protein-2
Isopentenyl diphosphate
  carbon-14 compound preparation, 152
  isoprenoid pathway precursor, 152
  tritiated compound preparation
    materials, 153–154
    overview, 152–153
    storage, 16
    synthesis, 154–156
Isopentenyl pyrophosphate, radiolabeling precursor for coenzyme Q labeling, 3

K

Keap1
  binding of Nrf2
    DNA microarray analysis using knockout mice, 283
    knockout mouse studies, 276–277, 280
    Nrf2 phosphorylation effects on binding, 296–297
    reporter transfection analysis, 282–283
    yeast two-hybrid system, 280–281
  oxidative stress sensing, 277
  structure, 275–276

L

β-Lapachone
  antitumor activity, 67–69
  redox cycling
    CG 10–248 analog redox cycling by cytosol fraction, 86
    cytochrome P450 reductase
      aminopyrine *N*-demethylase assay of quinone effects, 75–77
      aniline hydroxylase assay of quinone effects, 77–79
      inhibitory activity of β-lapachone and derivatives, 79–80
      one-electron reduction, 68
      semiquinone radical formation, 71–73
    DT-diaphorase
      semiquinone radical formation, 80–81
      two-electron reduction, 68–69
    electron paramagnetic resonance
      semiquinone radical formation by cytochrome P450 reductase, 71–73
      semiquinone radical formation by DT-diaphorase, 80–81
      spectra acquisition, 71
    microsomal and cytosol preparations, 70
    oxygen uptake studies of analog effects in liver cytosol fraction, 81–83
    quinone preparations, 70–71
    superoxide production studies, 74, 84
  structure and analogs, 69–70
Lipoamide dehydrogenase, ubiquinone reduction studies
  extraction and high-performance liquid chromatography of ubiquinone-6, ubiquinone-10, and ubiquinol-10, 137–138
  materials, 135–136
  overview, 133–135
  reaction conditions, 136

M

MAPK, *see* Mitogen-activated protein kinase
Melanin
  biosynthetic pathways
    balance between eumelanogenesis and pheomelanogenesis, 104–105
    overview, 88, 90
    ortho-quinone chemistry

Melanin *(cont.)*
addition reactions, 91–93
cross-linking, 106–107
pulse radiolysis studies, 102–104
redox exchange, 93–94
tautomerization, 94–95
therapeutic implications, 108–109
Raper–Mason pathway, 95–97
tyrosinase
assays, 100–102
oxidation of phenolamines and catecholamines, 97–100
quantitative analysis, 88–89
structure, 107
types, 88
Menadione
CD45 protein tyrosine phosphatase inhibition
assay, 264
epidermal growth factor receptor activation studies, 264–266
connexin phosphorylation induction
immunohistochemistry analysis, 270–272
overview, 259, 268–269
Western blot analysis, 269–270
cytotoxicity, 259
metabolism, 258
mitogen-activated protein kinase activation, 259, 266–268
reactive oxygen species production, 321–324
therapeutic prospects, 272
tyrosine kinase activation studies
epidermal growth factor receptor tyrosine kinase, 260–264
materials, 260–261
platelet-derived growth factor receptor B tyrosine kinase, 261–263
Western blot and immunoprecipitation, 261–264
Mevalonate, radiolabeling precursor for coenzyme Q labeling, 3–4
Mitogen-activated protein kinase, menadione activation, 259, 266–268

N

NAD(P)H:quinone oxidoreductase 1, *see* DT-diaphorase
NF-E2
redox state sensing, 222
related factors, *see* Nrfs
Nitric oxide synthase, quinone reduction, 319
NOS, *see* Nitric oxide synthase
Nrfs
antioxidant response element binding, 229, 242, 286–287
antioxidant response element-mediated DT-diaphorase gene expression studies
luciferase reporter assay, 232–234
plasmids, 231
transfection, 231–232
antioxidant response signaling of detoxifying enzyme synthesis, 235–238
Bach family homology, 275
inhibitor of Nrf2, 234
Keap1 binding of Nrf2
DNA microarray analysis using knockout mice, 283
knockout mouse studies, 276–277, 280
oxidative stress sensing, 277
reporter transfection analysis, 282–283
structure of Keap1, 275–276
yeast two-hybrid system, 280–281
knockout and transgenic mouse studies of Nrf2
DNA microarray analysis of *tert*-butylhydroquinone induction of genes
cortical culture, 251–252)
liver, 245, 248, 251
males versus females, 249–250
overview, 243–244, 274
Nrf3
knockout mouse studies, 284–285
MARE binding sequence, 283–284
tissue distribution, 284
phosphatidylinositol 3-kinase in antioxidant response element activation, Nrf2 effect studies
degradation inhibition, 255–257
nuclear translocation, 254–255, 258
protein kinase C regulation of Nrf2
antioxidant response element binding effects, 297–298
cell culture, 291
cell treatment with *tert*-butylhydroquinone, 291–292
cellular phosphorylation state assay, 294

Keap1 association effects, 296–297
nuclear translocation studies, 292–294
phosphorylation site identification, 288–289, 296
phosphorylation, *in vitro,* 294–296
reporter gene constructs and transfection, 291, 298–299
stabilization effects, 289–290, 299–300
Western blot of Nrf2 levels, 300–301
structure, 230–231, 242–243, 273–274
subcellular localization of Nrf2, 242–243
types and functions, 229–230, 278, 285–286
NRH:quinone oxidoreductase 2
antioxidant response signaling of detoxifying enzyme synthesis, 235–238
expression analysis
antioxidant treatment, 225
calcium phosphate transfection, 225
chloramphenicol acetyltransferase reporter assay, 225–226
human hepatoblastoma cell culture, 224
Northern blot, 224
promoter elements and mutagenesiss, 226–229
xenobiotic treatment, 225
knockout mouse, 223–224

O

Oxidative stress
antioxidant enzyme induction, 222–223, 273
antioxidant response signaling of detoxifying enzyme synthesis, 235–238
reactive oxygen species toxicity, 221
sensing by transcription factors, 222
OxyR, redox state sensing, 222

P

p53, polyaromatic hydrocarbon *o*-quinone mutagenesis assay, 64–65
PAHs, *see* Polycyclic aromatic hydrocarbons
PCR, *see* Polymerase chain reaction
Phosphatidylinositol 3-kinase, antioxidant response element activation
apoptosis inhibition, 257
inhibitor studies, 253–254
Nrf2 studies
degradation inhibition, 255–257
nuclear translocation, 254–255, 258
overview, 252–253
PKC, *see* Protein kinase C
Plastoquinone
biosynthesis in plants
assay, 127–129
chromatography of intermediates, 127–129
enzymes, 125
intracellular localization, 125–126, 129–130
prospects for study, 130–131
radiolabeled precursors, 126–127
functions, 124
half-life estimation, 130
membrane structure, 124
Platelet-derived growth factor receptor B tyrosine kinase, menadione activation, 261–263
Polycyclic aromatic hydrocarbons
activation pathways, 33–35
dihydrodiol oxidation, *see* Aldo-keto reductases
Polymerase chain reaction, curcumin and phase II gene induction studies with real-time polymerase chain reaction
curcumin-induced genes, 312–313
overview, 308–310
primers, 311–312
reverse transcription, 310–311
RNA preparation, 310
Polyprenyl-*p*-hydroxybenzoate transferase, inhibitor effects on coenzyme Q synthesis, 8
Protein kinase C, Nrf2 regulation studies
antioxidant response element binding effects, 297–298
cell culture, 291
cell treatment with *tert*-butylhydroquinone, 291–292
cellular phosphorylation state assay, 294
Keap1 association effects, 296–297
nuclear translocation studies, 292–294
phosphorylation site identification, 288–289, 296
phosphorylation, *in vitro,* 294–296
reporter gene constructs and transfection, 291, 298–299
stabilization effects, 289–290, 299–300

Protein kinase C, Nrf2 regulation studies (*cont.*)
  Western blot of Nrf2 levels, 300–301
Pyrroloquinoline quinone, structure, 18

Q

Quinone oxidoreductases
  external NADH oxidase proteins
    age-related protein, 195–196
    clock function, 198–199
    constitutive proteins, 194, 199
    growth functions, 196–199
    oscillatory patterns of activity, 192–194
    properties, 188–189
    protein disulfide thiol interchange, 190–191
    substrate specificity, 189–190
    terminal oxidases in transplasma membrane electron transport, 188
    transplasma membrane electron transport role, 183–185, 196–198
    tumor-associated proteins, 194–195, 199
  transplasma membrane electron transport chain
    cofactors, 184
    DT-diaphorase, 185–186
    lipophilic quinones, 185
    NADH-quinone reductases, 186–188
    overview, 183–185
  transplasma membrane electron transport
    ferricyanide reductase, 181–182
    NADH:external acceptor [quinone] reductase, 182
    overview, 179–181

R

Raloxifene, *see* Selective estrogen receptor modulators
Raper–Mason pathway, melanogenesis, 95–97

S

Selective estrogen receptor modulators
  DNA adduct formation with quinoids
    deoxynucleoside reaction with *o*-quinones, 114–115
    raloxifene adducts, 118–119
    tamoxifen adducts, 116–118
    toremifene adducts, 117–118
  indications, 110–111
  quinoid preparation
    microsomal oxidation, 113–114
    quinone methides, 113
    o-quinones, 113
SERMs, *see* Selective estrogen receptor modulators
SoxRS, redox state sensing, 222
Sphingomyelinase
  ceramide signaling, 200, 205
  coenzyme Q studies of regulation
    assay, 204–205
    inhibition of neutral magnesium-dependent enzyme, 201–202
  forms, 200
  inhibitors, 205
  purification from plasma membrane
    anion-exchange chromatography, 204
    gel filtration, 203
    heparin affinity chromatography, 203–204
    hydrophobic interaction chromatography, 204
    plasma membrane purification by two-phase partition, 202–203
Superoxide
  aldo-keto reductase production, 52
  coenzyme Q relationship studies in generation, 142–143
  coenzyme Q–vitamin E
    interactions and anion radical generation, 149–151
  doxorubicin induction and hydroethidine detection, 372, 374
  β-lapachone effects on production, 74, 84

T

Tamoxifen, *see* Selective estrogen receptor modulators
Thioredoxin reductase, ubiquinone reduction studies
  extraction and high-performance liquid chromatography of ubiquinone-6, ubiquinone-10, and ubiquinol-10, 137–138
  materials, 135–136
  overview, 133–135
  reaction conditions, 136

α-Tocopherol, *see* Vitamin E
Topaquinone
  enzymes, *see* Copper amine oxidase
  structure, 18
Toremifene, *see* Selective estrogen receptor modulators
Transferrin receptor, doxorubicin response assay, 369–371, 381–382
Transplasma membrane electron transport, *see* Quinone oxidoreductases
Tyrosinase
  assays, 100–102
  balanid adhesion, 106
  melanogenesis, 97–100

U

Ubiquinone
  antioxidant action of ubiquinol, 132
  biosynthesis in plants
    assay, 127–129
    chromatography of intermediates, 127–129
    enzymes, 125
    intracellular localization, 125–126, 129–130
    prospects for study, 130–131
    radiolabeled precursors, 126–127
  flavoenzyme reduction studies
    cell line extracts, 136–137
    extraction and high-performance liquid chromatography of ubiquinone-6, ubiquinone-10, and ubiquinol-10, 137–138
    glutathione reductase, 136
    lipoamide dehydrogenase, 136
    materials, 135–136
    overview, 132–135
    thioredoxin reductase, 136
  functions, 124
  half-life estimation, 130
  membrane structure, 124

V

Vitamin C, *see* Ascorbic acid
Vitamin E
  coenzyme Q interactions
    inner mitochondrial membrane, 147–149
    reduction of α-tocopheryl radicals, 147
    superoxide generation studies, 149–151
  high-performance liquid chromatography with electrochemical detection of coenzyme Q forms and vitamin E
    advantages, 161–162
    chromatography and detection, 159–160
    extraction, 159
    materials, 158
    mitochondrial sample preparation and treatment, 158–159
    overview, 145, 156–158
    standards, 158
  high-performance liquid chromatography with electrochemical detection of coenzyme Q forms and vitamin E
    advantages, 161–162
    chromatography and detection, 159–160
    extraction, 159
    materials, 158
    mitochondrial sample preparation and treatment, 158–159
    overview, 145, 156–158
    standards, 158
  lipid peroxy radical scavenging, 146
Vitamin K3, *see* Menadione

W

Western blot
  iron regulatory protein-2, 345–346
  menadione phosphorylation induction studies
    connexins, 269–270
    receptor tyrosine kinase activation studies, 261–264
  Nrf2 levels, 300–301
  transferrin receptor response to doxorubicin, 369–371

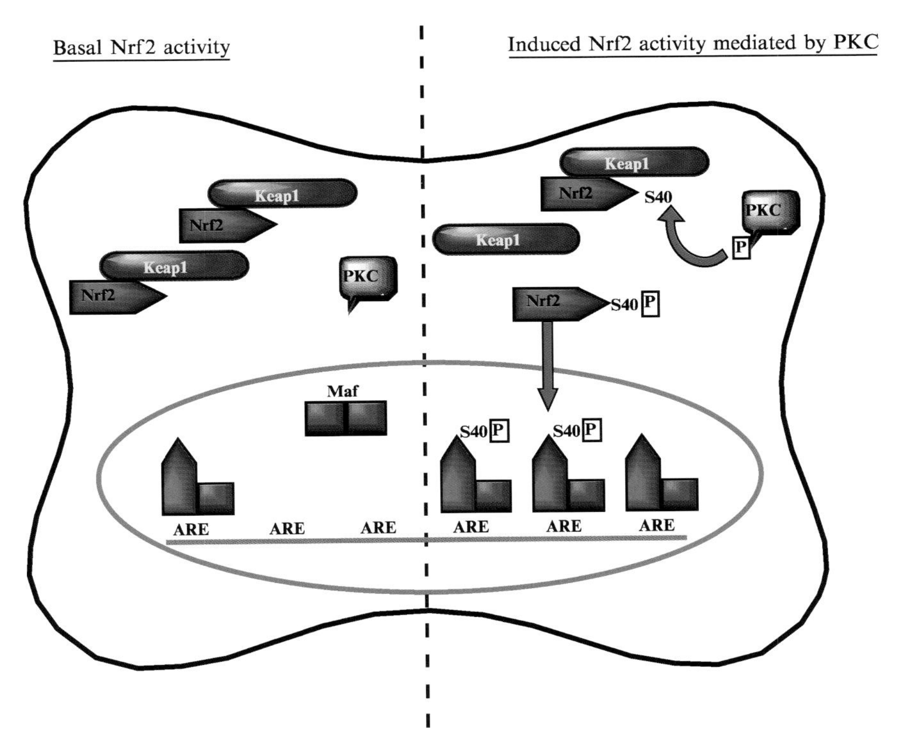

SHERRATT *ET AL.*, CHAPTER 22, FIG. 1. Under nonstimulated conditions, Nrf2 is repressed by Keap1, which localizes it to the cytoplasm of the cell. There is, however, a low level of basal ARE activity, because not all Nrf2 exists in the repressed state. Following induction by compounds such as *t*BHQ or PMA, PKC isoenzymes are activated and can then directly phosphorylate Nrf2 at serine 40. Once phosphorylated, Keap1 is less able to bind to and repress Nrf2, which subsequently accumulates in the nucleus of the cell to increase transactivation of the ARE.

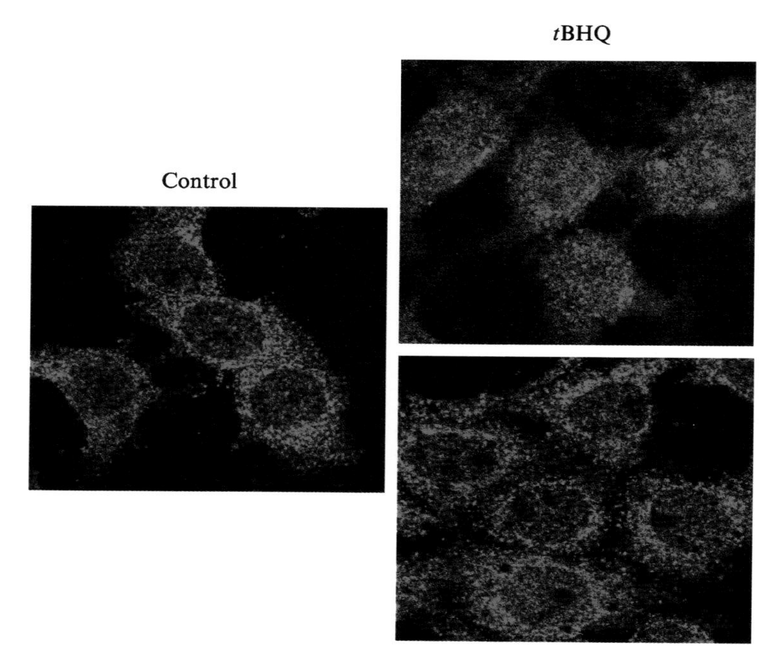

Sherratt *et al.*, Chapter 22, Fig. 3. Immunocytochemistry was used to demonstrate the accumulation of endogenous Nrf2 in the nucleus of cells following exposure to *t*BHQ. This can be impaired by pretreatment of the cells with PKC inhibitors. Rat hepatoma H4IIEC3 cells trypsinized onto coverslips were exposed to solvent (DMSO), *t*BHQ alone for 4 h, or pretreated with staurosporine for 1 h before treatment with *t*BHQ. *t*BHQ was used at a final concentration of 100 μM, and staurosporine was used at 15 n*M*.

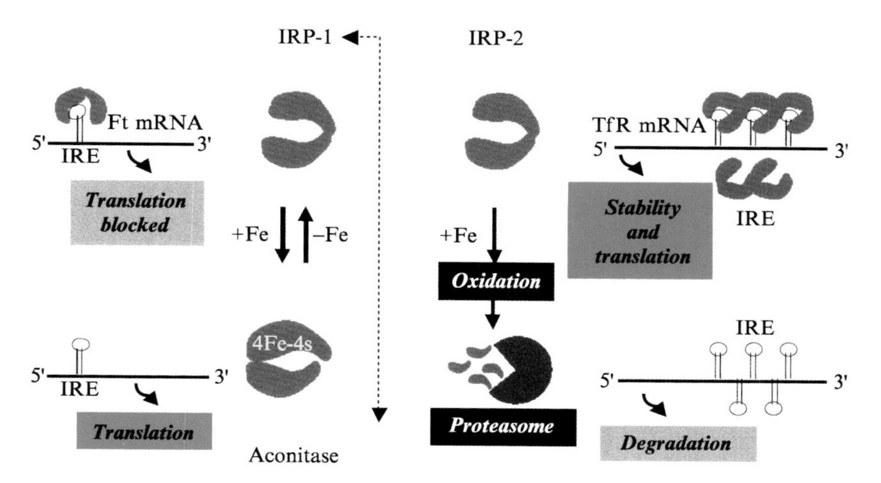

MINOTTI *ET AL.*, CHAPTER 25, FIG. 1. Simplified representation of iron-mediated switch between aconitase and IRP-1 mechanisms underlying Fe-dependent cluster assembly/disassesmbly, leading to aconitase↔IRP-1 switches, or Fe-induced IRP-2 degradation.

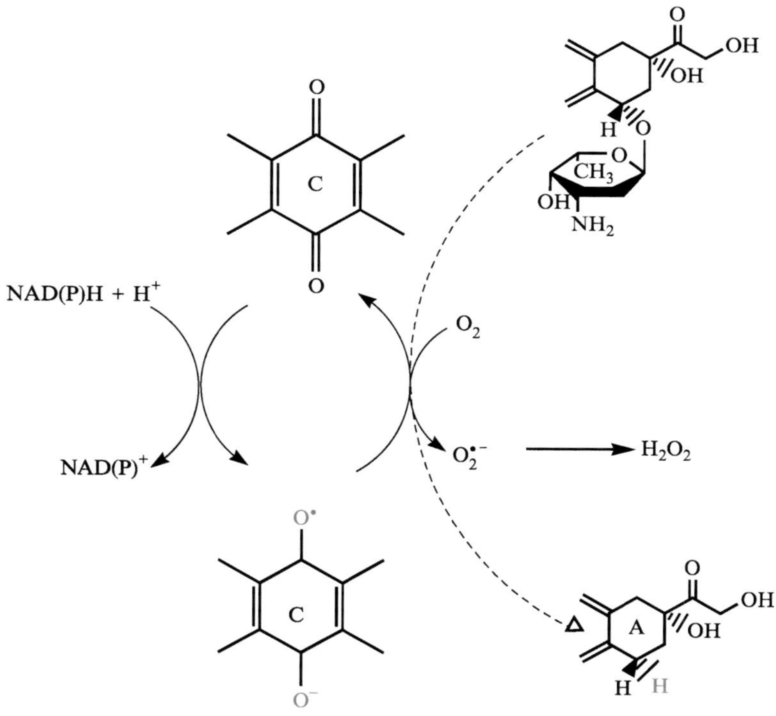

O
OH
OH
H
O
O
CH_3
OH
NH_2
O
C
O
NAD(P)H + H^+
O_2
NAD(P)$^+$
$O_2^{\bullet-}$
H_2O_2
$O^{\bullet}$
C
O^-
O
OH
A
OH
H H

MINOTTI *ET AL.*, CHAPTER 25, FIG. 3. Quinone-dependent DOX metabolism. Simplified representation of NAD(P)H-dependent $Q^{\bullet-}$ formation, leading to $O_2^{\bullet-}$ and H_2O_2 or reductive deglycosidation.

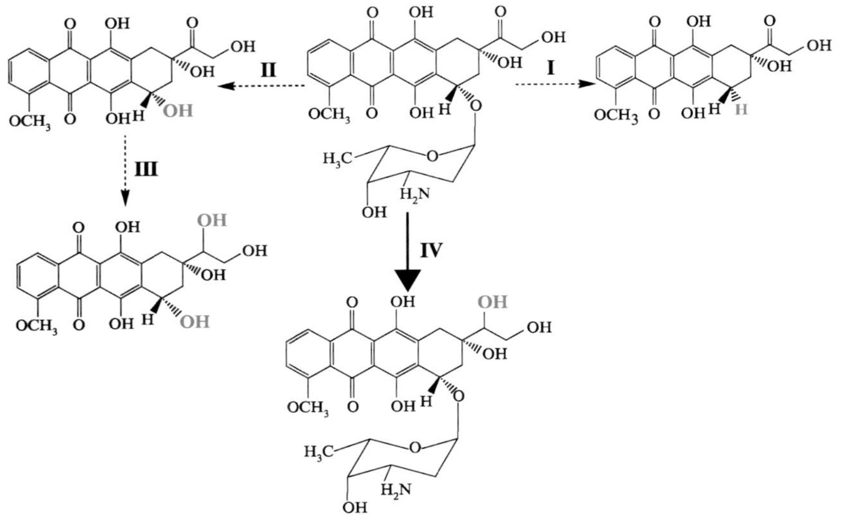

MINOTTI *ET AL.*, CHAPTER 25, FIG. 4. Quinone-independent DOX metabolism. Pathways of Q-independent reductive deglycosidation, leading to 7-deoxydoxorubicinone (I); hydrolase-type deglycosidation followed by carbonyl reduction, leading to doxorubicinone (II) and doxorubicinolone (III); direct carbonyl reduction, leading to DOXol (IV). Based on Licata *et al.*

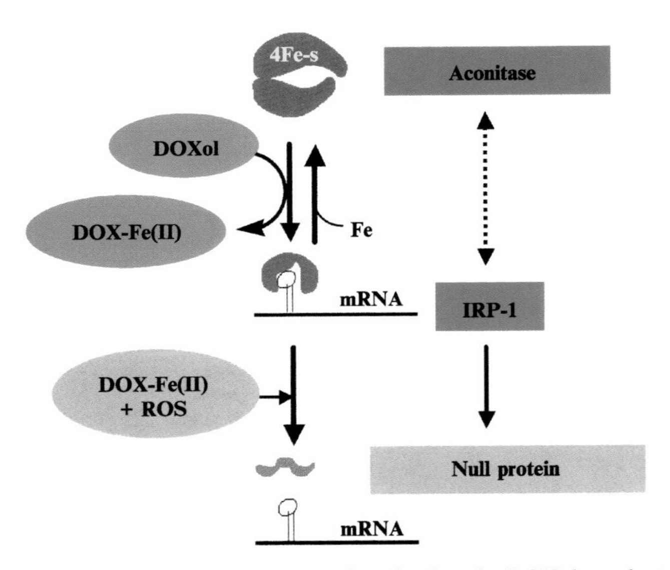

MINOTTI *ET AL.*, CHAPTER 25, FIG. 5. Proposed mechanisms for DOX-dependent conversion of aconitase/IRP-1 into a null protein during chronic cardiotoxicity. Simplified representation of (1) DOXol oxidation with Fe-S clusters, leading to reversible conversion of aconitase to IRP-1 and concomitant formation of DOX-Fe(II); (2) synergistic interaction of DOX-Fe(II) with ROS, leading to IRP-1 inactivation and formation of a null protein.